MW01345028

DISCRETE MATHEMATICS: MODELING OUR WORLD

Discrete Mathematics: Modeling Our World

Fourth Edition

Nancy Crisler
Gary Froelich

The Consortium for Mathematics and Its Applications (COMAP)
175 Middlesex Turnpike, Suite 3B
Bedford, MA 01730

Published and distributed by

ISBN 1-933223-48-0

Printed in the United States of America

Fourth Printing 2017

Contents

Preface

The first edition of Discrete Mathematics Through Applications appeared in 1994, the second in 2000, and the third in 2006. This new fourth edition is the first published by The Consortium for Mathematics and Its Applications (COMAP), a non-profit organization that has for decades advocated for modeling in mathematics education. This fourth edition shares a major goal with its predecessors: to introduce students to discrete mathematics and its importance in modeling today's world. We have retitled this fourth edition to reflect our new publisher and an increased emphasis on modeling.

We continue to hear from teachers who use the book. Their comments remain overwhelmingly positive. They tell us that their students find the book interesting, readable, and challenging. We are delighted to hear their stories of students who are thriving in the mathematics classroom for the first time.

When we began work on the first edition, our target audience was based on our own classroom experience. Twenty years later, we still believe we were close to the mark. Most users of the book describe their students as college-bound with plans to major in areas such as business, education, social science, and law. But they also say that students interested in mathematics, science, engineering, and computer science find the course helpful.

Many people continue to ask us, "What is discrete mathematics?" A good short description offers a comparison—one that contrasts discrete and continuous mathematics. The real numbers, for example, are a continuous set because between any two of them there are infinitely many others. Thus, the study of functions in algebra, trigonometry, and calculus is not discrete mathematics because most functions are defined on the real numbers. On the other hand, discrete topics such as permutations and combinations, which treat only whole numbers, can be found in traditional courses but in relative isolation.

The growing emphasis on discrete topics can be attributed to the changing nature of human society. Today's mathematics is being applied to important problems in the social sciences, to new problems in the biological sciences, and, of course, to problems that arise in the design of computer systems. Moreover, educators find many discrete topics accessible to a broad range of students because an understanding of important problems requires little specialized background.

Support for the implementation of discrete mathematics in schools in the United States has come from a number of organizations. In 1989, the National Council of Teachers of Mathematics (NCTM) gave considerable attention to discrete topics in its Curriculum and Evaluation Standards for School Mathematics. In 2000, the NCTM's Principles and Standards for School Mathematics (PSSM) reiterated the earlier document's call: "discrete mathematics should be an integral part of the school mathematics curriculum," and discrete topics should "span the years from prekindergarten through grade 12."

The authors of this book were members of an NCTM task force that developed guidelines for the implementation of discrete topics. Indeed, the content of this book was inspired by the work of that task force. And it was the generous offer of financial support for the development of a suitable text by COMAP that enabled us to write the first edition. We are delighted that COMAP finds our book worthy of a fourth edition.

As teachers, we found great satisfaction in teaching discrete mathematics. As writers, we are pleased to hear that other teachers are having similar experiences. The applicability of the subject matter in this book is so obvious to students that they rarely ask, When will we ever use this?

Features of this Book

Teachers who use this book often tell us that our approach is working well for them. Some of the features that they mention are:

- Exploration activities are often used to introduce important topics. We believe that exploration of a problem is critical to its understanding and solution and to student engagement.
- The text is written in an informal style that is easily read by students.
- Exercises often contain new ideas. Students learn by doing. In most cases, students are expected to do all the exercises.
- Chapter Extensions discuss topics related to the content of each chapter. These can be used as starting points for projects and class activities.
- Project suggestions are provided for students who want to go beyond the text.
- The themes of mathematical modeling, appropriate use of technology, and decision making are consistently emphasized.
- The strands of algorithmic thinking, recursive thinking, and mathematical induction are woven throughout the book.

New to this Edition

As in the past, we chose to keep the things that have worked well for us and other users of the book. But we chose to update and add content when appropriate. Among our changes are:

- An increased emphasis on mathematical modeling
- The first full-color edition
- There are two new chapters. Chapter 9 explores the topic of codes, and Chapter 10 consists of a series of miscellaneous topics (logic, set notation, bin packing, and linear programming)
- Updated news articles and data throughout
- Updated bibliographies

Supplements

There are several supplements to the text that teachers have found helpful.

- **Instructor's CD: ISBN: 1-933223-49-9**

  This CD includes our suggestions for teaching individual lessons, comments on exercises, masters for creating transparencies and handouts, test questions, computer/calculator software, and a complete answer key. PDFs of the student text are also included.

- **Video DVD: ISBN: 1-933223-51-0**

  This DVD includes most of the support videos that are listed in the teaching suggestions for each chapter.

- **A Supplemental Chapter on Statistics**
  Chapter 11: Measurement/Data Analysis

  This supplement is available in electronic format and can be found on the instructor's CD. The topics include graphical and numerical descriptions of data, normal distributions, and control charts.

Information on these supplements is available from COMAP. Contact COMAP by sending an email to info@comap.com or phoning (781) 862-7878 or (800) 772-6627.

Acknowledgments

Once again, we would like to thank COMAP for financial and technical support of all four editions. The faith in our abilities shown by Dr. Solomon Garfunkel, COMAP's executive director, is particularly appreciated.

Our thanks also go to George Ward, Anne Sterling, and John Tomicek of the COMAP staff for their work on this edition. We would also like to thank Sue Ann McGraw, Robert W. Owens, Sally Ziemba, and Kathy Berlin for their work on the test questions and answers.

We would like to thank the students who have used the book in their study of discrete mathematics and all those who will do so in the future. We cannot help but mention again the students we specifically thanked in the preface to the first three editions—Kari Bauer, Dan Froelich, and Tom Hehre—for writing computer programs that accompany the text. Kari, Dan, and Tom have since graduated from high school and college and are now pursuing their own careers.

Last, but certainly not least, we want to thank Patience Fisher, our first- and second-edition co-author. In addition to her writing, Dr. Fisher contributed teacher expertise, advice, and lots of love to the first two editions. We strove to maintain the many positive aspects of her work in the third and fourth editions.

We have been privileged to share our interest in discrete mathematics with other teachers by writing this book and by participating in institutes, seminars, and workshops around the country. We know that in so doing, we have helped you incorporate discrete topics into your own courses and curricula and to thereby better serve your students. That knowledge has made the many weeks and months spent writing or on the road worthwhile, and we hope that, in turn, the new edition of this book proves useful to you in your efforts.

Foreword

In the foreword to the third edition of this text, John Dossey sagely wrote, "Discrete mathematics and its applications is one of the most rapidly expanding areas in the mathematical sciences. The modeling and understanding of finite systems is central to the development of the economy, computer science, the natural and physical sciences, and mathematics itself. This rapid growth of discrete topics applications has made the definition and development of coursework in discrete mathematics a more difficult task than the development of materials and courses of study in many other areas of the mathematical sciences." Several changes and additions in the text before you reflect this fact.

First, you will note the change in the title with the added words – Modeling Our World. This reflects the added emphasis on modeling as both a process and an essential part of doing mathematics. Historically, students were not introduced to mathematical modeling until they had seen a significant amount of higher-level mathematics, in particular calculus and differential equations. But we feel strongly that this needs to change for two compelling reasons.

First, as we've already said more and more important areas of application lend themselves to analysis by discrete models. And secondly, and most importantly, we see modeling as a life skill. We know that math is everywhere. We want our students to know this and incorporate the use of mathematics in their careers, their future schooling, and their everyday lives. Working with discrete models as early as possible is one way to start this process. Moreover, modeling is now recognized as a practice in many state standards and will be emphasized and tested all through secondary school.

This text has always been a leader in the presentation of discrete mathematics in our nation's high schools. In this edition we have expanded our topic coverage and as always updated the data and examples to keep the applications current and relevant. COMAP is extremely proud to take over the publication of this text. We believe that it will continue to lead the way in establishing a firm base for student understanding of the power and applicability of mathematics.

Enjoy.

Sol Garfunkel
Executive Director, COMAP

Election Theory: Modeling the Voting Process

CHAPTER 1

Throughout your life you are faced with decisions. As a student, you must decide which courses to take and how to divide your time among homework, activities, social events, and, perhaps, a job. As an adult, you will be faced with many new decisions, including how to vote in elections.

The decisions that people make can have important consequences. For example, Nielsen Media Research regularly polls individuals to learn what television programs they decide to watch. These viewer decisions determine whether a show will survive to another season. Because of the consequences of their work, organizations like Nielsen have a formidable responsibility: to combine the preferences of all the individuals in their survey into a single result and to do so in a way that is fair to all television programs.

- How are the wishes of many individuals combined to yield a single result?
- Do the methods for doing so always treat each choice fairly?
- If not, is it possible to improve on these methods?

This chapter examines a process that is fundamental to any democratic society: group decision making. As you progress through the chapter, you will consider several models that can be used by a group of people to conduct a vote and reach a decision. Election theory, the study of the voting process, is a recent example of the use of mathematical modeling in the social sciences.

Lesson 1.1

An Election Activity

Every democratic institution must have a process by which the preferences of individuals are combined to produce a group decision. For example, the preferences of individual voters must be combined in a fair way in order to select government officials.

An excellent way to begin an exploration of group decision making is to give the process a try. Therefore, in this lesson you will develop a model for combining preferences of the individuals in your class into a single result. Before you begin, a word of reassurance and a preview of things to come: Many important problems in election theory (and other topics in discrete mathematics) can be understood and solved without a lot of background knowledge, and mathematicians know that there is no single right solution for many of these problems.

Explore This

On a piece of paper write the names of the following soft drinks, in the order given:

Coke
Dr. Pepper
Mountain Dew
Pepsi
Sprite

Rank the soft drinks. That is, beside the name of the soft drink you like best, write "1." Beside the name of your next favorite soft drink, write "2." Continue until you rank all five.

As directed by your instructor, collect the ballots from all members of your class and share the results by, for example, writing all ballots on a chalkboard. Since everyone has written the soft drinks in the same order, you should be able to record quickly only the digits from each ballot.

Your task in this activity is to devise a method of combining the rankings of all the individuals in your class into a single ranking for the entire class. Your method should produce a first-, second-, third-, fourth-, and fifth-place soft drink.

If you are working in a small group, your group should agree on a single method. After everyone finishes, each group or individual should present the ranking to the class and describe the method used to obtain it. Clear communication of the method used to obtain a result is important in mathematics, so everyone should strive for clarity when making the presentation.

As each group (or individual) makes its presentation, record the ranking in your notebook for use in the following exercises.

Howard Bans Distribution of Soft Drinks on County Property

Baltimore Sun
Dec. 11, 2012

Howard County Executive Ken Ulman moved Tuesday to ban the sale of high-sugar drinks such as soda in parks, libraries and other county properties and at county-sponsored events.

The sales and distribution ban — which mirrors efforts nationally and that may be adopted by Baltimore City — is aimed at reducing childhood obesity and raising awareness among parents and adults about the health hazards of sugary drinks.

Nationally, one of the most high-profile efforts aimed at tackling childhood obesity was the so-called super-size ban by the New York City Board of Health. That city plans to prohibit the sale of sugary drinks larger than 16 ounces by restaurants and other food vendors, starting in March.

Exercises

1. Do all the group rankings produced in your class have the same soft drink ranked first? If not, which soft drink is ranked first most often?
2. Repeat Exercise 1 for the soft drink ranked second.
3. Repeat Exercise 1 for the soft drink ranked third.
4. Repeat Exercise 1 for the soft drink ranked fourth.
5. Repeat Exercise 1 for the soft drink ranked fifth.
6. Write a description of the method you used to achieve a group ranking. Make it clear enough that another person can use the method. You may want to break down the method into numbered steps.
7. Did anyone in your class use a method similar to yours? Explain why you think they are similar.
8. Did your method result in any ties? How can your method be modified to break ties?
9. Mathematicians often find it convenient to represent a situation in a compact way. A good representation conveys the essential information about a situation. In election theory, a **preference schedule** is sometimes used to represent the preferences of one or more individuals. The following preference schedule displays four choices, called A, B, C, and D. It indicates that the individual whose preference it represents ranks B first, C second, D third, and A fourth.

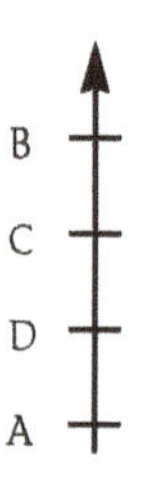

Since there are often several people who have the same preferences, mathematicians write the number of people or the percentage of people who expressed that preference under the schedule. The preferences in a group of 26 people are represented by the preference schedules at the top of the next page.

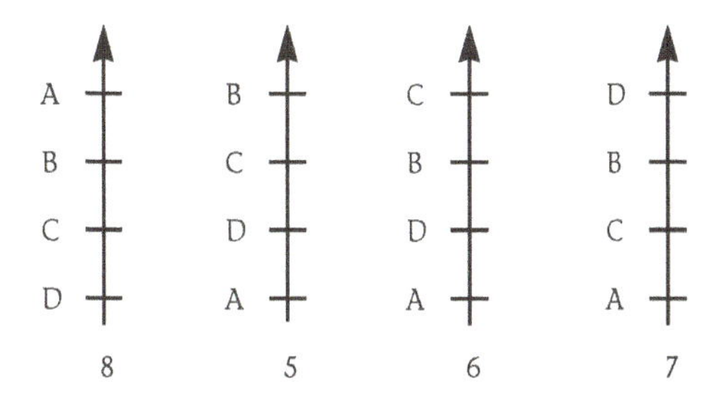

Total number of voters:
8 + 5 + 6 + 7 = 26.

a. Apply the method you used to determine your class's soft drink ranking to this set of preferences. List the first-, second-, third-, and fourth-place rankings that your method produces. If your method cannot be applied to this set of preferences, then explain why it cannot and revise it so that it can be used here.

b. Do you think the ranking your method produces is fair? If you worked in a group, do all members of your group think the result is fair? In other words, do the first-, second-, third-, and fourth-place rankings seem reasonable, or are there reasons that one or more of the rankings seem unfair? Explain.

c. Would preference schedules be a useful way to represent the individual preferences for soft drinks among the members of your class? Explain.

10. When your class members voted on soft drinks, they ranked them from first through fifth. A ballot that allows voters to rank the choices is called a **preferential ballot**. In most elections in the United States, preferential ballots are not used. Do you think preferential ballots are a good idea? Explain.

11. There are three choices in a situation that uses preferential ballots. Call the choices A, B, and C. The figure at the top of the next page gives the six possible preferences that a voter can express.

Academy to Use Preferential Voting to Pick Best Picture

Los Angeles Times
August 31, 2009

The Academy of Motion Picture Arts and Sciences today took the long expected step to ensure that this year's best picture winner won't be hated by 90% of its members by going with a preferential voting system for members.

In a preferential voting system, votes for the least popular first choice movie are eliminated and those members' second choices are taken into account. The process continues until a nominee receives more than 50% of the votes.

Under the old system, members simply voted for their first choice. With 10 nominees, that would mean a movie with one vote more than 10% could theoretically be named best picture.

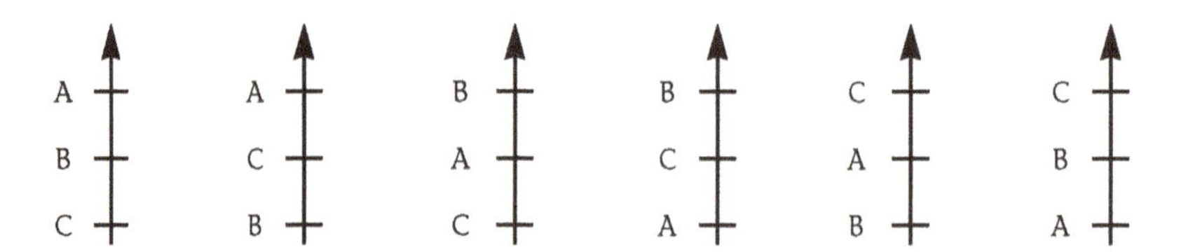

A fourth choice, D, enters the picture. If D is attached to the bottom of each of the previous schedules, there are six schedules with D at the bottom. Similarly, there are six schedules with D third, six with D second, and six with D first, or a total of $4(6) = 24$ schedules. Thus, the total number of schedules with four choices is four times the total number of schedules with three choices.

a. There are 24 possible schedules with four choices. How many are there with five choices? With six choices?

b. Mathematicians use symbols to represent this relationship. The symbol S_n represents the number of schedules when there are n choices. You have seen that $S_n = nS_{n-1}$. Write an English translation of the mathematical sentence $S_n = nS_{n-1}$.

12. The mathematical sentence in Exercise 11b is a **recurrence relation**, a verbal or symbolic statement that describes how one number in a list is derived from the previous number (or numbers). Since recurrence relations are an important part of discrete mathematics, your experience with them begins in this lesson.

For example, suppose the first number in a list is 7 and a recurrence relation states that to obtain any number in the list, add 4 to the previous number. Then the second number is $7 + 4$, or 11. This recurrence relation is stated symbolically as $T_n = 4 + T_{n-1}$.

Another example of a recurrence relation is $T_n = n + T_{n-1}$. Complete the following table for the recurrence relation $T_n = n + T_{n-1}$.

n	1	2	3	4	5
T_n	3	$2 + 3 = 5$			

13. Complete the following table for the recurrence relation $A_n = 3 + 2A_{n-1}$.

n	1	2	3	4	5
A_n	4	$3 + 2(4) = 11$			

14. State the recurrence relation $A_n = 3A_{n-1} - 4$ in words.

Lesson 1.2

Group-Ranking Models

If the soft drink data for your class are typical, you know that the problem of establishing a group ranking is not without controversy. Even among professionals, there isn't agreement on the best way to do so. This lesson examines several common models for determining a group ranking from a set of voter preferences. As you examine these models, consider whether any of them are similar to the ones devised by members of your class in Lesson 1.1.

Consider the preferences of Exercise 9 of the previous lesson, which are shown again in Figure 1.1.

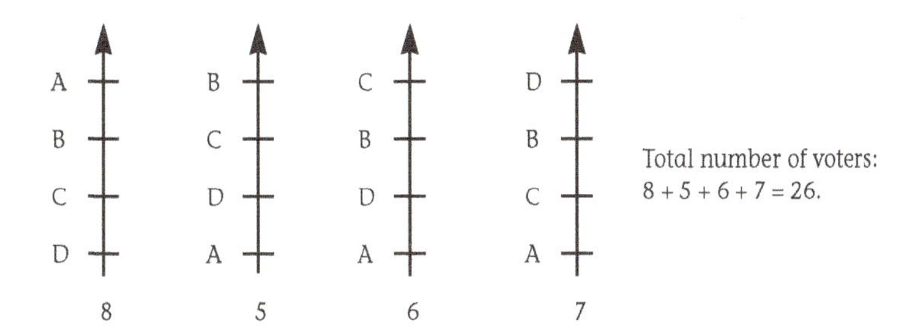

Figure 1.1. Preferences of 26 voters.

Many voting situations, such as elections in which there is only one office to fill, require the selection of a single winner. Although most such elections in the United States do not use a preferential ballot, they could. For example, in the set of preferences shown in Figure 1.1, choice A is ranked first on eight schedules, more often than any other choice.

If A wins on this basis, A is called the **plurality winner**. A plurality winner is based on first-place rankings only. The winner is the choice that receives the most votes. Note, however, that A is first only on about 30.8% of the schedules. If A is first on over half the schedules, A is a **majority winner**.

Mathematician of Note

Jean-Charles de Borda (1733–1799)

A French cavalry officer, naval captain, mathematician, and scientist, he preferred methods that assign points to the rankings of individuals because he was dissatisfied with plurality methods.

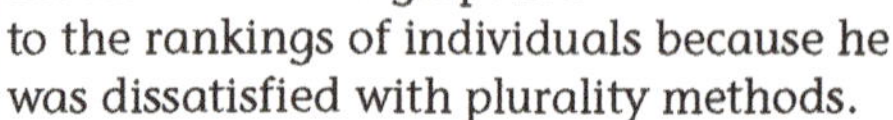

Borda Models

Did anyone in your class determine the soft drink ranking by assigning points to the first, second, third, and fourth choice of each individual's preference and obtaining a point total? If so, these groups used a type of **Borda count**.

The most common way that Borda methods are applied to a ranking of n choices is by assigning n points to a first-place ranking, $n - 1$ to a second-place ranking, $n - 2$ to a third-place ranking, . . . and 1 point to a last-place ranking. The group ranking is established by totaling each choice's points.

In the example of Figure 1.1, A is ranked first by 8 people and fourth by the remaining 18, so A's point total is $8(4) + 18(1) = 50$. Similar calculations give totals of 83, 69, and 58 for B, C, and D, respectively, as summarized below.

A: $8(4) + 5(1) + 6(1) + 7(1) = 50$

B: $8(3) + 5(4) + 6(3) + 7(3) = 83$

C: $8(2) + 5(3) + 6(4) + 7(2) = 69$

D: $8(1) + 5(2) + 6(2) + 7(4) = 58$

In this case, the plurality winner does not fare well under a Borda system.

Technology Note

Borda counts can be done quickly on a graphing calculator: after the first calculation is typed and entered, it is replayed and edited.

```
8*4+5*1+6*1+7*1
               50
8*3+5*4+6*3+7*3
```

Runoff Models

Some elections in the United States and other countries require a majority winner. If there is no majority winner, a runoff election between the top two candidates is held. Runoff elections are expensive because of the cost of holding another election and time-consuming for voters because they require a second trip to the polls. However, if voters use a preferential ballot, both disadvantages can be avoided.

To conduct a runoff, determine the number of firsts for each choice. In the example of Figure 1.1, A is first eight times, B is first five times, C is first six times, and D is first seven times.

> **High Cost of Runoff Elections Has Some in Albany Thinking It's Time to End Them**
>
> *New York Daily News*
> Oct. 4, 2009
>
> Alabama will spend $3 million to conduct a runoff election Tuesday that's expected to attract only one-fifth of the state's registered voters.
>
> Turnout for the primary June 4 was 34 percent. The runoff is for races where one candidate didn't get at least one vote over 50 percent in the primary.

Eliminate all choices except the two with more first-place votes than the others: Choices B and C are eliminated; A and D are retained.

Now consider each preference schedule on which the eliminated choices are ranked first. Choice B is first on the second schedule. Of the two remaining choices, A and D, D is ranked higher than A, so these 5 votes are transferred to D. Similarly, the 6 votes from the third schedule are transferred to D. The totals are now 8 for A and 7 + 5 + 6 = 18 for D, and so D is the runoff winner (see Figure 1.2).

The runoff method eliminates all choices except the two with the most firsts:
A: 8 B: 5 C: 6 D: 7
(Eliminate B & C.)

The five votes for B are transferred to D, and the six votes for C are transferred to D.
A: 8 D: 7 + 5 + 6 = 18.

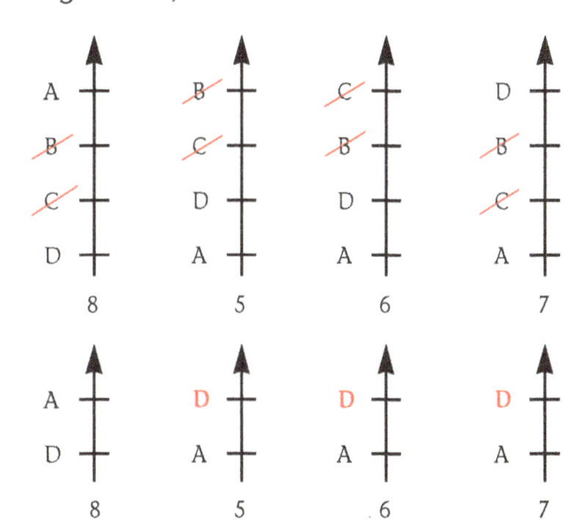

Figure 1.2. A runoff.

Sequential Runoff Models

Some elections are conducted by a runoff variant that eliminates only one choice at a time. However, if there are several choices and voters must vote again when one is eliminated, sequential runoff methods compound the disadvantages of runoff methods. As with the runoff method, if voters use a preferential ballot, they need vote only once.

In the example of Figure 1.1, B is eliminated first because it is ranked first the fewest times. The 5 first-place votes for B are transferred to C. The point totals are now 8 for A, 5 + 6 = 11 for C, and 7 for D.

There are three choices remaining. Now D's total is the smallest, so D is eliminated next. The 7 votes are transferred to the remaining choice that is ranked highest by these 7. Thus, C is given an additional 7 votes and so defeats A by 18 to 8 (see Figure 1.3).

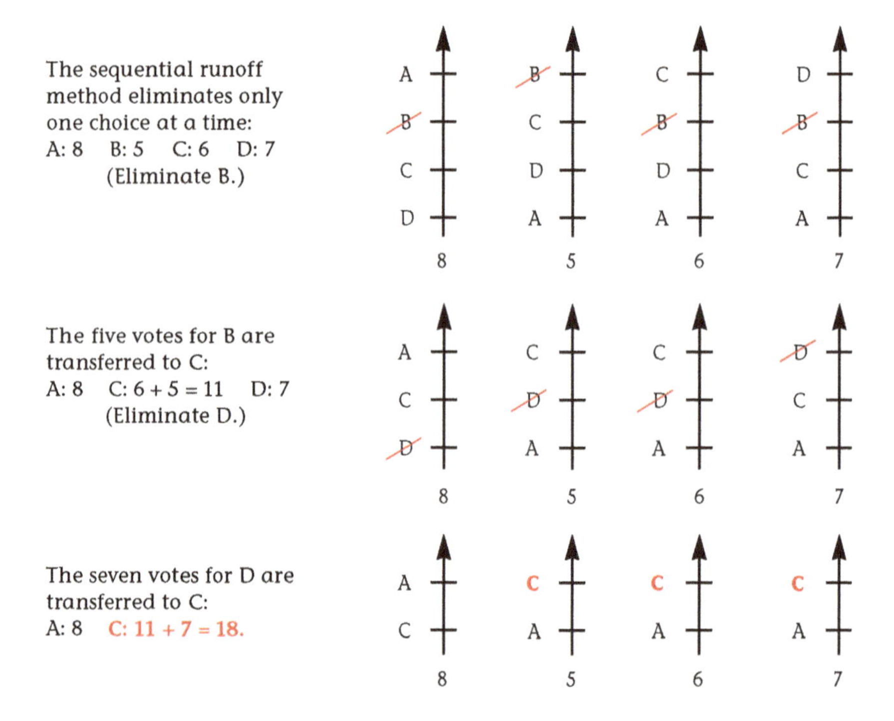

Figure 1.3. A sequential runoff.

When a preferential ballot is used and computers process the data, sequential runoff results are obtained almost instantaneously. Thus, the term **instant runoff** is often applied to sequential runoffs.

Exercises

For Exercises 1–4, apply the methods as they are described in this lesson.

1. Find a plurality winner for the soft drink voting in your class. Is it also a majority winner? Explain.
2. Find a Borda winner for the soft drink voting in your class.
3. Find a runoff winner for the soft drink voting in your class.
4. Find a sequential runoff winner for the soft drink voting in your class.
5. The International Olympic Committee uses sequential runoff voting to choose Olympic sites. Since the committee is relatively small, separate rounds of voting are used rather than a single round with preferential ballots. There were three cities competing to host the 2014 winter games. This table summarizes the voting.

Sochi	34	51
Pyeongchang	36	47
Salzburg	25	

   a. Write a short summary of the voting, including the order in which the cities were eliminated.

   b. Five cities competed to host the 2016 summer games. Read the news article about the voting and construct a similar table.

Rio to Stage 2016 Olympic Games

BBC
Oct. 2, 2009

Brazil will become the first South American country to host the Olympics after the city of Rio de Janeiro was chosen to stage the 2016 Games.

Chicago's early exit was a surprise, after bookmakers made them favorites. Chicago received only 18 of the 94 votes available in the first round poll of IOC delegates. Madrid came out top with 28, followed by Rio with 26.

In the second round, however, Rio almost secured the absolute majority needed to win outright, with 46 of the 95 votes cast. Madrid came a distant second with 29, while Tokyo was eliminated after receiving 20.

The final ballot saw Rio win by a comprehensive margin of 66 votes to 32.

6. For the example of Figure 1.1, find the percentage of voters that rank each choice first and last.

   a. Enter the results in a table like this:

Choice	First	Last
A		
B		
C		
D		

   b. On the basis of these percentages only, which choice do you think is most objectionable to voters? Least objectionable? Explain your answers.

   c. Which choice do you think most deserves to be ranked first for the group? Explain your reasoning.

   d. Give at least one argument against your choice.

7. The 2010 race for governor of Minnesota had three strong candidates. The following are the results from the general election.

Mark Dayton	919,232
Tom Emmer	910,462
Tom Horner	251,487
Others	25,840

   a. What percentage of the vote did the winner receive? Is the winner a majority winner?

   b. What is the smallest percentage that a plurality winner can receive in a race with three candidates? Explain.

8. Determine plurality, Borda, runoff, and sequential runoff winners for the following set of preferences. Apply the methods as they are described in this lesson.

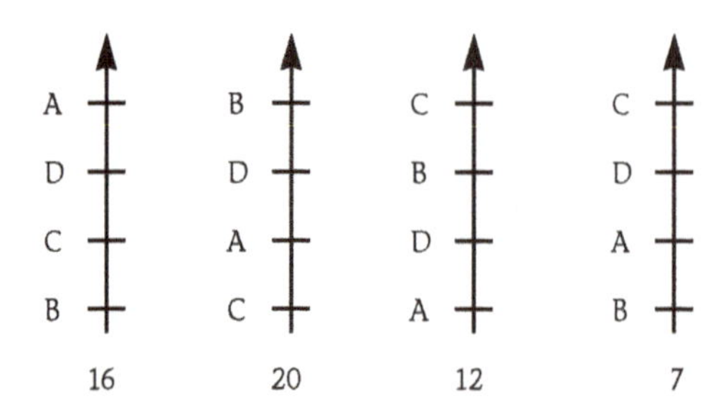

9. Borda models produce a complete group ranking, but the other modelss examined in this lesson determine a winner only. However, each of them can be extended to produce a complete group ranking.

   a. Describe how a plurality model could be extended to determine a second, third, and so forth. Apply your extension to the example in Figure 1.1 and list the second, third, and fourth that it produces.

   b. Describe how runoff models could be extended to determine a second, third, and so forth. Apply your extension to the example in Figure 1.1 and list the second, third, and fourth that it produces.

   c. Describe how a sequential runoff model could be extended to determine a second, third, and so forth. Apply your extension to the example in Figure 1.1 and list the second, third, and fourth that it produces.

10. Each year the Heisman Trophy recognizes one of the country's outstanding college football players. In 2012, Texas A & M freshman quarterback Johnny Manziel received the award. The results of the voting follow. Each voter selects a player to rank first, another to rank second, and another to rank third.

3-2-1

	1st	2nd	3rd	Points
1. Johnny Manziel	474	252	103	2,029
2. Manti Te'o	321	309	125	1,706
3. Collin Klein	60	197	320	894
4. Marqise Lee	19	33	84	207
5. Braxton Miller	3	29	77	144
6. Jadeveon Clowney	4	13	23	61
7. Jordan Lynch	3	8	27	52
8. Tavon Austin	6	4	21	47
9. Kenjon Barner	1	12	15	42
10. Jarvis Jones	1	10	18	41

   a. How many points are awarded for a first-place vote? For a second-place vote? For a third-place vote? 3-2-1

   b. Does the ranking produced by this model differ if a plurality model is used? Explain. Starting @ 5th place

11. In most American runoff elections, voters do not use a preferential ballot and therefore must return to the polls to vote in the runoff. Voters in some countries, such as Ireland, use a preferential ballot and therefore go to the polls only once. Examine the vote totals in the two runoffs shown below. What do the totals tell you about the merits of preferential ballots in runoff elections?

President of Ireland: 2011 Results

	General Election	Runoff
Michael Higgins	701,101	1,007,104
Seán Gallagher	504,964	628,114
Martin McGuinness	243,030	
Gay Mitchell	113,321	
David Norris	109,469	
Dana Scallon	51,220	
Mary Davis	48,657	

U.S. House Louisiana District 3: 2012 Results

	General Election	Runoff
Charles Boustany	139,123	58,820
Jeff Landry	93,527	37,764
Ron Richard	67,070	
Bryan Barrilleaux	7,908	
Jim Stark	3,765	

Michael Higgins Becomes Irish President

The Guardian
Oct. 29, 2011

The Irish Labour party's Michael D Higgins has been confirmed as the ninth president of the republic after winning a landslide victory in the most fractious campaign in the country's history.

The poet and campaigner gained a 56.8% share of the vote, putting him comfortably ahead of his rivals – Dragons' Den star Seán Gallagher, who came second, and former IRA commander Martin McGuinness, who ended in third place.

The result capped a two-day count of ballots to determine who would succeed Mary McAleese as Ireland's head of state.

12. In sequential runoffs, the number of choices on a given round is 1 less than the number of choices on the previous round. Let C_n represent the number of choices after n rounds and write this as a recurrence relation.

13. A procedure for solving a problem is called an **algorithm**. This lesson discusses various algorithms for obtaining a group decision from individual preferences. Algorithms are often written in numbered steps in order to make them easy to apply. The following is an algorithmic description of a runoff model as discussed in this lesson.

    1. For each choice, determine the number of preference schedules on which the choice is ranked first.
    2. Eliminate all choices except the two that are ranked first most often.
    3. For each preference schedule, transfer the vote total to the remaining choice that ranks highest on that schedule.
    4. Determine the vote total for the preference schedules on which each of the remaining choices is ranked first.
    5. The winner is the choice ranked first on the most schedules.

    a. Write an algorithmic description of the sequential runoff model discussed in this lesson.

    b. Write an algorithmic description of the Borda model discussed in this lesson.

14. The number of first-, second-, third-, and fourth-place votes for each choice in an election can be described in a table, or **matrix**, as shown below.

The preferences:

A	B	C	C
D	D	B	D
C	A	D	A
B	C	A	B
20	10	12	15

The matrix:

	A	B	C	D
1st	20	10	27	0
2nd	0	12	0	45
3rd	25	0	20	12
4th	12	35	10	0

The number of points that a choice receives for first, second, third, and fourth place can be written in a matrix, as shown below.

	1st	2nd	3rd	4th
Points	4	3	2	1

A new matrix that gives Borda point totals for each choice can be computed by writing this matrix alongside the first, as shown below.

$$[4\ \ 3\ \ 2\ \ 1]\begin{bmatrix} 20 & 10 & 27 & 0 \\ 0 & 12 & 0 & 45 \\ 25 & 0 & 20 & 12 \\ 12 & 35 & 10 & 0 \end{bmatrix}$$

Entries for a new matrix can be computed in this way: Multiply each entry of the first matrix by the corresponding entry in the first column of the second matrix and find the sum of these products.

$$4(20) + 3(0) + 2(25) + 1(12) = 142$$

This number is the first entry in a new matrix that gives Borda point totals for choices A, B, C, and D:

$$\begin{array}{rcccc} & A & B & C & D \\ \text{Point totals:} & [142 & \_\_ & \_\_ & \_\_] \end{array}$$

a. Calculate the remaining entries of the new matrix.

b. If you have this new matrix but not the voter preference schedules, by which methods is it possible to determine the winner? Explain.

Computer/Calculator Explorations

15. Enter the soft drink preferences of your class members into the election machine computer program that accompanies this book. Compare the results given by the computer to your answers to the first four exercises of this lesson. Resolve any discrepancies.

Projects

16. Write a short report on the history of any of the models discussed in this lesson. Look into the lives of people who were influential in developing the model. Discuss factors that led them to propose the model.

17. Find at least two examples of group-ranking models that are currently used somewhere in the world but not discussed in this lesson. Describe how the group ranking is determined. Compare each new model with those described in this lesson. What are some advantages and disadvantages of each new model?

18. Select one or more countries that are not discussed in this lesson and report on the models they use to conduct elections.

Bayern Munich Stays Best Team in AP Global Poll

Associated Press
April 30, 2013

After dominating the first legs of their Champions League semifinals, it's no surprise to see Bayern Munich and Borussia Dortmund leading the way in the latest Associated Press global football poll.

Bayern pipped Dortmund to be voted the world's top team for the fifth straight week, with its 4-0 hammering of Barcelona impressing the AP's panel of 15 journalists.

AP Global Football Rankings for the week ending April 29.

Based on 15 voters, using 10 points for first, nine for second, one for bottom place. Previous rankings in parentheses.

1. Bayern Munich (1), 147
2. Borussia Dortmund (7), 136
3 . Juventus (2), 71
5. Chelsea, 58
6. Liverpool, 50
7. Real Madrid (4), 36
8. Paris Saint-Germain (6), 30
9. Real Sociedad
10. Barcelona (5), 24

Lesson 1.3

More Group-Ranking Models and Paradoxes

Different models for finding a group ranking can give different results. This fact led the Marquis de Condorcet to propose that a choice that could obtain a majority over every other choice should be ranked first for the group.

Again consider the set of preference schedules used in the previous lesson (see Figure 1.4).

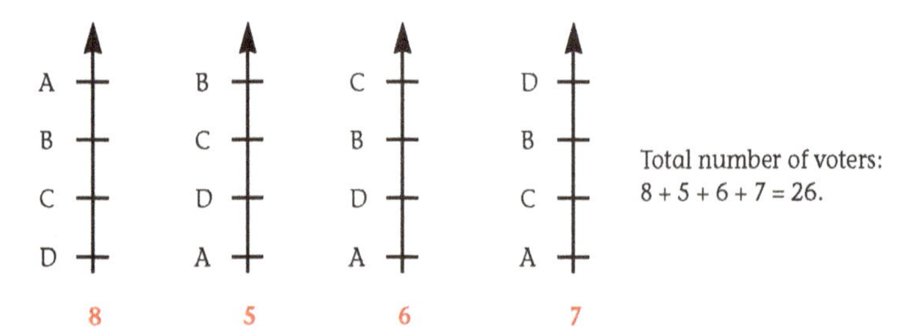

Figure 1.4. Preferences of 26 voters.

To examine these data for a Condorcet winner, compare each choice with every other choice. For example, begin by comparing A with B, then with C, and finally with D. Notice in Figure 1.4 that A is ranked higher than B on 8 schedules and lower on 18. (An easy way to see this is to cover C and D on all the schedules.) Because A cannot obtain a majority against B, A cannot be a Condorcet winner. Therefore, there is no need to check to see if A can beat C or D.

Now consider B. You have already seen that B beats A, so begin by comparing B with C. B is ranked higher than C on 8 + 5 + 7 = 20 schedules and lower than C on 6.

Now compare B with D. B is ranked higher than D on 8 + 5 + 6 = 19 schedules and lower than D on 7. Therefore, B has a majority over each of the other choices and so is a Condorcet winner.

Mathematician of Note

Marquis de Condorcet (1743–1794)

A French mathematician, philosopher, and economist, he shared an interest in election theory with his friend, Jean-Charles de Borda.

Since B is a Condorcet winner, it is unnecessary to make comparisons between C and D. Although all comparisons do not always have to be made, it can be helpful to organize them in a table:

	A	B	C	D
A		L	L	L
B	W		W	W
C	W	L		W
D	W	L	L	

To see how a choice does in one-on-one contests, read across the row associated with that choice. You see that A, for example, loses in one-on-one contests with B, C, and D.

Although Condorcet's model may sound ideal, it sometimes fails to produce a winner. Consider the set of schedules shown in Figure 1.5.

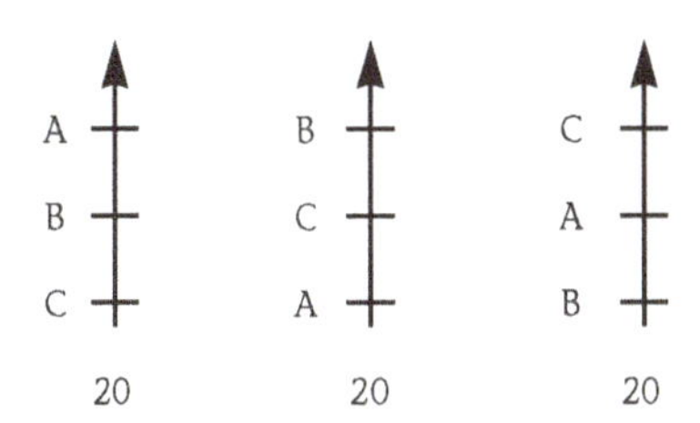

Figure 1.5. Preferences of 60 voters.

Notice that A is preferred to B on 40 of the 60 schedules but that A is preferred to C on only 20. Although C is preferred to A on 40 of the 60, C is preferred to B on only 20. Therefore there is no Condorcet winner.

You might expect that if A is preferred to B by a majority of voters and B is preferred to C by a majority of voters, then a majority of voters prefer A to C. But the example shows that this need not be the case.

In other mathematics classes you have learned that many relationships are transitive. The relation "greater than" (>), for example, is transitive because if $a > b$ and $b > c$, then $a > c$.

You have just seen that group-ranking models may violate the transitive property. Because this intransitivity seems contrary to intuition, it is known as a **paradox**. This particular paradox is sometimes referred to as the **Condorcet paradox**. There are other paradoxes that can occur with group-ranking models, as you will see in this lesson's exercises.

Exercises

1. Find a Condorcet winner in the soft drink ballot your class conducted in Lesson 1.1.

2. Propose a method for resolving situations in which there is no Condorcet winner.

3. In a system called **pairwise voting**, two choices are selected and a vote taken. The loser is eliminated, and a new vote is taken in which the winner is paired against a new choice. This process continues until all choices but one have been eliminated. An example of the use of pairwise voting occurs in legislative bodies in which bills are considered two at a time. The choices in the set of preferences shown in the following figure represent three bills being considered by a legislative body.

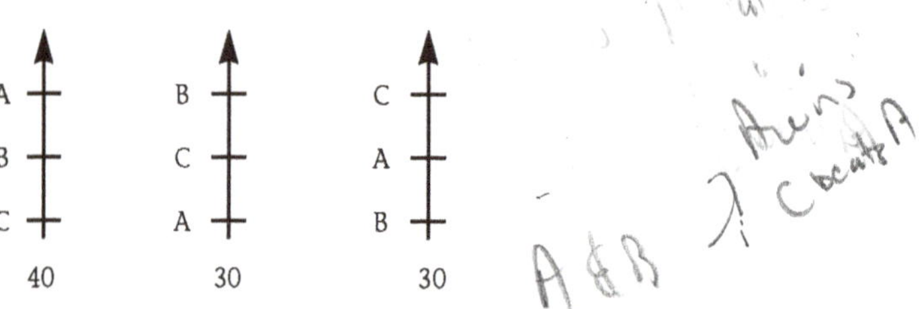

a. Suppose you are responsible for deciding which two bills appear on the agenda first. If you strongly prefer bill C, which two bills would you place on the agenda first? Why?

b. Is it possible to order the voting so that some other choice wins? Explain.

4. A panel of sportswriters is selecting the best football team in a league, and the preferences are distributed as follows.

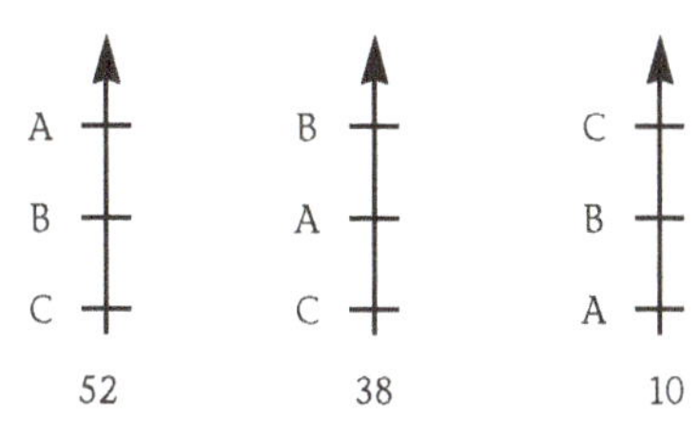

   a. Determine a best team using a 3-2-1 Borda count.

   b. The 38 who rank B first and A second decide to lie in order to improve the chances of their favorite and so rank C second. Determine the winner using a 3-2-1 Borda count.

5. When people decide to vote differently from the way they feel about the choices, they are said to be *voting insincerely*. People are often encouraged to vote insincerely because they have some idea of an election's result beforehand. Explain why such advance knowledge is possible.

6. Many political elections in the United States are decided with a plurality model. Construct a set of preferences with three choices in which a plurality model could encourage insincere voting. Identify the group of voters that might be encouraged to vote insincerely and explain the effect of their insincere voting on the election outcome.

7. Many people consider plurality models flawed because they can produce a winner that a majority of voters do not like.

   a. What percentage of voters ranks the plurality winner last in the preferences shown below?

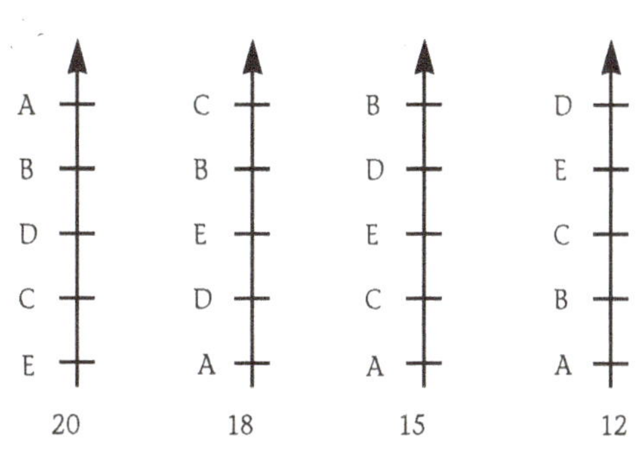

b. Runoffs are sometimes used to avoid the selection of a controversial winner. Is a runoff winner an improvement over a plurality winner in this set of preferences? Explain.

c. Do you consider a sequential runoff winner an improvement over plurality and runoff winners? Explain.

8. a. Use a runoff to determine a winner in the following set of preferences.

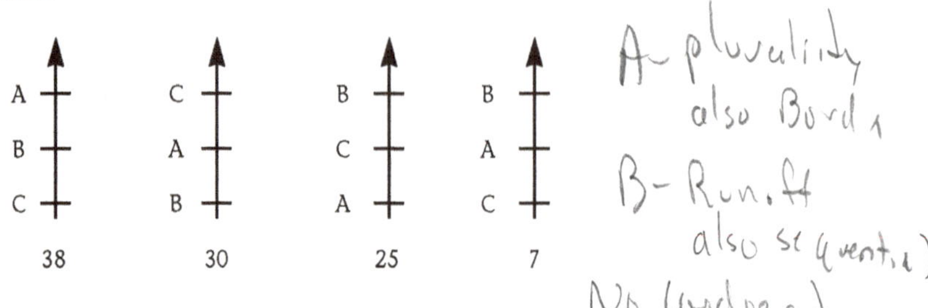

b. In some situations, votes are made public. For example, people have the right to know how their elected officials vote on issues. Suppose these schedules represent such a situation. Because they expect to receive some favors from the winner and because they expect A to win, the seven voters associated with the last schedule decide to change their preferences from

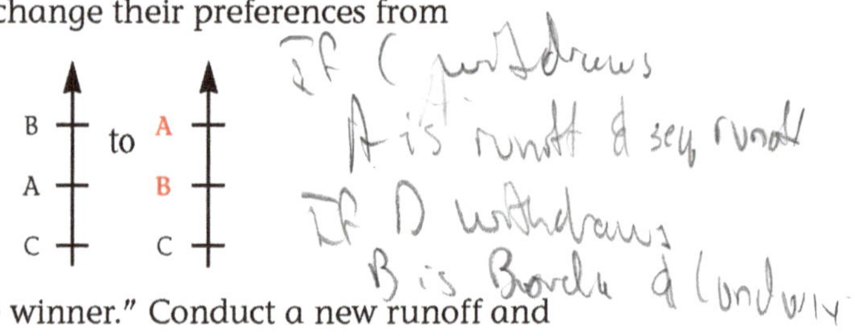

and to "go with the winner." Conduct a new runoff and determine the winner.

c. Explain why the results are a paradox.

9. a. Use a 4-3-2-1 Borda count to find a group ranking for the following set of preferences.

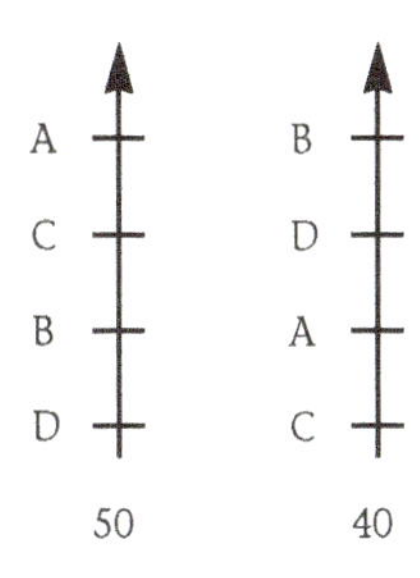

b. These preferences represent the ratings of four college athletic teams, and team C has been disqualified because of a recruiting violation. Write the schedules with team C removed and use a 3-2-1 Borda count to determine a group ranking.

c. Explain why these results are a paradox.

10. For each voting model discussed in this and the previous lesson (plurality, Borda, runoff, sequential runoff, and Condorcet), write a brief summary. Include at least one example of why the model can lead to unfair results.

11. In theory, Condorcet models require that each choice be compared with every other one, although in practice many of the comparisons do not have to be made in order to determine the winner. Consider the number of comparisons when every possible comparison is made.

Mathematicians sometimes find it helpful to represent the choices and comparisons visually. If there are only two choices, a single comparison is all that is necessary. In the diagram that follows, a point, or *vertex*, represents a choice, and a line segment, or *edge*, represents a comparison.

a. Add a third choice, C, to the diagram. Connect it to A and to B to represent the additional comparisons. How many new comparisons are there? What is the total number of comparisons?

b. Add a fourth choice, D, to the diagram. Connect it to each of A, B, and C. How many new comparisons are there, and what is the total number of comparisons?

c. Add a fifth choice to the diagram and repeat. Then add a sixth choice and repeat. Complete the following table.

Number of Choices	Number of New Comparisons	Total Number of Comparisons
1	0	0
2	1	1
3		
4		
5		
6		

12. Let C_n represent the total number of comparisons necessary when there are n choices. Write a recurrence relation that expresses the relationship between C_n and C_{n-1}.

13. U.S. College Hockey Online (USCHO) has several ranking systems. In a system called pairwise ranking, USCHO compares each team to every other team. In each comparison, the team that compares favorably to the other is awarded a point. The team with the most points is ranked first. Consider a simple version of this system in a league with 6 teams, A, B, C, D, E, and F. The following table shows the results of the comparisons. An X in a team's row indicates that it won the comparison with the team at the top of the column.

	A	B	C	D	E	F
A		X	X	X	X	
B			X	X		
C				X	X	
D					X	X
E		X				X
F	X	X	X			

Top Ten Pairwise Rankings

U.S. College Hockey Online
May 1, 2013

Rank	Team	Points
1	Quinnipiac	30
2	Minnesota	29
3	Massachusetts-Lowell	27
4	Notre Dame	26
5t	Miami	25
5t	Boston College	25
7	New Hampshire	24
8	North Dakota	23
9t	Denver	20
9t	Niagara	20

a. Find two pairwise comparisons in this table that demonstrate the transitive property. Find two comparisons that demonstrate a violation of the transitive property.

b. If each team receives a point for each comparison that it wins, find a group ranking for these teams.

c. Suggest a modification to the point system that is advantageous to team F.

d. Suppose team D drops out of the league. What effect does this have on the rankings you found in part b?

Computer/Calculator Explorations

14. Use the preference schedule program that accompanies this book to find a set of preferences with at least four choices that demonstrates the same paradox found in Exercise 8, but when sequential runoff is used.

15. Use the preference schedule program to enter several schedules with five choices. Use the program's features to alter your data in order to produce a set of preferences with several different winners. Can you find a set of preferences with five choices and five different winners? If so, what is the minimum number of schedules with which this can be done? Explain.

Projects

16. Research and report on paradoxes in mathematics. Try to determine whether the paradoxes have been satisfactorily resolved.

17. Research and report on paradoxes outside mathematics. In what way have these paradoxes been resolved?

18. Select an issue of current interest in your community or school that involves more than two choices. Have each member of your class vote by writing a preference schedule. Compile the preferences and determine winners by five different methods.

19. Investigate the contributions of Charles Dodgson (Lewis Carroll) to election theory. Was he responsible for any of the group-ranking procedures you have studied? What did he suggest doing when Condorcet models fail to produce a winner?

20. Investigate the system your school uses to determine academic rankings of students. Is it similar to any of the group-ranking procedures you studied? If so, could it suffer from any of the same problems? Propose another system and discuss why it might be better or worse than the one currently in use.

21. Investigate elections in your school (class officers, officers of organizations, homecoming royalty, and so forth). Report on the type of voting and the way winners are chosen. Recommend alternative methods and explain why you think the methods you recommend are fairer

Switzerland Jumps to #1 as Brand US Falls Further into Decline

PRNewswire
Oct. 24, 2012

Today, FutureBrand reveals its 8th annual ranking of the world's leading country brands - moving Switzerland to the #1 seat in the consultancy's 2012-13 Country Brand Index, a preeminent global study of country brands.

In keeping with past year's studies, the 2012-13 CBI ranks the world's countries - from their cultures, to their industries, to their economic vitality and public policy initiatives - based on global perceptions. Drawing insights from a 3,600 opinion-formers and frequent international travelers from 18 countries, FutureBrand used its proprietary Hierarchal Decision Model to determine how key audiences - residents, investors, tourists and foreign governments - see the world's country brands.

Top 25 country brands of 2012–13:

1. Switzerland (+1 from 2012)
2. Canada (–1)
3. Japan (+1)
4. Sweden (+3)
5. New Zealand (–2)
6. Australia (–1)
7. Germany (+4)
8. United States (–2)
9. Finland (–1)
10. Norway (+2)
11. United Kingdom (+2)
12. Denmark (+3)
13. France (–4)
14. Singapore (+2)
15. Italy (–5)
16. Maldives (+2)
17. Austria (0)
18. Netherlands (+5)
19. Spain (–5)
20. Mauritius (+2)
21. Ireland (–1)
22. Iceland (–3)
23. United Arab Emirates (+2)
24. Bermuda (–3)
25. CostaRica (–1)

Lesson 1.4

Arrow's Conditions and Approval Voting

Paradoxes, unfair results, and insincere voting are some of the problems that have caused people to look for better models for reaching group decisions. In this lesson you will learn of some recent and important work that has been done in attempts to improve the group-ranking process. First, consider an example involving pairwise voting.

Ten representatives of the language clubs at Central High School are meeting to select a location for the clubs' annual joint dinner. The committee must choose among a Chinese, French, Italian, or Mexican restaurant (see Figure 1.6).

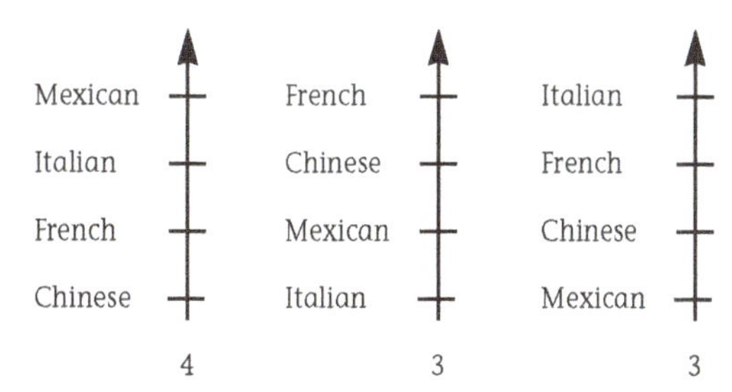

Figure 1.6. Preferences of 10 students.

Racquel says that because the last two dinners were at Mexican and Chinese restaurants, this year's dinner should be at either an Italian or a French restaurant. The group votes 7 to 3 in favor of the Italian restaurant.

Martin, who doesn't like Italian food, says that the community's newest Mexican restaurant has an outstanding reputation. He proposes that the group choose between Italian and Mexican. The other members agree and vote 7 to 3 to hold the dinner at the Mexican restaurant.

Sarah, whose parents own a Chinese restaurant, says that she can obtain a substantial discount for the event. The group votes between the Mexican and Chinese restaurants and selects the Chinese by a 6 to 4 margin.

Look carefully at the group members' preferences. Note that French food is preferred to Chinese by all, yet the voting selected the Chinese restaurant!

In 1951, paradoxes such as this led Kenneth Arrow, a U.S. economist, to formulate a list of five conditions that he considered necessary for a fair group-ranking model. These fairness conditions today are known as **Arrow's conditions**.

Mathematician of Note

Kenneth Arrow (1921–)

Kenneth Arrow received a degree in mathematics before turning to economics. His use of mathematical methods in election theory brought him worldwide recognition.

One of Arrow's conditions says that if every member of a group prefers one choice to another, then the group ranking should do the same. According to this condition, the choice of the Chinese restaurant when all members rated French food more favorably than Chinese is unfair. Thus, Arrow considers pairwise voting a flawed group-ranking method.

Arrow inspected common models for determining a group ranking for adherence to his five conditions. He also looked for new models that would meet all five. After doing so, he arrived at a surprising conclusion.

In this lesson's exercises, you will examine a number of group-ranking models for their adherence to Arrow's conditions. You will also learn Arrow's surprising result.

Arrow's Conditions

1. Nondictatorship: The preferences of a single individual should not become the group ranking without considering the preferences of the others.
2. Individual Sovereignty: Each individual should be allowed to order the choices in any way and to indicate ties.
3. Unanimity: If every individual prefers one choice to another, then the group ranking should do the same. (In other words, if every voter ranks A higher than B, then the final ranking should place A higher than B.)
4. Freedom from Irrelevant Alternatives: If a choice is removed, the order in which the others are ranked should not change. (The choice that is removed is known as an irrelevant alternative.)
5. Uniqueness of the Group Ranking: The method of producing the group ranking should give the same result whenever it is applied to a given set of preferences. The group ranking should also be transitive.

Exercises

1. Your teacher decides to order soft drinks for your class on the basis of the soft drink vote conducted in Lesson 1.1 but, in so doing, selects the preference schedule of a single student (the teacher's pet). Which of Arrow's conditions are violated by this method of determining a group ranking?
2. Instead of selecting the preference schedule of a favorite student, your teacher places all the individual preferences in a hat and draws one. If this method were repeated, would the same group ranking result? Which of Arrow's conditions does this method violate?

3. Do any of Arrow's conditions require that the voting process include a secret ballot? Is a secret ballot desirable in all group-ranking situations? Explain.

4. Examine the paradox demonstrated in Exercise 9 of Lesson 1.3 on page 23. Which of Arrow's conditions are violated?

5. Construct a set of preference schedules with three choices, A, B, and C, showing that the plurality method violates Arrow's fourth condition. In other words, construct a set of preferences in which the outcome between A and B depends on whether C is on the ballot.

6. You have seen situations in which insincere voting occurs. Do any of Arrow's conditions state that insincere voting should not be part of a fair group-ranking model? Explain.

7. Suppose that there are only two choices in a list of preferences and that the plurality method is used to decide the group ranking. Which of Arrow's conditions could be violated? Explain.

8. A group of voters have the preferences shown in the following figure.

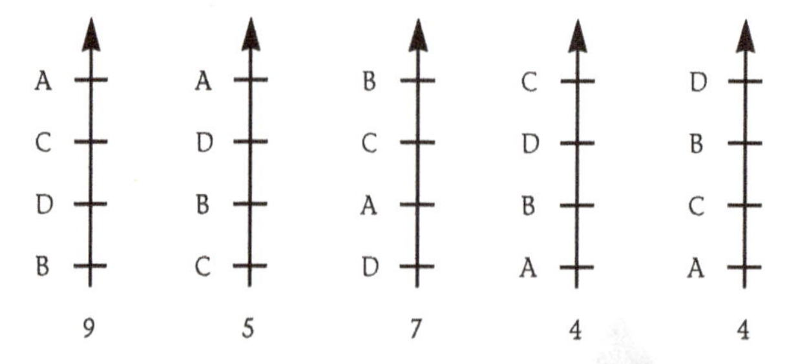

a. Use plurality, Borda, runoff, sequential runoff, and Condorcet models to find winners.

b. Investigate this set of preferences for violation of Arrow's fourth condition. That is, can a choice change a winner by withdrawing?

9. Read the news article about the Google search engine.

   a. Does the transitive property apply to individual Google voting? That is, if site A casts a Google vote for site B and site B casts a Google vote for site C, then must site A cast a Google vote for site C?

   b. Does the transitive property apply to the Google ranking system? That is, if site A ranks higher than site B and site B ranks higher than site C, then must site A rank higher than site C? Explain.

Is Google Page Rank Still Important?

Search Engine Journal
October 6, 2004

Since 1998 when Sergey Brin and Larry Page developed the Google search engine, it has relied on the Page Rank Algorithm. Google's reasoning behind this is, the higher the number of inbound links 'pointing' to a website, the more valuable that site is, in which case it would deserve a higher ranking in its search results pages.

If site 'A' links to site 'B', Google calculates this as a 'vote' for site B. The higher the number of votes, the higher the overall value for site 'B'. In a perfect world, this would be true. However, over the years, some site owners and webmasters have abused the system, implementing some 'link farms' and linking to websites that have little or nothing to do with the overall theme or topic presented in their sites.

10. After failing to find a group-ranking model for three or more choices that always obeyed all his fairness conditions, Arrow began to suspect that such a model does not exist. He applied logical reasoning and proved that no model, known or unknown, can always obey all five conditions. In other words, any group-ranking model violates at least one of Arrow's conditions in some situations.

    Arrow's proof demonstrates how mathematical reasoning can be applied to areas outside mathematics. This and other achievements earned Arrow the 1972 Nobel Prize in economics.

Although Arrow's work means that a perfect group-ranking model will never be devised, it does not mean that current models cannot be improved. Recent studies have led some experts to recommend **approval voting**.

In approval voting, you may vote for as many choices as you like, but you do not rank them. You mark all those of which you approve. For example, if there are five choices, you may vote for as few as none or as many as five.

a. Write a soft drink ballot like the one you used in Lesson 1.1. Place an "X" beside each of the soft drinks you find acceptable. At the direction of your instructor, collect ballots from the other members of your class. Count the number of votes for each soft drink and determine a winner.

b. Determine a complete group ranking.

c. Is the approval winner the same as the plurality winner in your class?

d. How does the group ranking in part b compare with the Borda ranking that you found in Lesson 1.1?

11. Examine Exercise 4 of Lesson 1.3 on page 21. Would any members of the panel of sportswriters be encouraged to vote insincerely if approval voting were used? Explain.

12. What is the effect on a group ranking of casting approval votes for all choices? Of casting approval votes for none of the choices?

13. The voters whose preferences are represented below all feel strongly about their first choices but are not sure about their second and third choices. They all dislike their fourth and fifth choices. Since the voters are unsure about their second and third choices, they flip coins to decide whether to give approval votes to their second and third choices.

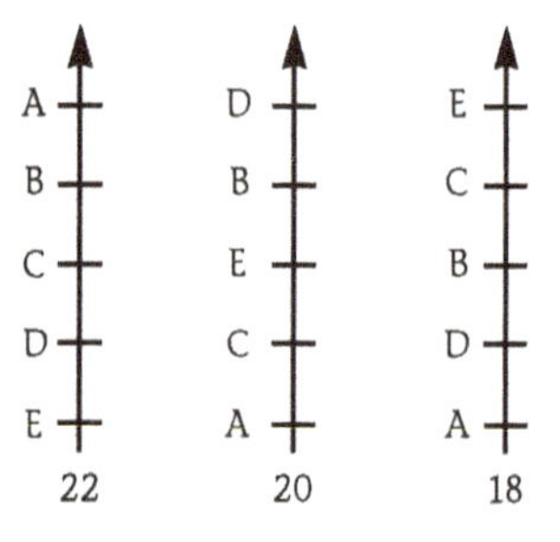

a. Assuming the voters' coins come up heads half the time, how many approval votes would you expect each of the five choices to get? Explain your reasoning.

b. Do the results seem unfair to you in any way? Explain.

14. Approval voting offers a voter many choices. If there are three candidates for a single office, for example, the plurality system offers the voter four choices: vote for any one of the three candidates or for none of them. Approval voting permits the voter to vote for none, any one, any two, or all three.

    To investigate the number of ways in which you can vote under approval voting, consider a situation with two choices, A and B. You can represent voting for none by writing { }, voting for A by writing {A}, voting for B by writing {B}, and voting for both by writing {A, B}.

    a. List all the ways of voting under an approval system when there are three choices.

    b. List all the ways of voting under an approval system when there are four choices.

    c. Generalize the pattern by letting V_n represent the number of ways of voting under an approval system when there are n choices and writing a recurrence relation that describes the relationship between V_n and V_{n-1}.

15. Listing all the ways of voting under the approval system can be difficult if not approached systematically. The following algorithm describes one way to find all the ways of voting for two choices. The results are shown applied to a ballot with five choices, A, B, C, D, and E.

    1. List all choices in order in List 1.
    2. Draw a line through the first choice in List 1 that doesn't already have a line drawn through it. Write this choice as many times in List 2 as there are choices in List 1 without lines through them.
    3. Beside each item you wrote in List 2 in step 2, write a choice in List 1 that does not have a line through it.
    4. Repeat steps 2 and 3 until each choice has a line through it. The items in the second list show all the ways of voting for two items.

List 1	List 2
A B C D E	
~~A~~ B C D E	A
	A
	A
	A
	A B
	A C
	A D
	A E

Write an algorithm that describes how to find all the ways of voting for three choices. You may use the results of the previous algorithm to begin the new one.

16. Many patterns can be found in the various ways of voting when the approval system is used. The following table shows the number of ways of voting for exactly one item when there are several choices on the ballot. For example, in Exercise 14, you listed all the ways of voting when there are three choices on the ballot. Three of these, {A}, {B}, and {C}, are selections of one item.

Number of Choices on the Ballot	Number of Ways of Selecting Exactly One Item
1	1
2	2
3	3
4	—
5	—

Complete the table.

17. Let $V1_n$ represent the number of ways of selecting exactly one item when there are n choices on the ballot and write a recurrence relation that expresses the relationship between $V1_n$ and $V1_{n-1}$.

18. The following table shows the number of ways of voting for exactly two items when there are from one to five choices on the ballot. For example, your list in Exercise 14 shows that when there are three choices on the ballot, there are three ways of selecting exactly two items: {A, B}, {A, C}, and {B, C}.

Number of Choices on the Ballot	Number of Ways of Selecting Exactly Two Items
2	1
3	3
4	—
5	—

Complete the table.

19. Let $V2_n$ represent the number of ways of selecting exactly two items when there are n choices on the ballot and write a recurrence relation that expresses the relationship between $V2_n$ and $V2_{n-1}$. Can you find more than one way to do this?

Computer/Calculator Explorations

20. Design a computer program that lists all possible ways of voting when approval voting is used. Use the letters A, B, C, . . . to represent the choices. The program should ask for the number of choices and then display all possible ways of voting for one choice, two choices, and so forth.

Projects

21. Investigate the number of ways of voting under the approval system for other recurrence relations (see Exercises 16 through 19). For example, in how many ways can you vote for three choices, four choices, and so forth?

22. Arrow's result is an example of an impossibility theorem. Investigate and report on other impossibility theorems.

23. Research and report on Arrow's theorem. The theorem is usually proved by an indirect method. What is an indirect method? How is it applied in Arrow's case?

24. In approval voting voters apply an approve or disapprove rating to each choice. Thus, approval voting is a rating system--not a ranking system. In 2007, Michel Balinski and Rida Laraki proposed another type of rating system called majority judgment, in which voters are allowed more than two ratings. Research and report on majority judgment. What are its advantages and disadvantages over other voting models?

Lesson 1.5

Weighted Voting and Voting Power

The first four lessons of this chapter examined situations in which all voters are considered equals. This lesson examines situations in which some voters have more votes than other voters.

Public Hearing Set for Madison County Weighted Voting

Oneida Daily Dispatch
April 11, 2013

The Madison County Board of Supervisors will hear public comments over upgrading weighted voting totals.

In Madison County, supervisors vote based on a weighted formula calculated by the population of their towns. That voting configuration hasn't been updated since 2002. Using population data from the 2010 Census, the number has been recalculated.

Under the proposed law, there are to be 1500 votes total. The votes are not divisible and are to be cast as one lump sum by each town. The county uses the Banzhaf power index to determine how many votes each supervisor has.

Members of legislative bodies such as the United States Congress and county boards represent districts. The constitutional principle of one person, one vote requires that such districts be approximately the same size. Thus, every 10 years, district lines are redrawn to reflect changes in population. But redrawing boundaries can be difficult for many reasons. Some legislative bodies try to resolve difficulties by adopting a system in which some votes carry more weight than others.

Consider a simple example. A small high school has 110 students. Because of recent growth in the size of the community, the sophomore class is quite large. It has 50 members, and the junior and senior classes each have 30 members.

The school's student council is composed of a single representative from each class. Each of the three members is given a number of votes proportionate to the size of the class represented. Accordingly, the sophomore representative has five votes, and the junior and senior representatives each have three. The passage of any issue that is before the council requires a simple majority of six votes.

The student council's voting model is an example of **weighted voting**. Weighted voting occurs whenever some members of a voting body have more votes than others have.

In recent years, several people have questioned whether weighted voting is fair. Among them is John Banzhaf III, a law professor at George Washington University who has initiated several legal actions against weighted voting procedures used in local government.

To understand Banzhaf's objection to weighted voting, consider the number of ways that voting on an issue could occur in the student council example.

It is possible that an issue is favored by none of the members, one of them, two of them, or all three. In which cases does an issue pass? The following list gives all possible ways of voting for an issue and the associated number of votes.

Mathematician of Note

John Banzhaf III (1940–)

John Banzhaf, a law professor who also holds an engineering degree, is a well-known consumer rights advocate.

{; 0} {So; 5} {Jr; 3} {Sr; 3} {So, Jr; 8} {So, Sr; 8} {Jr, Sr; 6} {So, Jr, Sr; 11}

For example, {Jr, Sr; 6} indicates that the junior and senior representatives vote for an issue and that they have a total of six votes between them.

Each of these collections of voters is called a *coalition*. Those with enough votes to pass an issue are **winning coalitions**. The winning coalitions in this example are those with six or more votes and are listed below along with their respective vote totals.

{So, Jr; 8} {So, Sr; 8} {Jr, Sr; 6} {So, Jr, Sr; 11}

The last winning coalition is different from the other three in one important respect: If any one of the members decides to vote differently, the coalition still wins. No single member is essential to the coalition. Banzhaf argued that the only time a voter has power is when the voter belongs to a coalition that needs the voter in order to pass an issue. The coalitions for which at least one member is essential are

{So, Jr; 8} {So, Sr; 8} {Jr, Sr; 6}.

Notice that the sophomore representative is essential to two of the coalitions, which is also true of the junior and senior representatives. In other words, about the same number of times, each of the representatives can be expected to cast a key vote in passing an issue.

A paradox: Although the votes have been distributed to give greater power to the sophomores, the outcome is that all members have the same power!

Since distributing the votes in a way that reflects the population distribution does not always result in a fair distribution of power, mathematical procedures can be used to develop ways to measure actual power in weighted voting situations.

A measure of the power of a member of a voting body is called a **power index**. In this lesson, a voter's power index is the number of winning coalitions in which the voter is essential. For example, in the student council situation, the sophomore representative is essential to two winning coalitions and thereby has a power index of 2, as do the junior and senior representatives.

A Power Index Algorithm

1. List all coalitions of voters that are winning coalitions.
2. Select any voter, and record a 0 for that voter's power index.
3. From the list in step 1, select a coalition of which the voter selected in step 2 is a member. Subtract the number of votes the voter has from the coalition's total. If the result is less than the number of votes required to pass an issue, add 1 to the voter's power index.
4. Repeat step 3 until all coalitions of which the voter chosen in step 2 is a member are checked.
5. Repeat steps 2 through 4 until all voters are checked.

Exercises

1. Consider a situation in which A, B, and C have 3, 2, and 1 votes respectively, and in which 4 votes are required to pass an issue.
   a. List all possible coalitions and all winning coalitions.
   b. Determine a power index for each voter.
   c. If the number of votes required to pass an issue is increased from 4 to 5, determine a power index for each voter.
2. In a situation with three voters, A has 7 votes, B has 3, and C has 3. A simple majority is required to pass an issue.
   a. Determine a power index for each voter.
   b. A *dictator* is a member of a voting body who has all the power. A *dummy* is a member who has no power. Are there any dictators or dummies in this situation?
3. The student council example in this lesson depicts a situation with three voters that results in equal power for all three. In Exercises 1 and 2, power is distributed differently. Find a distribution of votes that results in a power distribution among three voters that is different from the ones you have already seen. How many new power distributions in situations with three voters can you find?

4. In this lesson's student council example, can the votes be distributed so that the members' power indices are proportionate to the class sizes? Explain.

5. In this lesson's student council example, suppose that the representatives of the junior and senior classes always differ on issues and never vote alike. Does this make any practical difference in the power of the three representatives? Explain.

6. (See Exercise 14 of Lesson 1.4 on page 33.) Let C_n represent the number of coalitions that can be formed in a group of n voters. Write a recurrence relation that describes the relationship between C_n and C_{n-1}.

7. One way to determine all winning coalitions in a weighted voting situation is to work from a list of all possible coalitions. Use A, B, C, and D to represent the individuals in a group of four voters and list all possible coalitions.

8. Weighted voting is commonly used to decide issues at meetings of corporate stockholders. Each member has one vote for each share of stock held.

   a. A company has four stockholders: A, B, C, and D. They own 26%, 25%, 25%, and 24% of the stock, respectively, and more than 50% of the vote is needed to pass an issue. Determine a power index for each stockholder. (Use your results from Exercise 7 as an aid.)

   b. Another company has four stockholders. They own 47%, 41%, 7%, and 5% of the stock. Find a power index for each stockholder.

   c. Compare the percentage of stock owned by the smallest shareholder in parts a and b. Do the same for the power index of the smallest stockholder in each case.

9. A landmark court decision on voting power involved the Nassau County, New York, Board of Supervisors. In 1964, the board had six members. The number of votes given to each was 31, 31, 21, 28, 2, and 2.

   a. Determine a power index for each member.

   b. The board was composed of representatives of five municipalities with these populations:

Hempstead	728,625
North Hempstead	213,225
Oyster Bay	285,545
Glen Cove	22,752
Long Beach	25,654

   The members with 31 votes represented Hempstead. The others each represented the municipality listed in the same order as in the table. Compare the power indices of the municipalities with their populations.

10. A *minimal winning coalition* is one in which all voters are essential.

    a. Give an example of a weighted voting situation with a winning coalition for which at least one but not all of the voters is essential. Identify the minimal winning coalitions in this situation.

    b. Is defining a voter's power index as the number of minimal winning coalitions to which the voter belongs equivalent to the definition used in this lesson? Explain.

Nassau Districting Ruled Against Law

The New York Times
January 15, 1970

Albany—The Court of Appeals ruled today that the present "weighted voting" plan of the Nassau County Supervisors was unconstitutional but that a new plan was not necessary until after the 1970 federal census.

In a unanimous opinion, the state's highest court said the county's present charter provision is a clear violation of the one-man-one-vote principle, in that it specifically denied the town of Hempstead representation that reflected its population.

The town, the court pointed out, constituted 57.12 percent of the county's population, but because of the weighted voting plan its representatives on the board could cast only 49.6 percent of the board's vote.

11. The president of the United States is chosen in the Electoral College, a system that can be considered a form of weighted voting among the states. The number of electors given to each state (and the District of Columbia) is equal to its representation in Congress. That is, the number of electors equals the number of members of the House of Representatives plus two (the number of senators). In 2000, Albert Gore won the popular vote by about half a million votes over George Bush, but lost in the Electoral College.

    a. In the 2000 census, the population of California (the most populous state) was 33,871,648. The population of Wyoming (the least populous state) was 493,782. California has 52 representatives in the U.S. House. Wyoming has 1. Use these data to construct an argument that the electoral vote distribution is weighted in favor of small states.

    b. Some reformers have proposed removing the electoral weighting. They suggest equating the number of electors with the membership in the U.S. House only. The following table shows the Electoral College votes in the 2000 election. Would this reform proposal change the 2000 election results? Explain.

	Bush	Gore		Bush	Gore		Bush	Gore
Alabama	9		Kentucky	8		North Dakota	3	
Alaska	3		Louisiana	9		Ohio	21	
Arizona	8		Maine		4	Oklahoma	8	
Arkansas	6		Maryland		10	Oregon		7
California		54	Massachusetts		12	Pennsylvania		23
Colorado	8		Michigan		18	Rhode Island		4
Connecticut		8	Minnesota		10	South Carolina	8	
Delaware		3	Mississippi	7		South Dakota	3	
D. C.		3	Missouri	11		Tennessee	11	
Florida	25		Montana	3		Texas	32	
Georgia	13		Nebraska	5		Utah	5	
Hawaii		4	Nevada	4		Vermont		3
Idaho	4		New Hampshire	4		Virginia	13	
Illinois		22	New Jersey		15	Washington		11
Indiana	12		New Mexico		5	West Virginia	5	
Iowa		7	New York		33	Wisconsin		11
Kansas	6		North Carolina	14		Wyoming	3	

c. The Electoral College is mandated in the United States Constitution. Why do you think the founders of the country instituted a system weighted in favor of small states? (Do some research if you are not sure.)

Computer Explorations

12. Use the weighted voting software that accompanies this book to experiment with different weighted voting systems when there are three voters. Change the number of votes given to each voter and the number of votes required to pass an issue. How many different power distributions are possible? Do the same for weighted voting systems with four voters.

Projects

13. The Security Council of the United Nations is composed of five permanent members and ten others who are elected to two-year terms. For a measure to pass, it must receive at least nine votes that include all five of the permanent members. Determine a power index for a permanent member and for a temporary member. (Assume that all members are present and voting.)

Majority of Member States Back Pesticide Ban

European Voice
April 29, 2013

A majority of member states today voted to approve a European Commission proposal to ban the use of neonicotinoid seed treatment pesticides, thought to be harmful to bees.

The outcome was close, following a first vote last month which was inconclusive, with neither a majority for or against. Crucially, Germany changed its position in today's vote to approve the ban, after having abstained last month. The 15 member states voting in favor gave 189 weighted votes, versus 125 weighted votes against, including those of the UK. Abstentions amounted to 33 weighted votes.

14. Research and report on other power indices. What, for example, is the Shapley-Shubik power index?
15. What is the effect of the Electoral College system on the power of individual voters in selecting the president? Research the matter and report on the relative power of voters in different states.
16. Research and report on court decisions about weighted voting.

U.K., U.S. Wield Most Cyber Power

National Journal
January 13, 2012

The United Kingdom and United States lead other developing countries in their ability to withstand cyberattacks and develop strong digital economies, according to a new study by Booz Allen Hamilton and the Economist Intelligence Unit.

The U.K. tops the rest of the Group of 20 nations, including the U.S., in the "Cyber Power Inex." The European Union, considered the 20th member of the G20, was not included.

The index, which put the U.S. in the No. 2 spot, rates the countries on their legal and regulatory frameworks; economic and social issues; technology infrastructure; and industry.

"Overall, the top five countries exhibiting cyber power, as measured by the index--the U.K.; the U.S.; Australia; Germany; and Canada--illustrate that developed Western countries are leading the way into the digital era," the companies said in a statement.

Chapter Extension

Proportional Representation

Democracies are founded on the principle that all people should have representation in government. Most democratic countries have minority populations who feel that they should be represented by one of their own members, and courts have agreed.

However, ensuring minority representation in a legislative body such as the United States House of Representatives is not always easy. If, for example, a state has five representatives in the U.S. House and a minority is 40% of the state's population, it seems reasonable that the minority should hold $0.4 \times 5 = 2$ of the seats. But, depending on how the boundaries of the state's five congressional districts are drawn, the minority might hold no seats.

The task of ensuring minority representation in the U.S. House must be undertaken every ten years when district boundaries are redrawn after a census. In states with a significant minority population, districts are established in which the minority has over half the population. However, this practice sometimes produces districts with a shape so unusual that courts reject them.

How, then, does a democracy provide fair representation in government? Many democracies use some form of **proportional representation**. Although there are several proportional representation models in use, they all have a common goal: to ensure that minorities and/or political parties have representation in government proportionate to their numbers in the general population.

One form of proportional representation is achieved through a practice called **cumulative voting**. In this model, several representatives are elected from a single region (e.g., state). For example, if the region has three representatives, each voter has three votes. The voter can split the three votes in any way, and can even cast more than one vote for a single candidate. Cumulative voting is used in some local elections in the United States.

Another model, the **party-list model**, is used in some European countries. In this model, each party has a list of candidates on the ballot. Each voter votes for one of the parties. When the election is over, the party receives a number of seats proportionate to the vote it received. The seats are usually assigned to names on the party's ballot by taking them in order from the top of the ballot until the correct number is obtained.

The **mixed member model** has voters vote for a party and a candidate. A portion of the seats is assigned to candidates and another portion to parties. All individual winning candidates receive seats. The remaining seats are awarded to members of parties that do not have a number of individual seats proportionate to the vote they received. In 1994, New Zealand voters abandoned a plurality model like the one currently used in the United States in favor of the mixed member model, which is also used in several European countries.

In the **preference vote model**, voters rank the candidates. A threshold is established, and all candidates with a vote total over the threshold are elected. Remaining seats are distributed by conducting a form of sequential runoff among the remaining candidates.

Canadians Favor Proportional Representation

Winnipeg Free Press
April 8 2013

A new poll released today by Fair Vote Canada shows a majority of Canadians, of all political stripes, favor changing our voting system to incorporate proportional representation.

That would mean instead of just being elected based on winning the most votes, at least some MPs would be elected based on the share their party gets of the popular vote. Canada's first- past-the-post system is one of the last of its kind in the world, with most other developed nations using some form of proportional representation.

The poll found 70 per cent of Canadians strongly (24 per cent) or somewhat (46 per cent) support moving towards proportional representation.

Just as a reminder of why there is a push to amend our system: In the 2011 federal election, the Conservatives won 53.8 per cent of the seats and a majority government with few checks on its power, with just 39.6 per cent of the votes. Much the same as the Liberals in 2000 won 57 per cent of the seats with 40.8 per cent of the vote.

Chapter 1 Review

1. Write a summary of what you think are the important points of this chapter.
2. Consider the following set of preferences.

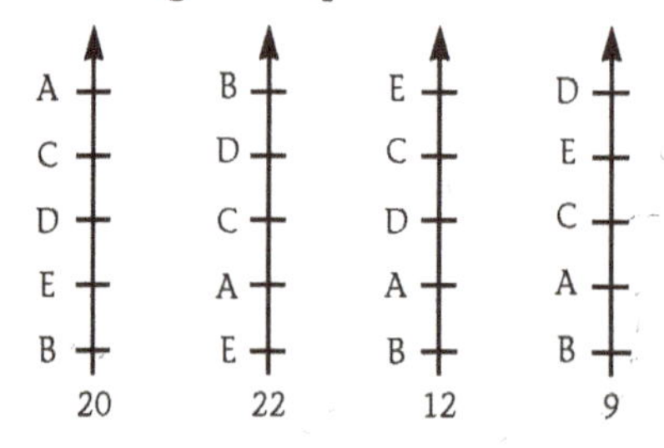

   a. Determine a winner using a 5-4-3-2-1 Borda count.

   b. Determine a plurality winner.

   c. Determine a runoff winner.

   d. Determine a sequential runoff winner.

   e. Determine a Condorcet winner.

   f. Suppose that this election is conducted by an approval model and all voters approve of the first two choices on their preference schedules. Determine an approval winner.

3. Complete the following table for the recurrence relation $B_n = 2B_{n-1} + n$.

n	B_n
1	3
2	$2(3) + 2 = 8$
3	
4	
5	

4. In this chapter, you encountered several paradoxes involving group-ranking models.
   a. One of the most surprising paradoxes occurs when a winning choice becomes a loser when the choice's standing actually improves. In which group-ranking model(s) can this occur?
   b. Discuss at least one other paradox that occurs with group-ranking models.
5. In the 1912 presidential election, polls showed that the preferences of voters were as follows.

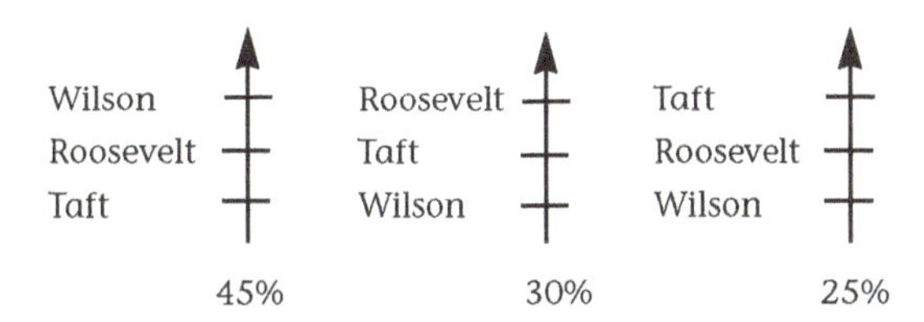

   a. Who won the election? Was he a majority winner?
   b. How did the majority of voters feel about the winner?
   c. How could one of the groups of voters have changed the results of the election by voting insincerely?
   d. Discuss who might have won the election if a different model had been used.
6. Your class is ranking soft drinks and someone suggests that the names of the soft drinks be placed in a hat and the group ranking be determined by drawing them from the hat. Which of Arrow's conditions does this method violate?
7. After their final round of skating in the 1995 World Figure Skating Championship, Chen Lu of China, Nicole Bobek of the United States, and Surya Bonaly of France were in first, second, and third place, respectively, with little chance of any remaining skater passing them. However, when American Michelle Kwan skated, she did well enough to move into fourth place. But something else quite surprising happened. Kwan's scores reversed the positions of Bobek and Bonaly. Which of Arrow's conditions did the scoring system violate in this case? Explain.

8. State Arrow's theorem. In other words, what did Arrow prove?

9. Can the point system used to do a Borda count affect the ranking (for example, a 5-3-2-1 system instead of a 4-3-2-1 system)? Construct an example to support your answer.

10. The 1992 presidential election was unusual because of a strong third-party candidate. In that election Bill Clinton received 43% of the popular vote, George Bush 38%, and Ross Perot 19%.

    Steven Brams and Samuel Merrill III used polling results to estimate the percentage of those voting for one candidate who also approved of another.

    - Approximately 15% of Clinton voters approved of Bush and approximately 30% approved of Perot.
    - Approximately 20% of Bush voters approved of Clinton and approximately 20% approved of Perot.
    - Approximately 35% of Perot voters approved of Clinton and approximately 30% approved of Bush.

    a. Estimate the percentage of approval votes each candidate would have received if approval voting had been used in the election.

    b. Find the total of the three percentages you gave as answers in part a. Explain why the total is not 100%.

11. Choose an election model from those you have studied in this chapter that you think best to use to determine a winner for the following preferences. Explain why you think your choice of method is best.

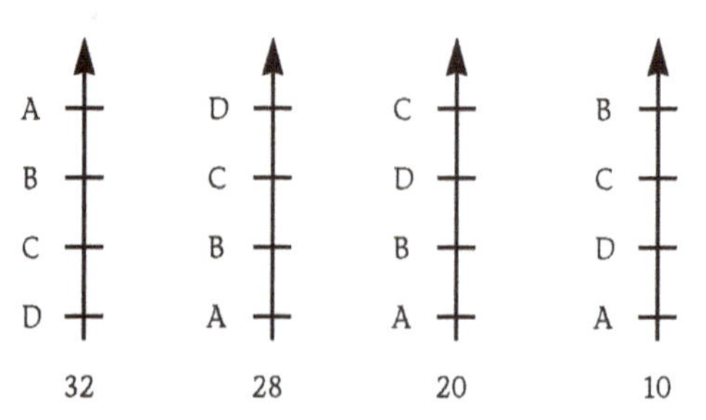

12. In 2012, Washington Nationals outfielder Bryce Harper won the National League Jackie Robinson Rookie of the Year Award. (There are two winners each year: one in the National League and one in the American League.) The results of the National League Voting are shown in the following table.

Player	First-Place Votes	Second-Place Votes	Third-Place Votes	Total
Bryce Harper	16	8	8	112
Wade Miley	12	13	6	105
Todd Frazier	3	7	9	45
Wilin Rosario	1	2	1	12
Norichika Aoki		2	5	11
Yonder Alonso			1	1
Matt Carpenter			1	1
Jordan Pachecho			1	1

What type of voting model is used to select rookie of the year?

13. There are conditions other than Arrow's that some experts consider important to a fair group-ranking model. For example, Donald Saari thinks that a good method should not retain the same winner if the voters reverse their preferences.

    a. To see how this reversal effect works, find a plurality winner for the following set of voter preferences. Then reverse the order of the rankings on each schedule and find a plurality winner.

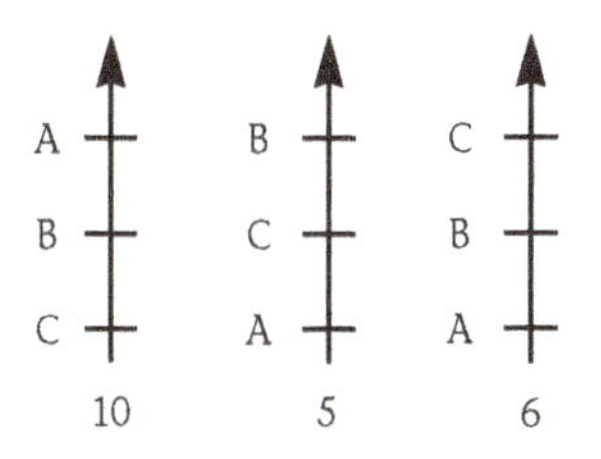

Mathematician of Note

Donald Saari (1940–)

Professor Saari is Distinguished Professor of Mathematics and Economics at University of California Irvine.

b. Do any other models that you studied in this chapter demonstrate a reversal effect in the set of preferences in part a? That is, do any other models leave the winner unchanged when preferences are reversed? Explain.

c. Examine the following set of preferences for reversal effects.

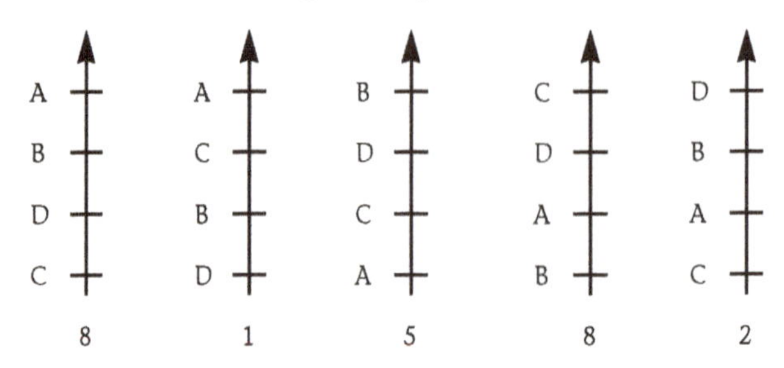

14. Consider a situation in which voters A, B, C, and D have 4, 3, 3, and 2 votes, respectively, and 7 votes are needed to pass an issue.

    a. List all winning coalitions and their vote totals.

    b. Find a power index for each voter.

    c. Do the power indices reflect the distribution of votes? Explain.

    d. Suppose the number of votes necessary to pass an issue increases from 7 to 8. How does this change the voters' power indices?

15. A county planning commission has five members. Each member's vote is weighted to reflect the population of the community the member represents. Member A has 1 vote, B has 1 vote, C has 2 votes, D has 5 votes, and E has 6 votes. A simple majority of the vote total is required to pass an issue. Do any of the members seem to have considerably more or less power than intended? Explain.

Bibliography

Amy, Douglas J. 2002. *Real Choices/New Voices:* New York: Columbia University Press.

Amy, Douglas J. 2000. *Behind the Ballot Box: A Citizen's Guide to Voting Systems*. Westport CT: Praeger.

Balinski, MIchel and Rida Laraki. 2011. *Majority Judgment: Measuring, Ranking, and Electing.* Cambridge/New York: MIT Press.

Brams, Steven J. 1995. "Approval Voting on Bills and Propositions." *The Good Society* 5(2): 37–39.

Brams, Steven J. 2002. *Paradoxes in Politics*. New York: Free Press.

Brams, Steven J., and Peter Fishburn. 2007. *Approval Voting*. New York: Springer.

Bunch, Bryan. 1997. *Mathematical Fallacies and Paradoxes*. New York: Dover.

COMAP. 2013. *For All Practical Purposes: Introduction to Contemporary Mathematics.* 9th ed. New York: W. H. Freeman.

Cole, K. C. 1999. *The Universe and the Teacup: The Mathematics of Truth and Beauty*. New York: Mariner.

Davis, Morton. 1980. *Mathematically Speaking*. New York: Harcourt Brace Jovanovich.

Falletta, Nicholas. 1990. *The Paradoxican*. New York: Wiley.

Gardner, Martin. 1980. "From Counting Votes to Making Votes Count: The Mathematics of Elections." *Scientific American* 251(4): 16.

Hoffman, Paul. 1988. *Archimedes' Revenge*. New York: W. W. Norton & Company.

Lijphart, Arend. 2012. *Patterns of Democracy: Government Forms and Performance in Thirty-Six Countries*. New Haven: Yale University Press.

Lucas, William F. 1992. *Fair Voting: Weighted Votes for Unequal Constituencies*. Lexington, MA: COMAP, Inc.

Lum, Lewis, and David C. Kurtz. 1989. *Voting Made Easy: A Mathematical Theory of Election Procedures*. Greensboro, NC: Guilford Press.

Niemi, Richard G., and William H. Riker. 1976. "The Choice of Voting Systems." *Scientific American* 234(6): 21.

Pakhomov, Valery. 1993. "Democracy and Mathematics." *Quantum* January-February: 4-9.

Rush, Mark E., and Richard L. Engstrom. 2001. *Fair and Effective Representation: Debating Electoral Reform and Minority Rights.* Lanham, MD: Rowman & Littlefield.

Saari, Donald. 2001. *Chaotic Elections!: A Mathematician Looks At Voting.* Providence, RI: American Mathematical Society.

Saari, Donald. 2001. *Decisions and Elections: Explaining the Unexpected.* New York: Cambridge University Press.

Saari, Donald. 2008. *Disposing Dictators, Demystifying Voting Paradoxes: Social Choice Analysis.* New York: Cambridge University Press.

Szpiro, George G. 2010. *Numbers Rule: The Vexing Mathematics of Democracy.* Princeton, NJ: Princeton University Press.

Taylor, Alan D. 2008. *Mathematics and Politics: Strategy, Voting, Power and Proof.* New York: Springer.

Taylor, Alan D. 2005. *Social Choice and the Mathematics of Manipulation.* New York: Cambridge University Press.

Fair Division

CHAPTER

2

Whether the parties involved are individuals, organizations, communities, states, or nations; a joint endeavor raises questions of fairness. Massive undertakings like monitoring the health of the earth raise many fairness issues. For example, how can the cost be shared equitably among all nations? What is a fair way to share the data that is generated?

Fair division questions arise in the simplest of situations. For example, you probably experienced unfairness as a child when you felt that someone received a piece of cake or portion of ice cream that was better than yours.

- How can a portion of food be divided fairly among two or more children?
- Is the meaning of fairness when food is divided among children different from the meaning of fairness when an estate is divided among heirs or when seats in Congress are divided among states?
- Are the methods that are commonly used to divide food, estates, and legislatures necessarily the fairest methods?

Discrete mathematics plays an important role in answering these questions.

In this chapter, you will consider several fairness issues and examine mathematical models for resolving them. Most of these fair division models are algorithms that can be expressed in a few numbered steps.

Lesson 2.1

A Fair Division Activity

Nations Seek Flexible Climate Approach

Reuters
May 3, 2013

New, more flexible ways to fight climate change were sketched out on Friday at the end of a week of talks between 160 nations.

The meeting of senior officials in Bonn, Germany, aired formulas to resolve disputes between rich and poor on sharing out the cost of curbing greenhouse gas emissions as part of a new U.N. deal, a successor to the 1997 Kyoto Protocol.

Attempts to reach agreement have foundered above all on a failure to agree on the contribution developing countries should make to curbing the industrial emissions responsible for global warming.

The United Nations said there was a broad agreement among delegates that any new accord should have flexibility to ratchet up curbs on emissions, without a need for further negotiations, if scientific findings about floods, droughts and rising sea levels worsen in coming years.

There are many situations in which the division of an object in a fair way is important to those involved. Three of the most common are the division of food among children, a house in an estate among heirs, and the seats in a governmental body among districts. Each has characteristics that make it different from the others.

In this lesson, you will consider an example of each of these three situations and propose solution models of your own. As is the case in election theory, fair division is an area of discrete mathematics in which important problems can be understood and solved without having considerable background knowledge.

Explore This

Below are three fair division problems. You will find it helpful to discuss one or more of them with a few other people, as time permits. Here is a way to divide the problems among small groups in your class. At the direction of your instructor, divide your class into groups of three people. Write the numbers 1, 2, or 3 on each of several slips of paper. Have a person from each group draw one of the slips from a bag or box. Each group should discuss the problem that corresponds to the number drawn by the group's representative. Allow about 15 minutes for the discussions.

After all groups have finished their discussions, a spokesperson for each group should present the group's result to the class. Each group that discussed Problem 1 should report first, and so forth.

In your notebook, record the method used by each group. You will need that information for this lesson's exercises.

1. Martha and Ray are siblings who want to divide the last piece of the cake that their mother baked yesterday. Propose a way to divide the last piece of cake that will seem fair to both Martha and Ray.
2. Juan and Mary are the only heirs to their mother's estate. The only object of significant value is the house in which they were raised. Propose a way to resolve the issue of dividing the house that will seem fair to both Juan and Mary.
3. The sophomore, junior, and senior classes of Central High School have 333, 288, and 279 members, respectively. The school's student council is composed of 20 members divided among the three classes. Determine a fair number of seats on the council for each class.

The members of your class may not have reached consensus on the best way to solve each of the three problems. In this lesson's exercises you will consider some of the important fairness issues in each of the three problems.

Exercises

1. a. Did any groups resolve Problem 1 by relying on the mother's authority? In what way or ways was this done?

   b. Does the resolution of such a problem by the mother or other authority figure always produce a solution that seems fair to both children? Explain.

   c. Cite at least two examples of situations in which fair division problems are resolved by an authority.

2. a. Did any of the groups use a random event such as a coin flip to resolve Problem 1? In what way or ways was this done?

   b. Does the use of randomness in such a problem always produce a solution that seems fair to both children? Explain.

   c. Cite at least two examples of situations in which randomness is used to resolve an issue.

3. a. Did any of the groups use a means of measuring the piece of cake to resolve Problem 1? In what way or ways was this done?

   b. Does the use of measurement in such problems always produce a solution that seems fair to both children? Explain.

   c. Give an example of a situation in which measurement is likely to result in an agreeable solution to a fair division problem.

4. A common way of resolving Problem 1 is to have one of the children cut the cake into two pieces and to have the other choose first.

   a. If Martha cuts the cake, how does she feel about the two pieces?

   b. If Ray chooses one of the pieces that Martha cut, how does he feel about the two pieces?

   c. If you were one of the participants in this scheme, would you rather be the cutter or the chooser? Why?

5. Write a description of what you consider to be desirable results of a model that fairly divides a cake among any number of people.

6. Did any of the groups resolve Problem 2 by selling the house and dividing the cash? Why might the results of such a model be unsatisfactory to one or more of the heirs?
7. Did any of the groups use a model that considers the possibility that the heirs might not agree on the value of the house? In what way or ways was this done?
8. Suppose that Juan thinks the house is worth $100,000 and Mary feels it is worth $120,000.
   a. Who do you think should receive the house? Explain.
   b. How might the person who doesn't get the house be compensated?
9. Write a description of what you consider to be desirable results of a model that fairly divides a house among several heirs.
10. How might the possibility of lying about the value of the house affect the result of a division process?
11. Did all the groups that discussed Problem 3 divide the seats among the classes in the same way? If not, describe the differences.
12. If some of the groups that discussed Problem 3 obtained different results, which of the models do you think is the fairest? If all groups produced the same result, do you agree that the result is fair? Why or why not?
13. Write a description of what you consider to be desirable results of a model that fairly divides the seats in a student council among a school's classes.
14. Summarize the similarities and differences in the meaning of fairness in this lesson's three problems. For each of the three problems, explain why you think it is or is not possible to achieve fairness.
15. Read the news article on the right.
    a. How is the problem of fairly sharing the cost of something different from the three problems you considered in this lesson?

County, Feds Share Cost of Road Widening

Galesburg Register-Mail
April 14, 2012

The Knox County Board met Wednesday to approve or fund a series of projects, ranging from road expansion to courtroom renovations.

Board members approved an agreement between the federal government and the county for work to be done on County Highway 4. Widening the highway is slated to cost $5 million, with $3 million in funding coming from the federal government. The county will pay $1.7 million for the project, while the state will chip in $306,000.

Final construction plans for the project have been submitted to the Illinois Department of Transportation, with work slated to begin in June

b. Which of the three problems of this lesson do you think is most like the problem of sharing the cost of something? Describe the characteristics that are similar and different.

Projects

16. Pick a situation in which individuals or groups have developed a model for settling fair division problems other than the division of food, an estate, or legislative seats. Report on the model used. Compare it with models developed in this lesson and later in this chapter.

New Airwave-Sharing Scheme Will Launch a Wireless Revolution

MIT Technology Review
December 11, 2012

Aiming to boost wireless bandwidth and innovation, the U.S. Federal Communications Commission is poised to recommend the biggest regulatory change in decades: one that allows a newly available chunk of wireless spectrum to be leased by different companies at different times and places, rather than being auctioned off to one high bidder.

The move in effect allocates spectrum for another Wi-Fi—a technology that has had tremendous impact. But it is the sharing approach that represents a dramatic change in unleashing bandwidth.

The move spells the beginning of the end of a system in which spectrum is either exclusively owned by a private company, walled off for government use, or unlicensed and crowded.

Under the new approach, a startup that wants cheap spectrum to test a breakthrough idea would no longer have to rely on cluttered unlicensed bands—such as those that handle baby monitors and garage-door openers—to experiment with its idea in the real world. Rather, it could use a slice of choice spectrum at zero or low cost—and for a short time period if desired.

Lesson 2.2

Estate Division

A fair division problem can be either *discrete* or *continuous*. The problem of dividing a house among heirs and that of dividing a student council among classes are examples of the discrete case. Discrete division occurs whenever the objects of the division cannot be meaningfully separated into pieces. Dividing a cake is an example of the continuous case because the cake can be divided into any number of pieces.

This lesson considers fair division of an estate among heirs. In your discussions in Lesson 2.1, you may have decided that fairness is difficult to define in some situations because different people place different values on the same object. However, it is sometimes possible to use such differences of opinion to the advantage of all those involved. The estate division model on the next page is an algorith that produces an appealing paradox: Each of the heirs receives a share that is larger than he or she thinks is fair.

An Algorithm for Dividing an Estate

1. Each heir submits a bid for each item in the estate. (Bids are not made on cash in the estate because it can be divided equally without controversy.)
2. A fair share is determined for each heir by finding the sum of his or her bids and dividing this sum by the number of heirs. (If there is cash in the estate, its total is included in each heir's sum.)
3. Each item in the estate is given to the heir who bid the highest on it.
4. Each heir is given an amount of cash from the estate that is equal to his or her fair share (from step 2) less the amount the heir bid on the objects he or she receives. If this amount is negative, the heir pays that amount into the estate's cash.
5. Any remaining cash in the estate is divided equally among the heirs.

Wellington's Millions

mlive.com
May 24, 2011

Trustees for the $110 million estate of Wellington R. Burt, the 19th-century lumber baron who froze out family from his fortunes for nearly a century, distributed the millions of dollars to 12 Burt heirs by a Monday deadline.

The estate's distribution closes the book on Burt's storied estate, which first made headlines when the lumber baron died in 1919. That's when the public learned his will featured an odd stipulation freezing the bulk of his fortunes from family until 21 years after his last living grandchild's death. His last living grandchild died in 1989.

This month, the dozen heirs ironed out a settlement. They have Burt's blood in common, but many share little else. The youngest is 19; the oldest, 94. They're scattered across eight states—as far west as California and as far east as Connecticut. One resides in Michigan.

The first four steps of the algorithm give each heir goods and cash whose total value equals what the heir feels is a fair share of the estate. Extra cash awarded in step 5 is a bonus.

Whenever you encounter a new model, it is helpful to give it a try. After considering the following example, you will have the opportunity to use the model in this lesson's exercises.

Estate Division Example

Amanda, Brian, and Charlene are heirs to an estate that includes a house, a boat, a car, and $150,000 in cash.

Step 1

Each heir submits bids for the house, boat, and car. The bids are summarized in the following table, or matrix.

	House	Boat	Car
Amanda	$80,000	$5,000	$8,000
Brian	$70,000	$9,000	$11,000
Charlene	$76,000	$7,000	$13,000

For example, the entries in Amanda's row indicate the value to Amanda of each item in the estate.

Step 2

A fair share is determined for each heir.

Amanda: ($80,000 + $5,000 + $8,000 + $150,000) ÷ 3 = $81,000

Brian: ($70,000 + $9,000 + $11,000 + $150,000) ÷ 3 = $80,000

Charlene: ($76,000 + $7,000 + $13,000 + $150,000) ÷ 3 = $82,000

Step 3

The house is given to Amanda, the boat to Brian, and the car to Charlene.

Step 4

Each heir receives cash equal to the difference between the fair share and the value of the awarded items.

Amanda: $81,000 – $80,000 = $1,000

Brian: $80,000 – $9,000 = $71,000

Charlene: $82,000 – $13,000 = $69,000

Technology Note

The graphing calculator features used to do Borda counts can be used to determine fair shares. Replay the calculation for the first heir and edit it by moving the cursor and typing new digits when necessary. Delete and Insert features can be used when digits must be eliminated or introduced.

```
(80000+5000+8000
+150000)/3
           81000
(70000+9000+1100
0+150000)/3
```

Step 5

The cash that is given to the heirs totals \$141,000, which leaves \$150,000 – \$141,000 = \$9,000 in the estate. Each heir receives a bonus of \$9,000 ÷ 3 = \$3,000.

This example's results can be summarized in a matrix:

	Amanda	Brian	Charlene
Total of bids and cash	\$243,000	\$240,000	\$246,000
Fair share	\$81,000	\$80,000	\$82,000
Items received	House	Boat	Car
Value of items received	\$80,000	\$9,000	\$13,000
Initial cash received	\$1,000	\$71,000	\$69,000
Share of remaining cash	\$3,000	\$3,000	\$3,000

For each heir, totaling the last three rows of the matrix gives the value the heir attaches to the items and cash received. For example, Amanda feels the value of her share of the estate is \$80,000 + \$1,000 + \$3,000 = \$84,000, which is more than the \$81,000 that Amanda feels is a fair share.

The final settlements for each heir are:

Amanda: the house and \$4,000

Brian: the boat and \$74,000

Charlene: the car and \$72,000

James Brown's Heirs Settle Estate Feud

IMDb.com
May 27, 2009

James Brown's heirs have ended their bitter feud over his estate. A South Carolina judge has signed off on a settlement reached by the Godfather of Soul's wife and children.

His wife, Tomi Rae Hynie, and six adult children have fought over his riches since his death in 2006. His children were named in his will, but his wife and their son James Brown II were not. The will was signed before their marriage and before James II was born.

Under the terms of the settlement, half the legend's assets will go to his charitable trust to pay for his grandchildren's education and help poor children in South Carolina and Georgia, a quarter will go to Hynie and James II, and the rest to his six children.

James Brown

Exercises

1. The success of a fair division model requires certain assumptions, or *axioms*. For example, this lesson's algorithm works successfully only if each heir is capable of placing a value on each of the estate's objects. If any heir considers an object priceless or is otherwise incapable of placing a dollar value on an object, the algorithm fails. State at least one other axiom that you think is necessary for the success of this algorithm.

2. Calvin and Hobbes are using this lesson's algorithm to divide an estate composed of a wagon, a sled, and \$20 in cash. Calvin bids \$20 for the wagon and \$10 for the sled. Hobbes bids \$10 for the wagon and \$15 for the sled.

   a. What does Calvin think is a fair share of the estate? Hobbes?

   b. What value does Calvin place on his final settlement? Hobbes? Explain.

3. Garfield and Marmaduke are heirs to an estate that contains only a summer cottage. Garfield bids $80,000. Marmaduke bids $70,000.

   a. What does Garfield feel is a fair share? Marmaduke?

   b. What is the difference between Garfield's bid and Garfield's fair share?

   c. Because the value Garfield assigned to the cottage is more than Garfield's fair share, Garfield must pay cash into the estate. How much cash must Garfield pay?

   d. Marmaduke is given an amount of cash from Garfield's payment equal to Marmaduke's fair share. How much does Marmaduke receive? If the remaining cash is divided equally, what is the final value of Marmaduke's settlement? Of Garfield's settlement?

   e. Garfield must borrow money in order to pay into the estate, and the interest on this loan is $2,000. Do you think this should be considered when arriving at a settlement? If so, suggest how the settlement can be revised.

   f. If the division between Garfield and Marmaduke is settled by another model that is frequently used to divide estates, Marmaduke receives half of Garfield's bid. Compare the final settlements for Garfield and Marmaduke by this model with the settlements in part d. Of the two results, which do you think is fairest? Explain.

4. Amy, Bart, and Carol are heirs to an estate that consists of a valuable painting, a motorcycle, a World Series ticket, and $5,000 in cash. Their bids are shown in the following matrix.

	Painting	Motorcycle	Ticket
Amy	$2,000	$4,000	$500
Bart	$5,000	$2,000	$100
Carol	$3,000	$3,000	$300

   a. Use this lesson's algorithm to divide the estate among the heirs. For each heir, state the fair share, the items received, the amount of cash, and the final settlement. Summarize your results in a matrix.

A helpful tip: It is relatively easy to lose track of the estate's cash if several payments are made into and out of the estate. Errors can be avoided by tracking the cash with a small table designed for that purpose:

Cash in the estate	$5,000
Received from Amy	______
Received from Bart	______
Paid to Carol	______
Cash remaining	______

b. It is common for one or more heirs to pay into an estate. This lesson's estate division model fails if an heir who must pay into the estate cannot do so. Suggest a way to modify the model for situations in which one or more heirs cannot raise the cash needed to complete the division.

5. Suppose that in the division of Exercise 4, Amy received previous financial support from the estate in the form of a loan to pay college tuition. A will states that she is to receive only 20% of the estate, whereas Bart and Carol are to receive 40% each. Adapt the model of this lesson to this situation and describe a fair division of the estate.

6. If two heirs submit identical highest bids for an item, how would you resolve the tie?

7. Alan, Betty, and Carl are heirs to an estate. They submit the bids shown in the following table.

	House	Boat	Car
Alan	$115,000	$6,000	$13,000
Betty	$120,000	$7,000	$11,000
Carl	$117,000	$6,000	$12,000

The awarding of items in the estate can be represented in a matrix, as shown below.

	Alan	Betty	Carl
House	0	1	0
Boat	0	1	0
Car	1	0	0

The entries in Alan's column show which items he receives. For example, the 1 in Alan's column and the car's row indicates that Alan receives the car. Each of the other two entries in Alan's column is a 0; these zeros indicate that Alan receives neither the house nor the boat.

A new matrix can be computed by writing the second matrix beside the first, as shown below.

$$\begin{bmatrix} \$115{,}000 & \$6{,}000 & \$13{,}000 \\ \$120{,}000 & \$7{,}000 & \$11{,}000 \\ \$117{,}000 & \$6{,}000 & \$12{,}000 \end{bmatrix} \begin{bmatrix} 0 & 1 & 0 \\ 0 & 1 & 0 \\ 1 & 0 & 0 \end{bmatrix}$$

One of the entries in the new matrix is computed by multiplying each entry in the first row of the first matrix by the corresponding entry in the first column of the second matrix and then finding the sum of these products.

$$\$115{,}000(0) + \$6{,}000(0) + \$13{,}000(1) = \$13{,}000$$

Because the result, \$13,000, is obtained from the first row of the first matrix and the first column of the second matrix, it is written in the first row and the first column of the new matrix.

$$\begin{bmatrix} \$13{,}000 & \underline{\qquad} & \underline{\qquad} \\ \underline{\qquad} & \underline{\qquad} & \underline{\qquad} \\ \underline{\qquad} & \underline{\qquad} & \underline{\qquad} \end{bmatrix}$$

The entry in the first row and the second column of the new matrix is found by performing a similar calculation with the first row of the first matrix and the second column of the second matrix.

$$\$115{,}000(1) + \$6{,}000(1) + \$13{,}000(0) = \$121{,}000$$

a. Calculate the remaining entries of the new matrix.

b. The \$13,000 in the first row and the first column of the new matrix can be interpreted as the value to Alan of the items he received. Write an interpretation of the number in the first row and the second column of the new matrix.

c. Write an interpretation of the number in the second row and the second column of the new matrix.

8. Could this lesson's estate division model encourage insincere bidding by one or more of the heirs? Explain.

9. In 1998, the U.S. Supreme Court settled a dispute between New York and New Jersey over control of Ellis Island by dividing the island between the two states. Is the problem of how to divide an island among two or more parties a continuous or discrete problem? Explain.

The Battle Over Ellis Island

Government Technology Magazine
January, 2000

Between 1892 and 1954, Ellis Island was the gateway to America for some 17 million people on their way to a new life. Many Americans today have parents or grandparents who arrived there as immigrants.

Today, Ellis Island is a national monument that draws 5 million visitors annually. Although historically associated with New York, about 83 percent of the island was recently declared by the U.S. Supreme Court to be within the jurisdiction of New Jersey.

In 1993, New Jersey initiated a lawsuit against New York in an effort to resolve a territorial dispute over the island that had dragged on for more than 160 years.

Since the island is owned by the United States, the court's ruling is unlikely to have significance for anyone but the litigants.

10. Two friends plan to share an apartment in order to obtain one that is nicer than either could afford individually. They choose a two-bedroom apartment that rents for $1,300 monthly, including utilities. One bedroom is larger and sunnier than the other. Propose a model for deciding which of the friends gets the nicer bedroom.

Computer/Calculator Exploration

11. It can be instructive to examine the results of an estate division when one or more of the bids changes. However, it is tedious to redo all the calculations several times over. Fortunately, this lesson's estate division algorithm can be implemented on a spreadsheet, which simplifies changing values, and inspecting the results. Use a computer spreadsheet to perform this lesson's estate division algorithm. A sample output and the formulas that generate it are shown below and on the next page. In this case, the results are those of Exercise 4. Once your spreadsheet is complete, use it to answer the questions that follow.

	A	B	C	D	E	F
1		Estate Division Spreadsheet				
2						
3		Amy	Bart	Carol		
4	Painting	2000.00	5000.00	3000.00		
5	Motorcycle	4000.00	2000.00	3000.00		
6	Ticket	500.00	100.00	300.00		
7						
8		Amy	Bart	Carol		Cash:
9	Bid total	11500.00	12100.00	11300.00		5000.00
10	Share	0.333333	0.333333	0.333333		666.67
11	Fair share	3833.33	4033.33	3766.67		966.67
12	Object value	4500.00	5000.00	0.00		-3766.67
13	Cash received	-666.67	-966.67	3766.67		
14	Extra cash	955.56	955.56	955.56		2866.67
15						
16	Final total	4788.89	4988.89	4722.23		
17						

	A	B	C	D	E	F
1		Estate Division Spreadsheet				
2						
3		Amy	Bart	Carol		
4	Painting	2000.00	5000.00	3000.00		
5	Motorcycle	4000.00	2000.00	3000.00		
6	Ticket	500.00	100.00	300.00		
7						
8		Amy	Bart	Carol		Cash:
9	Bid total	=SUM(B4:B6)+$F$9	=SUM(C4:C6)+$F$9	=SUM(D4:D6)+$F$9		5000
10	Share	=1/3	=1/3	=1/3		=–B13
11	Fair share	=B9*B10	=C9*C10	=D9*D10		=–C13
12	Object value	=B5+B6	=C4	=0		=–D13
13	Cash received	=B11–B12	=C11–C12	=D11–D12		
14	Extra cash	=$F$14*B10	=$F$14*C10	=$F$14*D10		=SUM(F9:F12)
15						
16	Final total	=SUM(B12:B14)	=SUM(C12:C14)	=SUM(D12:D14)		

a. What if the amount of cash in the estate is 0? Change the amount in cell F9 to 0 and see what happens. Describe the result.

b. What would happen if Bart lies about the value he places on the motorcycle and says he feels it is worth $5,000? Change the amount in cell C5 to 5,000 and see. (Change the cash back to 5,000 before doing this one.) Describe the result.

c. What happens if Bart really does feel that the motorcycle is worth $5,000 but accepts a $2,000 bribe from Amy to bid $2,000? How does this collusion between Bart and Amy change the value of the final settlements for Bart and Amy?

d. Explain how to change the formulas in the spreadsheet to account for the situation in Exercise 5.

Projects

12. Matrix calculations like the multiplication shown in Exercise 7 are useful in programming computers to do tedious calculations. Research and report on the use of matrix applications in computer science.

13. Research division models that are used at auctions. What are Dutch and English auctions? Why do some auctions award the contract to the second-highest bidder? When are closed and open bids used?

Lesson 2.3

Apportionment Models

The problem of dividing an estate fairly involves discrete objects, but also involves cash. When a fair division problem is strictly discrete, it can be impossible to solve in a way that treats all parties fairly.

A variety of situations involve fair allocation of discrete objects. For example, your school's administrators must decide a fair way to allocate teaching positions to the school's various departments and equipment such as computers to classrooms. One of the most politically charged fair allocation problems in the United States involves the apportionment of seats in the U.S. House of Representatives among the states. (The House is reapportioned every ten years after a new census is completed.)

Unlike the discrete objects in estate division situations, the value of a seat in the U.S. House is not subjective. Thus, fairness has a simple definition, one that is mandated by the Constitution: that the seats be distributed among the states according to population.

Although the definition of fairness that is used to apportion seats in the U.S. House is uncontroversial, the method of apportionment can be. The first veto by a U.S. president occurred in 1792 when George Washington rejected an apportionment bill advocated by Alexander Hamilton in favor of a method championed by Thomas Jefferson. This lesson considers the two models for apportioning seats in a governmental body that were at the center of the Hamilton-Jefferson dispute.

Alexander Hamilton.

Since the apportioning of the U.S. House of Representatives involves 435 seats and 50 states, you will feel more comfortable starting with a simpler situation. (Although some of this lesson's examples may seem artificial, they are designed to reflect the large population differences among the 50 states.)

Thomas Jefferson.

Central High School has sophomore, junior, and senior classes of 464, 240, and 196 students, respectively. The 20 seats on the school's student council are divided among the classes according to population. Since there are 900 students in the school and since 900 ÷ 20 = 45, ideally each member of the council would represent 45 students. In other words, the **ideal ratio** of students to seats is 45.

$$\text{Ideal ratio} = \frac{\text{Total population}}{\text{Number of seats}}$$

In cases of political representation, the ideal ratio is often called the *ideal district size.* If, for example, the population of the United States is 280 million, then the ideal district size is 280,000,000 ÷ 435, or about 645,000. Ideally, each member of the U.S. House would represent 645,000 people. This ideal cannot be achieved because district boundaries cannot cross state lines.

Because Central High's sophomore class has 464 members, it deserves 464 ÷ 45 = 10.31 seats. Accordingly, 10.31 is called the sophomore class **quota**. Similarly, the junior quota is 5.33, and the senior quota is 4.36.

The Quotas	
Sophomores:	10.31
Juniors:	5.33
Seniors:	4.36

$$\text{Quota} = \frac{\text{Class size}}{\text{Ideal ratio}}$$

In the case of the U.S. House, a state's quota is determined by dividing its population by the ideal district size. For example, a state with 4 million people deserves 4,000,000 ÷ 645,000, or about 6.2 seats.

It isn't possible to split a seat in Central High's council and give 0.36 of it to the seniors, 0.33 of it to the juniors, and 0.31 of it to the sophomores. The school must decide a fair way to award this seat to one of the classes.

The model favored by Hamilton has something in common with the one favored by Jefferson: Each begins by ignoring the quota's decimal part and assigning a number of seats equal to the whole number part. Regardless of whether the quota is 10.31 or 10.91, both Hamilton and Jefferson begin by awarding 10 seats. Ignoring the decimal part of a number in this way is called *truncating.*

The total of the truncated quotas is $10 + 5 + 4 = 19$ seats. The difference between the Hamilton and Jefferson models is in the way they award the remaining seat.

The **Hamilton model** gives the remaining seat to the class whose quota has the largest decimal part. Since the decimal part of the senior quota, 0.36, is larger than either of the other decimal parts, the senior class gets the extra seat. The Hamilton results are summarized in the following table.

Class Size	Quota	Hamilton Apportionment
464	10.31	10
240	5.33	5
196	4.36	5

The Hamilton model seems reasonable. Perhaps some members of your class proposed a similar model in Lesson 2.1. However, the Hamilton model has fallen out of favor in the United States for reasons you will consider in this lesson's exercises, after you have examined Jefferson's approach.

You might think of the **Jefferson model** as conducting a race to see which quota can get to the next whole number first. That is, whether the sophomore quota can increase to 11 or the junior quota can increase to 6 or the senior quota can increase to 5 first. Here is how this race is conducted.

Since a quota is found by dividing the class size by the ideal ratio, the quota becomes larger when the ideal ratio becomes smaller. For example, consider what happens if the ideal ratio is decreased from 45 students per seat to 40 students per seat. The results are summarized in the following table.

Class Size	Quota with Ideal Ratio of 45	Quota with Ideal Ratio of 40
464	$464 \div 45 = 10.31$	$464 \div 40 = 11.6$
240	$240 \div 45 = 5.33$	$240 \div 40 = 6.0$
196	$196 \div 45 = 4.36$	$196 \div 40 = 4.9$
Seats	$10 + 5 + 4 = 19$	$11 + 6 + 4 = 21$

For example, the sophomore class receives 10 seats when the ideal ratio is 45 and 11 seats when the ratio is 40. Therefore, there must be a ratio between 45 and 40 that causes the sophomore apportionment to be exactly 11. It can be found by dividing the sophomore class size by 11: $464 \div 11 \approx 42.18$. The number 42.18 is called a **Jefferson adjusted ratio**.

$$\text{Jefferson adjusted ratio} = \frac{\text{Class size}}{\text{Truncated quota} + 1}$$

Similarly, the junior quota passes 6 when the ratio drops below $240 \div 6 = 40$. The senior quota passes 5 when the ratio drops below $196 \div 5 = 39.2$.

Proponents of the Jefferson model argue that since the ideal ratio does not apportion all of the seats, it should be replaced with a ratio that does give a complete apportionment and that this new ratio should be as close as possible to the ideal ratio. If the ratio gradually decreases from the ideal 45, it reaches a value at which the sophomores receive another seat before it reaches a value at which either of the other classes receives one, as shown in the following table.

Ideal Ratio	Sophomore Seats	Junior Seats	Senior Seats
45	10	5	4
↓			
42.18	11	5	4
↓			
40	11	6	4
↓			
39.2	11	6	5

Greater detail can be seen in the following table, in which the adjusted ratio is decreased in steps of 0.2. Note that the sophomore class quota passes the next integer before either the junior or the senior quota does.

Adjusted Ratio	Sophomore	Junior	Senior
45.00	10.31	5.33	4.36
44.80	10.36	5.36	4.38
44.60	10.40	5.38	4.39
44.40	10.45	5.41	4.41
44.20	10.50	5.43	4.43
44.00	10.55	5.45	4.45
43.80	10.59	5.48	4.47
43.60	10.64	5.50	4.50
43.40	10.69	5.53	4.52
43.20	10.74	5.56	4.54
43.00	10.79	5.58	4.56
42.80	10.84	5.61	4.58
42.60	10.89	5.63	4.60
42.40	10.94	5.66	4.62
42.20	**11.00**	5.69	4.64
42.00	11.05	5.71	4.67
41.80	11.10	5.74	4.69
41.60	11.15	5.77	4.71
41.40	11.21	5.80	4.73
41.20	11.26	5.83	4.76
41.00	11.32	5.85	4.78
40.80	11.37	5.88	4.80
40.60	11.43	5.91	4.83
40.40	11.49	5.94	4.85
40.20	11.54	5.97	4.88
40.00	11.60	**6.00**	4.90
39.80	11.66	6.03	4.92
39.60	11.72	6.06	4.95
39.40	11.78	6.09	4.97
39.20	11.84	6.12	**5.00**
39.00	11.90	6.15	5.03

The Jefferson model can be summarized as an algorithm:

1. Divide the total population by the number of seats to obtain the ideal ratio.
2. Divide the population of each class (state, district, etc.) by the ideal ratio to obtain the class quota.

3. Assign a number of seats to each class that equals its truncated quota.
4. If the number of seats assigned matches the total number of seats to be apportioned, then stop.
5. If the number of seats assigned is smaller than the total number of seats to be apportioned, then divide the size of each class by one more than the number of seats assigned to it in step 3, to obtain an adjusted ratio.
6. Give an extra seat to the class with the largest adjusted ratio. (In other words, to the class whose adjusted ratio is closest to the ideal ratio.)

This algorithm applies only to situations in which the total of the truncated quotas falls one short of the number of seats to be apportioned. In some cases, the shortfall after truncation is more than one seat. This lesson's exercises consider what to do in such cases.

Exercises

1. The Central High council has had trouble deciding a number of issues because of disagreements between the sophomore representatives and the other representatives. The vote has been a 10-10 tie. In order to avoid future ties, the council decides to add one seat.
   a. On the basis of the data in this lesson, which class do you think should get the extra seat? Why?
   b. Find the new ideal ratio of students per seat for the 21-seat council.
   c. Use this ideal ratio to find the quota for each of the three classes.
   d. Use the Hamilton model to apportion the 21 seats on the new council among the three classes.
   e. Compare the Hamilton apportionment for the 21-seat council to that of the 20-seat council and explain why the results constitute a paradox.

2. A senior council member who recently studied apportionment in the school's discrete mathematics course is unhappy about the 21-seat Hamilton apportionment and proposes the Jefferson model.

   a. Find an adjusted ratio for each class as described in the Jefferson model's algorithm of this lesson.

   b. Decrease the 21-seat ideal ratio until all 21 seats are apportioned. State the number of seats that the Jefferson model gives each class.

   c. Compare the 21-seat Jefferson apportionment with the 20-seat Jefferson apportionment. Does the Jefferson model produce a paradox similar to the one in Exercise 1?

3. Revise the Jefferson apportionment algorithm given in this lesson to account for situations in which more than one seat remains after truncation.

4. The paradox you observed in Exercise 1 occurs because increases in a divisor do not produce equal changes in all quotients. When the size of a representative assembly increases and the total population remains the same, the ideal ratio decreases. As an example, consider two classes with populations of 100 and 230. An increase in the size of the council causes the ideal ratio to decrease from 22 to 21.

   a. Complete the following table. Then explain why it could result in the shift of a council seat from one class to the other.

Class Size	Quota with Ideal Ratio of 22	Quota with Ideal Ratio of 21
100		
230		

   b. Does the paradox in Exercise 1 result in the loss of a seat for a small class or a large class? Why?

   c. Why do you think Thomas Jefferson opposed the Hamilton model? (If you're not sure, look up the 1790 census. Jefferson was from Virginia.)

5. The student council members at Central High, aware of the strange results that slight differences can make, decide to monitor the council's apportionment. At the end of the first quarter of the school year, the class numbers have changed somewhat:

Sophomores	Juniors	Seniors
459	244	197

   a. Use the Hamilton model to divide the council's 21 seats among the classes.

   At the end of the first semester the classes have changed again:

Sophomores	Juniors	Seniors
460	274	196

   At the council's first meeting of the new semester, the members are surprised when one of the representatives of the senior class, the only class that has decreased in size, demands a reapportionment.

   b. Use the Hamilton model to reapportion the council.

   c. Explain why the results constitute a paradox.

   d. Use an analysis similar to that of Exercise 4 to explain why this type of paradox occurs. Does it have an adverse effect on small classes or large classes?

6. The cartoon on page 74 has been described as "an indiscreet attempt to apply a continuous division procedure to a discrete problem." Explain what this means.

7. Read the news article on the following page about the 2010 census results. In the 2010 census, the populations of Texas, New York, and California were 25,268,418, 19,421,055, and 37,341,989, respectively.

   a. Find the quota for each of these three states.

   b. Compare the quotas to the actual apportionment mentioned in the news article.

   c. Do you think the apportionment treated any of these three states unfairly? Explain.

2010 Census: Slowest Growth Since Great Depression

USA Today
February 3, 2011

The U.S. population grew 9.7% in the past decade to 308,745,538, according to the first results of the 2010 Census — the slowest growth since the Great Depression for a nation hard hit by a recession and housing bust.

Census population counts taken every 10 years are used to distribute $400 billion a year in federal funds and reallocate seats in the U.S. House of Representatives.

The shifting of House seats affects 18 states. Arizona, Georgia, Nevada, South Carolina, Utah and Washington all gain one seat. Florida gains two, and Texas gains four. Illinois, Iowa, Louisiana, Massachusetts, Michigan, Missouri, New Jersey and Pennsylvania all lose a seat. Both Ohio and New York lose two.

Texas is the big winner. It gained more people than any other state for the first time, outpacing California, which had dominated for almost a century. Texas' four new seats will give it 36, still behind California's 53. New York's loss of two seats and Florida's gain of two will give each state 27.

d. The 2010 census total for Louisiana was 4,553,962. Minnesota's total was 5,314,879. The apportionment gave Louisiana 6 seats and Minnesota 8. Louisiana felt that the apportionment was unfair and went to court, but lost. Do you think the apportionment was less fair to Louisiana than to Minnesota? Explain.

8. A county is composed of four districts: A, B, C, and D. Their populations are 210; 1,082; 311; and 284; respectively. The county commission has 18 seats.

   a. What is the ideal ratio?

   b. Find each district's quota.

   c. Find the Hamilton apportionment.

   d. Does the Jefferson model apportion all of the seats initially? Explain.

   e. Find the final Jefferson apportionment.

   f. This exercise shows why some do not like the Jefferson model. Explain.

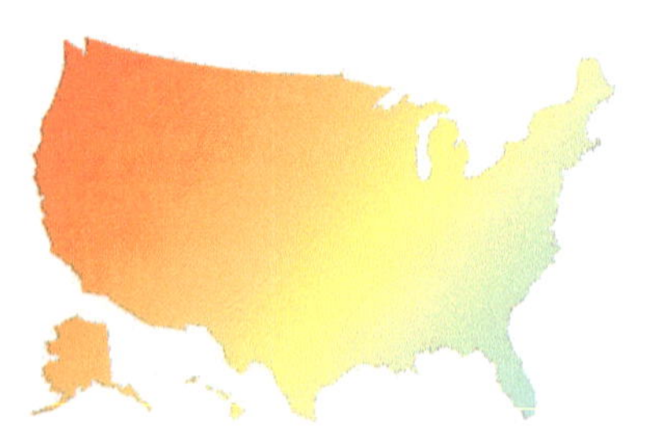

Computer/Calculator Explorations

9. Develop a spreadsheet to do the Jefferson apportionment for the three classes in this lesson's example. When finished, the spreadsheet should be similar to the following.

	A	B	C	D
1		Population	Quota	Seats
2	Sophomore	464	10.311111	10
3	Junior	240	5.333333	5
4	Senior	196	4.355556	4
5	Total	900		19
6	Seats	20		
7	Ideal ratio	45		

The values in columns C and D and in cells B5 and B7 should be calculated with formulas. The formulas in column D require the use of your spreadsheet's truncation function. On many spreadsheets, this function is abbreviated TRUNC. For example, TRUNC(C2,0) truncates the value in cell C2 so it has 0 decimal places.

When your spreadsheet is done, show how the proper apportionment can be found by changing the value in cell B7 (the ideal ratio).

Projects

10. Research and report on models that have been used to apportion the U.S. House of Representatives and the controversies that have arisen. Why has the apportionment model been changed? Why has the size of the House been changed? Did paradoxes occur with any of the models?

11. The president of the United States is chosen in the Electoral College. The number of electoral votes a state has is determined by the size of its congressional delegation. Thus, apportionment can affect the Electoral College vote. Research and report on the impact of apportionment on the election of the president. Has the apportionment model ever made a difference in the person the Electoral College chose?

Lesson 2.4

More Apportionment Models and Paradoxes

Dissatisfaction with paradoxes that can occur with the Hamilton model led to its abandonment as a method of apportioning the U.S. House of Representatives. The Jefferson model is disliked by small states because it favors large ones. This lesson considers alternatives to the Jefferson and Hamilton models and some recent developments in the debate over which model is fairest.

The Jefferson model is one of several *divisor models*. The term *divisor* is used because these models determine quotas by dividing the population by an ideal ratio or an adjusted ratio. This ratio is the divisor. The Hamilton model is not a divisor model.

There are two divisor models given considerable attention today. One is named for Daniel Webster; another is named for Joseph Hill, an American statistician. The Webster and Hill models differ from the Jefferson model in the way they round quotas. Recall that the Jefferson model truncates a quota and apportions a number of seats equal to the integer part of the quota. Quotas of 11.06 and 11.92 both receive 11 seats under the Jefferson model.

The **Webster model** uses the rounding method with which you are familiar: A quota above or equal to 11.5 receives 12 seats, and a quota below 11.5 receives 11 seats. The number 11.5 is sometimes called the **arithmetic mean** of 11 and 12. The arithmetic mean of two numbers is the number halfway between them. It can be calculated by dividing the sum of the two numbers by 2.

The **Hill model** rounds by using the geometric mean instead of the arithmetic mean. The **geometric mean** of two numbers is the square root of their product. If the quota exceeds the geometric mean of the integers directly above and below the quota, the quota is rounded up; otherwise it is rounded down. For example, a quota between 11 and 12 must exceed $\sqrt{11 \times 12} \approx 11.4891$ to receive 12 seats (see Figure 2.1).

Population Change Could Alter Makeup of Cumberland-North Yarmouth School Board

The Forecaster
February 5, 2013

The Cumberland-North Yarmouth school board voted 7-0 Monday to ask the Maine commissioner of education to determine if the board is apportioned according to the principle of one person, one vote.

The Town Council voted unanimously Dec. 10, 2012, to request the board's action "to determine the necessity for reapportionment based upon the 2010 census."

Stephen Moriarty, at the time chairman of the Cumberland Town Council, explained to Superintendent of Schools Robert Hasson in a letter last December that Cumberland's population in the 2010 census was 7,211, nearly 67 percent of the district's total population; North Yarmouth's population was 3,565.

The School Board now has five members from Cumberland and three from North Yarmouth.

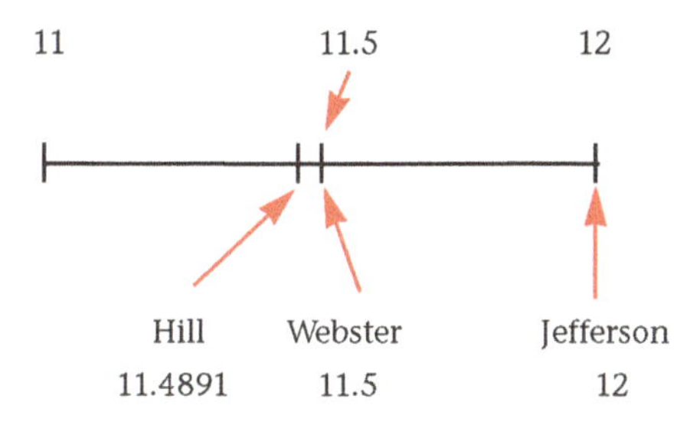

Figure 2.1.
The Hill, Webster, and Jefferson roundup points for quotas between 11 and 12.

The Hill model can be confusing because the quota for which it awards an extra seat must be calculated. For example, when the Webster model is used, you know immediately that a quota of 7.4903 yields 7 seats because 7.4903 is slightly below 7.5. However, you do not know whether the Hill model awards 7 or 8 seats until you calculate $\sqrt{7 \times 8} \approx 7.4833$. Since the quota (7.4903) is larger than 7.4833, the Hill model awards 8 seats.

The following table summarizes the apportionment of the 20-seat Central High student council from Lesson 2.3 by the models discussed in this and the previous lesson.

Class Size	Initial Apportionment				
	Quota	Hamilton	Jefferson	Webster	Hill
464	10.31	10	10	10	10
240	5.33	5	5	5	5
196	4.36	5	4	4	4

The Jefferson, Webster, and Hill models all fail to assign one of the seats and so require an adjusted ratio. The following table lists the adjusted ratio necessary for each class to gain a seat by each method.

Class Size	Adjusted Ratio for		
	Jefferson	Webster	Hill
464	$464 \div 11 = 42.1818$	$464 \div 10.5 = 44.1905$	$464 \div \sqrt{10 \times 11} = 44.2407$
240	$240 \div 6 = 40.0000$	$240 \div 5.5 = 43.6364$	$240 \div \sqrt{5 \times 6} = 43.8178$
196	$196 \div 5 = 39.2000$	$196 \div 4.5 = 43.5556$	$196 \div \sqrt{4 \times 5} = 43.8269$

Recall that the adjusted Jefferson ratio for the sophomore class is found by dividing the class size by 11. The adjusted Webster ratio for the sophomore class is found by dividing the class size by 10.5. The adjusted Hill ratio for the sophomore class is found by dividing the sophomore class size by $\sqrt{10 \times 11} \approx 10.4881$.

Each of these models requires that the ideal ratio of 45 decrease until it is smaller than exactly one of the adjusted ratios. To compare the models, it can be helpful to list the adjusted ratios in decreasing order:

Jefferson: 42.1818 (sophomores), 40.0000 (juniors), 39.2000 (seniors)

Webster: 44.1905 (sophomores), 43.6364 (juniors), 43.5556 (seniors)

Hill: 44.2407 (sophomores), 43.8269 (seniors), 43.8178 (juniors)

For all three models, the sophomore class is first when the adjusted ratios are listed in decreasing order. Therefore, all three models award the extra seat to the sophomores. Note, however, that the Hill model lists the senior class second rather than third.

Because the Webster and Hill models do not truncate quotas, they sometimes apportion too many seats rather than too few. In this lesson's exercises you will consider how to apply the Webster and Hill models in such cases and learn of some surprising results.

Exercises

1. a. Complete the following apportionment table for the 21-seat Central High student council described in Exercise 1 of Lesson 2.3 (see page 77).

Initial Apportionment

Class Size	Quota	Hamilton	Jefferson	Webster	Hill
464	___	___	___	___	___
240	___	___	___	___	___
196	___	___	___	___	___

b. The Jefferson model distributes only 19 seats. You can use your results from Exercise 1 of Lesson 2.3 to complete the Jefferson apportionment. Give the final Jefferson apportionment.

Both the Webster and Hill models apportion 22 seats. Therefore, the ideal ratio must increase until one of the classes loses a seat. The sophomore class, for example, loses a seat under the Webster model if its quota drops below 10.5. This requires an adjusted ratio of $464 \div 10.5 \approx 44.1905$. For the sophomore class to lose a seat under the Hill model, its quota must drop below $464 \div \sqrt{10 \times 11} \approx 44.2407$.

c. Complete the following table of adjusted ratios for the Webster and Hill models.

Class Size	Adjusted Ratio for Webster	Adjusted Ratio for Hill
464	$464 \div 10.5 = 44.1905$	$464 \div \sqrt{10 \times 11} = 44.2407$
240	___	___
196	___	___

d. List the adjusted ratios for the Webster model in increasing order. The ideal ratio must increase until it passes the first ratio in your list. The class whose adjusted ratio is passed first loses one seat. Give the final Webster apportionment.

e. List the adjusted ratios for the Hill model in increasing order. The ideal ratio must increase until it passes the first ratio in your list. The class with this ratio loses one seat. Give the final Hill apportionment.

f. For each of the Central High classes, which model do you think the class favors? Explain.

2. Since the number of seats assigned to a class rarely equals its quota exactly, apportionments are seldom completely fair. This fact has led people to ask whether one apportionment is less fair than another. A way to measure the unfairness of an apportionment is to total the discrepancies between the quota and the number of seats assigned to each class. For example, if the quota is 11.25 and 11 seats are apportioned, the unfairness is 0.25 seats. If 12 seats are apportioned, the unfairness is 0.75 seats.

a. Use the apportionments for the 20-seat Central High student council to measure the discrepancy for each class by means of each model. Record the results in the following table.

		Amount of Discrepancy			
Class Size	Quota	Hamilton	Jefferson	Webster	Hill
464	10.31	___	___	___	___
240	5.33	___	___	___	___
196	4.36	___	___	___	___
Total discrepancy		___	___	___	___

b. Which model has the smallest total discrepancy?

c. Do you think smallest total discrepancy is a good criterion for choosing an apportionment model? Explain.

3. Another way to measure an apportionment's unfairness is to compare the representation of two classes (states, districts, etc.) as a percentage. For example, if a class with 250 students has 5 seats, then each seat represents 250 ÷ 5 = 50 people. If another class has 270 students and 6 seats, the representation is 270 ÷ 6 = 45 people

per seat. The representation is unfair to the first class by 50 – 45 = 5 people per seat, which is 5 ÷ 50 = 0.10, or 10% of its representation.

a. The initial Hill apportionment in Exercise 1 apportions one seat too many. Compare the representation of the junior and senior classes if one seat is taken from the juniors. What is the unfairness percentage to the juniors?

b. Compare the representation of the junior and senior classes if one seat is taken from the seniors. What is the unfairness percentage to the seniors?

c. By this percentage measure, is it fairer to take a seat from the juniors or to take a seat from the seniors? Which apportionment in Exercise 1 agrees with your answer?

4. None of the divisor models is plagued by the paradoxes that caused the demise of the Hamilton model. Divisor methods, however, can cause their own problems. As an illustration, consider the case of South High School. The freshman class has 1,105 members, the sophomore class has 185, the junior class 130, and the senior class 80. The 30 members of the student council are apportioned among the classes by the Webster model.

   a. What is the ideal ratio?

   b. Complete the following Webster apportionment table.

Class Size	Quota	Initial Webster Apportionment
1,105	22.1	22
185		
130		
80		

   c. Because the Webster model apportions too many seats, the ideal ratio must decrease. Calculate the adjusted ratio necessary for each class to lose a seat and enter the results in the following table.

Class Size	Adjusted Ratio
1,105	1,105 ÷ 21.5 = 51.3953
185	
130	
80	

d. Determine the final apportionment. Explain your method.

e. Explain why the freshman class would consider the final apportionment unfair.

The Webster apportionment in Exercise 4 demonstrates a **violation of quota**. This occurs whenever a class (district, state, etc.) is given a number of seats that does not equal either the integer directly below its quota or the one directly above. It can occur with any divisor model and is considered a flaw of divisor models.

The following table shows how the quotas for each class change as the ideal ratio is gradually increased. The freshman quota drops much more quickly than do the others.

Ratio	Freshman	Sophomore	Junior	Senior
50.00	22.10	3.70	2.60	1.60
50.20	22.01	3.69	2.59	1.59
50.40	21.92	3.67	2.58	1.59
50.60	21.84	3.66	2.57	1.58
50.80	21.75	3.64	2.56	1.57
51.00	21.67	3.63	2.55	1.57
51.20	21.58	3.61	2.54	1.56
51.40	21.50	3.60	2.53	1.56
51.60	21.41	3.59	2.52	1.55
51.80	21.33	3.57	2.51	1.54
52.00	21.25	3.56	2.50	1.54

Mathematicians of Note

Michel L. Balinski is a director of the Laboratoire d'Econometrie and the Ecole Polytechnique in Paris.

H. Peyton Young is a senior fellow at the Brookings Institution.

Around 1980 Michel L. Balinski and H. Peyton Young proved an important impossibility theorem: Any apportionment model sometimes produces at least one of these undesirable results: violation of quota and the two paradoxes that occur with the Hamilton model (see Exercises 1 and 5 of Lesson 2.3).

5. Every person in a small community belongs to exactly one of the community's four political parties. Membership is distributed as shown in the following table.

Party	Membership
A	561
B	200
C	100
D	139

The 20 seats in the community's council are apportioned by the Jefferson model.

a. Determine the Jefferson apportionment.

b. Parties C and D decide to join together to form a single party. Determine the new apportionment.

c. Would the apportionment that results from the merger of parties C and D occur with any of the other models you have studied? Explain.

6. A county is divided into districts A, B, C, and D with populations of 400, 652, 707, and 1,644, respectively. There are 14 seats on the county commission. Examine various apportionment models that could be used by the county. Which districts do you think would favor a particular model? Explain.

7. Read the news article on page 83. Compare Cumberland's quota after the 2010 census with the number of members it has on the board. How does this situation compare with the weighted voting situations you studied in Lesson 1.5?

8. Compare the Hill round-off point with the Webster round-off point for quotas of various sizes. How, for example, do they compare for quotas between 2 and 3, between 10 and 11, between 100 and 101? Can you prove any relationship between the two?

Computer/Calculator Explorations

9. Extend the spreadsheet you made in Exercise 9 of Lesson 2.3 to include columns for the Webster and Hill models. You will need to use your spreadsheet's rounding function to do the Webster apportionment. For the Hill apportionment you will need to use the spreadsheet's logical functions. Logical functions on a spreadsheet

are similar to those on a calculator. For example, the calculation A(B < 5) + (A + 1)(B > 5) produces the value of A when B < 5 and the value of A + 1 when B > 5 because an equation or inequality has a value of 0 when it is false and 1 when it is true.

Projects

10. Research and report on the work of Balinski and Young. How did they prove their result? What apportionment model did they recommend for the U.S. House of Representatives?
11. Research divisor models for other ways in which violation of quota can occur. For example, can you find four classes for which the number of apportioned seats falls two short of the size of the council and results in awarding both seats to the same class?
12. Investigate measures of fairness such as those given in Exercises 2 and 3. For each apportionment model discussed in this and the previous lesson, which measure of fairness produces the same apportionment?
13. Research and report on the process used by the U.S. Census Bureau to apportion U.S. House seats after each census. How does the bureau reach its final count? How does it implement the Hill model?
14. Research and report on apportionment models used in other countries. Which models are used and why?
15. Research and report on the way presidential primary results are used to apportion convention delegates among the candidates.

Lesson 2.5

Fair Division Models: The Continuous Case

The problem of dividing a cake fairly is similar in some ways to estate division. Like the cash in an estate, a cake can be divided in any number of ways. Moreover, the participants in a cake division may not agree on the value of a particular slice of cake, just as the heirs to an estate may place different values on a house. However, unlike estate division and legislative apportionment, a cake division problem is strictly continuous and involves no items such as cash and legislative seats that are of indisputable value.

The problem of dividing a continuous medium fairly is not limited to the activities of children. States and countries, for example, must agree on how to share water resources.

Before beginning your exploration of cake division, you may want to review the work that you did in Lesson 2.1. In that lesson, you concluded that the resolution of a cake division problem by an outside authority might not yield a solution that is fair in the eyes of both individuals. You also concluded that the use of a random event such as a coin toss might not result in a division that is fair in the eyes of both parties. For two people to feel that a piece of cake has been divided fairly, each must feel that he or she received at least half the cake.

Historic U.S.-Mexico Water Sharing Agreement Signed

Coronado Path,
November 20, 2012

Secretary of the Interior Ken Salazar traveled to Coronado to sign the treaty to manage use of Colorado River water.

Dignitaries from agencies on both sides of the U.S.-Mexican border joined Tuesday to sign what one called "the most important" water agreement between the two countries since the original 1944 treaty that determined how water from the Colorado River would be allocated.

"This serves as an international model on how to resolve water issues," said U.S. Secretary of the Interior Ken Salazar.

Yet Salazar described the pact as significant, calling it "a collaborative effort, remarkable in the diversity of its benefits."

It adjusts water usage in times of surplus and shortages, and recognizes as Salazar said, that the Colorado River serves as "the lifeblood of local communities."

Facing reduced supplies in the years ahead, and a common need for water, the agreement adjusts usage in times of drought and seeks cooperation in sharing responsibility for water shortages and surpluses.

Delegations from the U.S. and Mexico worked on the pact for three years.

A good model for the problem of dividing a piece of cake fairly is impossible without a *definition of fairness.* Therefore, this lesson uses the following definition: A division among n people is called fair if each person feels that he or she receives at least $1/n$th of the cake (or other object).

Lesson 2.1 also showed that an appeal to a measurement scheme such as weighing may not be adequate because an individual's evaluation of a piece of cake can be based on more than just size. Cake icing, for example, could be more valuable to one person than to another. Along with a definition of fairness, a good model for the cake division problem requires a few realistic *assumptions* about the participants.

1. Each individual is capable of dividing a portion of the cake into several portions that the individual feels are equal.
2. Each individual is capable of placing a value on any portion of the cake. The total of the values a person places on all portions of a cake is 1, or 100%.
3. The value that each individual places on a portion of the cake can be based on more than just the size of the portion.

The two-person cake division problem is often solved first by having one person cut the cake into two pieces. Then the other person chooses one of them. It is instructive to examine the role of the above definition and assumptions in this model.

The requirement that one person (the cutter) cut the cake into two portions that the person feels are equal is the first assumption. The second assumption ensures that the second person (the chooser) places values that total 1 on the two pieces. However, because of the third assumption, the chooser need not place the same value on both pieces. Even if the chooser feels that his or her piece is more than half the cake, the division is fair because the definition of fairness requires only that each person feel that his or her piece is at least half the cake.

No model for the problem of dividing a cake fairly among three or more people is adequate unless it adheres to the definition of a fair cake division. For example, a fair division among three people must result in each person receiving a piece the person feels is at least one-third of the cake.

In mathematical modeling, often there is more than one way to solve a problem. That is the case with the problem of dividing a cake fairly among three or more people. Sometimes a model can be based on one for a simpler problem. The following model for the three-person problem is based on the two-person model.

Call the three individuals Ava, Bert, and Carlos. The model is described in algorithmic form:

1. Ava cuts the cake into two pieces that she feels are equal.
2. Bert chooses one of the pieces; Ava gets the other.
3. Ava cuts her piece into three pieces that she considers equal; Bert does the same with his.
4. Carlos chooses one of Ava's three pieces and one of Bert's three pieces.

To see that this is indeed a solution to the three-person problem, you must be satisfied that the model adheres to the definition of fairness. That is, you must believe that each person receives a portion that she or he feels is at least one-third of the cake.

Consider Ava. In step 1, Ava feels that each piece is one-half of the cake. She therefore feels that the piece she received in step 2 is half the cake. She feels that each piece she cut in step 3 is one-third of half the cake, or one-sixth of the cake. She therefore feels that she receives two-sixths, or one-third, of the cake.

Bert's case is similar except that he might feel that the portion he chose in step 2 is more than half the cake. Thus, he may feel that each piece he cut in step 3 is more than one-sixth of the cake and that his final share is more than two-sixths, or one-third.

Carlos' case is different from Ava's or Bert's. He might feel that the two pieces Ava cut in step 1 are unequal. He might, for example, consider one piece 0.6 of the cake and the other 0.4. Likewise, he might not feel that the cuts made in step 3 produced equal pieces. He could decide that the piece he valued at 0.6 is divided into pieces he values at 0.3, 0.2, and 0.1. Similarly, he could decide that the piece he values at 0.4 is divided into pieces he values at 0.2, 0.1, and 0.1. However, because he chooses first, he picks the largest piece from each: 0.3 from the piece he values at 0.6 and 0.2 from the piece he values at 0.4. Thus, in this example he feels that the value of the portion he receives is $0.3 + 0.2 = 0.5$.

The previous example gives Carlos a portion he values at more than one-third, but it is only one example. To see that Carlos always gets a portion he values as at least one-third, use the variable x to represent the value Carlos places on one of the pieces cut by Ava in step 1. His value for the other piece cut by Ava must be $1 - x$. Although Carlos may not feel that the piece he values at x is divided into equal thirds, he chooses the largest of the three pieces and values it as at least one-third of x or $\frac{1}{3}x$. Similarly, he values the piece he chooses from the part he values as $1 - x$ as at least $\frac{1}{3}(1 - x)$. Thus, the total value of his two pieces is at least $\frac{1}{3}x + \frac{1}{3}(1 - x) = \frac{1}{3}x + \frac{1}{3} - \frac{1}{3}x = \frac{1}{3}$.

This model is referred to as the **cut-and-choose model**.

This lesson's exercises consider several issues related to cake division and several models for solving problems with three or more participants.

Exercises

1. a. In the division among Ava, Bert, and Carlos, who values his or her share as exactly one-third?

   b. Who might feel that he or she received more than one-third? Explain.

2. a. Does the division among Ava, Bert, and Carlos result in three portions or three pieces?

   b. Does your answer to part a violate the definition of fairness or any of the three assumptions?

3. For each of the four steps of the division among Ava, Bert, and Carlos, state which of the three fairness assumptions is applied.

In Exercises 4 and 5, suppose Carlos feels that Ava's initial division in step 1 is even, that Ava's subdivision in step 3 is also even, but that Bert's subdivision in step 3 is not. (Give your answers as fractions or as decimals rounded to the nearest 0.1.)

4. What value does Carlos place on the piece he takes from Ava? Explain.

5. Although Carlos feels that the piece Bert divided is half the cake, he does not feel that Bert subdivided it equally. He could, for example, place values of 0.3, 0.1, and 0.1 or values of 0.4, 0.06, and 0.04 on the three pieces.

   a. Explain why the largest value Carlos can place on any of Bert's three subdivisions is 0.5.

   b. What is the smallest value Carlos can place on the piece he takes from Bert? Explain.

   c. What is the largest total value Carlos can place on the two pieces he takes from Ava and Bert?

   d. What is the smallest value Carlos can place on the two pieces he takes from Ava and Bert?

6. Alisha and Balavan are dividing a cake by the cut-and-choose model. After the cake is cut, Alisha thinks that the values of the two pieces are 60% and 40%. Balavan thinks that the values are 50% and 50%. Who did the cutting? Explain.

7. Is the problem of dividing cookies fairly among three children discrete or continuous? Explain.

Peanuts © reprinted by permission of United Feature Syndicate, Inc.

8. In mathematics, a fundamental principle of counting states: If there are m ways of performing one task and n ways of performing another, then there are $m \times n$ ways of performing both. For example, a tossed coin can fall in two ways, and a rolled die can fall in six ways. Together they can fall in a total of $2 \times 6 = 12$ ways.

   a. If two people each have a piece of cake and each person cuts his or her piece into three pieces, show how to use the fundamental counting principle to determine the number of pieces that result.

   b. If k people each have a piece of cake and each cuts his or her piece into $k + 1$ pieces, what is an expression for the total number of pieces that result? Show how to use the distributive property to write an equivalent expression.

   c. If $k + 1$ boxes each contain $k + 5$ toothpicks, what are two equivalent expressions for the total number of toothpicks?

d. Two offices are being filled in an election: mayor and governor. If there are three candidates for governor and four for mayor and conventional voting procedures are used, in how many ways can a person vote?

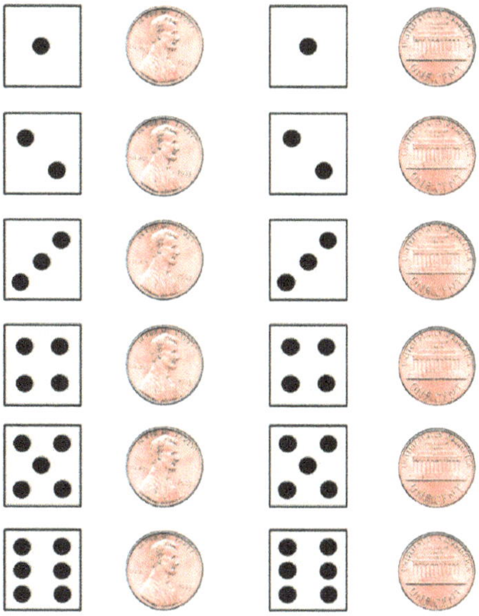

9. Consider the following division of a cake among three people: Arnold, Betty, and Charlie. Arnold cuts the cake into three pieces. Betty chooses one of the pieces, and Charlie chooses either of the remaining two. Arnold gets the third piece.

   a. Does Arnold feel he receives at least one-third of the cake? Might he feel he receives more than one-third?

   b. Does Betty feel she receives at least one-third of the cake? Might she feel she receives more than one-third?

   c. Does Charlie feel he receives at least one-third of the cake? Might he feel he receives more than one-third?

10. Arnold, Betty, and Charlie decide to divide a cake in the following way: Arnold cuts a piece that he considers one-third of the cake. Betty inspects the piece. If she feels it is more than one-third of the cake, she must cut enough from it so that she feels it is one-third of the cake. The removed portion is returned to the cake. Charlie now inspects Betty's piece and has the option of doing similarly if he thinks that it is more than one-third of the cake. The piece of cake is given to the last person who cuts from it.

    Next, one of the remaining two people slices a piece that he or she feels is half of the remaining cake. The other person inspects the piece with the option of removing some of it if he or she feels that it is more than half the remaining cake.

    a. Does the person who receives the first piece feel that it is at least one-third of the cake? Could he or she feel it is more than one-third? Explain your answers.

b. Does the person who receives the second piece feel that it is at least one-third of the cake? Could he or she feel it is more than one-third? Explain your answers.

c. Does the person who receives the third piece feel that it is at least one-third of the cake? Could he or she feel it is more than one-third? Explain your answers.

This model is called the **inspection model**.

11. The definition of a fair cake division used in this lesson does not state that each person must feel his or her portion of the cake is at least as large as the portion that each of the other participants receives.

a. Construct an example for the three-person cut-and-choose division among Ava, Bert, and Carlos to show that Carlos may feel that the piece received by one of the other two is better than his.

b. Could the definition of fair cake division result in jealously? Explain.

Computer/Calculator Exploration

12. Use the moving knife program that accompanies this book to divide a cake of any shape amongst yourself and two other people by means of the **moving knife model**.

a. Explain why each of the people in your group feels that he or she received at least one-third of the cake.

b. Might any of the people feel jealous of the share received by another? Explain.

13. A model for dividing a cake fairly can be considered "optimal" if it fulfills some criterion. For example, a model might be called optimal if it results in exactly the same number of pieces as participants. By this criterion, is the cut-and-choose model, the inspection model, or the moving knife model optimal? Explain.

14. Could the cut-and-choose, the inspection, or the moving knife model be extended to divide a cake among four people? Explain how to do this.

Chimps Possess a Sense of 'Fairness'

ABC News
January 15, 2013

Sharing and caring Chimpanzees, in a test of their willingness to share with other chimps, displayed a surprising sense of fairness, scientists say, debunking the idea that only humans boast that quality.

The researchers say they are the first to show chimpanzees have a sense of fairness and that this may have evolved over time to aid their survival.

The experiments were carried out by scientists at the Yerkes National Primate Research Center at Emory University near Atlanta, Georgia, working with colleagues at Georgia State University.

They taught the chimps to play the game 'Ultimatum' in which "one individual needs to propose a reward division to another individual, and then have that individual accept the proposition before both can obtain the rewards," says researcher Frans de Waal.

Researchers ran the test separately on six adult chimpanzees and 20 human children between the ages of two and seven.

The game was simple: One individual chose between two differently colored tokens that, with his or her partner's cooperation, could be exchanged for rewards - a snack item for the chimps, or stickers for the children.

A token of one color offered equal rewards to both players, while the other gave the individual making the choice a much larger share of the spoils.

The results showed no differences between human and chimp behavior: If the partner's cooperation was required, the chimpanzees and children split the rewards equally, researchers write.

The scientists note that from an evolutionary standpoint, chimpanzees are highly cooperative in the wild and would need to be sensitive to reward distributions for survival.

Lesson 2.6

Mathematical Induction

The cut-and-choose model, the inspection model, and the moving knife model all produce a fair division of a cake among two or three people. It appears that these models can be extended to larger groups, but you or others in your class may not be convinced. This lesson considers a method mathematicians use to prove that a discrete process works indefinitely. The method is called **mathematical induction**.

The principle behind mathematical induction is relatively simple because there only two essential parts:

- Show that the process in question works for at least one *base case*.
- Use a general argument to show that the process can be extended indefinitely.

For example, consider the cut-and-choose model. Since mathematical induction requires the establishment of at least one base case, it is important to review the situations for which the cut-and-choose model is known to work. It was first offered as a means by which two people can divide a cake fairly. It was extended to three people by requiring that two of them first apply the two-person model. Then each of the two cut his or her piece into three pieces that he or she considered equal. The third person selected a piece from each. Therefore, the cut-and-choose model is known to work in two base cases: two and three people.

Quite often a general argument can be modeled after a specific one. So, before trying to develop a general argument, consider how the three-person case can be extended to a four-person case.

Call the four people Ann, Bart, Carl, and Daiva. First, have Ann, Bart, and Carl apply the three-person model. Next, have each of them divide his or her portion into four portions that he or she considers equal. Daiva chooses one from each. To be sure this division is fair, you must show that each of the three cutters and the chooser feel that the received share is at least one-fourth of the cake.

Consider one of the cutters, Ann. The three-person model guarantees that she feels she has at least one-third of the cake. Therefore, Ann divides what she considers a third into four equal pieces. Since she retains three of these pieces, the value she attaches to her portion is at least $\frac{3}{4} \times \frac{1}{3} = \frac{1}{4}$.

Daiva may not feel that the three-person model produced three equal portions, but she must feel that the total value of the three portions is 1. Suppose she attaches values of p_1, p_2, and p_3 to the three portions, where $p_1 + p_2 + p_3 = 1$. Because Daiva is given first choice of a portion from each of the cutters, she places a value of at least $\frac{1}{4}p_1 + \frac{1}{4}p_2 + \frac{1}{4}p_3 = \frac{1}{4}(p_1 + p_2 + p_3) = \frac{1}{4}(1)$ on the resulting portion. Therefore, Daiva considers the division fair.

It appears that this division can be extended to a five-person situation, to a six-person situation, and, in general, to an n-person situation. If, for example, a cake is to be divided fairly among 10 people, the division can be based on a nine-person division, which, in turn, is based on an eight-person division, which is based on . . . .

Mathematical induction generalizes this pattern of solutions by using a variable instead of a specific number. In other words, assuming you know how to divide a cake fairly among k people, you must show that it is possible to use that knowledge to divide a cake fairly among $k + 1$ people.

The proof begins with the assumption that k people can divide the cake fairly. When these k people finish the division, each of them divides his or her share into $k + 1$ portions that he or she feels are equal. The $(k + 1)$th person then selects one portion from each.

It is now necessary to show that each of the $k + 1$ people places a value of at least $\frac{1}{k+1}$ on his or her share.

In the case of the k cutters, the k-person solution guarantees that each has a portion valued as at least $\frac{1}{k}$ of the cake. This portion is divided into $k + 1$ equal portions, so the value of each is at least $\frac{1}{k(k+1)}$. Since k of the $k + 1$ portions are retained by the cutter, the total value attached to them is

$$k\left(\frac{1}{k(k+1)}\right) = \frac{1}{k+1}.$$

Although the chooser may not feel that each of the k portions from the k-person solution is $\frac{1}{k}$ of the cake, he or she must feel that the total value is 1. Suppose this person assigns values of $p_1, p_2, \ldots, p_k$ to the k portions. Then $p_1 + p_2 + \ldots + p_k = 1$. Because the chooser is given first choice of a portion from each of the cutters, the chooser places a value of at least

$$\frac{1}{k}p_1 + \frac{1}{k}p_2 + \ldots + \frac{1}{k}p_k = \frac{1}{k}(p_1 + p_2 + \ldots + p_k) = \frac{1}{k}(1)$$

on the resulting portion. Therefore, the chooser considers the division fair.

The proof is complete because it demonstrates that whenever a cake can be divided fairly among k people, it can also be divided fairly among $k + 1$ people. In other words, the cut-and-choose model can be extended indefinitely. Note that the mathematical induction argument is merely a generalization of the one given to justify the extension to four people.

Keep in mind that it is senseless to attempt to show that a process can be extended indefinitely unless it is known to work in at least one base case. Your work in Lesson 2.5 showed that the cut-and-choose model works with groups of two and three.

Verifying a Formula with Mathematical Induction

One of the most common uses of mathematical induction is to verify that a formula that describes a numerical pattern does so indefinitely. The identification and verification of numerical patterns helps people analyze and predict trends in science, business, economics, world affairs, and many other areas. By confirming that a formula accurately describes a pattern, mathematical induction helps avoid erroneous predictions that can waste time and money.

For example, consider a situation in which Luis and Britt are investigating the number of handshakes that are made in a group of people if every person shakes hands with every other person. Luis notes that if there is only one person in the group, no handshakes are made and that if there are two people, one handshake is made. Britt draws a geometric model (see Figure 2.2) in which the vertices represent people, and the segments represent handshakes. She concludes that a group of three people requires a total of three handshakes.

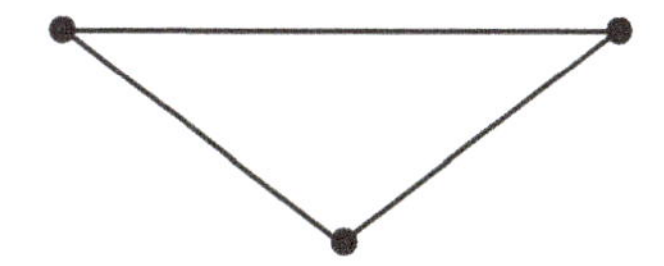

Figure 2.2. Diagram representing handshakes among three people.

Luis suggests organizing the data into a table.

Number of People in the Group	Number of Handshakes
1	0
2	1
3	3

Britt and Luis must now determine a formula that predicts the number of handshakes in a group from the number of people in the group. They need mathematical induction to prove that any conjectured formula works for a group of any size. In the first four of this lesson's exercises you will help Britt and Luis complete their work. After that you will try some similar problems on your own.

Exercises

1. To use mathematical induction, you must be able to express numerical patterns in symbols. You will use some of the expressions you write in this exercise in the mathematical induction proof of a formula for the handshake problem.

    a. If there are three people in a group and another person joins the group, then there are four people in the group. If a person leaves the original group, there are two. Write an expression for the number of people if there are k people in a group and another person joins. Do the same if a person leaves the group of k people.

    b. Repeat part a if the original group has $k + 1$ people and if the original group has $2k$ people.

2. Draw a graph like Britt's and make a table like Luis's.

   a. Add another vertex to the graph to represent a fourth person, and draw segments to represent the additional handshakes that result from the addition of the fourth person. In your table, write the total number of handshakes in a group of four people.

   b. Add a fifth vertex to represent a fifth person and draw segments to represent the additional handshakes. In your table, write the total number of handshakes in a group of five people.

3. a. Suppose that there are seven people in a group and each of them has shaken hands with each of the others. If an eighth person enters the group, how many additional handshakes must be made? Explain.

   b. Suppose that there are k people in a group and each of them has shaken hands with every other person. If a new person enters the group, how many additional handshakes must be made? Explain.

   c. If H_n represents the number of handshakes in a group of n people, write a recurrence relation that expresses the relationship between H_n and H_{n-1}. Write a recurrence relation that expresses the relationship between H_{n+1} and H_n.

4. After studying the data, Britt suggests that the number of handshakes in a group can be found by multiplying the number of people by the number that is 1 less than that and then dividing by 2. She and Luis establish three base cases by checking her formula for 1, 2, and 3 people using the data in Luis's table.

   a. If her guess is correct, how many handshakes are there in a group of 10 people?

   b. On the basis of Britt's guess, write an expression for the number of handshakes if there are k people in a group. Do the same for a group of $2k$ people. Do the same for a group of $k + 1$ people.

Britt's formula, if correct, is sometimes called a solution of the recurrence relation you wrote in part c of Exercise 3. One of the solution's advantages is that it allows you to determine the number of handshakes in a group directly (that is, without knowing the number of handshakes in a smaller group).

To prove that Britt's formula always works, you must show that it is possible to extend it to a group that is one larger. In other words, whenever the conjecture works for a group of k people, it must work for a group of $k + 1$ people.

c. Assume that Britt's formula works for a group of k people. Write her formula for the number of handshakes in a group of k people.

d. If an additional person enters a group of k people, how many new handshakes are necessary? Explain.

e. You must show that Britt's formula works for a group of $k + 1$ people. Write down what this means. That is, write her formula for the number of handshakes in a group of $k + 1$ people.

An expression for the total number of handshakes in a group of $k + 1$ people can be found by adding the expression for the number of handshakes in a group of k people (part c) to the number of additional handshakes when another person enters the group (part d): $\frac{k(k-1)}{2} + k$.

f. You can conclude that Britt's formula works indefinitely if this expression is equivalent to the one you wrote in part e. Use algebraic procedures to transform the expression $\frac{k(k-1)}{2} + k$ until it matches the one you wrote in part e.

5. Although Britt's formula finds the number of handshakes in a group of people, it could also be used to find the number of potential two-party conflicts in a group.

a. Use the formula to compare the number of potential conflicts when the size of a group doubles. Does the number of potential conflicts also double? Explain.

b. Why do the results of Exercise 4 suggest that some of the costs associated with government, such as that of maintaining a law enforcement system, may outpace the growth of population?

In Exercises 1 through 4, you supplied many of the steps of a mathematical induction proof. In Exercise 6, which considers a formula for the relationship between the number of candidates on a ballot and the number of ways of casting an approval vote, you will again supply many of the steps of an induction proof. However, before beginning the proof, you must do several preliminary steps that lead to the conjecture of a formula. The preliminary steps are summarized here.

Preliminary Steps

Do the following before using mathematical induction to prove that a formula describes a relationship between two quantities.

1. Organize a table of data for several small values. For example, how many ways of voting are there with 1, 2, 3, or 4 choices on the ballot?
2. Study the table and describe the data with a recurrence relation. For example, how many additional ways of voting are there when another choice is added to the ballot?
3. Conjecture a formula that predicts the outcome for a collection of k items. For example, what is a formula that predicts the number of ways of voting when there are k choices on the ballot?
4. Verify that your formula works for the values in your table.

6. In Exercise 14 of Lesson 1.4 (page 33), you considered the number of ways of voting under an approval system. The data you gathered in Lesson 1.4 are reproduced below. The goal of this exercise is to first conjecture a formula for the number of ways of voting when there are n choices on the ballot and then to use mathematical induction to prove that the formula is correct.

Number of Choices on the Ballot	Number of Possible Ways of Voting
1	2
2	4
3	8
4	16

a. Collecting these data completes the first of the preliminary steps. Next, you must find a recurrence relation that describes the relationship between the number of ways of voting when there are $k + 1$ choices on the ballot (V_{k+1}) and the number of ways of voting when there are k choices on the ballot (V_k). Do so, but be careful; if you do not establish the recurrence relation properly, the proof that comes later will fail. Here's a hint.

With Three Choices There are Eight Ways		The New Ways When a Choice Is Added
{ }		{D}
{A}		{A, D}
{B}	Append D	{B, D}
{C}	to each	{C, D}
{A, B}	→	{A, B, D}
{A, C}		{A, C, D}
{B, C}		{B, C, D}
{A, B, C}		{A, B, C, D}

b. You are ready for the third and fourth preliminary steps. Study the data carefully. Notice that the values in the second column are all powers of 2. What formula does this suggest for the number of ways of voting when there are n choices? Check the formula with each pair of values in the table.

Parts a and b complete the preliminary process. You are now ready to prove that the formula you conjectured in part b of Exercise 6 always works.

The Proof

After you have conjectured a formula and shown that it works for a few base cases, you can prove that it always works:

1. State the meaning of your formula for a collection of size k. This is the assumption or hypothesis. (It is similar to the "given" in a geometric proof.)
2. State the meaning of your formula for a collection of size $k + 1$. You can do this by replacing k with $k + 1$ in the formula you wrote in step 1. This is the goal. (It is similar to the "prove" in a geometric proof.)
3. Use your recurrence relation to describe the effect of an additional object on the formula you stated in step 1.
4. Use algebraic procedures to manipulate the expression you wrote in step 3 until it matches the one you wrote in step 2.

c. Assume that the formula you conjectured in part b works for a ballot with k choices: $V_k = 2^k$. You must show that the formula works for a ballot with $k + 1$ choices. Write the formula for a ballot with $k + 1$ choices to complete the first two steps of the proof process.

d. Write an expression for the total number of ways of voting when a ballot has $k + 1$ choices by applying the recurrence relation you gave in part a to the formula you stated in part b. Use algebraic procedures to show that the result is equivalent to your answer to part c. This completes steps 3 and 4 of the proof process, and your induction proof is complete.

In Exercises 7 through 13, collect and organize data into a table, examine the data, conjecture a formula, and use mathematical induction to prove that the formula is correct. If you need help with any of the steps, use the summaries of the preliminary steps and the proof in Exercise 6.

7. Dominoes come in sets of different sizes. A double-six set, for example, contains dominoes that pair each number of spots from 0 through 6 with itself and with every other number of spots. Find a formula for the number of dominoes in a double-k set. After you have found and proved your formula, use it to determine the number of dominoes in the image of Carl Gauss.

Mathematician and domino artist Robert Bosch created this likeness of the famous mathematician Carl Gauss from 48 sets of double-nine dominoes.

8. Bowling pins are normally set in a triangular configuration. Find a formula for the number of pins in a triangular configuration of k rows.

9. An ancient legend says that the inventor of the game of chess was offered a reward of his own choosing for the delight the game gave the king. The inventor asked for enough grains of wheat to be able to place one grain on the first square of the chessboard, two on the second, four on the third, and so forth, doubling the number of grains each time. Find a formula for the total number of grains on a chessboard after the kth square has been filled.

10. It takes four toothpicks to make a 1×1 square. It takes 12 toothpicks to make a 2×2 square that is subdivided into 1×1 squares. Find a formula for the number of toothpicks needed to make a $k \times k$ square that is subdivided into 1×1 squares.

11. In Exercises 18 and 19 of Lesson 1.4 (page 34), you found a recurrence relation for the number of ways of selecting exactly two items when there are several choices on a ballot. Find a formula for the number of ways of selecting exactly two items when there are k choices on the ballot.

12. In the popular song "The Twelve Days of Christmas," one gift is given on the first day, one plus two on the second, and so on. Find a formula for the total number of gifts given on the kth day.

13. Find a formula for the number of building blocks in a set of $k \times k$ steps.

A 2×2 set of steps has three building blocks

A 3×3 set of steps has six building blocks.

14. In the 1950s while a student at Harvard, Solomon Golomb coined the term polyominoes for generalized dominoes. For example, this is called a pentomino because it is formed from five squares, as shown:

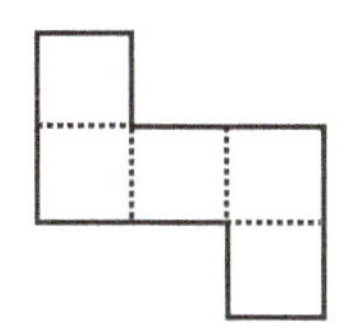

Mathematician of Note

Solomon Golomb (1932-)

Dr. Golomb is professor of communications at the University of Southern California. He received the National Medal of Science in 2013

Numerous puzzles and games are based on Golomb's polyominoes, including the computer game Tetris.

A polyomino formed from three squares is called a tromino. There are two tromino shapes. One is straight. The other is shaped like the letter L:

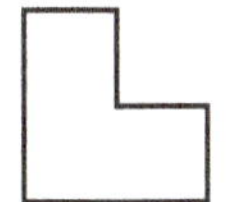

Golomb uses mathematical induction to prove various theorems about polyominoes. One of these theorems is that any $2^k \times 2^k$ ($k = 1$, 2, 3, . . .) square with one corner square removed can be tiled with L-shaped trominoes.

a. Explain why the following figure establishes a base case.

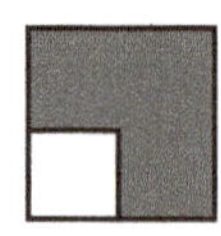

b. Verify the theorem for another base case, $k = 2$.

c. You can use your answer to part b to show that the theorem is true for the next case, $k = 3$ by dividing this square into smaller squares. Explain how to tile a $2^3 \times 2^3$ square. (Hint: draw the $2^3 \times 2^3$ square on a piece of graph paper and cut it into four smaller squares.)

d. Complete the mathematical induction proof of the theorem. That is, show how to use the assumption that a $2^k \times 2^k$ square with a corner removed can be tiled to explain how a $2^{k+1} \times 2^{k+1}$ square with a corner removed can be tiled.

15. Mathematicians often use mathematical induction to prove facts about numbers.

    a. Use mathematical induction to prove that a formula for the kth odd integer is $2k - 1$.

    b. Find and prove a formula for the sum of the first k odd integers. (Hint: The formula in part a may be of help.)

16. a. Find and prove a formula for the kth even integer.

    b. Find and prove a formula for the sum of the first k even integers.

Projects

17. Research and report on the use of mathematical induction in computer science.

18. Research and report on the history of mathematical induction as a method of proof. When was it first used? How have its uses evolved over time?

Chapter Extension

Envy-Free Division

The first mathematician to take a serious interest in the problem of dividing a cake fairly may have been Hugo Steinhaus, a Polish mathematician who studied the problem during World War II. Steinhaus appears to have been motivated only by the challenge the problem posed. In the last half-century the world has served up an increasing number of problems that resemble cake division. Mathematicians and social scientists have risen to the challenge with a variety of new models.

Steinhaus used the fairness definition that you used in this chapter: A cake is divided fairly among n people if each person feels that he or she receives at least $1/n$th of the cake. The cleverness of the people who have tackled fair division problems is demonstrated by the existence of an estate division model that can result in everyone's getting more than a fair share by this definition.

Mathematician of Note

Hugo Steinhaus (1887–1972)

Professor Steinhaus did important mathematical research at universities in Lvov and Wroclaw.

However, the fairness definition is deficient in at least one important way. Fair division situations are sometimes charged with emotion. Not all participants behave as rationally as might be desired. For example, according to this definition, a participant in a four-person estate division should be happy if the value placed on his or her received share is, say, 0.3. But what if this individual feels someone else's share has a value of 0.35? Feeling that your share is fair is not the same as feeling that no one else has a larger share.

For example, consider three individuals, A, B, and C, who divide a cake by the cut-and-choose model. Suppose that A and B are the cutters and that C places a value of 0.9 on A's piece and 0.1 on B's. Following the subsequent division of A's and B's pieces, C values A's three pieces at 0.3 each. C now chooses a piece with value 0.3 from A, leaving A with two pieces C values at 0.3 each. Even if C gets all of B's pieces, C can do no better than 0.4 against A's 0.6.

The existence of situations in which envy arises suggests an alternative definition of fairness: Each person must feel that the received portion is at least as big as every other person's. The two-person cut-and-choose model is free of envy, but the three-person cut-and-choose model is not.

In the 1960s, several mathematicians independently devised a three-person envy-free model. It begins by having A cut the cake into three pieces. If B feels the division is unfair, B trims the best piece to make it the same value as the second-best. C gets first choice of one of the three pieces, B second choice, and A third. However, when B chooses, B must take the trimmed piece if C did not choose it. The trimmings can be disposed of by, say, feeding them to a dog. If the trimmings are substantial, they can be divided in a second round of cutting and choosing (although the roles of A, B, and C change).

Since C chooses first, C's piece seems at least as good to C as either of the others. After C chooses, B is left at least one of the two pieces B considers better. Since B is forced to pick the trimmed piece if it is available, A is left one of the two pieces that A cut on the first round.

Mathematicians of Note

Steven Brams (1940–)

Steven Brams is a political scientist at New York University.

Alan Taylor (1947–)

Alan Taylor is a mathematician at Union College in Schenectady, New York.

None of the mathematicians who discovered this three-person envy-free model succeeded in extending it to four. Then, in 1992, Steven Brams and Alan Taylor discovered an algorithm that produces an envy-free solution for any number of people.

Although the model of Brams and Taylor applies to a divisible object such as a cake, the pair quickly adapted it to situations in which only a portion of the goods is divisible (i.e., the cash in an estate). In a pair of recent books on fair division, Brams and Taylor adapt the model to a wide range of real-world fair division problems, including situations, such as the division of chores, in which the "goods" are really "bads."

Brams: Negotiate Mideast Peace With Point System

Pacific Standard
December 27, 2011

Steven Brams has examined the current peace agreement between Egypt and Israel, negotiated at Camp David MD in 1978, as a way to test his fairness theories.

When Brams thinks about dividing up goods, he has several goals. One is efficiency, where no other allocation is better for one player and at least as good for all the others. Another is envy-freeness, where each player thinks it receives at least a tied-for-largest portion.

What he recommends (when what is being negotiated is divisible) is a technique called adjusted winner.

In this system, each side is given 100 points to distribute across two or more goods. After the sides assign points independently, the goods are allocated based on the highest bids. If one side gets goods worth more points than the other side overall, an adjustment is made, via negotiations, to even the point totals.

Such a system doesn't eliminate negotiations but requires them to be structured so that both sides avoid entangling details. As Brams notes, most negotiations get hung up on the procedural issues before arriving at substantive matters.

Here's an example Brams gives based on the Camp David negotiations. Assume that Israel, anticipating that Egypt would put very high points on return of the Sinai, reduced Israel's own points on that issue from 35 to 20. And then Israel guesses Egypt won't value Palestinian rights too highly, so Israel bumps up its points on that, to 20.

Israel may win most of the issues, but it could find that it is tied with Egypt on Palestinian rights and that both countries have assigned 20 points to it.

If Egypt's total points won are below Israel's, the tied issue of Palestinian rights will be awarded to Egypt.

If that puts Egypt ahead on total points won, another adjustment is needed. But by allocating its points in an untruthful way, Israel will end up with fewer points for what it really feels is important in each subsequent adjustment.

The only way gaming this system can work is if one side had precise information about how the other side will distribute its points, and then the manipulator could optimally allocate its points to exploit the knowledge.

Brams' assessment of Camp David, which has stood the test of time, is that his adjusted winner system could have cut through a lot of the preliminaries and produced a less-crisis-driven process than what occurred.

Chapter 2 Review

1. Write a summary of what you think are the important points of this chapter.
2. Joan, Henry, and Sam are heirs to an estate that includes a vacant lot, a boat, a computer, a stereo, and $11,000 in cash. Each heir submits bids as summarized in the following table.

	Joan	Henry	Sam
Vacant lot	$8,000	$7,500	$6,200
Boat	$6,500	$5,700	$6,700
Computer	$1,340	$1,500	$1,400
Stereo	$800	$1,100	$1,000

   For each heir, find the fair share, the items received, the amount of cash, and the final settlement. Summarize your results in a matrix.

3. Anne, Beth, and Jay are heirs to an estate that includes a computer, a used car, and a stereo. Each heir submits bids for the items in the estate as summarized in the following table.

	Anne	Beth	Jay
Computer	$1,200	$900	$1,050
Car	$7,600	$7,400	$7,000
Stereo	$800	$600	$1,000

   For each heir, find the fair share, the items received, the amount of cash, and the final settlement. Summarize your results in a matrix.

4. Lynn, Pauline, and Tim have just learned that they are the heirs to the estate of their recently deceased Uncle George. The only items of value in the estate are a rare guitar, a car, a kayak, and an expensive watch. Lynn, Pauline, and Tim submit bids as shown in the following table.

	Lynn	Pauline	Tim
Guitar	$1,500	$2,500	$2,200
Car	$6,000	$5,500	$5,500
Kayak	$700	$200	$600
Watch	$250	$400	$350

Lynn was always George's favorite, and so his will states that Lynn should receive half of his estate and that Pauline and Tim should each receive a quarter. For each heir, find the fair share, the items received, the amount of cash, and the final settlement. Summarize your results in a matrix.

5. States A, B, and C have populations of 647, 247, and 106, respectively. There are 100 seats to apportion among them.

   a. What is the ideal ratio?

   b. Find the quota for each state.

   c. Apportion the 100 seats among the three states by the Hamilton model.

   d. What is the initial Jefferson apportionment?

   e. Find the Jefferson adjusted ratio for each state.

   f. Apportion the 100 seats by the Jefferson model.

   g. What is the initial Webster apportionment?

   h. Find the Webster adjusted ratio for each state.

   i. Apportion the 100 seats by the Webster model.

   j. What is the initial Hill apportionment?

   k. Find the Hill adjusted ratio for each state.

   l. Apportion the 100 seats by the Hill model.

m. Suppose the populations of the states change to 650, 255, and 105, respectively. Reapportion the 100 seats by the Hamilton model.

n. Explain why the results in part m constitute a paradox.

6. States A, B, C, and D have populations of 156, 1,310, 280, and 254, respectively. There are 20 seats to apportion among them. Two of the states have strong opinions about the model used, and the other two do not care. Determine which states care and explain why.

7. Discuss the theorem proved by Michel Balinski and H. Peyton Young. That is, what did they prove?

8. Arnold, Betty, and Charlie are dividing a cake in the following way.

   - Arnold divides the cake into what he considers six equal pieces.
   - The pieces are then chosen in this order: Betty, Charlie, Betty, Charlie, Arnold, Arnold.

   Who is guaranteed a fair share by his or her own assessment?

9. Four people have divided a cake into four pieces that each considers fair, and then a fifth person arrives. Describe a way to divide the four existing pieces so that each of the five people receives a fair share.

In Exercises 10 and 11, collect and organize data into a table. Examine the data and conjecture a formula. Then use mathematical induction to prove that your formula is correct.

10. A basketball league schedules two games between each pair of teams. Find a formula for the number of games if the league has k teams.

11. In a set of concentric circles, a ring is any region that lies between any two of the circles. Find a formula for the number of rings in a set of k concentric circles.

12. On the basis of the enrollment in each of a high school's courses, the administration must decide the number of sections that are offered. A number of factors affect the decision. For example, financial considerations require about 25 students in each section and a maximum of 300 sections for all courses. Develop a model the school might use to divide 300 sections fairly among all courses on the basis of the enrollment in those courses.

13. Discuss how fair division models you studied in this chapter might change to accommodate a situation in which the objects being divided are undesirable (i.e., the division of household chores among children).

Bibliography

Balinski, Michel, and H. Peyton Young. 2001. *Fair Representation: Meeting the Ideal of One Man, One Vote.* 2nd ed. Washington: Brookings Institution Press.

Berlinski, David. 2001. *The Advent of the Algorithm.* New York: Mariner.

Bradberry, Brent A. 1992. "A Geometric View of Some Apportionment Paradoxes." *Mathematics Magazine* 65(1): 3–17.

Brams, Steven J., and Alan D. Taylor. 1996. *Fair Division: From Cake-Cutting to Dispute Resolution.* Cambridge/New York: Cambridge University Press.

Brams, Steven J., and Alan D. Taylor. 2000. *The Win-Win Solution: Guaranteeing Fair Shares to Everybody.* New York: W. W. Norton & Company.

Bruce, Colin. 2002. *Conned Again, Watson.* New York: Basic Books.

Bunch, Bryan. 1997. *Mathematical Fallacies and Paradoxes.* Mineola, NY: Dover Publications.

Burrows, Herbert, et al. 1989. *Mathematical Induction.* Lexington, MA: COMAP, Inc.

Cole, K. C. 1999. *The Universe and the Teacup: The Mathematics of Truth and Beauty.* New York: Mariner.

COMAP. 2011. *For All Practical Purposes: Introduction to Contemporary Mathematics.* 9th ed. New York: W. H. Freeman.

Eisner, Milton P. 1982. *Methods of Congressional Apportionment.* Lexington, MA: COMAP, Inc.

Falletta, Nicholas. 1990. *The Paradoxican.* New York: John Wiley & Sons.

Goetz, Albert. "Cost Allocation: An Application of Fair Division." *Mathematics Teacher* 93(7): 600–603.

Golomb, Solomon. 1996. *Polyominoes: Puzzles, Patterns, Problems, and Packings*. 2nd ed. Princeton, NJ: Princeton University Press.

Hively, Will. 1995. "Dividing the Spoils." *Discover* (March): 49.

Lambert, J. P. 1988. *Voting Games, Power Indices, and Presidential Elections.* Lexington, MA: COMAP, Inc.

Peterson, Ivars. 1996. "Formulas for Fairness." *Science News* 149(18).

Robertson, Jack, and William Webb. 1998. *Cake-Cutting Algorithms: Be Fair If You Can.* Wellesley, MA: A K Peters Ltd.

Steinhaus, Hugo. 2011. *Mathematical Snapshots.* 3rd ed. Mineola, NY: Dover Publications.

CHAPTER

3

Matrix Operations and Applications

It has been said that sports fans are the nation's foremost consumers of statistics and that baseball fans are the most prominent among them. Whether it's in the information conveyed by a scoreboard or in the current information on league leaders, baseball records are full of numbers.

- How can large collections of data be organized and managed in an efficient way?
- What calculations provide meaningful information to people who use the data?
- How can computers and calculators assist them?

Baseball statisticians, business executives, and wildlife biologists are among the diverse groups of people who turn to the mathematics of matrices for answers to these questions.

Lesson 3.1

Addition and Subtraction of Matrices

In Lesson 1.2, you were introduced to matrices as a natural way to organize, manipulate, and display information. As you have seen, matrices can be used to represent sets of discrete data that can be described with two characteristics. One characteristic can be represented by the rows of the matrix and the other by the columns. In this lesson you will be introduced to some of the terminology and notation used when working with matrices. Matrix addition and subtraction will also be explored.

Matrix Terminology

Suppose that you and a few of your friends are planning a pizza and video party. You decide to call several pizza houses to ask about prices for large single-topping pizzas, liter containers of cold drinks, and family-sized salads with house dressing. You could record your information in a table such as the following.

	Gina's	Vin's	Toni's	Sal's
Pizza	$12.16	$10.10	$10.86	$10.65
Drinks	$1.15	$1.09	$0.89	$1.05
Salad	$4.05	$3.69	$3.89	$3.85

Or you might choose to organize your data in matrix form by writing the numbers in a rectangular array and enclosing them in brackets or parentheses.

$$\begin{array}{l} \\ \text{Pizza} \\ \text{Drinks} \\ \text{Salad} \end{array} \begin{array}{c} \begin{array}{cccc} \text{Gina's} & \text{Vin's} & \text{Toni's} & \text{Sal's} \end{array} \\ \begin{bmatrix} \$12.16 & \$10.10 & \$10.86 & \$10.65 \\ \$1.15 & \$1.09 & \$0.89 & \$1.05 \\ \$4.05 & \$3.69 & \$3.89 & \$3.85 \end{bmatrix} \end{array}$$

When writing this price matrix, you could omit the row and column labels and dollar signs and write only the values. However, if you delete the labels you will have to remember that the rows represent the prices for pizzas, drinks, and salads, while the columns represent the various pizza houses.

$$\begin{bmatrix} 12.16 & 10.10 & 10.86 & 10.65 \\ 1.15 & 1.09 & 0.89 & 1.05 \\ 4.05 & 3.69 & 3.89 & 3.85 \end{bmatrix}$$

This matrix has 3 rows and 4 columns. So its **order** or **dimension** is 3 by 4 (written as 3 × 4). Notice that when you give the order of a matrix, you write the number of rows followed by the number of columns. Each of the individual entries in the matrix is called an **element** or a **component** of the matrix.

In general, a matrix with *m* rows and *n* columns is called an *m* by *n* matrix.

After looking over your data, you might decide to drop Gina's options since they are more expensive than any of the other pizza houses. If you do this, you will be left with a 3 × 3 **square matrix.** Notice that in a square matrix the number of rows equals the number of columns or $m = n$.

$$\begin{array}{l} \\ \text{Pizza} \\ \text{Drinks} \\ \text{Salad} \end{array} \begin{array}{c} \begin{array}{ccc} \text{Vin's} & \text{Toni's} & \text{Sal's} \end{array} \\ \begin{bmatrix} \$10.10 & \$10.86 & \$10.65 \\ \$1.09 & \$0.89 & \$1.05 \\ \$3.69 & \$3.89 & \$3.85 \end{bmatrix} \end{array}$$

If a matrix has only one column, it is called a **column matrix**. For example, if you list only the prices for Sal's offerings, the result will be a column matrix of order 3 × 1.

$$\begin{matrix} & \text{Sal's} \\ \begin{matrix}\text{Pizza}\\ \text{Drinks}\\ \text{Salad}\end{matrix} & \begin{bmatrix}\$10.65\\ \$1.05\\ \$3.85\end{bmatrix}\end{matrix}$$

Point of Interest

The simplest order for a matrix would be 1 × 1. A **1 × 1 matrix** such as [10.65] contains only one element.

If you choose to look at the pizza prices alone, they can be represented with a 1 × 3 **row matrix**.

$$\begin{matrix} & \text{Vin's} & \text{Toni's} & \text{Sal's} \\ \text{Pizza} & [\$10.10 & \$10.86 & \$10.65] \end{matrix}$$

Exercises

1. Give an example of a 2 × 5 matrix, a row matrix with two elements, and a square matrix.
2. How many elements does a 2 × 8 matrix have? A 7 × 1 matrix? An $m \times n$ matrix?
3. A trendy garment company receives orders from three clothing shops. The first shop orders 25 jackets, 75 shirts, and 75 pairs of pants. The second shop orders 30 jackets, 50 shirts, and 50 pairs of pants. The third shop orders 20 jackets, 40 shirts, and 35 pairs of pants. Display this information in a matrix. Let the rows represent the shops and the columns represent the type of garment ordered. Label the rows and columns of your matrix accordingly.
4. Although matrices contain many data values, they can also be thought of as single entities. This feature allows us to refer to a matrix with a single capital letter.

$$\begin{matrix} & & \text{Vin's} & \text{Toni's} & \text{Sal's} \\ A = & \begin{matrix}\text{Pizza}\\ \text{Drinks}\\ \text{Salad}\end{matrix} & \left[\begin{matrix}\$10.10\\ \$1.09\\ \$3.69\end{matrix}\right. & \begin{matrix}\$10.86\\ \$0.89\\ \$3.89\end{matrix} & \left.\begin{matrix}\$10.65\\ \$1.05\\ \$3.85\end{matrix}\right] \end{matrix}$$

Individual entries in a matrix are identified by row number and column number, in that order. For example, the value $10.65 is the entry in row 1 and column 3 of matrix A and is referenced as A_{13}. Entry A_{13} represents or *is interpreted as* the cost of a pizza at Sal's. Notice that A_{31} is not the same as A_{13}. Entry A_{31} has the value $3.69 and represents the cost of a salad at Vin's.

a. What is the value of A_{21}? Of A_{12}? Of A_{32}?

b. Write an interpretation for entry A_{21}. For entry A_{12}. For entry A_{32}.

5. In the row matrix S, the entries are referenced as S_1, S_2, and S_3.

	Vin's	Toni's	Sal's
S = Pizza	[$10.10	$10.86	$10.65]

Write an interpretation of S_3.

6. For breakfast Yoko had cereal, a banana, a cup of milk, and a slice of toast. She recorded the following information in her food journal. Cereal: 165 calories, 3 g fat, 33 g carbohydrate, and no cholesterol. Banana: 120 calories, no fat, 26 g carbohydrate, and no cholesterol. Milk: 120 calories, 5 g fat, 11 g carbohydrate, and 15 mg cholesterol. Toast: 125 calories, 6 g fat, 14 g carbohydrate, and 18 mg cholesterol.

a. Write this information in a matrix N whose rows represent the foods. Label the rows and columns of your matrix.

b. State the values of N_{23}, N_{32}, and N_{42}.

c. Write an interpretation of N_{23}, N_{32}, and N_{42}.

7. As you continue to plan your pizza party, you discover that the local supermarket has a sale on 2-liter bottles of soft drinks. You decide not to order drinks from a pizza house after all. Write and label a 2×3 matrix that represents the prices for just pizza and salad at Vin's, Toni's, and Sal's.

8. Suppose that when you were calling the pizza houses about prices, you also collected the following information about the cost of additional toppings and salad dressings.

	Vin's	Toni's	Sal's
Additional toppings	$1.15	$1.10	$1.25
Additional dressings	$0.00	$0.45	$0.50

Represent the information from this table in another 2×3 matrix whose rows represent the additional toppings and dressings and whose columns represent the three pizza houses. Label the rows and columns of your matrix.

9. Suppose you want to find the cost of ordering pizzas with two toppings and salads with a choice of two salad dressings. This can be done by adding corresponding elements of your two price matrices from Exercises 7 and 8. If you let A represent the basic price matrix and B represent the matrix of additional costs, then you can add A and B to get a third matrix C. Matrix C will represent the total prices for pizza and salads at each pizza house.

$$A = \begin{bmatrix} 10.10 & 10.86 & 10.65 \\ 3.69 & 3.89 & 3.85 \end{bmatrix}$$

and

$$B = \begin{bmatrix} 1.15 & 1.10 & 1.25 \\ 0.00 & 0.45 & 0.50 \end{bmatrix},$$

then

$$A + B = \begin{bmatrix} 10.10 + 1.15 & 10.86 + 1.10 & \_\_\_\_\_\_ \\ \_\_\_\_\_\_ & \_\_\_\_\_\_ & \_\_\_\_\_\_ \end{bmatrix}$$

$$C = \begin{bmatrix} \_\_\_\_\_\_ & \_\_\_\_\_\_ & \_\_\_\_\_\_ \\ \_\_\_\_\_\_ & \_\_\_\_\_\_ & \_\_\_\_\_\_ \end{bmatrix}.$$

Complete the addition.

10. In Exercise 9, the entries of matrix C represent the sum of the corresponding entries in matrices A and B. For example, C_{13}, which represents the cost of a pizza with an extra topping at Sal's, equals the sum of A_{13} and B_{13}.

    a. What is the value of A_{21}? Of B_{21}? Of C_{21}?

    b. Write an interpretation of A_{21}, B_{21}, and C_{21}.

In Exercises 9 and 10, you found that you could add two matrices, A and B, by adding the corresponding entries of each. In general, you can add or subtract matrices only if they have the same orders.

If A and B are matrices of the same orders, then the sum $A + B$ is formed by adding the corresponding entries of A and B. The difference $A - B$ is formed by subtracting the corresponding entries of B from the corresponding entries of A.

11. Find the value of each of the following expressions. If it is not possible, explain.

    a. $\begin{bmatrix} 2 & 4 \\ -3 & 5 \end{bmatrix} + \begin{bmatrix} 7 & -8 \\ -1 & 2 \end{bmatrix}$

    b. $\begin{bmatrix} 1 & 5 & 7 \\ 0 & 2 & 4 \end{bmatrix} + \begin{bmatrix} 1 & 0 \\ 5 & 2 \\ 7 & 4 \end{bmatrix}$

    c. $\begin{bmatrix} 1 & 0 \\ 3 & 0 \\ 5 & 0 \\ 0 & -4 \end{bmatrix} - \begin{bmatrix} 1 & 5 \\ -3 & 0 \\ 2 & -4 \\ 0 & -2 \end{bmatrix}$

    d. $\begin{bmatrix} 1 & 5 & 10 \end{bmatrix} - \begin{bmatrix} 1 \\ 5 \\ 10 \end{bmatrix}$

12. The National League batting leaders for 2011 had the following batting statistics.

	AB	R	H	HR	RBI	Avg
J. Reyes (New York)	537	101	181	7	44	.337
R. Braun (Milwaukee)	563	109	187	33	111	.332
M. Kemp (Los Angeles)	602	115	195	39	126	.324

The following statistics for the same three players were published at the end of the 2012 season.

	AB	R	H	HR	RBI	Avg
J. Reyes (Miami)	642	86	184	11	57	.287
R. Braun (Milwaukee)	598	108	191	41	112	.319
M. Kemp (Los Angeles)	403	74	122	23	69	.303

Source: www.espn.go.com

Find and label a matrix that displays the changes in these statistics from the 2011 season to the 2012 season. Notice that several of the statistics decreased from 2011 to 2012. How will you show this in your matrix?

13. The matrices that follow give the winning times in seconds for three track and field events in the 1964 and 2012 Olympic Games.

$$\begin{array}{l} \\ \\ \text{100-meter race} \\ \text{200-meter race} \\ \text{400-meter race} \end{array} \begin{array}{c} \text{1964} \\ \begin{array}{cc} \text{Men} & \text{Women} \end{array} \\ \begin{bmatrix} 10.0 & 11.4 \\ 20.3 & 23.0 \\ 45.1 & 52.0 \end{bmatrix} \end{array} \quad \begin{array}{c} \text{2012} \\ \begin{array}{cc} \text{Men} & \text{Women} \end{array} \\ \begin{bmatrix} 9.80 & 10.92 \\ 19.82 & 21.69 \\ 44.12 & 49.28 \end{bmatrix} \end{array}$$

a. Find and label the matrix that represents the change in times in seconds for each event from 1964 to 2012.

b. In which event and sex was there the greatest decrease in time? The smallest decrease?

14. In your study of algebra, you learned that the commutative and associative properties hold for addition over the set of real numbers, that is, for all real numbers a, b, and c, $a + b = b + a$ and $a + (b + c) = (a + b) + c$.

a. Do you think that the commutative and associative properties hold for addition of matrices? Why?

b. Use the following matrices to test your conjecture in part a.

$$A = \begin{bmatrix} 4 & -2 \\ 3 & 1 \end{bmatrix} \quad B = \begin{bmatrix} 1 & 3 \\ -2 & 5 \end{bmatrix} \quad C = \begin{bmatrix} 2 & 4 \\ 1 & -1 \end{bmatrix}$$

15. Do you think that the commutative and associative properties hold for subtraction of matrices? Why? Test your conjecture using matrices A, B, and C in Exercise 14b.

16. A matrix all of whose entries are the number zero is called a **zero matrix** and is denoted using a capital letter O alone or with subscripts $O_{m \times n}$.

a. Use matrix A from Exercise 14b to show that $A + O = O + A = A$ and that $A - A = O$.

b. Show that $A + (-A) = (-A) + A = O$, where the matrix $-A$, called the negative of A, is obtained by negating each entry in A.

17. The matrix M below shows the mileage between 5 major U.S. cities.

$M =$

	Atlanta	Boston	Chicago	Los Angeles	St. Louis
Atlanta	0	1,075	716	2,211	555
Boston	1,075	0	1,015	3,026	1,187
Chicago	716	1,015	0	2,034	297
Los Angeles	2,211	3,026	2,034	0	1,842
St. Louis	555	1,187	297	1,842	0

a. Entries that are located in row i, column j, where $i = j$, are said to be located on the **main diagonal** of the matrix. Examine the entries on the main diagonal of M. What do you notice?

b. A square matrix R with order $n \times n$ is **symmetric** if $R_{ij} = R_{ji}$, where i and $j = 1, 2, 3, \ldots, n$. Is matrix M symmetric? Explain.

c. Give an example of a 3×3 matrix that is symmetric.

d. Could a matrix that is not square be symmetric? Why?

Point of Interest

In a symmetric matrix you need to know only the values along the main diagonal and either the triangle above the main diagonal (the upper triangle) or below it (the lower triangle). Because of this feature, symmetric matrices are often written with blanks in either the upper or lower triangle.

18. In statistics, a correlation matrix is a matrix whose entries represent the degree of relationship between variables. The values in a correlation matrix range from –1 to 1, where 0 indicates that there is no relationship, a negative value indicates that as one variable increases the other decreases, and a positive value indicates that as one variable increases the other one also increases. In a study of the relationship between ACT test scores, high school class rank, and college grade point average, the following correlation matrix was generated. Notice that the row labels and the column labels are the same in a correlation matrix.

	ACT Comp.	ACT Eng.	ACT Math	ACT S.S.	ACT Sci.	H. S. Rank	Coll. GPA
ACT Composite	1.00	0.80	0.79	0.81	0.82	0.59	0.51
ACT English	0.80	1.00	0.54	0.58	0.55	0.53	0.48
ACT Math	0.79	0.54	1.00	0.42	0.52	0.57	0.42
ACT Social Studies	0.81	0.58	0.42	1.00	0.61	0.39	0.39
ACT Science	0.82	0.55	0.52	0.61	1.00	0.44	0.36
High School Rank	0.59	0.53	0.57	0.39	0.44	1.00	0.45
College GPA	0.51	0.48	0.42	0.39	0.36	0.45	1.00

Source: Aksamit, Mitchell, and Pozehl, 1986.

a. Why do you think the values along the main diagonal of a correlation matrix are all 1s?

b. Why are the values in a correlation matrix symmetric about the main diagonal?

c. Which variable had the highest correlation with college GPA?

d. Which subject area test had the highest correlation with high school rank?

Project

19. Bring to class at least two matrices from newspapers or magazines. Be prepared to share your matrices with other class members.

a. What are the dimensions of each of your matrices?

b. What is represented by the rows and columns of your matrices?

Lesson 3.2

Multiplication of Matrices, Part 1

In the previous lesson, matrix addition and subtraction were defined by looking at several matrix models of real-world situations. In this lesson, matrix multiplication is approached in a similar manner.

Multiplying a Matrix by a Scalar

Return to the pizza problem from the last lesson. In that lesson you used the following price matrix to represent the costs of two-topping pizzas and salads with a choice of two dressings from Vin's, Toni's, and Sal's.

$$C = \begin{array}{r} \\ \text{Pizzas} \\ \text{Salads} \end{array} \begin{array}{c} \begin{array}{ccc} \text{Vin's} & \text{Toni's} & \text{Sal's} \end{array} \\ \begin{bmatrix} 11.25 & 11.96 & 11.90 \\ 3.69 & 4.34 & 4.35 \end{bmatrix} \end{array}$$

Suppose you wish to order four of each of the pizzas and four of each of the salads. To do this, multiply each element in matrix C by 4 to get a new matrix that is equal to $4C$.

$$4C = 4 \times \begin{bmatrix} 11.25 & 11.96 & 11.90 \\ 3.69 & 4.34 & 4.35 \end{bmatrix} = \begin{bmatrix} 4(11.25) & 4(11.96) & 4(11.90) \\ 4(3.69) & 4(4.34) & 4(4.35) \end{bmatrix}$$

$$= \begin{bmatrix} 45.00 & 47.84 & 47.60 \\ 14.76 & 17.36 & 17.40 \end{bmatrix}$$

If you call the new matrix T and label the rows and columns of the matrix, you have

$$T = \begin{array}{c} \\ \text{Pizza} \\ \text{Salad} \end{array} \begin{array}{c} \begin{array}{ccc} \text{Vin's} & \text{Toni's} & \text{Sal's} \end{array} \\ \begin{bmatrix} \$45.00 & \$47.84 & \$47.60 \\ \$14.76 & \$17.36 & \$17.40 \end{bmatrix} \end{array}.$$

When working with matrices, a real number is often called a **scalar**.

In general, if k is a real number and A is a matrix, the matrix kA is formed by multiplying each entry in the matrix A by k.

Multiplying a Column Matrix by a Row Matrix

Consider the following situation.

Suppose that Ruben, a student at Washington High, goes to a nearby store to buy some food to stock up his locker for between-class snacks. He chooses four small bags of chips, five candy bars, a box of cheese crackers, three packs of sour drops, and two bags of cookies.

Ruben's purchases can be represented by a row matrix Q.

$$Q = \begin{array}{c} \begin{array}{ccccc} \text{Chips} & \text{Candy} & \text{Crackers} & \text{Drops} & \text{Cookies} \end{array} \\ \begin{bmatrix} 4 & 5 & 1 & 3 & 2 \end{bmatrix} \end{array}$$

Suppose further that chips cost 30 cents a bag, candy bars cost 35 cents each, crackers cost 50 cents a box, sour drops cost 20 cents a pack, and cookies sell for 75 cents a bag. These prices can be represented in column matrix P.

$$P = \begin{array}{c} \\ \text{Chips} \\ \text{Candy} \\ \text{Crackers} \\ \text{Drops} \\ \text{Cookies} \end{array} \begin{array}{c} \text{Cents} \\ \begin{bmatrix} 30 \\ 35 \\ 50 \\ 20 \\ 75 \end{bmatrix} \end{array}$$

Now the obvious question to ask is, "How much did Ruben pay for all these snacks?" You can answer this question by multiplying the price matrix P by the quantity matrix Q.

$$Q \times P = [4 \;\; 5 \;\; 1 \;\; 3 \;\; 2] \begin{bmatrix} 30 \\ 35 \\ 50 \\ 20 \\ 75 \end{bmatrix}$$

$$= 4(30) + 5(35) + 1(50) + 3(20) + 2(75)$$

$$= 120 + 175 + 50 + 60 + 150$$

$$= 555 \text{ cents} = \$5.55$$

This matrix computation is, of course, exactly what the clerk at the store would do in figuring Ruben's bill. The price of each item is multiplied by the number purchased and the products are summed.

In order to do this computation, the number of entries in each matrix must be the same. Items and prices must also correspond.

In general, if Q is a row matrix and P is a column matrix, each having the same number of entries, then the product QP is defined. QP can be determined by multiplying the corresponding entries and summing the results.

Example

Suppose a second student, Terri, goes along with Ruben to the store. Her purchases are a bag of chips, two candy bars, two packs of gum that cost 25 cents each, and a medium drink for 75 cents. Find the total cost of Terri's purchases.

Solution:

To solve this problem, use a row matrix to represent the quantity of each item and a column matrix to represent the price of each item.

Let Q represent the quantity matrix.

$$\begin{array}{cccc} \text{Chips} & \text{Candy} & \text{Gum} & \text{Drink} \end{array}$$
$$Q = \begin{bmatrix} 1 & 2 & 2 & 1 \end{bmatrix}$$

Let P represent the item-price matrix.

$$P = \begin{matrix} \text{Chips} \\ \text{Candy} \\ \text{Gum} \\ \text{Drink} \end{matrix} \overset{\text{Cents}}{\begin{bmatrix} 30 \\ 35 \\ 25 \\ 75 \end{bmatrix}}$$

To find the total cost of the purchase find the product QP.

$$Q \times P = [1 \quad 2 \quad 2 \quad 1] \begin{bmatrix} 30 \\ 35 \\ 25 \\ 75 \end{bmatrix}$$

$$= 1(30) + 2(35) + 2(25) + 1(75)$$

$$= 30 + 70 + 50 + 75$$

$$= 225 \text{ cents} = \$2.25$$

As you can see from these examples, if a $(k \times 1)$ column matrix P is multiplied by a $(1 \times k)$ row matrix Q, the result is a (1×1) single-value matrix.

Multiplying a Matrix with More than One Column by a Row Matrix

Now that you can multiply a column matrix by a row matrix, you can use that process to multiply a multidimensional matrix by a row matrix.

Return to the pizza problem. Suppose your group decides to order five pizzas and three salads and you want to calculate the total cost at each of the pizza houses. If you do the calculations without using matrices, you multiply the pizza price by 5 and add the result to 3 times the salad price for each pizza house.

Cost at Vin's: 5($11.25) + 3($3.69) = $56.25 + $11.07 = $67.32

Cost at Toni's: 5($11.96) + 3($4.34) = $59.80 + $13.02 = $72.82

Cost at Sal's: 5($11.90) + 3($4.35) = $59.50 + $13.05 = $72.55

To use matrix multiplication to solve this problem, set up a 1×2 row matrix A to represent the number of pizzas and salads you plan to order.

$$A = \begin{array}{cc} \text{Pizzas} & \text{Salads} \\ \left[\ 5 \right. & \left. 3\ \right] \end{array}$$

Set up a 2×3 matrix C to represent the prices for the pizzas and salads at each of the three pizza houses.

$$C = \begin{array}{c} \\ \text{Pizzas} \\ \text{Salads} \end{array} \begin{array}{c} \begin{array}{ccc} \text{Vin's} & \text{Toni's} & \text{Sal's} \end{array} \\ \begin{bmatrix} 11.25 & 11.96 & 11.90 \\ 3.69 & 4.34 & 4.35 \end{bmatrix} \end{array}$$

Now when matrix C is multiplied by the row matrix A, the expected result is another matrix whose entries will give the total cost for five pizzas and three salads at each pizza house. To accomplish this, it makes sense to multiply each of the columns in matrix C by the row matrix A.

$$A \times C = [5 \quad 3] \times \begin{bmatrix} 11.25 & 11.96 & 11.90 \\ 3.69 & 4.34 & 4.35 \end{bmatrix}$$

$$= [5(11.25) + 3(3.69) \quad 5(11.96) + 3(4.34) \quad 5(11.90) + 3(4.35)]$$

$$= [56.25 + 11.07 \quad 59.80 + 13.02 \quad 59.50 + 13.05]$$

$$= [67.32 \quad 72.82 \quad 72.55]$$

The last step is to label the entries in the final product:

$$\begin{array}{ccc} \text{Vin's} & \text{Toni's} & \text{Sal's} \\ \left[\$67.32\right. & \$72.82 & \left.\$72.55\right] \end{array}$$

Notice that in carrying out this matrix computation, the process was exactly the same as that of calculating the costs without the use of matrices. In the matrix multiplication, each of the entries in the columns of matrix C was multiplied by the corresponding entry in the row matrix A. These products were then summed to give the entries of the final product matrix.

You can see from this model that multiplying a multidimensional matrix by a row matrix can only be defined if the number of entries in the row matrix equals the number of rows in the multidimensional matrix.

In general, the product of a $(1 \times k)$ row matrix A and a $(k \times n)$ matrix C is a $(1 \times n)$ row matrix P.

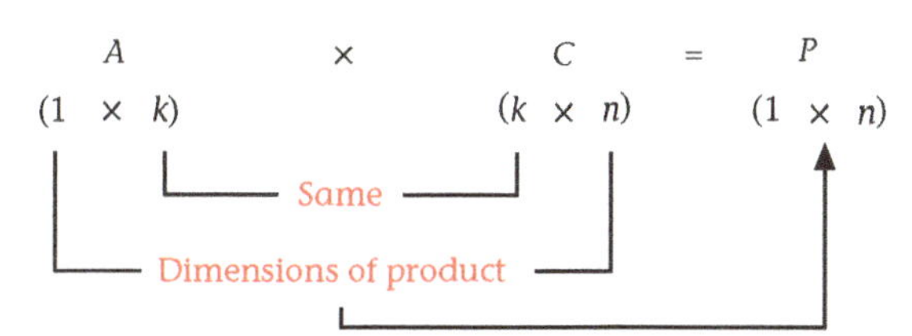

Example

Find the product of V and W.

$$V = [2 \;\; 4 \;\; 7 \;\; 0 \;\; 1] \qquad W = \begin{bmatrix} 1 & 2 \\ 3 & 8 \\ 2 & 5 \\ 2 & 1 \\ 1 & 1 \end{bmatrix}$$

Solution:

The product of the two matrices will be a 1×2 matrix.

$$VW = [2(1) + 4(3) + 7(2) + 0(2) + 1(1) \quad 2(2) + 4(8) + 7(5) + 0(1) + 1(1)]$$

$$= [2 + 12 + 14 + 0 + 1 \quad 4 + 32 + 35 + 0 + 1]$$

$$= [29 \quad 72]$$

Exercises

1. Recall that a matrix T was found by multiplying 4 times the price matrix that represented the costs of two-topping pizzas and salads with a choice of two dressings from Vin's, Toni's, and Sal's. (See page 130.)

$$T = \begin{array}{c} \\ \text{Pizzas} \\ \text{Salads} \end{array} \begin{array}{c} \begin{array}{ccc} \text{Vin's} & \text{Toni's} & \text{Sal's} \end{array} \\ \begin{bmatrix} \$45.00 & \$47.84 & \$47.60 \\ \$14.76 & \$17.36 & \$17.40 \end{bmatrix} \end{array}$$

   a. What does matrix T represent?

   b. What is the cost of four pizzas at Sal's?

   c. Interpret T_{12} and T_{21}.

2. A jeweler has a small shop where she makes and sells four different kinds of jewelry: earrings (e), pins (p), necklaces (n), and bracelets (b). She makes each item out of either pearls or jade beads. The following matrix represents the jeweler's sales for May.

$$M = \begin{array}{c} \\ \text{Pearl} \\ \text{Jade} \end{array} \begin{array}{c} \begin{array}{cccc} e & p & n & b \end{array} \\ \begin{bmatrix} 8 & 4 & 6 & 5 \\ 20 & 10 & 12 & 9 \end{bmatrix} \end{array}$$

The jeweler hopes to sell twice as many of each piece in June.

a. Calculate a matrix J, where $J = 2M$ to represent the number of each item the jeweler will sell in June if she reaches her goal.

b. Label the rows and columns of matrix J.

c. How many jade necklaces does the jeweler expect to sell in June?

d. Interpret J_{21} and J_{12}.

3. Matt reads on the side of his cereal box that each ounce of cereal contains the following percentages of the minimum daily requirements of:

Vitamin A 25%
Vitamin C 25%
Vitamin D 10%

If Matt eats 3 ounces of cereal for breakfast, what percentages of each vitamin will he get? Show the matrices and matrix operation involved in your calculation. Label your matrices.

4. The regents at a state university recently announced a 7% raise of tuition rates per semester hour. The current rates per semester hour are shown in the following table.

	Undergraduate	Graduate
Resident	\$75.00	\$99.25
Nonresident	\$204.00	\$245.25

a. Write and label a matrix that represents this information.

b. Find a new matrix that represents the tuition rates per semester hour after the 7% raise goes into effect. Label your matrix.

c. Find a matrix that represents the dollar increase for each of the categories. Label your matrix.

5. For each of the following, state whether the matrix product QP is defined. If so, give the order of the product.

a. Order of Q: 1×4. Order of P: 4×2.

b. Order of Q: 1×3. Order of P: 2×3.

c. Order of Q: 1×5. Order of P: 5×4.

d. Order of Q: 1×2. Order of P: 4×2.

e. Order of Q: $1 \times m$. Order of P: $m \times 1$.

6. A local credit union has investments in three states—Massachusetts, Nebraska, and California. The deposits in each state are divided between consumer loans and bonds.

   The amount of money (in thousands of dollars) invested in each category is displayed in the following table.

	MA	NE	CA
Loans	230	440	680
Bonds	780	860	940

   The current yields on these investments are 6.5% for consumer loans and 7.2% for bonds. Use matrix multiplication to find the total earnings for each state. Label your matrices.

7. A carpenter makes a trip to the lumber company to pick up ten 2 × 6s, four 4 × 6s, and two 5 × 5s. In 8-foot lengths, 2 × 6s cost $3.00, 4 × 6s cost $8.50, and 5 × 5s cost $9.50.

   a. Write and label a row matrix and a column matrix to represent the information in this problem.

   b. Will everyone necessarily write the same row and column matrices? Explain your answer.

   c. Perform a matrix multiplication to find the total cost of the carpenter's purchases.

8. Use the following matrices to compute the given expressions.

$$A = \begin{bmatrix} 3 & 8 & -1 \\ 2 & 0 & 4 \end{bmatrix} \quad B = [2\ \ 4] \quad C = \begin{bmatrix} 2 \\ 4 \end{bmatrix}$$

   a. $3A$

   b. BA

   c. BC

   d. $-2C$

9. You have $10,000 in a 12-month CD at 7.3% (annual yield); $17,000 in a credit union at 6.5%; and $12,000 in bonds at 7.5%. Use multiplication of a column matrix by a row matrix to find your earnings for a year. Label your matrices.

10. The **transpose** (A^T) of a matrix A is the matrix obtained by interchanging the rows and columns of matrix A.

    a. Describe the transpose of a row matrix.

    b. Describe the transpose of a column matrix.

    c. Matrix M from Exercise 2 is given below.

$$M = \begin{array}{c} \\ \text{Pearl} \\ \text{Jade} \end{array} \begin{array}{c} \begin{array}{cccc} \text{e} & \text{p} & \text{n} & \text{b} \end{array} \\ \begin{bmatrix} 8 & 4 & 6 & 5 \\ 20 & 10 & 12 & 9 \end{bmatrix} \end{array}$$

    Write and label the transpose (M^T) of matrix M.

11. Refer to Exercise 2. Suppose it takes the jeweler 2 hours to make a pair of earrings, 1 hour to make a pin, 2.5 hours to make a necklace, and 1.5 hours to make a bracelet.

    a. Write and label a row matrix that represents this information.

    b. Use matrix multiplication to find a matrix that represents the total hours the jeweler spends making each type of jewelry (pearls or jade) for the month of May. (Hint: Use the transpose of matrix M that you found in the previous exercise.)

    c. Label your product matrix.

    d. Interpret each of the entries in the product matrix.

12. A hobby shop has three different locations, North, South, and East. The store's sales for July are shown in the following table.

	North	South	East
Model trains	10	8	12
Model cars	6	5	4
Model planes	3	2	2
Model trucks	4	3	2

Suppose that model trains sell for $40 each, cars for $35, planes for $80, and trucks for $45. Use matrix multiplication to find the shop's total sales at each location. Label your matrices.

13. During the first week of a recent fundraiser, math club students sold the following number of calendars.

	Mon	Tues	Wed	Thurs	Fri
Calendars	10	15	20	30	50

a. Write this information in a column matrix C. Label your matrix.

b. Find a row matrix N such that the product N times C gives the total number of calendars that the students sold for the week. Then use matrix multiplication to find the total number of calendars.

c. Find a row matrix A such that the product A times C gives the average number of calendars that the students sold each day. (Hint: What fraction would you multiply the total number of calendars by to find the average?) Then use matrix multiplication to find the daily average

Lesson 3.3

Multiplication of Matrices, Part 2

In Lesson 3.2 you multiplied a row matrix times a 2 by 3 price matrix to determine the total cost of 5 pizzas and 3 salads at each pizza house. (See page 133.) But what if you want to compare the total costs of several different combinations of pizzas and salads?

One way to do this is to multiply each of the row matrices times a price matrix and then compare the products. Another way is to combine the combination options into a single matrix. Then you can multiply this new matrix times the price matrix. For example, let matrix *B* represent three different pizza/salad combinations and matrix *C* represent the price matrix for the pizza houses.

$$B = \begin{array}{c} \\ \text{Option 1} \\ \text{Option 2} \\ \text{Option 3} \end{array} \begin{array}{c} \begin{array}{cc} \text{Pizzas} & \text{Salads} \end{array} \\ \begin{bmatrix} 4 & 3 \\ 4 & 4 \\ 5 & 3 \end{bmatrix} \end{array} \qquad C = \begin{array}{c} \\ \text{Pizzas} \\ \text{Salads} \end{array} \begin{array}{c} \begin{array}{ccc} \text{Vin's} & \text{Toni's} & \text{Sal's} \end{array} \\ \begin{bmatrix} 11.25 & 11.96 & 11.90 \\ 3.69 & 4.34 & 4.35 \end{bmatrix} \end{array}$$

If you multiply matrix *B* times matrix *C*, the product will be a 3×3 matrix (call it *D*). The rows of *D* will represent the three options and the columns will represent the three pizza houses. The elements of *D* give the total cost for each of the three options at each of the three pizza houses.

Notice as you follow the steps of this matrix multiplication that the computations are exactly the same as making three separate calculations, one for each option. You expect, then, that row 1 of the product represents the cost of four pizzas and three salads, that row 2 of the product represents the cost of four pizzas and four salads, and that row 3 of the product represents the cost of five pizzas and three salads at each of the pizza houses.

$$D = \begin{array}{c} \\ \text{Option 1} \\ \text{Option 2} \\ \text{Option 3} \end{array} \begin{array}{c} \begin{array}{cc} \text{Pizzas} & \text{Salads} \end{array} \\ \begin{bmatrix} 4 & 3 \\ 4 & 4 \\ 5 & 3 \end{bmatrix} \end{array} \times \begin{array}{c} \\ \text{Pizzas} \\ \text{Salads} \end{array} \begin{array}{c} \begin{array}{ccc} \text{Vin's} & \text{Toni's} & \text{Sal's} \end{array} \\ \begin{bmatrix} 11.25 & 11.96 & 11.90 \\ 3.69 & 4.34 & 4.35 \end{bmatrix} \end{array}$$

or

$$D = \begin{bmatrix} 4 & 3 \\ 4 & 4 \\ 5 & 3 \end{bmatrix} \times \begin{bmatrix} 11.25 & 11.96 & 11.90 \\ 3.69 & 4.34 & 4.35 \end{bmatrix}$$

$$= \begin{bmatrix} 4(11.25) + 3(3.69) & 4(11.96) + 3(4.34) & 4(11.90) + 3(4.35) \\ 4(11.25) + 4(3.69) & 4(11.96) + 4(4.34) & 4(11.90) + 4(4.35) \\ 5(11.25) + 3(3.69) & 5(11.96) + 3(4.34) & 5(11.90) + 3(4.35) \end{bmatrix}$$

$$= \begin{bmatrix} 45.00 + 11.07 & 47.84 + 13.02 & 47.60 + 13.05 \\ 45.00 + 14.76 & 47.84 + 17.36 & 47.60 + 17.40 \\ 56.25 + 11.07 & 59.80 + 13.02 & 59.50 + 13.05 \end{bmatrix}$$

$$= \begin{bmatrix} 56.07 & 60.86 & 60.65 \\ 59.76 & 65.20 & 65.00 \\ 67.32 & 72.82 & 72.55 \end{bmatrix}$$

The labels of the product are

$$D = \begin{array}{c} \\ \text{Option 1} \\ \text{Option 2} \\ \text{Option 3} \end{array} \begin{array}{c} \begin{array}{ccc} \text{Vin's} & \text{Toni's} & \text{Sal's} \end{array} \\ \begin{bmatrix} \$56.07 & \$60.86 & \$60.65 \\ \$59.76 & \$65.20 & \$65.00 \\ \$67.32 & \$72.82 & \$72.55 \end{bmatrix} \end{array}.$$

In this matrix, D_{11} represents the cost of four pizzas and three salads at Vin's. How would you interpret D_{23} and D_{33}?

In order for the product of two matrices to be defined, the number of columns in the first matrix must equal the number of rows in the second matrix.

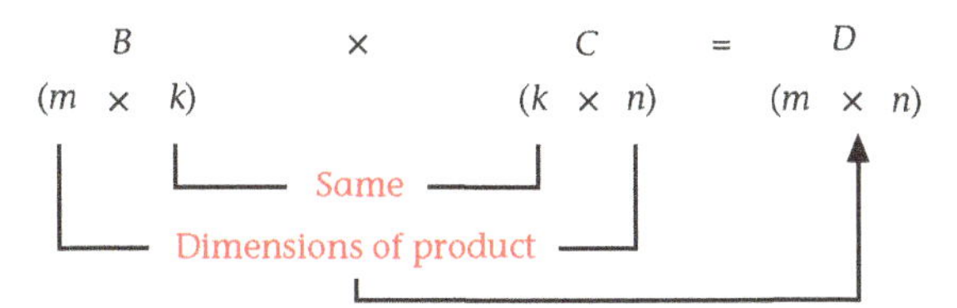

Notice that the number of rows of the first matrix and the number of columns of the second give the order of the product.

It is important to observe that the dimensions of these matrices can also be described using the row and column labels. Matrix B classifies the data according to Options (rows) and Foods (columns). Hence you can refer to matrix B as an Options by Foods matrix. Likewise you can describe C as a Foods by Houses matrix. The product B times C, in turn, results in a matrix of dimension Options by Houses. (See the following diagram.)

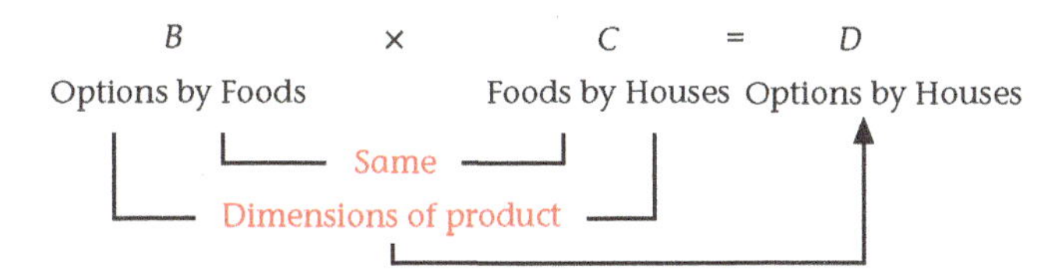

Using row and column labels in this manner helps determine whether a matrix multiplication will result in a meaningful interpretation or, indeed, whether it will give you the results that you want.

Example

M is a 2×3 matrix and N is a 4×2 matrix. Which of the products is defined, MN or NM? Explain.

Solution:

The product MN is not defined as the number of columns in the first matrix is not equal to the number of rows in the second matrix.

The product NM is defined as the number of columns in the first matrix is equal to the number of rows in the second matrix. The product matrix will be a 4 by 3 matrix.

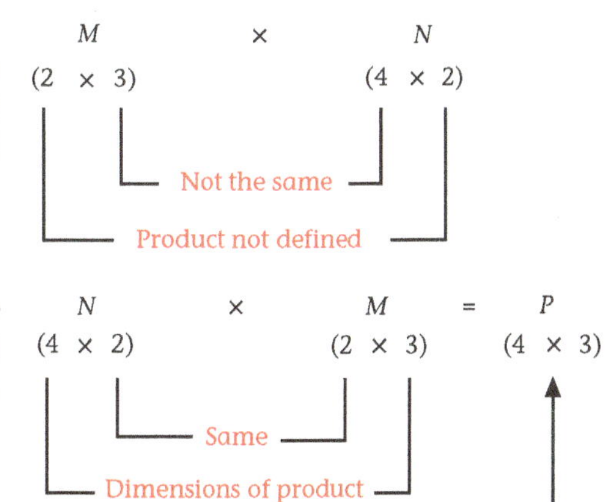

Exercises

1. Use the following matrices to compute the given expression. If the expression is not defined, give the reason.

$$A = \begin{bmatrix} 2 & 1 \\ -1 & 3 \end{bmatrix} \quad B = \begin{bmatrix} 2 & 1 & 0 \\ -4 & 3 & 5 \end{bmatrix} \quad C = \begin{bmatrix} 1 & 2 \\ -1 & -2 \\ 0 & 4 \end{bmatrix}$$

   a. AB

   b. BA

   c. BC

   d. CB

2. Mike, Liz, and Kate are heirs to an estate that consists of a condominium, a customized BMW, and choice season tickets to the Nebraska Cornhusker football games, and for the purposes of fair division, they have submitted the bids shown in matrix E.

$$E = \begin{array}{r} \\ \text{Mike} \\ \text{Liz} \\ \text{Kate} \end{array} \begin{array}{c} \begin{array}{ccc} \text{Condo} & \text{BMW} & \text{Tickets} \end{array} \\ \begin{bmatrix} \$185{,}000 & \$76{,}000 & \$250 \\ \$175{,}000 & \$60{,}000 & \$215 \\ \$180{,}000 & \$75{,}000 & \$325 \end{bmatrix} \end{array}$$

   The awarding of the items in the estate is indicated by matrix A.

$$A = \begin{array}{r} \\ \text{Condo} \\ \text{BMW} \\ \text{Tickets} \end{array} \begin{array}{c} \begin{array}{ccc} \text{Mike} & \text{Liz} & \text{Kate} \end{array} \\ \begin{bmatrix} 1 & 0 & 0 \\ 1 & 0 & 0 \\ 0 & 0 & 1 \end{bmatrix} \end{array}$$

   a. Find the matrix product $P = EA$. Label the rows and columns of P.

   b. Write an interpretation of the entries in matrix P. (Refer to Exercise 7 in Lesson 2.2, pages 67 and 68.)

3. Rosa and Max go out to eat at Sammy's Drive Inn. Rosa orders a Sammy's special, fries, and a shake. Max has a cheeseburger, a baked potato with sour cream, and a shake. The approximate numbers of calories, grams of fat, and milligrams of cholesterol in each of these foods are represented in the following table.

	Calories	Fat (g)	Cholesterol (mg)
Cheeseburger	450	40	50
Sammy's special	570	48	90
Potato/sour cream	500	45	25
French fries	300	30	0
Shake	400	22	50

a. Write a matrix Q that describes Rosa's and Max's orders, with the columns representing the foods. Label the rows and columns of this matrix.

b. Write a matrix C that represents the information in the preceding table with the rows representing the foods. Label the rows and columns of this matrix.

c. What are the dimensions of matrix Q and of matrix C?

d. What is the dimension of the product Q times C? Show why your answer is correct by using a diagram such as the one on page 141.

e. The dimension of matrix Q could be described as Persons by Foods. Describe the dimensions of matrices C and Q times C in a similar manner. Justify your answer for matrix Q times C with a diagram such as the one on page 141.

f. Multiply matrix Q times matrix C to get a matrix R. Label the rows and columns of matrix R.

g. Interpret R_{12}, R_{21}, and R_{23}.

4. a. What must be true about the dimensions of matrices A and B if the product $C = AB$ is defined?

   b. If the products AB and BA are both defined, what must be true about the dimensions of matrices A and B? Why?

   c. Find two nonsquare matrices A and B, where AB and BA are both defined. Compute AB and BA. Does $AB = BA$? Why?

   d. As illustrated by your answer in part c, if AB and BA are both defined, it does not necessarily follow that $AB = BA$ (i.e., *in general, matrix multiplication is not commutative*). Using 2×2 matrices, find examples in which $AB = BA$ and in which AB is not equal to BA.

5. An **identity matrix** is any matrix in which each entry along the main diagonal is 1 and all other entries are 0s. Identity matrices act in the same way for matrix products as the number 1 does for number products.

   Let A be any 3×3 matrix and let

   $$I = \begin{bmatrix} 1 & 0 & 0 \\ 0 & 1 & 0 \\ 0 & 0 & 1 \end{bmatrix}.$$

   Show that $IA = AI = A$.

6. Given the matrices A, B, and C.

   $$A = \begin{bmatrix} 1 & 0 & 1 \\ 0 & 1 & 1 \end{bmatrix} \quad B = \begin{bmatrix} 3 & 1 \\ 2 & 2 \\ -1 & 1 \end{bmatrix} \quad C = \begin{bmatrix} 1 & -1 & 0 \\ 2 & 1 & 1 \end{bmatrix}$$

   a. Do you think that $A(BC) = (AB)C$?

   b. Test your conjecture by computing the products $A(BC)$ and $(AB)C$.

   c. The computations in part b show one case in which matrix multiplication is associative. Do you think this property holds for all matrices A, B, and C for which the product $A(BC)$ is defined? Why or why not?

7. Find two (2×2) matrices A and B to demonstrate that $(A + B)(A - B)$ is not necessarily equal to $A^2 - B^2$.

In algebra you learned that two numbers whose product is 1 (the identity element for multiplication) are called inverses of each other. For example, 5 and $\frac{1}{5}$ (or 5^{-1}) are inverses of each other since $5\left(\frac{1}{5}\right) = \left(\frac{1}{5}\right)5 = 1$. Similarly, if A and B are two square matrices such that $AB = BA = I$, then A and B are called **inverses** of each other. The inverse of A is denoted A^{-1}.

8. a. Verify that the matrices A and B are inverses of each other by computing AB and BA.

   $$A = \begin{bmatrix} 2 & 3 \\ 1 & 2 \end{bmatrix} \quad B = \begin{bmatrix} 2 & -3 \\ -1 & 2 \end{bmatrix}$$

b. Not all square matrices will have an inverse. Use algebra to show that matrix C does not have an inverse.

$$C = \begin{bmatrix} 2 & 4 \\ 3 & 6 \end{bmatrix}$$

9. Carefully plot the points $A(0, 0)$, $B(6, 2)$, $C(8, 6)$, and $D(2, 4)$ on graph paper. Connect the points to form a polygon $ABCD$. You can represent this polygon with a matrix P as follows.

$$P = \begin{matrix} A & B & C & D \\ \end{matrix}$$
$$P = \begin{bmatrix} 0 & 6 & 8 & 2 \\ 0 & 2 & 6 & 4 \end{bmatrix}$$

a. Multiply the matrix that represents polygon $ABCD$ by the matrix

$$T_1 = \begin{bmatrix} -1 & 0 \\ 0 & 1 \end{bmatrix}.$$

b. Plot and label the four points represented in your new matrix as A', B', C', and D'. Connect the points to form polygon $A'B'C'D'$.

c. Describe the relationship between polygon $A'B'C'D'$ and polygon $ABCD$.

d. Multiply the matrix representing polygon $A'B'C'D'$ by the matrix

$$T_2 = \begin{bmatrix} 1 & 0 \\ 0 & -1 \end{bmatrix}.$$

e. Plot and label the four points represented in your new matrix as A'', B'', C'', and D''. Describe the relationship between polygon $A''B''C''D''$ and polygon $A'B'C'D'$.

f. Multiply T_2T_1 to get a new matrix R. Multiply R times the matrix P, that represents the original polygon $ABCD$, and plot the resulting points. What effect does multiplication by R have on $ABCD$? Do the following to test your conjecture: Use a blank sheet of unlined paper and trace both your axes and polygon $ABCD$. Leave your copy on top of the original polygon and place the point of your pencil on the origin. Now, holding the original paper in place, rotate the top sheet until your copy of $ABCD$ rests on top of polygon $A''B''C''D''$. Describe what happened to polygon $ABCD$.

g. Find a matrix T_3 that reflects polygon $A''B''C''D''$ about the y-axis into quadrant IV of your graph.

h. Find a matrix T_4 that rotates polygon $A'B'C'D'$ about the origin into quadrant IV. How does T_4 relate to T_2 and T_3?

For Exercises 10–13, you need either a graphing calculator or access to computer software that performs matrix operations.

10. A manufacturing company that makes fine leather bags has three factories—one in New York, one in Nebraska, and one in California. One of the bags they make comes in three styles—handbag, standard shoulder bag, and roomy shoulder bag. The production of each bag requires three kinds of work—cutting the leather, stitching the bag, and finishing the bag.

Matrix T gives the time (in hours) of each type of work required to make each type of bag.

$$T = \begin{array}{c} \text{Handbag} \\ \text{Standard} \\ \text{Roomy} \end{array} \overset{\begin{array}{ccc} \text{Cutting} & \text{Stitching} & \text{Finishing} \end{array}}{\begin{bmatrix} 0.4 & 0.6 & 0.4 \\ 0.5 & 0.8 & 0.5 \\ 0.6 & 1.0 & 0.6 \end{bmatrix}}$$

Matrix P gives daily production capacity at each of the factories.

$$P = \begin{array}{c} \text{New York} \\ \text{Nebraska} \\ \text{California} \end{array} \overset{\begin{array}{ccc} \text{Handbag} & \text{Standard} & \text{Roomy} \end{array}}{\begin{bmatrix} 10 & 15 & 20 \\ 25 & 15 & 12 \\ 20 & 12 & 10 \end{bmatrix}}$$

Matrix W provides the hourly wages of the different workers at each factory.

$$W = \begin{array}{c} \text{New York} \\ \text{Nebraska} \\ \text{California} \end{array} \overset{\begin{array}{ccc} \text{Cutting} & \text{Stitching} & \text{Finishing} \end{array}}{\begin{bmatrix} 7.50 & 8.50 & 9.00 \\ 7.00 & 8.00 & 8.50 \\ 8.40 & 9.60 & 10.10 \end{bmatrix}}$$

Matrix D contains the total orders received at each factory for the months of May and June.

$$D = \begin{array}{l} \\ \text{Handbag} \\ \text{Standard} \\ \text{Roomy} \end{array} \begin{array}{c} \begin{array}{cc} \text{May} & \text{June} \end{array} \\ \begin{bmatrix} 600 & 800 \\ 800 & 1{,}000 \\ 400 & 600 \end{bmatrix} \end{array}$$

a. Matrix T can be described as a Bag by Work matrix. Describe matrices P, W, and D in a similar manner.

For parts b–e, use the matrices above (or their transposes). Label the rows and columns of the matrix in each answer. Hint: The label dimensions from part a will help you decide what your matrix products should look like.

b. Find the hours of each type of work needed each month to fill all orders.

c. Find the production cost per bag at each factory.

d. Find the cost of filling all May orders at the Nebraska factory. (Hint: In this example the answer, a single value, is the product of a row matrix and a column matrix).

e. Find the daily hours of each type of work needed at each factory if production levels are at capacity.

11. (For students who have studied trigonometry.)

a. Plot the polygon $ABCD$ represented in Exercise 9.

b. Multiply the matrix P by the following transformation matrix.

$$T_1 = \begin{bmatrix} \cos 30^\circ & -\sin 30^\circ \\ \sin 30^\circ & \cos 30^\circ \end{bmatrix}$$

c. Plot the resulting polygon and label it $A'B'C'D'$. How does polygon $A'B'C'D'$ relate to polygon $ABCD$? Try repeating the transformation using 180° to test your conjecture.

d. Write a matrix that will rotate a polygon through 60°. Does this transformation matrix have the same effect as applying T_1 twice? Test your conjecture.

e. Find a matrix that rotates polygon $ABCD$ through 90° and another that rotates it through –90°. Find the product of these two transformation matrices. What is the relationship between these two matrices? Test your conjecture by finding the product of the matrices that will rotate the polygon through 60° and –60°.

12. The matrix A is called an upper-triangular matrix.

$$A = \begin{bmatrix} 1 & 1 & 1 \\ 0 & 1 & 1 \\ 0 & 0 & 1 \end{bmatrix}$$

a. Calculate A^2, A^3, and A^4.

b. Make a conjecture about the form of A^k.

c. Test your conjecture by computing additional powers of A.

d. Challenge: Prove your conjecture using mathematical induction.

13. Challenge: Refer to Exercise 12 and explore the following.

a. Replace the 1s in the upper-triangular matrix A with 2s, 3s, and 4s and repeat part a of Exercise 12 for each of your new upper-triangular matrices.

b. Use the results of part a to make a conjecture for A^k when the 1s in A are replaced by any natural number m.

c. Prove your conjecture in part b using mathematical induction.

Computer/Calculator Exploration

14. Write a program for the graphing calculator based on the method of Exercise 11 that will allow you to enter the coordinates of the vertices of a polygon and the angle of rotation. Design your program so that both the original polygon and the rotation will be displayed.

Modeling Project

15. Research and write a short report on modeling with matrices in trigonometry. Possible topics include the representation vectors and complex numbers as matrices.

Lesson 3.4

Population Growth: The Leslie Model, Part 1

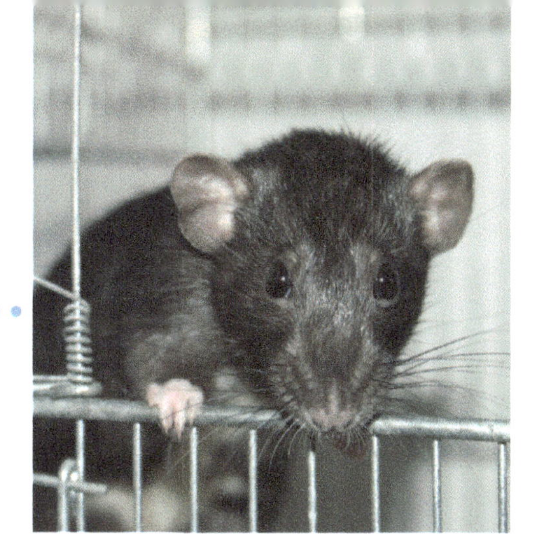

In previous lessons, you learned to add, subtract, and multiply matrices. In this lesson you will learn to use these skills to model the growth of a population.

Population growth is a topic that is of great concern to many people. For example, urban planners are interested in knowing how many people there will be in various age groups after certain periods of time have passed. Wildlife managers are concerned about keeping animal populations at levels that can be supported in their natural habitats.

If you know the age distribution of a population at a certain date and the birth and survival rates for age-specific groups, you can use this data to create a mathematical model. You can use your model to determine the age distributions of the survivors and descendants of the original population at successive intervals of time.

The problem used to illustrate this model was posed by P. H. Leslie. In his problem, the growth rate of a population of an imaginary species of small brown rats, *Rattus norvegicus*, is examined.

Mathematician of Note

P. H. Leslie

In 1945, P. H. Leslie of the Bureau of Animal Population at Oxford University in Oxford, England created a powerful mathematical model. This model, known as the **Leslie matrix model**, is an effective tool that can be used to determine the growth of a population.

In order to simplify the model, the following assumptions are made.

- Only the female population is considered.
- Birth rates and survival rates are held constant over time.
- The survival rate of a rat is the probability that it will survive and move into the next age group.
- The lifespan of these rodents is 15–18 months.
- The rats will have their first litter at approximately 3 months and continue to reproduce every 3 months until they reach the age of 15 months.

Birth rates and age-specific survival rates for 3-month periods are summarized in the following table.

Age (months)	Birth Rate	Survival Rate
0–3	0	0.6
3–6	0.3	0.9
6–9	0.8	0.9
9–12	0.7	0.8
12–15	0.4	0.6
15–18	0	0

Suppose the original female rat population is 42 animals with the age distribution shown in the following table.

Initial Female Rat Population

Age (months)	0–3	3–6	6–9	9–12	12–15	15–18
Number	15	9	13	5	0	0

You can use this table and the birth rate/survival rate information to find the total number of rats and their age distribution after 3 months.

To find this new distribution, you will need to find the number of new female babies introduced into the population. You will also need to determine the number of female rats that survive in each group and move up to the next age group.

To find the number of new births after 3 months (1 cycle), multiply the number of female rats in each age group times the corresponding birth rates and then find the sum.

$$15(0) + 9(0.3) + 13(0.8) + 5(0.7) + 0(0.4) + 0(0) = 0 + 2.7 + 10.4 + 3.5 + 0 + 0 = 16.6$$

So after 3 months, there will be about 17 female rats in the 0–3 age group.

The number of female rats who survive in each age group and move up to the next can be found as follows:

Age	No.	SR*	Number Moving Up to the Next Age Group
0–3	15	0.6	(15)(0.6) = 9.0 move up to the 3–6 age group.
3–6	9	0.9	(9)(0.9) = 8.1 move up to the 6–9 age group.
6–9	13	0.9	(13)(0.9) =11.7 move up to the 9–12 age group.
9–12	5	0.8	(5)(0.8) = 4.0 move up to the 12–15 age group.
12–15	0	0.6	(0)(0.6) = 0 move up to the 15–18 age group.
15–18	0	0	No rodent lives beyond 18 months.

* SR stands for survival rate.

The distribution of female rats after 3 months (1 cycle) is shown in the following table.

Female Rat Population after 3 Months

Age	0–3	3–6	6–9	9–12	12–15	15–18
Number	16.6	9.0	8.1	11.7	4.0	0

The sum of the number of female rats in each age group results in a total population of female rats equal to 16.6 + 9.0 + 8.1 + 11.7 + 4.0 + 0, or 49.4. So, after 3 months (1 cycle) the female rat population has grown from 42 to approximately 50.

Notice that in the table the number of female rats in each age group is not rounded to the nearest integer. This is because when the values are to be used for further analysis, rounding off can mean a significant difference in calculations over time even though it makes no sense to have a fractional part of a rat.

Exercises

1. Use the preceding distribution table (Female Rat Population after 3 Months on page 151) and the process introduced in this lesson to compute the following.

   a. Calculate the number of newborn rats (aged 0–3) after 6 months (2 cycles).

   b. Calculate the number of rats that survive in each age group after 6 months and move up to the next age group.

   c. Use the results to parts a and b to show the distribution of the rat population after 6 months. Approximately how many rats will there be after 6 months?

   d. Use your population distribution from part c to calculate the number of rats and the approximate number in each age group after 9 months (3 cycles). Continue this process to find the number of rats after 12 months (4 cycles).

   e. Compare the original number of rats with the number of rats after 3, 6, 9, and 12 months. What do you observe?

   f. What do you think might happen to this population if you extended the calculations to 15, 18, 21, . . . months?

For Exercises 2 and 3 use the following birth and survival rates for a certain species of deer.

Age (years)	Birth Rate	Survival Rate
0–2	0	0.6
2–4	0.8	0.8
4–6	1.7	0.9
6–8	1.7	0.9
8–10	0.8	0.7
10–12	0.4	0

2. a. The following table shows a distribution of an initial population of 148 deer.

Age (years)	0–2	2–4	4–6	6–8	8–10	10–12
Number	50	30	24	24	12	8

Find the number of newborn female deer after 2 years (1 cycle).

b. Calculate the number of deer that survive in each age group after 2 years and move up to the next age group.

c. Arrange the initial population distribution in a row matrix and the birth rates in a column matrix. Multiply the row matrix times the column matrix. Interpret this result.

3. Explore the possibility of multiplying the initial population distribution in a row matrix times some column matrix to find the number of deer after 2 years that move from:

   a. The 0–2 group to the 2–4 group. (Hint: the column matrix that you use will need to contain several zeros in order to produce the desired product.)

   b. The 2–4 to the 4–6 group.

   c. The 4–6 group to the 6–8 group.

   d. The 6–8 group to the 8–10 group.

   e. The 8–10 group to the 10–12 group.

4. Use the birth and survival rate information for *Rattus norvegicus* from this lesson (see the table on page 150) to find the population total and distribution after 3 months (1 cycle) for the following initial populations.

   a. $[35\ \ 0\ \ 0\ \ 0\ \ 0\ \ 0]$

   b. $[5\ \ 5\ \ 5\ \ 5\ \ 5\ \ 5]$

5. Assume an initial deer population of $[25\ \ 0\ \ 0\ \ 0\ \ 0\ \ 0]$. Use the birth and survival rate information for the deer population in Exercises 2 and 3 to find the population total and distribution after each of the following time spans.

   a. 2 years (1 cycle)

   b. 4 years (2 cycles)

   c. 6 years (3 cycles)

   d. 8 years (4 cycles)

Lesson 3.5

Population Growth: The Leslie Model, Part 2

In Lesson 3.4, you found that it was possible to use an initial population distribution along with birth and survival rates to predict population numbers at future times. As you explored your model, you found that you could look 2, 3, or even 4 cycles into the future. However, the arithmetic soon became cumbersome. What do wildlife managers and urban planners do if they want to look 10, 20, or even more cycles into the future?

In Lesson 3.4 (Exercises 2 and 3), you began to get a glimpse of the model that Leslie proposed. The use of matrices seems to hold the key. And with the aid of computer software or a calculator, looking ahead many cycles is not difficult. In fact, some very fascinating results are produced.

Return to the original rat model. If you multiply the original population distribution (P_0) times a matrix that we will call L, you can calculate the population distribution at the end of cycle 1 (P_1).

$$P_0L = [15 \quad 9 \quad 13 \quad 5 \quad 0 \quad 0]\begin{bmatrix} 0 & 0.6 & 0 & 0 & 0 & 0 \\ 0.3 & 0 & 0.9 & 0 & 0 & 0 \\ 0.8 & 0 & 0 & 0.9 & 0 & 0 \\ 0.7 & 0 & 0 & 0 & 0.8 & 0 \\ 0.4 & 0 & 0 & 0 & 0 & 0.6 \\ 0 & 0 & 0 & 0 & 0 & 0 \end{bmatrix}$$

$$= [15(0) + 9(0.3) + 13(0.8) + 5(0.7) + 0(0.4) + 0(0) \quad 15(0.6) \quad 9(0.9) \quad 13(0.9) \quad 5(0.8) \quad 0(0.6)]$$

$$= [16.6 \quad 9.0 \quad 8.1 \quad 11.7 \quad 4.0 \quad 0] = P_1$$

The matrix L is called the **Leslie matrix**. This matrix is formed by augmenting or joining the column matrix containing the birth rates of each age group and a series of column matrices that contain the survival rates. Notice that the survival-rate columns contain the survival rates as one entry and zeros everywhere else. The survival rates (of which there is one less than the actual number of survival rates since no animal survives beyond the 15–18 age group) lie along the **super diagonal** that is immediately above the main diagonal of the matrix.

When the matrix L is multiplied by a population distribution P_k, a new population distribution P_{k+1} results. To find population distributions at the end of other cycles, the process can be continued.

$P_1 = P_0L$

$P_2 = P_1L = (P_0L)L = P_0(LL) = P_0L^2$

In general, $P_k = P_0L^k$.

Example

Use the formula $P_k = P_0L^k$ to find the population distribution for the rats after 24 months (8 cycles) and the total population of the rats.

Solution:

The population distribution after 8 cycles is

$$P^8 = P_0L^8 = [15\ \ 9\ \ 13\ \ 5\ \ 0\ \ 0]\begin{bmatrix} 0 & 0.6 & 0 & 0 & 0 & 0 \\ 0.3 & 0 & 0.9 & 0 & 0 & 0 \\ 0.8 & 0 & 0 & 0.9 & 0 & 0 \\ 0.7 & 0 & 0 & 0 & 0.8 & 0 \\ 0.4 & 0 & 0 & 0 & 0 & 0.6 \\ 0 & 0 & 0 & 0 & 0 & 0 \end{bmatrix}^8$$

$$= [21.03\ \ 12.28\ \ 10.90\ \ 9.46\ \ 7.01\ \ 4.27]$$

Total population $= 21.03 + 12.28 + 10.90 + 9.46 + 7.01 + 4.27$
$= 64.95$, or approximately 65 rats.

Technology Note

You can perform the calculation P_0L^8 on a calculator with matrix features.

```
[A][B]^8
[[21.026 12.283...
```

Exercises

Note: For the following exercises, you need to have access to either a graphing calculator or computer software that performs matrix operations.

1. Use the original population distribution, [15 9 13 5 0 0], and the Leslie matrix from the *Rattus norvegicus* example to find the following.
   a. The population distribution after 15 months (5 cycles)
   b. The total population after 15 months (Hint: Multiply $P_0 L^5$ times a column matrix consisting of six 1s.)
   c. The population distribution and the total population after 21 months

2. Suppose the *Rattus norvegicus* start dying off from overcrowding when the total female population for a colony reaches 250. Find how long it will take for this to happen when the initial population is
   a. [18 9 7 0 0 0].
   b. [35 0 0 0 0 0].
   c. [5 5 5 5 5 5].
   d. [25 15 10 11 7 13].

3. a. Complete the table for the given cycles of *Rattus norvegicus* using the original population distribution of [15 9 13 5 0 0].

Cycle	Total Population	Growth Rate
Original	42	
1	49.4	$\frac{49.4-42}{42} = 0.176 = 17.6\%$
2	56.08	$\frac{56.08-49.4}{49.4} = 0.135 = 13.5\%$
3	57.40	$\frac{57.40-56.08}{56.08} = 0.024 = 2.4\%$
4		
5		
6		

b. What do you observe about the growth rates?

c. Calculate the total populations for P_{25}, P_{26}, and P_{27}. What is the growth rate between these successive years? Hint: To find the growth rate from P_{25} to P_{26}, subtract the total population for P_{25} from the total population for P_{26} and divide the result by the total population for P_{25}.

4. One characteristic of the Leslie model is that growth does stabilize at a rate called the **long-term growth rate** of the population. As you observed in Exercise 3, the growth rate of *Rattus norvegicus* converges to about 3.04%. This means that for a large enough k, the total population in cycle k will equal about 1.0304 times the total population in the previous cycle.

   a. Find the long-term growth rate of the total population for each of the initial population distributions in Exercise 2.

   b. How does the initial population distribution seem to affect the long-term growth rate?

5. Again, consider the deer species from Lesson 3.4. The birth and survival rates follow.

Age (years)	Birth Rate	Survival Rate
0–2	0	0.6
2–4	0.8	0.8
4–6	1.7	0.9
6–8	1.7	0.9
8–10	0.8	0.7
10–12	0.4	0

   a. Construct the Leslie matrix for this animal.

   b. Given that $P_0 = [50\ \ 30\ \ 24\ \ 24\ \ 12\ \ 8]$, find the long-term growth rate.

   c. Suppose the natural range for this animal can sustain a herd that contains a maximum of 1,250 females. How long before this herd size is reached?

   d. Once the long-term growth rate of the deer population is reached, how might the population of the herd be kept constant?

In his study of the application of matrices to population growth, P. H. Leslie was particularly interested in the special case in which the birth rate matrix has only one nonzero element. The following example falls into this special case.

6. Suppose there is a certain kind of bug that lives at most 3 weeks and reproduces only in the third week of life. Fifty percent of the bugs born in one week survive into the second week, and 70% of the bugs that survive into their second week also survive into their third week. On the average, six new bugs are produced for each bug that survives into its third week. A group of five 3-week-old female bugs decide to make their home in a storage box in your basement.

   a. Construct the Leslie matrix for this bug.

   b. What is P_0?

   c. How long will it be before there are at least 1,000 female bugs living in your basement?

7. Exercise 6 is an example of a population that grows in waves. Will the population growth for this population stabilize in any way over the long run? To explore this question, make a table of the population distributions P_{22} through P_{30}.

   a. Examine the population change from one cycle to the next. Can you find a pattern in the population growth?

   b. Examine the population change from P_{22} to P_{25}, P_{23} to P_{26}, P_{24} to P_{27}, P_{25} to P_{28}, P_{26} to P_{29}, and P_{27} to P_{30}. Are you surprised at the results? Why?

8. a. Change the initial population in Exercise 6 to $P_0 = [4 \ \ 4 \ \ 4]$ and repeat the instructions in Exercise 7 looking at the total population growth for each cycle.

   b. Examine the changes in successive age groups from P_{22} to P_{25}, P_{23} to P_{26}, P_{24} to P_{27}, P_{25} to P_{28}, P_{26} to P_{29}, and P_{27} to P_{30}. Make a conjecture based on your results.

9. Using mathematical induction, prove that $P_k = P_0L^k$ for any original population P_0 and Leslie matrix L, where k is a natural number.

Modeling Projects

10. Search the Web for applications of the Leslie matrix model in managing wildlife or domestic herds.

Chapter Extension

Harvesting Animal Populations

The following application of the Leslie matrix to a population mestic sheep in New Zealand was originally published in 1967. (See G. Caughley, "Parameters for Seasonally Breeding Populations," *Ecology* 48(1967):834-839. Anton and Rorres (1987) developed the problem in their text, *Elementary Linear Algebra with Applications*. More recent references can be found by searching the Web for Leslie matrix applications.

This application involves what is referred to as the harvesting of a population. The term harvesting is defined as removal of the animals from the population. This could entail slaughtering some of the animals as is the case of wild deer or caribou herds that grow too large to be supported by their habitat. It could also mean selling or relocating some of the animals to start a new herd or colony as is the case of domestic herds or wild colonies of animals such as beaver. The ultimate goal is to find a stable distribution from which the population growth can be harvested at regular intervals so that the population can be held constant.

This model begins with an initial population that undergoes a growth period that is described by the Leslie matrix. At the end of the growth period, however, a certain percentage of each age group in the distribution is harvested. This is usually done in such a way that the unharvested population has the same age distribution as the initial population. A plan for harvesting the same percentage of each age group on a regular basis so the population remains the same after each harvesting is called a sustainable harvesting policy

To describe this situation using matrices, suppose a Leslie matrix and a population distribution vector for n age groups of females in a population have been defined. If h_i, for $i = 1, 2, 3, \ldots, n$, represents the fraction offemales harvested for each of the age groups, an $n \times n$ diagonal matrix H called the harvesting matrix can be formed as follows.

$$H = \begin{bmatrix} h_1 & 0 & 0 & . & . & . & 0 \\ 0 & h_2 & 0 & . & . & . & 0 \\ 0 & 0 & h_3 & . & . & . & 0 \\ . & . & . & & & & . \\ . & . & . & & & & . \\ . & . & . & & & & . \\ 0 & 0 & 0 & . & . & . & h_n \end{bmatrix}$$

If all the h_i's are the same value, then the harvesting is called uniform. From the definition of a sustainable harvesting policy, it follows that

$$\begin{bmatrix} \text{Age distribution} \\ \text{at end of} \\ \text{growth period} \end{bmatrix} - [\text{harvest}] = \begin{bmatrix} \text{Age distribution} \\ \text{at beginning of} \\ \text{growth period} \end{bmatrix}.$$

Using matrix notation, where P represents the distribution at the beginning of the growth period, L represents the Leslie matrix, and H represents the harvesting matrix, this translates to

$$PL - PLH = P.$$

This model was applied in 1967 to a species of domestic sheep in New Zealand which has a lifespan of 12 years. The sheep were divided into 12 age groups with a growth period of 1 year. Birth and survival rates for each age group of sheep were found using demographics, and the following Leslie matrix was developed.

$$\begin{bmatrix}
0 & 0.845 & 0 & 0 & 0 & 0 & 0 & 0 & 0 & 0 & 0 & 0 \\
0.045 & 0 & 0.975 & 0 & 0 & 0 & 0 & 0 & 0 & 0 & 0 & 0 \\
0.391 & 0 & 0 & 0.965 & 0 & 0 & 0 & 0 & 0 & 0 & 0 & 0 \\
0.472 & 0 & 0 & 0 & 0.950 & 0 & 0 & 0 & 0 & 0 & 0 & 0 \\
0.484 & 0 & 0 & 0 & 0 & 0.926 & 0 & 0 & 0 & 0 & 0 & 0 \\
0.546 & 0 & 0 & 0 & 0 & 0 & 0.895 & 0 & 0 & 0 & 0 & 0 \\
0.543 & 0 & 0 & 0 & 0 & 0 & 0 & 0.850 & 0 & 0 & 0 & 0 \\
0.502 & 0 & 0 & 0 & 0 & 0 & 0 & 0 & 0.786 & 0 & 0 & 0 \\
0.468 & 0 & 0 & 0 & 0 & 0 & 0 & 0 & 0 & 0.691 & 0 & 0 \\
0.459 & 0 & 0 & 0 & 0 & 0 & 0 & 0 & 0 & 0 & 0.561 & 0 \\
0.433 & 0 & 0 & 0 & 0 & 0 & 0 & 0 & 0 & 0 & 0 & 0.370 \\
0.421 & 0 & 0 & 0 & 0 & 0 & 0 & 0 & 0 & 0 & 0 & 0
\end{bmatrix}$$

If left to reproduce without a harvesting policy in place, the sheep would eventually approach a stable growthrate. If the shepard allows this to happen without a harvesting policy, the income from selling the wool would not cover the cost of feeding the flock. The stable growth rate that can be found using the Leslie matrix can be used to approximate a uniform harvesting policy. In this case the uniform harvesting policy is one in which roughly 18% of the sheep from each of the 12 age groups is harvested each year.

Chapter 3 Review

1. Write a summary of what you think are the important points of this chapter.

For Exercises 2 and 3, use the following matrices.

$$A = \begin{bmatrix} 2 & 0 \\ 4 & 7 \\ -1 & 3 \end{bmatrix} \quad B = \begin{bmatrix} 4 & -1 & 2 \\ 1 & 6 & -3 \end{bmatrix} \quad C = \begin{bmatrix} 4 & -2 \\ -1 & 0 \\ 3 & -4 \end{bmatrix} \quad D = \begin{bmatrix} 1 & -3 \\ 2 & -2 \\ 3 & -1 \end{bmatrix}$$

2. a. How many elements does matrix B have?

   b. What is the value of C_{12}? Of C_{21}?

3. Find the value of each of the given expressions. If it is not possible, state the reason why.

   a. $A + C$

   b. $C - B$

   c. $(A + C) - D$

   d. $2A + D$

4. Your math club is planning a Saturday practice session for an upcoming math contest. For lunch the students ordered 35 Mexican lunches, 6 bags of corn chips, 6 containers of salsa, and 12 six-packs of cold drinks.

   a. Write this information in a row matrix L. Label your matrix.

   b. Interpret L_2 and L_4.

   c. Suppose that the club pays \$4.50 per lunch, \$1.97 per bag of corn chips, \$2.10 for each container of salsa, and \$2.89 for each six-pack of cold drinks. Use multiplication of a row and column matrix to find the total cost. Label your matrices.

5. A group of students is planning a retreat. They have contacted three lodges in the vicinity to inquire about rates. They found that Crystal Lodge charges $13.00 per person per day for lodging, $20.00 per day for food, and $5.00 per person for use of the recreational facilities. Springs Lodge charges $12.50 for lodging, $19.50 for meals, and $7.50 for use of the recreational facilities. Bear Lodge charges $20.00 per night for lodging, $18.00 a day for meals, and there is no extra charge for using the recreational facilities. Beaver Lodge charges a flat rate of $40.00 a day for lodging (meals included) and no additional fee for use of the recreational facilities.

   a. Display this information in a matrix C. Label the rows and columns.

   b. State the values of C_{22} and C_{43}.

   c. Interpret C_{13} and C_{31}.

6. Mr. Jones has been shopping for a vacuum-powered cleaning system. He found one at Z-Mart and another model at Base Hardware. The Z-Mart system cost $39.50, disposal cartridges were 6 for $24.50, and storage cases were $8.50 each. At Base Hardware the system cost $49.90, cartridges were 6 for $29.95, and cases were $12.50 each.

   a. Write and label a matrix showing the prices for the three items at the two stores.

   b. Mr. Jones decided to wait and see if the prices for the systems would be reduced during the upcoming sales. When he went back during the sales, the Z-Mart prices were reduced by 10% and the Base Hardware prices were reduced by 20%. Construct a matrix showing the sale prices for each of the three items at the two stores.

   c. Use matrix subtraction to compute how much Mr. Jones could save for each item at the two stores.

   d. Suppose Mr. Jones is interested in purchasing the systems for himself and three of his friends. Use multiplication of a matrix by a scalar to find how much he would pay for each of the three items at the two stores at the sale prices.

7. The dimensions of matrices P, Q, R, and S are 3×2, 3×3, 4×3, and 2×3, respectively. If matrix multiplication is possible, find the dimensions of the following matrix products. If it is not possible, state why.

   a. QP

   b. RQ

   c. QS

   d. RPS

8. An artist creates plates and bowls from small pieces of colored woods. She currently has orders for five plates, three large bowls, and seven small bowls. Each plate requires 100 pieces of ebony, 800 pieces of walnut, 600 pieces of rosewood, and 400 pieces of maple. It takes 200 ebony pieces, 1,200 walnut pieces, 1,000 rosewood pieces, and 800 pieces of maple to make a large bowl. A small bowl takes 50 pieces of ebony, 500 walnut pieces, 450 rosewood pieces, and 400 pieces of maple.

   a. Write a row matrix showing the current orders for this artist's work.

   b. Construct a matrix showing the number of pieces of wood used in an individual plate or bowl.

   c. Use matrix multiplication to compute the number of pieces of each type of wood the artist will need for the plates and bowls that are on order.

   d. Suppose it takes the artist 3 weeks to fashion a plate, 4 weeks to make a large bowl, and 2 weeks to complete a small bowl. Use matrix multiplication to show how long it will take the artist to fill all the orders for plates and bowls.

9. Tonya has money invested in three sports complexes in Smith City. Her return (annual) from a \$50,000 investment in a tennis club is 8.2%. She receives 6.5% from a \$100,000 investment in a golf club and 7.5% on a \$75,000 investment in a soccer club. Use matrix multiplication to find Tonya's income from her investments for one year. Label your matrices.

10. Three music classes at Central High are selling candy as a fundraiser. The number of each kind of candy sold by each of the three classes is shown in the following table.

	Jazz Band	Symphonic Band	Orchestra
Almond bars	300	220	250
Chocolate chews	240	330	400
Mint patties	150	200	180
Sour balls	175	150	160

The profit for each type of candy is sour balls, 30 cents; chocolate chews, 50 cents; almond bars, 25 cents; and mint patties, 35 cents. Use matrix multiplication to compute the profit made by each class on its candy sales.

11. Use the following matrices to find the value of each of the given expressions. If the expression is not defined, give the reason.

$$A = \begin{bmatrix} 2 & 0 \\ 4 & 7 \\ -1 & 3 \end{bmatrix} \quad B = \begin{bmatrix} 4 & -1 & 2 \\ 1 & 6 & -3 \end{bmatrix} \quad C = \begin{bmatrix} 1 & 2 \\ -1 & 3 \end{bmatrix} \quad D = [2 \ \ 1 \ \ 0] \quad E = [4 \ \ 3]$$

a. AB

b. BA

c. CA

d. $DA + E$

12. Write the transpose (A^T) of matrix A, where

$$A = \begin{bmatrix} 4 & 2 & 6 \\ 5 & 1 & 3 \end{bmatrix}.$$

13. Let matrix

$$M = \begin{bmatrix} 1 & 1 \\ 1 & 1 \end{bmatrix}.$$

a. Calculate M^2, M^3, and M^4.

b. Predict the components of M^5 and check your prediction.

c. Generalize to M^n, where n is a natural number.

d. Prove your conjecture in part c using mathematical induction.

e. Repeat parts a, b, c, and d for the matrix

$$M = \begin{bmatrix} 1 & 0 \\ 2 & 3 \end{bmatrix}.$$

14. Complete the following statement: If a square matrix A has an inverse A^{-1}, then the product AA^{-1} = the _____ matrix I, where I is a ________.

15. Which of the following matrices are inverses of each other? Explain your answers.

 a. $\begin{bmatrix} -1 & 3 \\ 2 & -5 \end{bmatrix}$ and $\begin{bmatrix} 5 & 3 \\ 2 & 1 \end{bmatrix}$

 b. $\begin{bmatrix} 1 & 0 \\ 0 & 1 \end{bmatrix}$ and $\begin{bmatrix} 1 & 0 \\ 0 & 1 \end{bmatrix}$

 c. $\begin{bmatrix} 2 & 1 & 0 \\ 3 & 2 & 1 \end{bmatrix}$ and $\begin{bmatrix} 1 & -1 \\ -1 & 2 \\ -1 & 2 \end{bmatrix}$

16. The students at Central High are planning to hire a band for the prom. Their choices are bands A, B, and C. They survey the Sophomore, Junior, and Senior classes and find the following percentages of students (regardless of sex) prefer the bands,

$$\begin{array}{c} \\ A \\ B \\ C \end{array}\begin{array}{c} \begin{array}{ccc} 10\text{th} & 11\text{th} & 12\text{th} \end{array} \\ \begin{bmatrix} 20\% & 35\% & 40\% \\ 30\% & 30\% & 25\% \\ 50\% & 35\% & 35\% \end{bmatrix} \end{array}.$$

The student population by class and sex is:

$$\begin{array}{c} \\ 10\text{th} \\ 11\text{th} \\ 12\text{th} \end{array}\begin{array}{c} \begin{array}{cc} \text{Male} & \text{Female} \end{array} \\ \begin{bmatrix} 235 & 225 \\ 205 & 215 \\ 175 & 190 \end{bmatrix} \end{array}.$$

Use matrix multiplication to find:

a. The number of males and females who prefer each band.

b. The total number of students who prefer each band.

17. The characteristics of the female population of a herd of small mammals are shown in the following table.

	Age Groups (months)					
	0–4	4–8	8–12	12–16	16–20	20–24
Birth Rate	0	0.5	1.1	0.9	0.4	0
Survival Rate	0.6	0.8	0.9	0.8	0.6	0

Suppose the initial female population for the herd is given by

$$P_0 = [22 \;\; 22 \;\; 18 \;\; 20 \;\; 7 \;\; 2].$$

a. What is the expected lifespan of this mammal?

b. Construct the Leslie matrix for this population.

c. Determine the long-term growth rate for the herd.

d. Suppose this mammal starts dying off from overcrowding when the total female population for the herd reaches 520. How long will it take for this to happen?

Bibliography

Anton, H., and C. Rorres. 2010. *Elementary Linear Algebra: Applications Version.* 10th ed. New York, NY: John Wiley & Sons.

Cozzens, M. B., and R. D. Porter. 1987. *Mathematics and Its Applications.* Lexington, MA: D.C. Heath and Company.

Kemeny, J. G., J. N. Snell, and G. L. Thompson. 1974. *Introduction to Finite Mathematics.* Englewood Cliffs, NJ: Prentice Hall.

Leslie, P. H. 1945. "On the Uses of Matrices in Certain Population Mathematics." *Biometrika* 33:183-212.

Maurer, S. B., and A. Ralston. 2004. *Discrete Algorithmic Mathematics.* 3rd ed. Wellesley, MA: A. K. Peters Ltd.

North Carolina School of Science and Mathematics. 1988. *New Topics for Secondary School Mathematics: Matrices*. Reston, VA: National Council of Teachers of Mathematics.

Ross, K. A., and C. R. B. Wright. 2003. *Discrete Mathematics*. 5th ed. Englewood Cliffs, NJ: Prentice Hall.

Tuchinsky, Philip M. 1986. *Matrix Multiplication and DC Ladder Circuits*. Lexington, MA: COMAP, Inc.

CHAPTER

4

Graphs as Models

When the boundaries or names of countries change, cartographers have to be prepared to provide the public with new maps. For years, mapmakers and mathematicians alike have wondered about the number of colors it takes to color a map.

- What is the minimum number of colors needed to color any map?
- Optimally, how do you color a map?
- What do coloring maps and scheduling meeting times for your school organizations have in common?

Whether you're a cartographer who must find a way to color a map, a businessperson who must determine whether a project can be completed on time, or a planner who wants to know the most efficient way to route a city's garbage trucks, you will find the answer in an area of mathematics known as graph theory.

Lesson 4.1

Modeling Projects

How does a building contractor organize all of the jobs needed to complete a project? How do your parents manage to get all the food for a Thanksgiving dinner done at the same time? Many people believe that planning is a simple activity. After all, everybody does it. Planning your day-to-day activities seems to be second nature. What most people fail to realize is that for people in the business world who must plan and control work on extensive projects, this haphazard manner of planning is not the most efficient way to complete a job. A more scientific, organized method must be used.

One way to model projects that consist of several different subprojects, or tasks, is through the use of a diagram, or **graph**, that is made up of points called **vertices** and connecting lines called **edges** (see Figure 4.1).

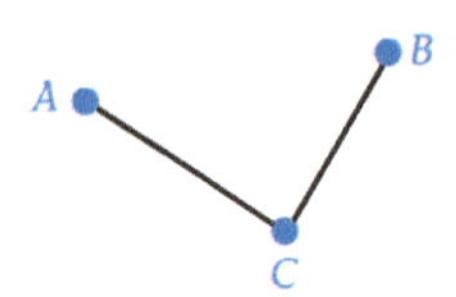

Figure 4.1. Graph with three vertices and two edges.

Explore This

The Central High yearbook staff has only 16 days left before the deadline for completing their book. They are running behind schedule and still have several tasks left to finish. The remaining tasks and time that it takes to complete each task are listed in the following table.

	Task	Time (days)
Start		0
A	Buy film	1
B	Load camera	1
C	Take photos of clubs	3
D	Take sports photos	2
E	Take photos of teachers	1
F	Develop film	2
G	Design layout	5
H	Print and mail pages	3

Is it possible for the yearbook to be completed on time if the tasks have to be done one after the other? If some tasks can be done at the same time as others, can the deadline be met?

As you may have noticed, some of the yearbook staff jobs can be done simultaneously, while several of them cannot be started until others have been completed. Assuming the following prerequisites, how soon can the project be completed?

	Task	Time (days)	Prerequisite Task
Start		0	—
A	Buy film	1	None
B	Load camera	1	*A*
C	Take photos of clubs	3	*B*
D	Take sports photos	2	*C*
E	Take photos of teachers	1	*B*
F	Develop film	2	*D*, *E*
G	Design layout	5	*D*, *E*
H	Print and mail pages	3	*G*, *F*

Using a graph to model this information makes it easier to see the relationships among the tasks. In the graph in Figure 4.2 on page 172, the tasks are represented by points (vertices). The arrows (directed edges) indicate which tasks must be finished before a new task can begin. Each edge also shows the number of days it takes to complete the preceding task. Note that tasks with the same prerequisites are aligned vertically. Although this is not necessary, it helps to make the graph easier to follow.

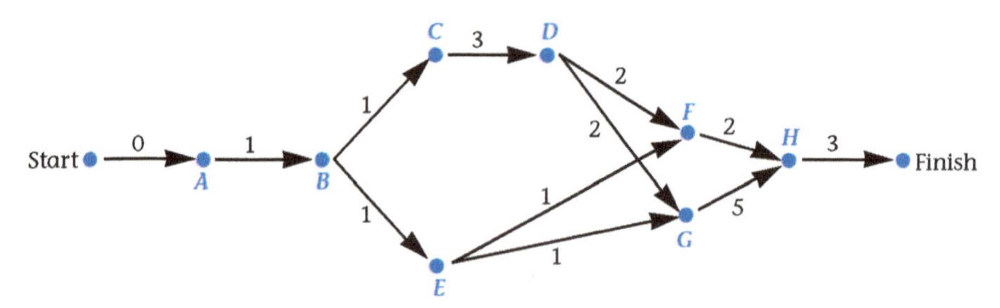

Figure 4.2. Diagram of the order of tasks necessary to complete the yearbook.

Exercises

In Exercises 1 through 4, use the task table to draw a graph with appropriately labeled vertices and edges.

1.

Task	Time	Prerequisites
Start	0	—
A	5	None
B	6	*A*
C	4	*A*
D	4	*B*
E	8	*B*, *C*
F	4	*C*
G	10	*D*, *E*, *F*
Finish		

2.

Task	Time	Prerequisites
Start	0	—
A	4	None
B	3	*A*
C	1	*A*
D	6	*A*
E	2	*B*
F	3	*C*, *D*
G	3	*E*
H	1	*E*, *F*
Finish		

3.

Task	Time	Prerequisites
Start	0	—
A	1	None
B	2	None
C	3	*A*, *B*
D	5	*B*
E	5	*C*
F	5	*C*, *D*
G	4	*D*, *E*
H	4	*E*, *F*
Finish		

4.

Task	Time	Prerequisites
Start	0	—
A	5	None
B	8	*A*, *D*
C	9	*B*, *I*
D	7	None
E	8	*B*
F	12	*I*
G	4	*C*, *E*, *F*
H	9	None
I	5	*D*, *H*
Finish		

5. To help organize the family dinner, Ms. Shu listed the following tasks.

Task	Time (min)	Prerequisite Task
Start	0	—
A Wash hands		
B Defrost hamburger		
C Shape meat into patties		
D Cook hamburgers		
E Peel and slice potatoes		
F Fry potatoes		
G Make salad		
H Set table		
I Serve food		

a. Complete the table by making reasonable time estimates in minutes for each of these tasks and indicating the prerequisites.

b. Construct a graph using the information from your table.

c. What is the least amount of time needed to prepare dinner?

6. Your best friend, Matt, has always been very disorganized. He is now preparing to leave for college and desperately needs your help.
   a. Create a table of at least six activities that will need to be completed before Matt can leave home. Give the times and prerequisites of these activities.
   b. Construct a corresponding graph.
   c. For your task list, what is the least amount of time it will take to get Matt off to school?
7. Consider the following graph.

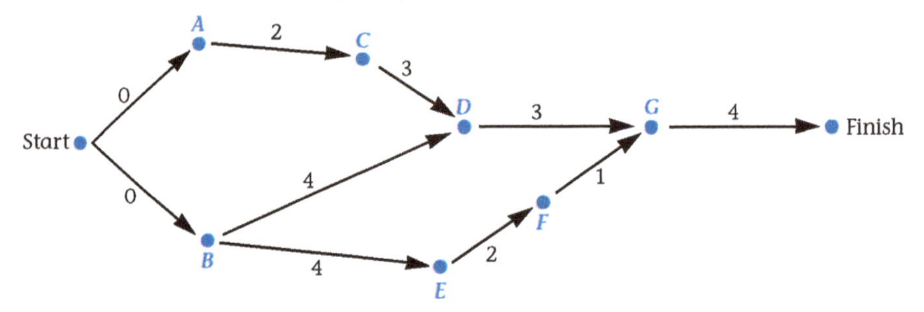

   a. Complete the following task table for this graph.

Task	Time	Prerequisite Task
Start	0	—
A		
B		
C		
D		
E		
F		
G		
Finish		

   b. What is the least amount of time it will take to complete all of the tasks in this graph? Explain why it cannot be completed in less time.

Lesson 4.2

Critical Paths

It is relatively easy to find the shortest time needed to complete a project if the project consists of only a few activities. But as the tasks increase in number, the problem becomes more difficult to solve by inspection alone.

In the 1950s the U.S. government was faced with the need to complete very complex systems such as the U.S. Navy Polaris Submarine project. In order to do this efficiently, a method was developed called PERT (Program Evaluation and Review Technique). This technique targeted tasks that were critical to the earliest completion of the project. The path of targeted tasks from the start to the finish of the project became known as the **critical path**.

Recall the graph in Lesson 4.1 that represented the Central High yearbook project. How might you go about finding a systematic way to identify the critical path for this project? To do this, an **earliest-start time** (EST) for each task must be found. The EST is the earliest that an activity can begin if all the activities preceding it begin as early as possible.

To calculate the EST for each task, begin at the start. Then label each vertex with the smallest possible time that is needed before the task can begin. The label for C in Figure 4.3 is found by adding the EST of B to the 1 day that it takes to complete task $B(1 + 1 = 2)$. Task G cannot be completed until both predecessors, D and E, have been completed. Hence, G cannot begin until 7 days have passed.

In the case of the yearbook staff, the earliest time in which the project can be completed is 15 days. As paradoxical as it may seem, the least amount of time that it takes to complete all of the tasks in the project corresponds to the time it takes to complete the longest path through the graph from start to finish.

A path with this longest time is the desired critical path. In Figure 4.3, the critical path is Start-*ABCDGH*-Finish.

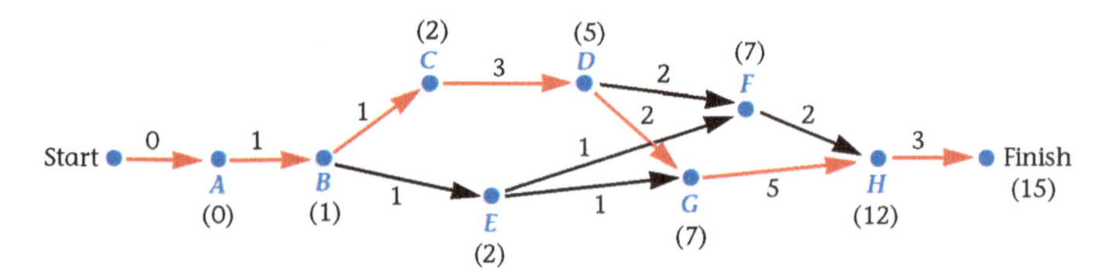

Figure 4.3. Yearbook diagram showing the earliest-start time for each task.

Example

1. Copy the graph and label the vertices with the EST for each task.

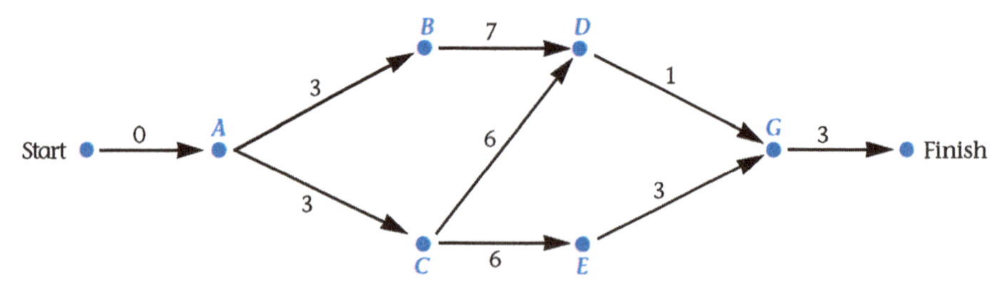

Then determine the earliest completion time for the project. All times are in minutes.

2. Identify the critical path.

Solution:

1.

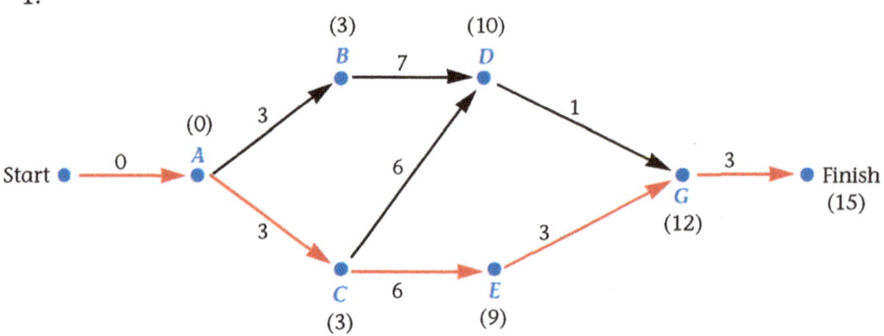

The earliest time in which the project can be completed is 15 minutes.

2. Since the critical path is the longest path from the start to the finish, the critical path is Start-*ACEG*-Finish.

If it is desirable to cut the completion time of a project, it can be done by shortening the length of the critical path once it is found. In the preceding example, one way to shorten the time it takes to complete the project is to cut the time it takes to complete task *E*. If task *E*'s time is cut from 3 minutes to 2 minutes, the completion time for the project is cut to 14 minutes.

The efficient management of large projects like the construction of a building requires the use of critical path analysis.

Exercises

1.

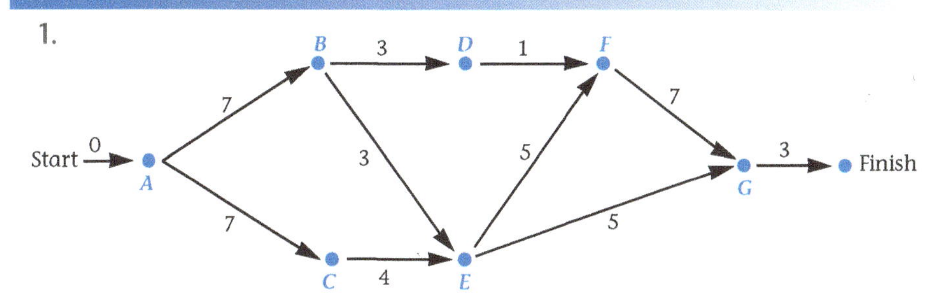

Complete the following.

Vertex	Earliest-Start Time
A	0
B	7
C	
D	
E	
F	
G	

Minimum project time =

Critical path(s) =

In Exercises 2 and 3, list the vertices of the graphs and give their earliest-start time, as in Exercise 1. Determine the minimum project time and all of the critical paths.

2.

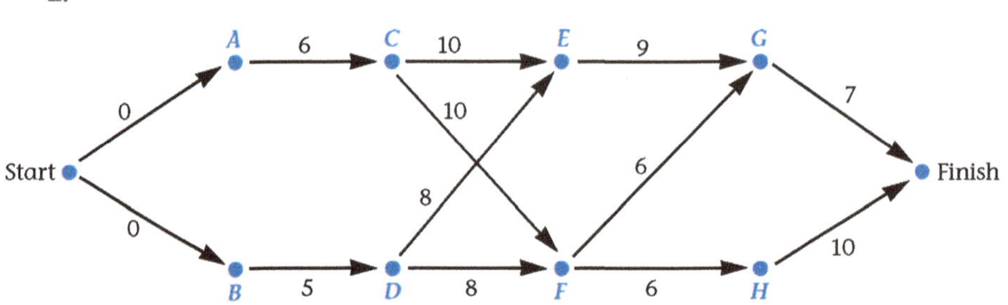

3.

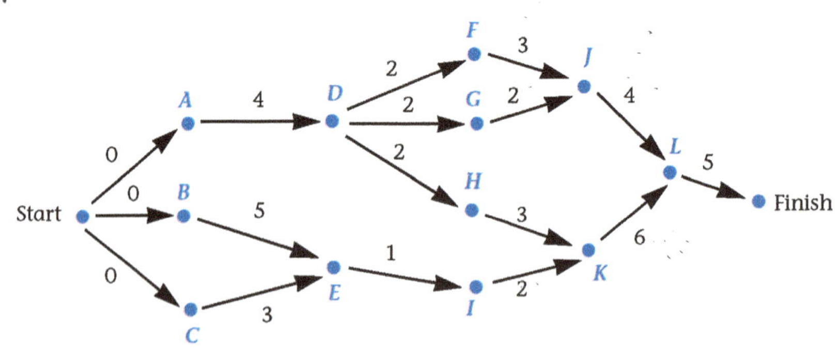

4. Using the information from the following table, construct a graph and label each of the vertices with its earliest-start time. Determine the minimum project time and critical path.

Task	Time	Prerequisites
Start	0	—
A	2	None
B	4	None
C	3	A, B
D	1	A, B
E	5	C, D
F	6	C, D
G	7	E, F

5.

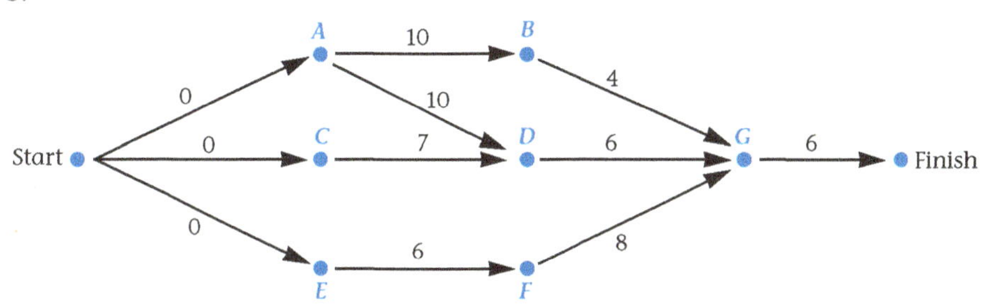

a. Copy the graph and label each vertex with its earliest-start time.

b. How quickly can the project be completed?

c. Determine the critical path.

d. What happens to the minimum project time if task *A*'s time is reduced to 9 days? To 8 days?

e. Will the project time continue to be affected by reducing the time of task *A*? Explain why or why not.

6. Construct a graph with three critical paths.

7. Determine the minimum project time and the critical path for the following graph.

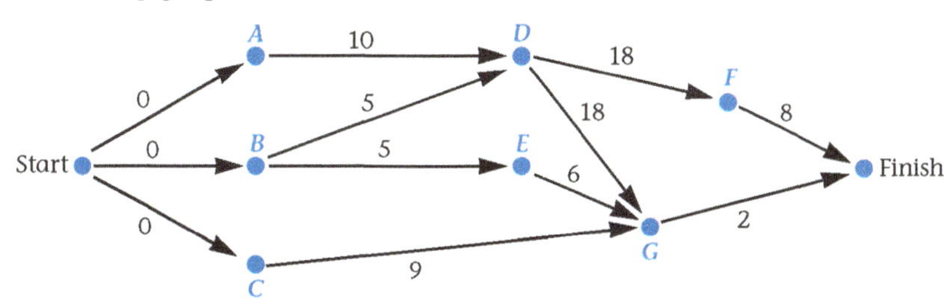

8.

Task	Time	Prerequisites
Start	0	—
A	13	None
B	10	None
C	4	*A*
D	8	*B*
E	6	*B*
F	7	*C, D, E*
G	5	*F*
H	8	*F*
Finish		

a. Draw a graph using the information in the table.

b. Label each vertex with its earliest-start time.

c. Determine the minimum project time.

d. Determine the critical path(s).

9. In the following graph, each vertex has been label with its EST, and the critical path is marked.

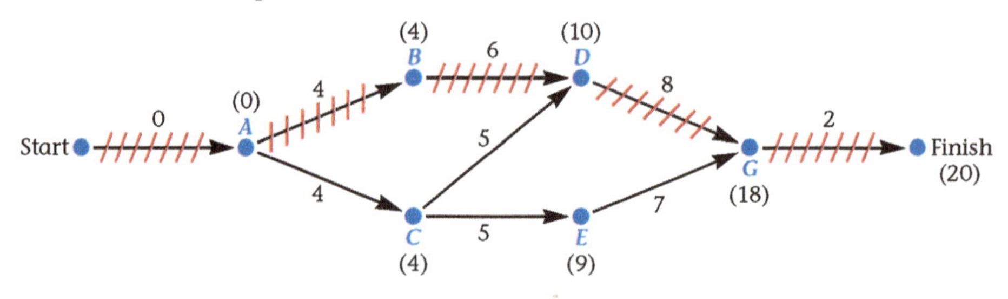

a. Task *E* can begin as early as day 9. If it begins on day 9, when will it be completed? If it begins on day 10? On day 11? What will happen if it begins on day 12?

b. What is the latest day on which task *E* can begin if task *G* is to begin on day 18?

If an activity is not on the critical path, it is possible for it to start later than its earliest-start time and not delay the project. The latest a task can begin without delaying the project's minimum completion time is known as the **latest-start time** (LST) for the task. For example, the LST for *E* is day 11.

c. In order to find the LST for vertex C, the times of the two vertices D and E need to be considered. Since vertex D is on the critical path, the latest it can start is day 10. For D to begin on time, what is the latest day on which C can begin? In part b, you found that the latest E can start is day 11. In that case, what is the latest C can begin? From this information, what is the latest (LST) that C can begin without delaying either task D or E?

10. To find the LST for each task, it is necessary to begin with the Finish and work through the graph in reverse order to the Start. Each of the vertices in the following graph is labeled with its EST. The LSTs for several of the tasks have been calculated and are shown below the ESTs on the vertices. Find the LSTs for the remaining tasks.

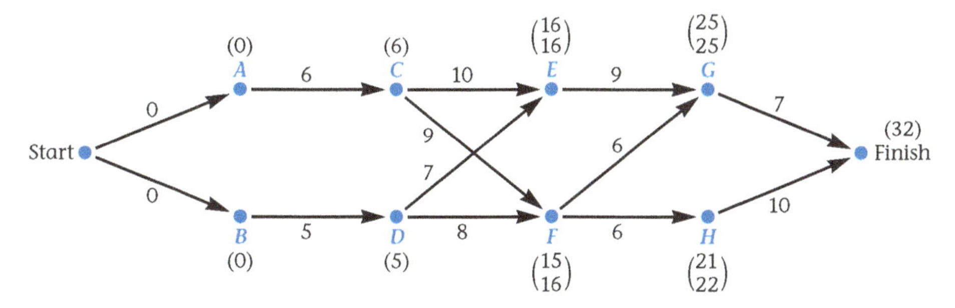

11. Write an algorithm to find the LSTs for the tasks in a graph. Test your algorithm on the graph in Exercise 1.

Project

12. Interview the yearbook sponsors in your school to find out how they organize the publication of your school's yearbook. Create a task table that shows the approximate times and prerequisite tasks that must be completed before your yearbook can go to the publisher. Design a graph with the EST for each task, and identify the critical path.

Modeling Project

13. Use the Internet or other sources to research and report on businesses or people who use PERT or similar evaluation techniques such as Gantt Charts to model project planning.

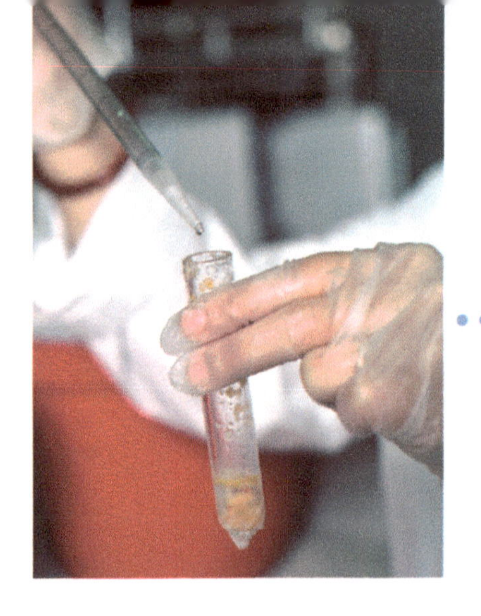

Lesson 4.3

The Vocabulary and Representations of Graphs

Graphs have many applications in addition to critical path analysis. They are frequently used as models in social science, computer science, chemistry, biology, transportation, and communications. In the following lessons several of these models are examined.

Recall that a graph is a set of points called vertices and a set of connecting lines called edges. Often graphs are used to model situations in which the vertices represent objects, and edges are drawn between the vertices on the basis of a particular relationship between the objects. Note that the important characteristics of a graph remain unchanged, whether the edges are curved or straight.

Explore This

Case 1

Suppose the vertices of Figure 4.4 represent the starting five players on a high school basketball team, and the edges denote friendships. This graph indicates that player *C* is friends with all of the other players and that *E* has only two friends, *C* and *B*. Notice that edge *CE* and edge *DB* intersect in this graph, but the intersection does not create a new vertex.

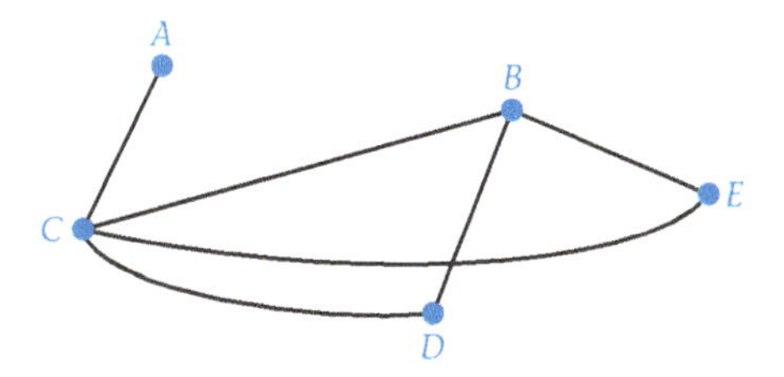

Figure 4.4. Graph showing five vertices and six edges.

1. Which player has only one friend?
2. How many friends does E have? Who are they?
3. Redraw the graph so that A has no friends.

Solution:

1. Player A has only one friend.
2. Player E has two friends. The friends of E are players C and B.
3. The graph in Figure 4.5 shows A with no friends.

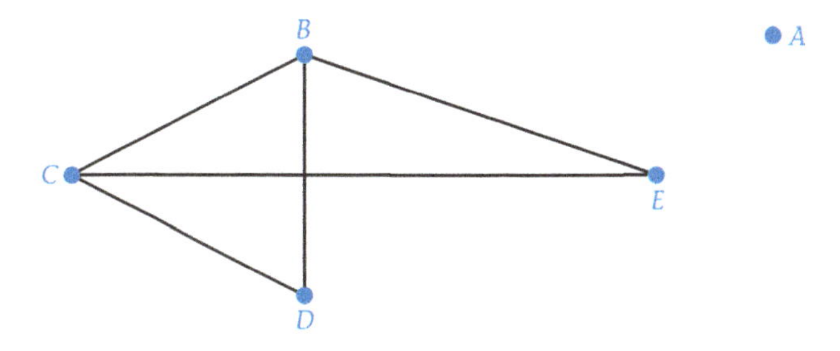

Figure 4.5. Graph that shows A with no friends.

A graph is **connected** if there is a path between each pair of vertices. The graph in Figure 4.5 is not connected because there is no path from A to any of the other vertices.

Case 2

Figure 4.4 on page 183 could represent many different things other than basketball players and their friendships. For example, let the vertices in the figure represent rooms in your school. The vertices are connected if there is a direct hallway between two rooms. Then, according to the graph in Figure 4.4, a student can get from room *C* directly to any of the other four rooms.

When two vertices are connected with an edge, they are said to be **adjacent**. In Figure 4.4, *C* is adjacent to *A, B, D,* and *E.* Although there is no single edge from *D* to *A,* it is possible to get from room *D* to room *A* by following a path that goes through room *C.* Although a path exists between *D* and *A,* they are not adjacent.

Try drawing a graph in which there is direct access from each room to every other room. Figure 4.6 shows two possible ways to represent this graph. Even though these graphs appear to be different, they are structurally the same. So they are considered the same graph. It is important to note that there is no single correct way to draw a graph to represent a given situation. A good graph is one that clearly represents the information needed to solve some problem.

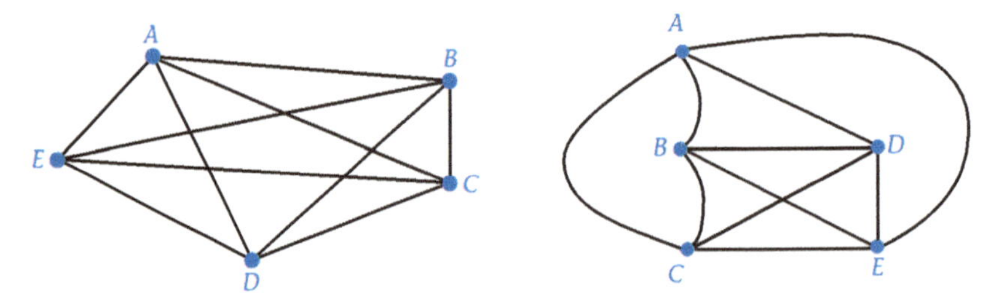

Figure 4.6. Two different representations of five rooms in a school, with each one having direct access to every other room.

Graphs such as the ones in Figure 4.6, in which every pair of vertices is adjacent, are called **complete** graphs. Complete graphs are often denoted by K_n, where *n* is the number of vertices in the graph. Figure 4.6 depicts a K_5 graph.

Alternative Representations

Graphs can be represented in several different ways. A diagram, such as the one in Figure 4.4, is just one of these ways. Another way to represent the information is to list the set of vertices and the set of edges. For example, the graph in Figure 4.4 on page 183 can be described by

Vertices = $\{A, B, C, D, E\}$ Edges = $\{AC, CB, CE, CD, BD, BE\}$.

A third way to represent this information is with an adjacency matrix. This type of representation can be used to depict the vertices and edges of the graph in a computer or calculator.

A 5×5 matrix in which both the rows and columns correspond to the vertices A, B, C, D, and E can be used to represent Figure 4.4. If an edge exists between vertices, a 1 appears in the corresponding position in the matrix; otherwise a 0 appears.

$$\begin{array}{c} \\ A \\ B \\ C \\ D \\ E \end{array} \begin{array}{c} \begin{array}{ccccc} A & B & C & D & E \end{array} \\ \begin{bmatrix} 0 & 0 & 1 & 0 & 0 \\ 0 & 0 & 1 & 1 & 1 \\ 1 & 1 & 0 & 1 & 1 \\ 0 & 1 & 1 & 0 & 0 \\ 0 & 1 & 1 & 0 & 0 \end{bmatrix} \end{array}$$

The entry in row 2, column 4 is a 1. This indicates that vertices B and D are adjacent; that is, an edge exists between them.

Exercises

1. Quin bought six different types of fish. Some of the fish can live in the same aquarium, but others cannot. Guppies can live with Mollies, Swordtails can live with Guppies, Plecostomi can live with both Mollies and Guppies, Gold Rams can live only with Plecostomi, and Piranhas cannot live with any of the other fish. Draw a graph to illustrate this.

2. Construct a graph for each of the following sets of vertices and edges.

   a. $V = \{A, B, C, D, E\}$
   $E = \{AB, AC, AD, AE, BE\}$

   b. $V = \{M, N, O, P, Q, R, S\}$
   $E = \{MN, SR, QS, SP, OP\}$

   c. $V = \{E, F, G, J, K, M\}$
   $E = \{EF, KM, FG, JM, EG, KJ\}$

   d. $V = \{W, X, Y, Z\}$
   $E = \{WX, XZ, YZ, XY, WZ, WY\}$

3. Which graphs in Exercise 2 are connected? Which are complete?

4. Draw a graph with vertices = {A, B, C, D, E, F} and edges = {AB, CD, DE, EC, EF}.

   a. Name two vertices that are not adjacent.

   b. F, E, C is one possible path from F to C. This path has a length of 2, since two edges were traveled to get from F to C. Name a path from F to C with a length of 3.

   c. Is this graph connected? Explain why or why not.

   d. Is this graph complete? Explain why or why not.

5. Draw a graph with five vertices in which vertex W is adjacent to Y, X is adjacent to Z, and V is adjacent to each of the other vertices.

6. Draw a graph with four vertices and three edges so that the graph is not connected.

7. Construct a graph for each adjacency matrix. Label the vertices $A, B, C, \ldots$.

   a. $\begin{bmatrix} 0 & 1 & 0 & 0 \\ 1 & 0 & 1 & 1 \\ 0 & 1 & 0 & 1 \\ 0 & 1 & 1 & 0 \end{bmatrix}$ b. $\begin{bmatrix} 0 & 1 & 0 & 0 & 1 \\ 1 & 0 & 1 & 0 & 1 \\ 0 & 1 & 0 & 1 & 1 \\ 0 & 0 & 1 & 0 & 1 \\ 1 & 1 & 1 & 1 & 0 \end{bmatrix}$ c. $\begin{bmatrix} 0 & 0 & 1 & 1 & 0 \\ 0 & 0 & 1 & 0 & 1 \\ 1 & 1 & 0 & 1 & 1 \\ 1 & 0 & 1 & 0 & 0 \\ 0 & 1 & 1 & 0 & 0 \end{bmatrix}$

8. Create an adjacency matrix for each of the following graphs:

   a.

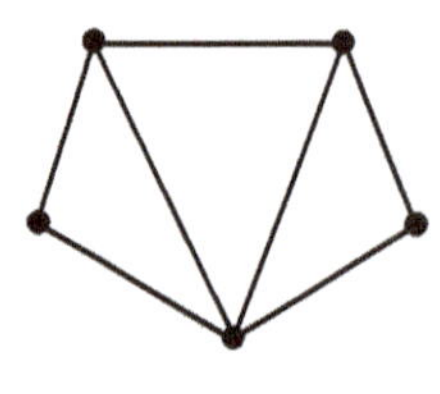

   b.

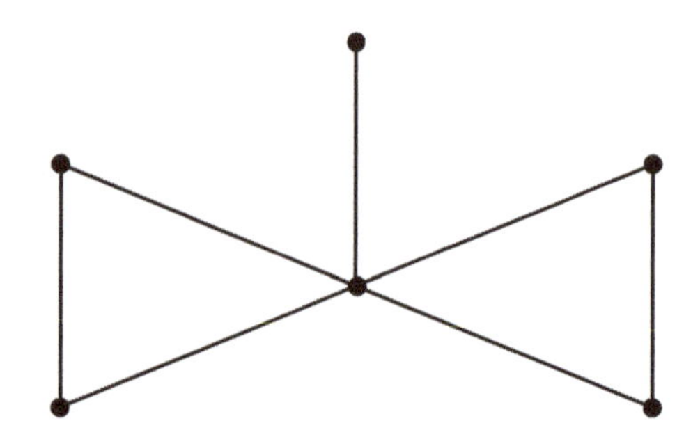

9. Give the adjacency matrix for the following graph.

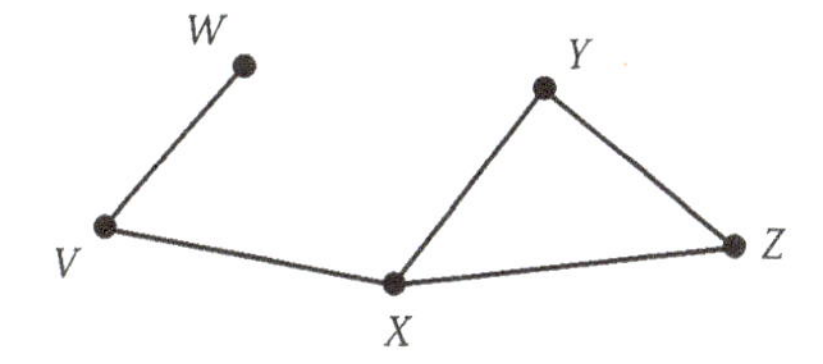

a. What do you notice about the main diagonal of the matrix?

b. Does your matrix possess symmetry? If so, where?

c. If an adjacency matrix has a 1 on the main diagonal, what would that indicate? What would a 2 in row 2, column 1 indicate?

10. Find the sum of each row of your matrix from Exercise 9. What do these sums tell you about the graph of the matrix?

11. In a graph, the number of edges that have a specific vertex as an endpoint is known as the **degree** or **valence** of that vertex. In the following graph, the degree of vertex W is 4. This is denoted by $\deg(W) = 4$. Find the degree of each of the other vertices.

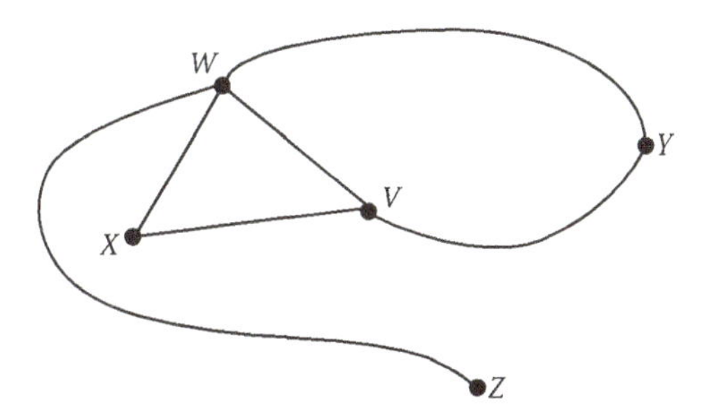

12. a. The degree of each vertex of a K_4 graph is ________.

b. The degree of each vertex of a K_n graph is ________.

13. An edge that connects a vertex to itself is called a **loop**. If a graph contains a loop or **multiple edges** (more than one edge between two vertices), the graph is known as a **multigraph**. When finding the degree of a vertex on which there is a loop, the loop is counted twice. For example, deg(A) = 3.

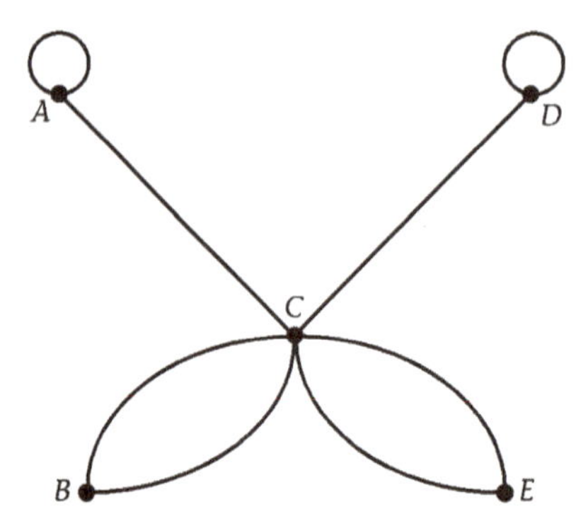

a. Find the degree of vertices B, C, D, and E.

b. Give the adjacency matrix for the above multigraph.

c. Compare an adjacency matrix for a graph and one for a multigraph. Without seeing the graph, can you tell which belongs to the graph and which belongs to the multigraph? Explain how you know.

14. Complete the following table for the sum of the degrees of the vertices in a complete graph.

Graph	Number of Vertices	Sum of the Degrees of All of the Vertices	Recurrence Relation
K_1	1	0	
K_2	2	2	$T_2 = T_1 + 2$
K_3	3	6	$T_3 = T_2 + 4$
K_4	___	___	___
K_5	___	___	___
K_6	___	___	___

Write a recurrence relation that expresses the relationship between the sum of the degrees of all the vertices for K_n and the sum for K_{n-1}.

15. Having completed the table in Exercise 14, what did you notice about the sum of the degrees of the vertices for any complete graph? Do you think this is true for any graph? If so, explain why this is true; if not, give a counterexample.

16. a. Try to construct a graph with four vertices, two of the vertices with degree 3 and two with degree 2. No loops or multiple edges may be used.

    b. Try to construct a graph with five vertices, three of the vertices with an odd degree and two with an even degree. No loops or multiple edges may be used.

    c. What do you think might be true about the number of vertices with even degree and the number of vertices with odd degree in any graph? (Hint: Try a few examples to check your hypothesis.)

17. Describe the adjacency matrix of a complete graph.

18. Complete the following table for the given complete graphs.

Graph	Number of Vertices	Number of Edges	Recurrence Relation
K_1	1	0	
K_2	2	1	$S_2 = S_1 + 1$
K_3	3	3	$S_3 = S_2 + 2$
K_4	___	___	___
K_5	___	___	___
K_6	___	___	___

Write a recurrence relation that expresses the relationship between the number of edges of K_n and the number of edges of K_{n-1}.

19. Central High School is a member of a five-team hockey league. Each team in the league plays exactly two games, which must be against different teams. Show that there is only one possible graph for this schedule.

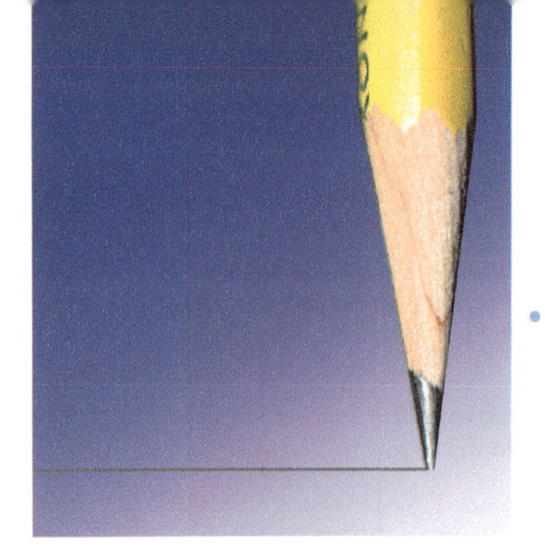

Lesson 4.4

Euler Circuits and Paths

Now that you are familiar with some of the concepts of graphs and the way graphs convey connections and relationships, it's time to begin exploring how they can be used to model many different types of situations.

Explore This

Consider the graph in Figure 4.7. Try to draw this figure without lifting your pencil from the paper and without tracing any of the lines more than once. Is this possible?

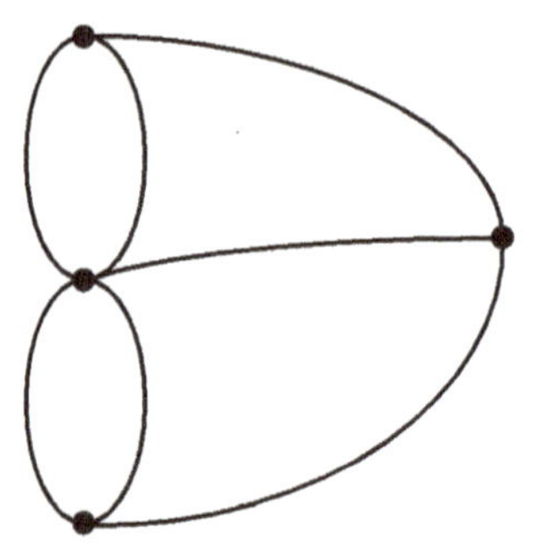

Figure 4.7. Graph.

The graph in Figure 4.7 represents an eighteenth-century problem that intrigued the famous Swiss mathematician Leonhard Euler (pronounced "oiler"). The problem was one that had been posed by the residents of Königsberg, a city in what was then Prussia but is now the

Russian city of Kaliningrad. In the 1700s, seven bridges connected two islands in the Pregel River to the rest of the city (see Figure 4.8). The people of Königsberg wondered whether it would be possible to walk through the city by crossing each bridge exactly once and return to the original starting point.

Mathematician of Note

Leonhard Euler (1707–1783)

Euler was an extraordinary mathematician who published over 500 works during his lifetime. Even total blindness for the last 17 years of his life did not stop his effectiveness and genius. He is often referred to as "the father of graph theory."

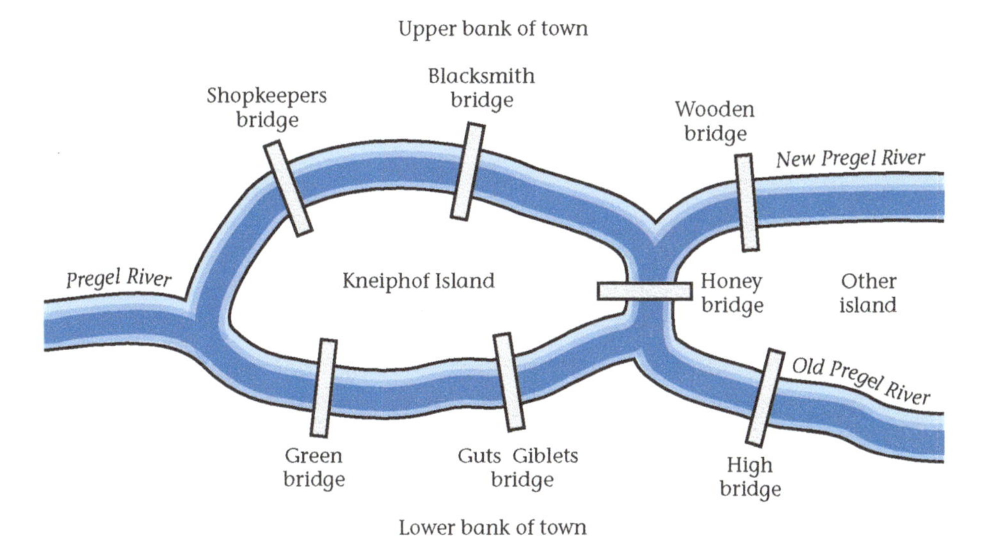

Figure 4.8. Representation of the seven bridges of Königsberg.

Using a graph like the one in Figure 4.7, in which the vertices represented the landmasses of the city and the edges represented the bridges, Euler found that it was not possible to make the desired walk through the city. In so doing, he also discovered a solution to problems of this general type.

What did Euler find? Try to reproduce the graphs in Figure 4.9 on the next page without lifting your pencil or tracing the lines more than once.

1. When can you draw the figure without retracing any edges and still end up at your starting point?
2. When can you draw the figure without retracing and end up at a point other than the one from which you began?
3. When can you not draw the figure without retracing?

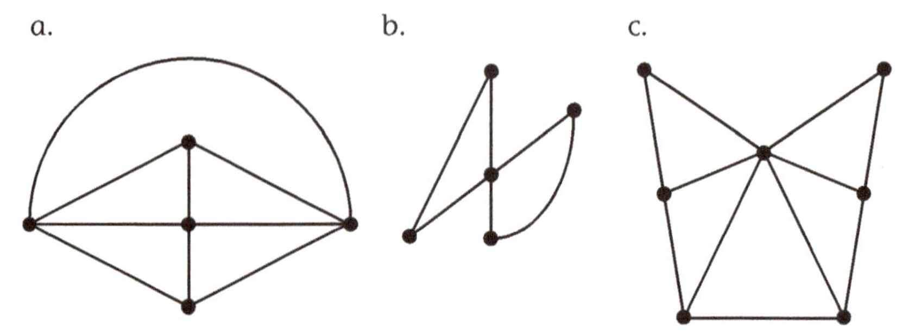

Figure 4.9. Graphs to trace.

Euler found that the key to the solution was related to the degrees of the vertices. Recall that the degree of a vertex of a graph is the number of edges that have that vertex as an endpoint. Find the degree of each vertex of the graphs in Figure 4.9. Do you see what Euler noticed?

Euler hypothesized and later proved that in order to be able to traverse each edge of a connected graph exactly once and to end at the starting vertex, the degree of each vertex of the graph must be even. See Figure 4.9b. In honor of Leonhard Euler, a path that uses each edge of a graph exactly once and ends at the starting vertex is called an **Euler circuit**.

Euler also noticed that if a connected graph had exactly two odd vertices, it was possible to use each edge of the graph exactly once but to end at a vertex different from the starting vertex. Such a path is called an **Euler path**. Figure 4.9a is an example of a graph that has an Euler path.

Figure 4.9c has four odd vertices. So it cannot be traced without lifting your pencil. It has neither an Euler circuit nor an Euler path.

An Euler circuit for a relatively small graph usually can be found by trial and error. However, as the number of vertices and edges increases, a systematic way of finding the circuit becomes necessary. The following algorithm gives a procedure for finding an Euler circuit for a connected graph with all vertices of even degree.

Example

Use the Euler circuit algorithm to find an Euler circuit for the following graph.

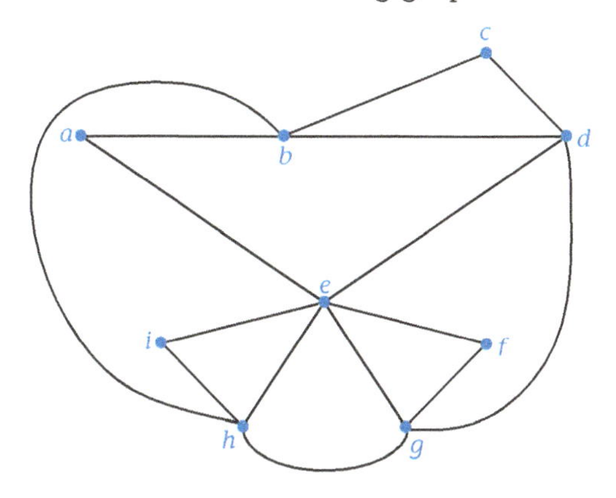

- Apply step 1 of the algorithm. Choose vertex *b*, and label it *S*.
- Let *C* be the circuit *S, c, d, e, a, S*.
- Circuit *C* does not contain all edges of the graph, so proceed to step 4 of the algorithm.
- Choose vertex *d*.
- Let C′ be the circuit *d, g, h, S, d*.
- Combine *C* and *C′* by replacing vertex *d* in the circuit *C* with the circuit *C′*. Let *C* now be the circuit *S, c, d, g, h, S, d, e, a, S*.
- Go to step 3 of the algorithm.
- Circuit *C* does not contain all edges of the graph, so again proceed to step 4.
- Choose vertex *g*.
- Let *C′* be the circuit *g, f, e, i, h, e, g*.
- Combine *C* and *C′* by replacing vertex *g* in the circuit *C* with the circuit *C′*. Let *C* now be the circuit *S, c, d, g, f, e, i, h, e, g, h, S, d, e, a, S*.
- Circuit *C* now contains all edges of the graph, so go to step 8 of the algorithm and stop. *C* is an Euler circuit for the graph.

Euler Circuit Algorithm

1. Pick any vertex, and label it *S*.
2. Construct a circuit, *C*, that begins and ends at *S*.
3. If *C* is a circuit that includes all edges of the graph, go to step 8.
4. Choose a vertex, *V*, that is in *C* and has an edge that is not in *C*.
5. Construct a circuit *C′* that starts and ends at *V* using edges not in *C*.
6. Combine *C* and *C′* to form a new circuit. Call this new circuit *C*.
7. Go to step 3.
8. Stop. *C* is an Euler circuit for the graph.

Edges with Direction

Many applications of graphs require that the edges have direction. A city with one-way streets is one such example. A graph that has directed edges, edges that can be traversed in only one direction, is known as a **digraph** (see Figure 4.10). The number of edges coming into a vertex is known as the **indegree** of the vertex, and the number of edges going out of a vertex is known as the **outdegree**.

Examine Figure 4.10. This digraph can be described by

Vertices = {A, B, C, D} Ordered edges = {AB, BA, BC, CA, DB, AD}.

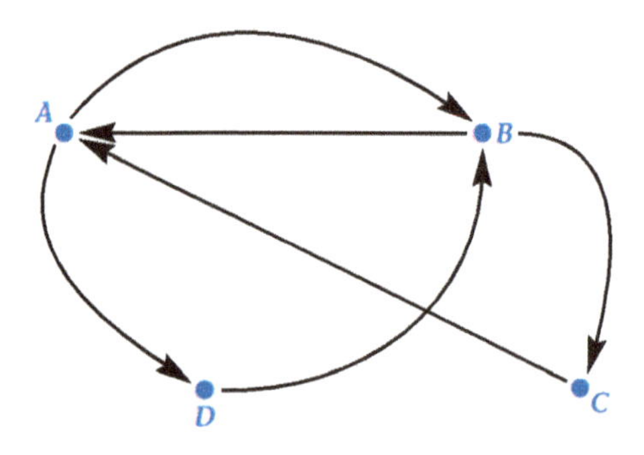

Figure 4.10. Digraph.

If you follow the indicated direction of each edge, is it possible to start at some vertex, draw the digraph, and end up at the vertex from which you started? That is, does this digraph have a directed Euler circuit?

Check the indegree and outdegree of each vertex. You will find that a connected digraph has an Euler circuit if the indegree and outdegree of each vertex are equal.

Exercises

1. State whether each graph has an Euler circuit, an Euler path, or neither. Explain why.

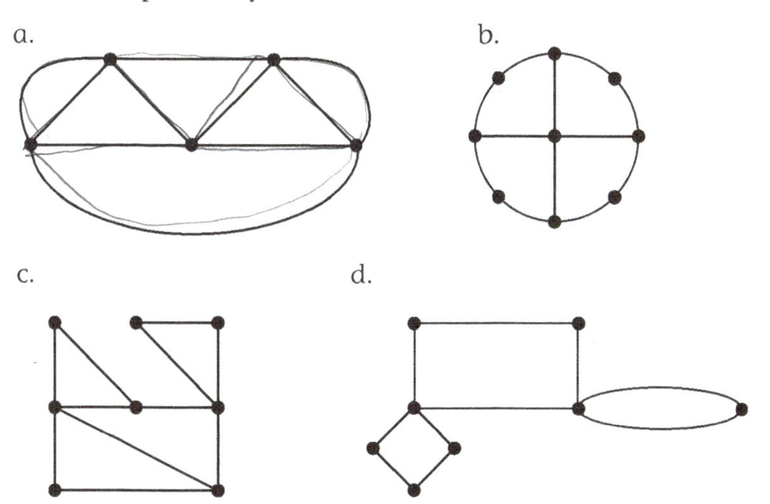

2. Draw a graph with six vertices and eight edges so that the graph has an Euler circuit.

3. Sally began using the Euler circuit algorithm to find the Euler circuit for the following graph. She started at vertex *d* and labeled it *S*. The first circuit she found was *S, e, f, a, b, c, S*. Using Sally's start, continue the algorithm and find an Euler circuit for the graph.

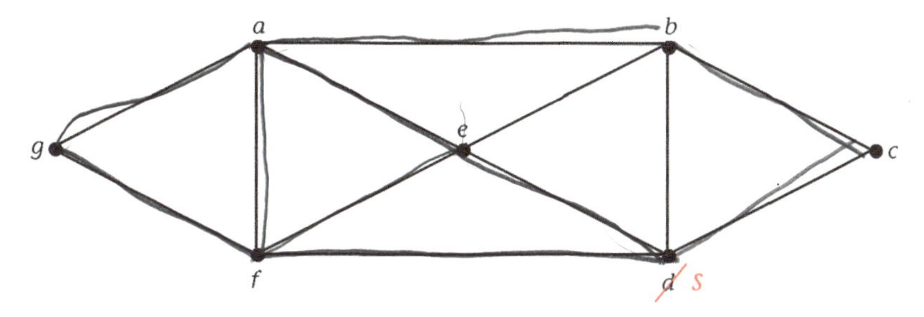

4. Use the Euler circuit algorithm to find the Euler circuit for the following graph.

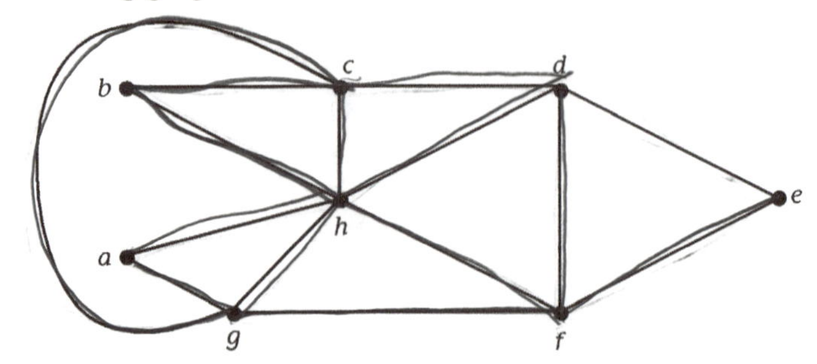

5. The text states that to apply the Euler circuit algorithm, the graph must be "connected with all vertices of even degree."
   a. Why is it necessary to state that the graph must be connected?
   b. Give an example of a graph with all vertices of even degree that does not have an Euler circuit.
   c. Draw a graph with exactly two vertices of odd degree that does not have an Euler path.
6. Will a complete graph with 2 vertices have an Euler circuit? With 3 vertices? With 4 vertices? With 5 vertices? With n vertices?
7. Suppose that the people of Königsberg built two more bridges across the river. If one bridge was added to connect the two banks on the river, A to B in the following figure, and another one was added to link the land to one of the islands, B to D, would it then be possible to make the famous walk and return to the starting point? Explain your reasoning.

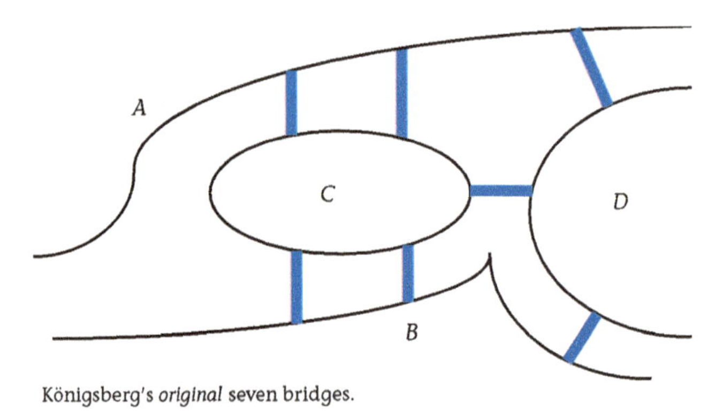

Königsberg's *original* seven bridges.

8. The street network of a city can be modeled with a graph in which the vertices represent the street corners, and the edges represent the streets. Suppose you are the city street inspector and it is desirable to minimize time and cost by not inspecting the same street more than once.

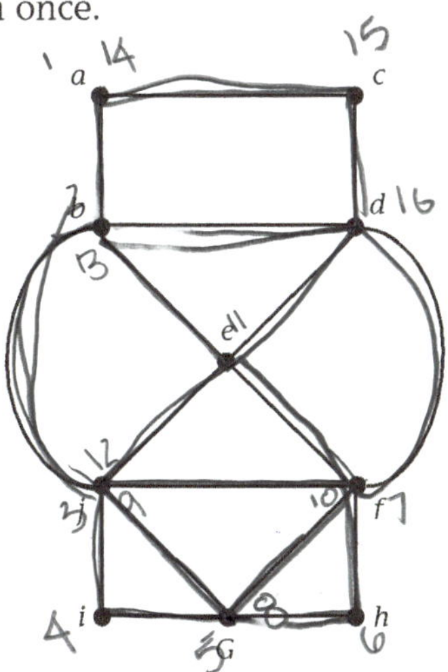

a. In this graph of the city, is it possible to begin at the garage (G) and inspect each street only once? Will you be back at the garage at the end of the inspection?

b. Find a route that inspects all streets, repeats the least number of edges possible, and returns to the garage.

9. Construct the following digraphs.

a. $V = \{A, B, C, D, E\}$
$E = \{AB, CB, CE, DE, DA\}$

b. $V = \{W, X, Y, Z\}$
$E = \{WX, XZ, ZY, YW, XY, YX\}$

10. a. Write a list of the set of vertices and the set of ordered edges that can be used to describe the following digraph.

b. Does the digraph have an Euler circuit? Explain.

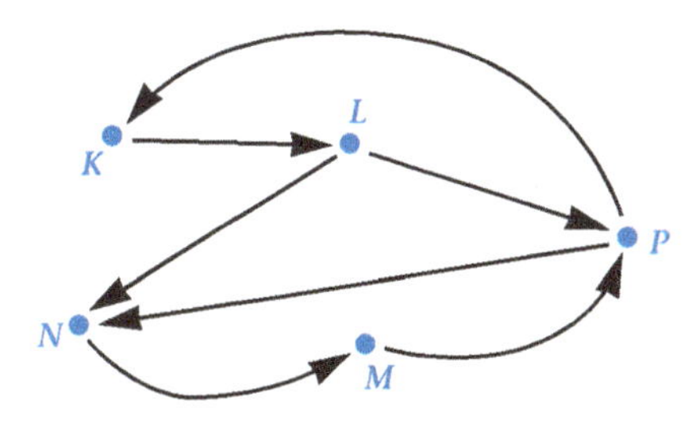

11. Determine whether the digraph has a directed Euler circuit.

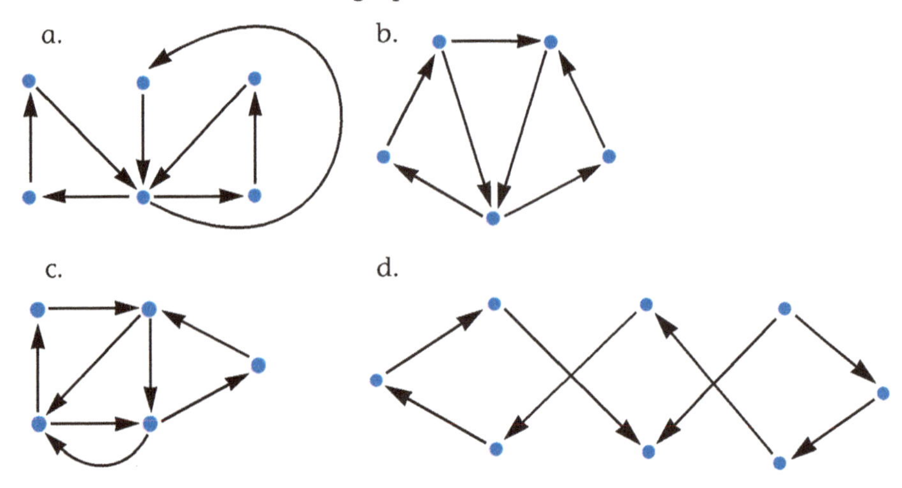

12. a. Does the following digraph have a directed Euler circuit? Explain why or why not.

    b. Does it have a directed Euler path? If it does, which vertices can be the starting vertex?

    c. Write a general statement explaining when a digraph has a directed Euler path.

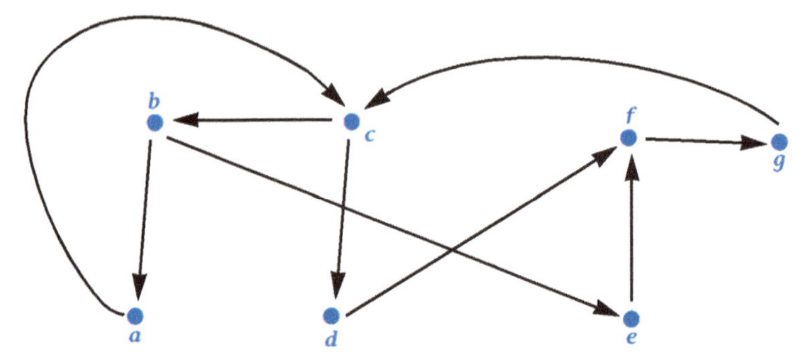

13. A digraph can be represented by an adjacency matrix. If there is a directed edge from vertex a to vertex b, then a 1 is placed in row a, column b of the matrix; otherwise a 0 is entered. Matrix M is the adjacency matrix for the following graph.

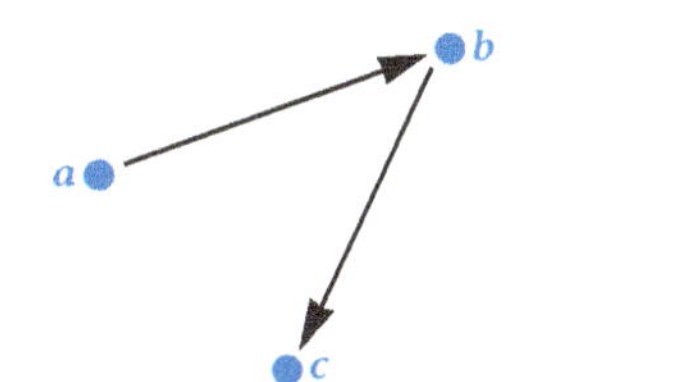

$$M = \begin{bmatrix} 0 & 1 & 0 \\ 0 & 0 & 1 \\ 0 & 0 & 0 \end{bmatrix}$$

Find the adjacency matrix for each of the following digraphs.

a.

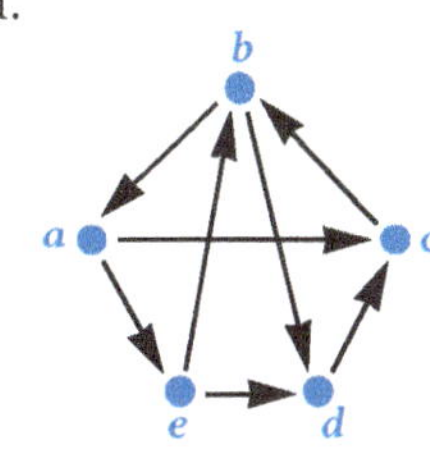

b.

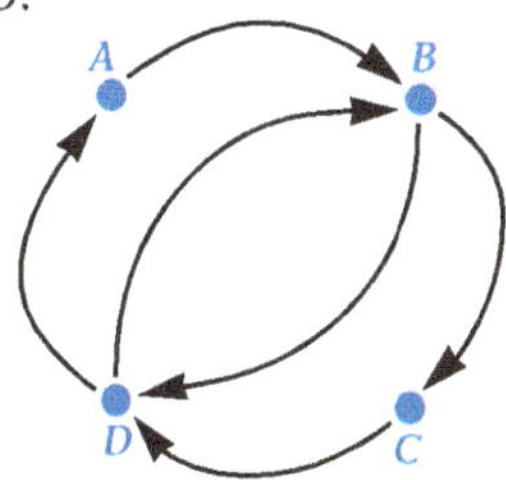

c.

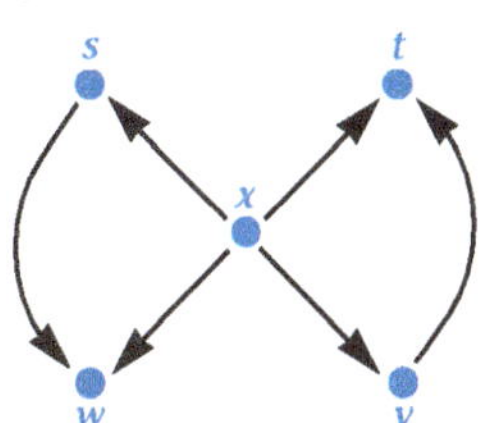

d.

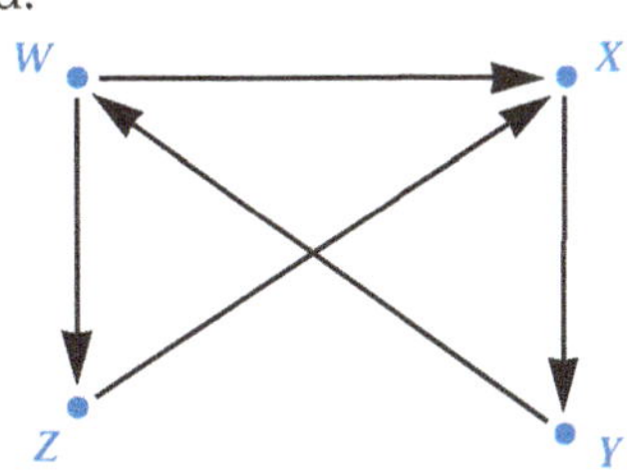

14. Use the following adjacency matrix to construct a digraph.

	A	B	C	D
A	0	1	1	0
B	1	0	0	1
C	0	1	0	0
D	0	0	1	0

15. a. Construct a digraph for the following adjacency matrix.

$$\begin{bmatrix} 0 & 1 & 0 & 1 & 0 \\ 1 & 0 & 1 & 0 & 1 \\ 0 & 0 & 0 & 1 & 0 \\ 0 & 1 & 1 & 0 & 0 \\ 1 & 0 & 0 & 1 & 0 \end{bmatrix}$$

b. Is there symmetry along the main diagonal of the adjacency matrix? Explain why or why not.

c. Find the sum of the numbers in the second row. What does that total indicate?

d. Find the sum of the numbers in the second column. What does that total indicate?

Computer/Calculator Explorations

16. Create a computer or calculator program that prompts the user to enter the adjacency matrix for a connected graph. Then the program should tell the user whether or not the graph has an Euler circuit.

Projects

17. Leonhard Euler was known for many accomplishments in addition to his discoveries related to graph theory. After researching Euler's achievements, create a "biographic poster" that illustrates the important milestones of his life.

18. Research and report on algorithms that determine Euler circuits for graphs that have them.

Lesson 4.5

Hamiltonian Circuits and Paths

Since its inception, graph theory has been closely tied to applications. In Lesson 4.4, you investigated situations in which you needed to traverse each edge of a graph. In this lesson, you will explore situations that can be modeled with graphs in which each vertex must be visited.

Explore This

Suppose once again that you are a city inspector, but instead of inspecting all of the streets in an efficient manner, you must inspect the fire hydrants that are located at each of the street intersections. This implies that you are searching for an optimal route that begins at the garage *G*, visits each intersection exactly once, and returns to the garage (see Figure 4.11).

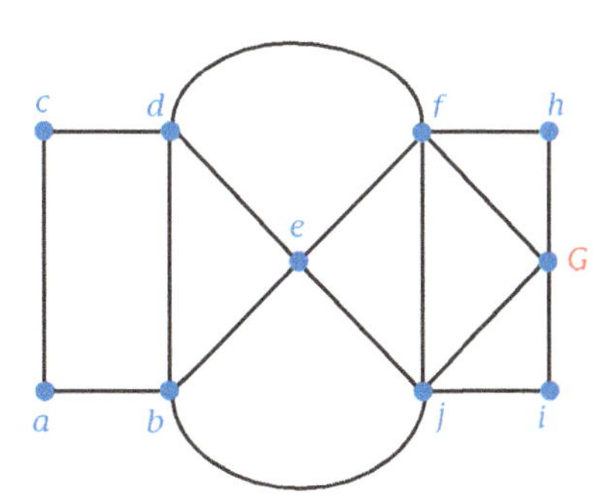

Figure 4.11. Street network.

One path that meets these criteria is *G, h, f, d, c, a, b, e, j, i, G*. Notice that it is not necessary to traverse every edge of the graph when visiting each vertex exactly once.

Mathematician of Note

Sir William Rowan Hamilton (1805–1865)

Hamilton, a leading nineteenth-century Irish mathematician, was appointed Astronomer Royal of Ireland at the age of 22 and knighted at 30. Shortly before his death, Hamilton received word that he had been elected the first foreign member of the National Academy of Sciences of the USA. Hamiltonian Mechanics is still used today to determine orbital trajectories of satellites.

In 1856, Sir William Rowan Hamilton used his mathematical knowledge to create a game called the Icosian game. The game consisted of a graph in which the vertices represented major cities in Europe. The object of the game was to find a path that visited each of the 20 vertices exactly once. In honor of Hamilton and his game, a path that uses each vertex of a graph exactly once is known as a **Hamiltonian path**. If the path ends at the starting vertex, it is called a **Hamiltonian circuit**.

Try to find a Hamiltonian circuit for each of the graphs in Figure 4.12.

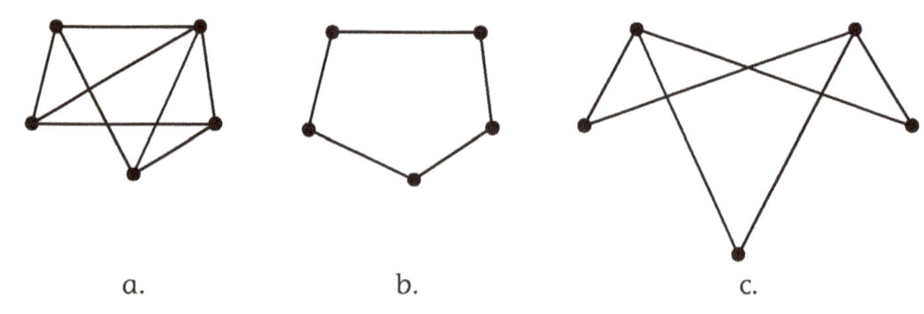

Figure 4.12. Graphs with possible Hamiltonian circuits.

Mathematicians continue to be intrigued with this type of problem because a simple test for determining whether a graph has a Hamiltonian circuit has not been found. It now appears that a general solution that applies to all graphs may be impossible. Fortunately, several theorems have been proved that help guarantee the existence of Hamiltonian circuits for certain kinds of graphs. The following is one such theorem.

If a connected graph has n vertices, where $n > 2$ and each vertex has degree of at least $n/2$, then the graph has a Hamiltonian circuit.

To apply the Hamiltonian circuit theorem to Figure 4.12a, check the degree of each vertex. Since each of the five vertices of the graph has degree of at least 5/2, the theorem guarantees that the graph has a Hamiltonian circuit. Unfortunately, it does not tell you how to find it.

The theorem has another downside as well. If a graph has a vertex with degree less than $n/2$, then the theorem simply does not apply to that graph. It may or it may not have a Hamiltonian circuit. Each of the graphs in parts b and c of Figure 4.12 has some vertices of degree less than 5/2, so no conclusions can be drawn. By inspection, Figure 4.12b has a Hamiltonian circuit, but Figure 4.12c does not.

Tournaments

As with Euler circuits, often it is useful for the edges of the graph to have direction. Consider a competition in which each player must play every other player. By using directed edges, it is possible to indicate winners and losers. To illustrate this, draw a complete graph in which the vertices represent the players, and a directed edge from vertex A to vertex B indicates that player A defeats player B. This type of graph is known as a tournament. A **tournament** is a digraph that results from giving directions to the edges of a complete graph. Figure 4.13 shows a tournament in which A beats B, C beats B, and A beats C.

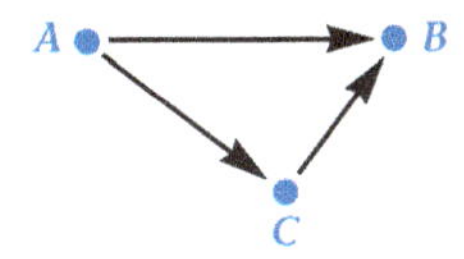

Figure 4.13. Tournament with three vertices.

One interesting property of such a digraph is that every tournament contains at least one Hamiltonian path. If there is exactly one such path, it can be used to rank the teams in order, from winner to loser.

Example

Suppose four teams play in the school soccer round-robin tournament. The results of the competition follow:

Game	*AB*	*AC*	*AD*	*BC*	*BD*	*CD*
Winner	*B*	*A*	*D*	*B*	*D*	*D*

Draw a digraph to represent the tournament. Find a Hamiltonian path and use it to rank the participants from winner to loser.

Solution:

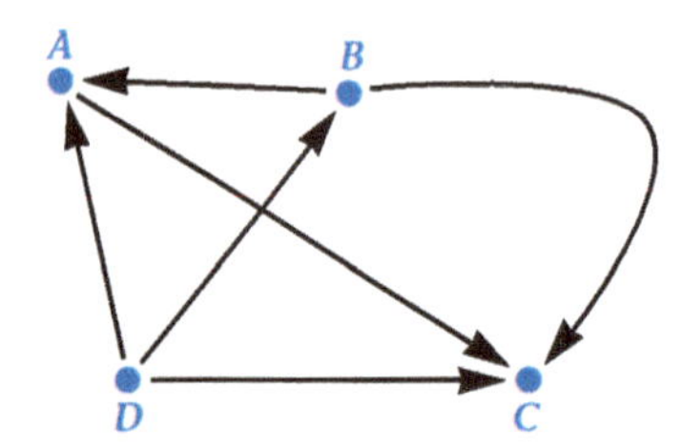

To determine a ranking, remember that a tournament results from a complete graph when direction is given to the edges. In this case, there is only one Hamiltonian path for the graph: *D, B, A, C*. Therefore, *D* finishes first, *B* is second, *A* is third, and *C* finishes fourth.

Exercises

1. Apply the theorem from page 202 to the following graphs. According to the theorem, which of the graphs have Hamiltonian circuits? Explain your reasoning.

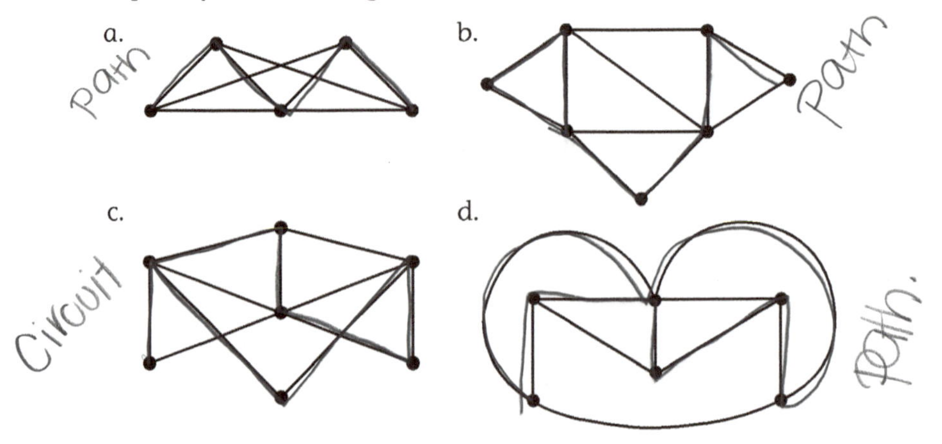

2. Give two examples of situations that could be modeled by a graph in which finding a Hamiltonian path or circuit would be of benefit.

3. a. Construct a graph that has both an Euler and a Hamiltonian circuit.

   b. Construct a graph that has neither an Euler nor a Hamiltonian circuit.

4. Hamilton's Icosian game was played on a wooden regular dodecahedron (a solid figure with 12 sides). Here is a planar representation of the game.

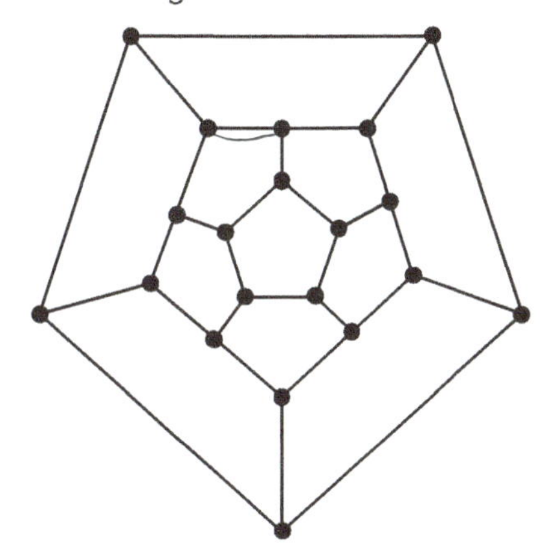

   a. Copy the graph onto your paper and find a Hamiltonian circuit for the graph.

   b. Is there only one Hamiltonian circuit for the graph?

   c. Can the circuit begin at any of the vertices or only some of them?

5. Find a Hamiltonian circuit for the following graph.

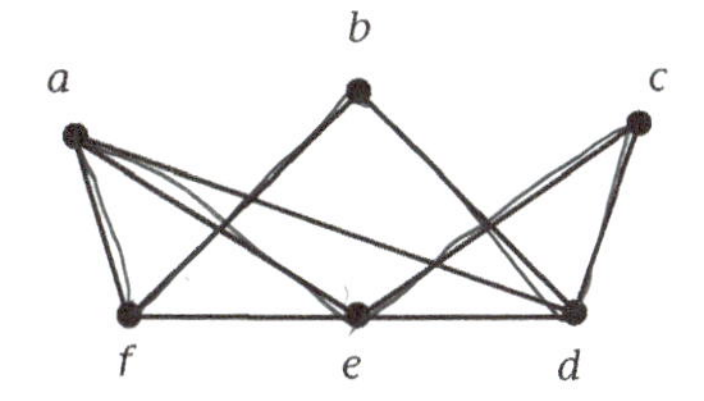

6. Draw a tournament with five players, in which player A defeats everyone, B defeats everyone but A, C is defeated by everyone, and D defeats E.

7. Find all the directed Hamiltonian paths for each of the following tournaments:

a.

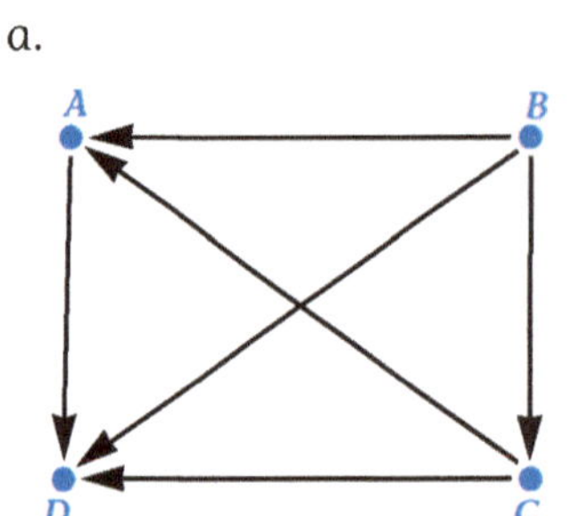

b.

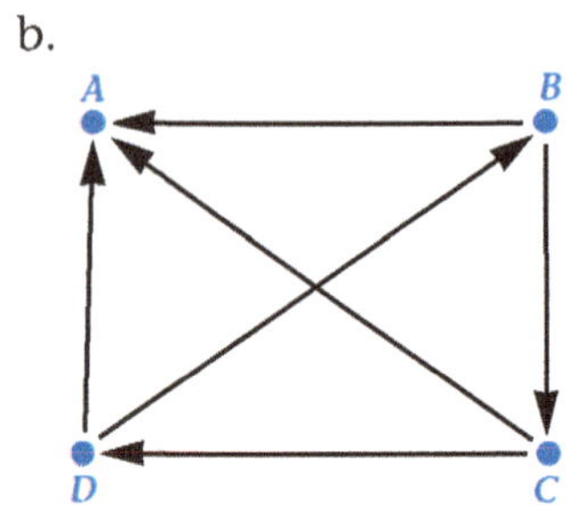

8. Draw a tournament with three vertices in which:

   a. One player wins all the games he or she plays.

   b. Each player wins exactly one game.

   c. Two players lose all of the games they play.

9. Draw a tournament with five vertices in which there is a three-way tie for first place.

10. When ties exist in a ranking for a tournament (e.g., more than one first place winner), there is more than one Hamiltonian path for the graph. Explain why this is so.

11. a. Write an algorithm that uses the outdegree of the vertices to find the Hamiltonian path for a tournament that has exactly one Hamiltonian path.

    b. Explain the difficulties that arise with your algorithm when the tournament has more than one Hamiltonian path.

12. Complete the following table for a tournament.

Number of Vertices	Sum of the Outdegrees of the Vertices
1	0
2	1
3	3
4	
5	
6	

Write a recurrence relation that expresses the relationships between S_n, the sum of the outdegrees for a tournament with n vertices, and S_{n-1}.

13. In a tournament a **transmitter** is a vertex with a positive outdegree and a zero indegree. A **receiver** is a vertex with a positive indegree and a zero outdegree. Explain why a tournament can have at most one transmitter and at most one receiver.

14. Use mathematical induction on the number of vertices to prove that every tournament has a Hamiltonian path.

    a. Begin the mathematical induction process by showing that every tournament with one vertex has a Hamiltonian path.

    b. Assume that a tournament of k vertices has a Hamiltonian path and use this assumption to prove that a tournament of $k + 1$ vertices has a Hamiltonian path.

15. Consider the set of preference schedules from Lesson 1.3:

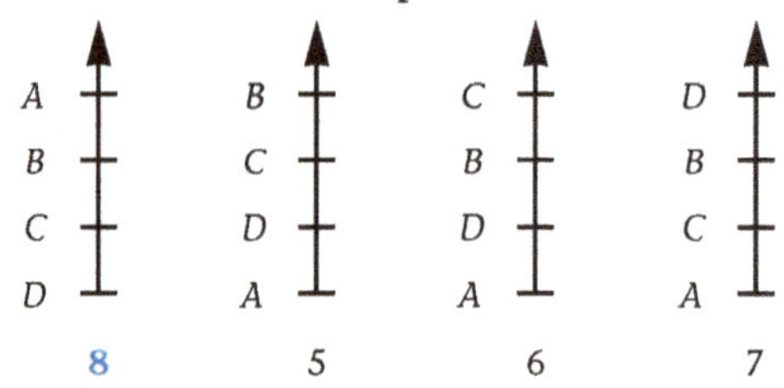

The first preference schedule could be represented by the following tournament.

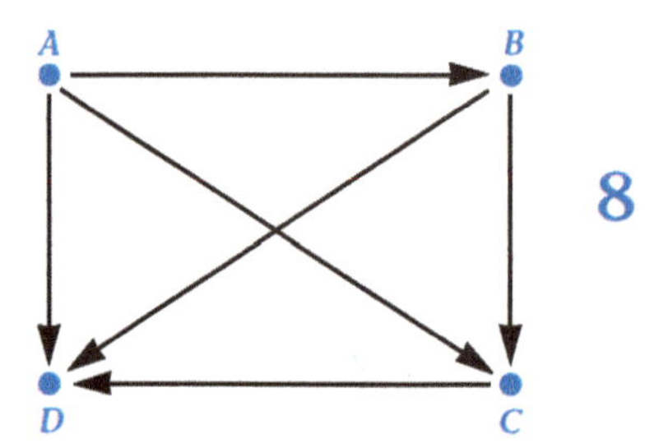

a. Construct tournaments for each of the three other preference schedules.

b. Construct a cumulative preference tournament that would show the overall results of the four individual preference schedules.

c. Is there a Condorcet winner in the election? (Recall from Lesson 1.3 that a Condorcet winner is one who is able to defeat each of the other choices in a one-on-one contest.)

d. Find a Hamiltonian path for the cumulative tournament. What does this path indicate?

16. a. Construct an adjacency matrix for the following digraph, and call the matrix M.

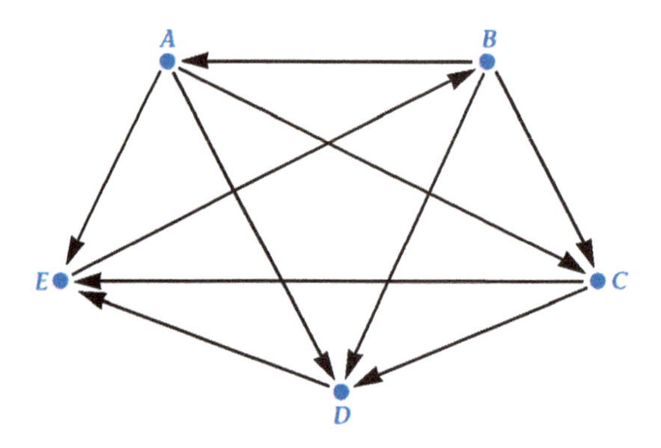

By summing the rows of M, you can see that a tie exists between A and B, each has three wins.

b. Square M. Notice that this new matrix M^2 gives the number of paths of length 2 between vertices. For example, the 3 in entry M_{25} indicates that 3 paths of length 2 exist between B and E. These paths are B, C, E; B, D, E; and B, A, E. In the case of a tournament, this means that B has three 2-stage wins, or **dominances**, over E.

c. Add M and M^2. Use the sum to determine the total number of ways that A and B can dominate in one and two stages. Who might now be considered the winner?

d. What would M^3 indicate? Find M^3 and see whether you are correct.

Project

17. Design and build a Planar Icosian Game by enlarging the graph in Exercise 4 and copying it onto a piece of plywood or heavy cardboard. Use tacks for the vertices. The game is then played by tying a piece of string (approximately 12 inches or longer) on one of the tacks (vertices) and attempting to wind the string from tack to tack (following the lines on the graph) until all of the tacks are touched and the player is back to the initial tack (in other words, until a Hamiltonian circuit is found). Try the game with younger children and adults. Who seems to find a Hamiltonian circuit the quickest?

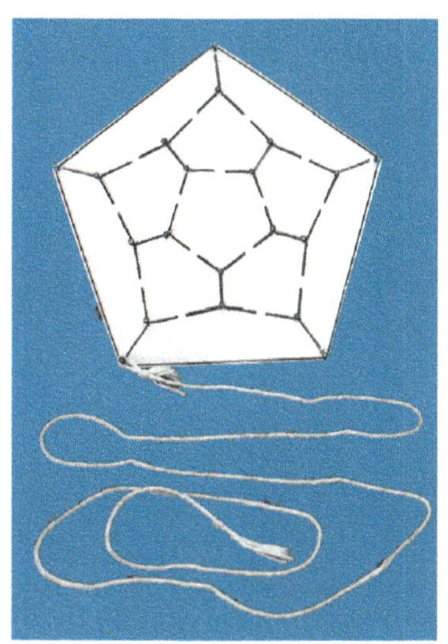
Student-Made Icosian Game

Lesson 4.6

Graph Coloring

When it is time to schedule meetings at school or register students into classes, scheduling conflicts often arise. Mathematicians have found that graphs are useful tools in helping to resolve conflicts of these types.

Explore This

Here is a table of clubs at Central High School and students who hold offices in these clubs.

	Math Club	Chess Club	Science Club	Art Club	Pep Club	Spanish Club
Nicole	X	X	X	—	—	—
Mattie	X	—	—	X	X	—
Quin	—	X	—	—	—	X
Emma	X	—	X	—	—	—
Alex	X	—	—	—	X	—

Each club at Central High wants to meet once a week. Since several students hold offices in more than one organization, it is necessary to arrange the meeting days so that no students are scheduled for more than one meeting on the same day. Is it possible to create such a schedule? What is the minimum number of days needed?

One possible solution to the problem is to use five days for the scheduling. The Math and Spanish Clubs could meet on Monday, and the remaining clubs could meet on the other four days. If the problem is to schedule the meetings in the fewest number of days, then the solution of using five days is not optimal. It is possible to create a schedule using only three days.

Finding such a schedule by trial and error is not difficult in this case, but a mathematical model would be helpful for more complicated problems. The first step in creating such a model might be to construct a graph in which the vertices represent clubs at Central High and the edges indicate conflict. If an edge joins two vertices, then those two clubs share an officer and cannot meet on the same day.

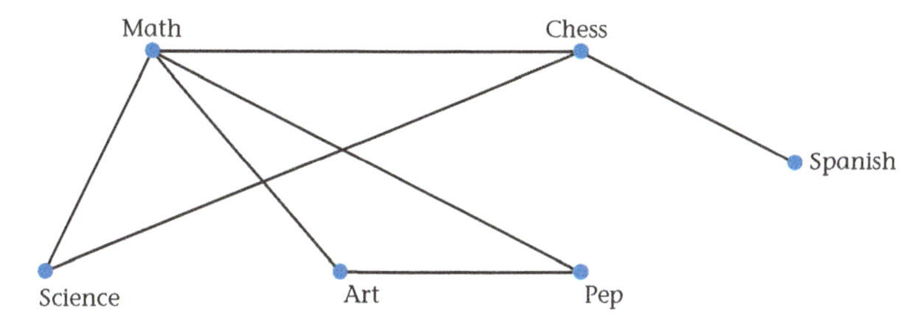

Figure 4.14. Graph showing the organizations of Central High that share an officer.

After completing the graph (see Figure 4.14), label the vertices with days of the week. When doing so, keep in mind that adjacent vertices must have different labels, because this is where the conflicts occur. One way of assigning days is to begin with the Math Club and label it with Monday. Look at the vertices not adjacent to the Math Club vertex. Since no officers belong to both the Math Club and the Spanish Club, also label the Spanish Club with Monday. Label the Chess Club with Tuesday. Label the Science Club Wednesday. The Pep or Art Club, but not both, can receive a Tuesday label. The one not labeled Tuesday is then labeled Wednesday, the third day. The resulting schedule is an optimal solution to the problem. But notice that this schedule is not unique.

Problems of this type are called *coloring* problems because historically the labels placed on the vertices of the graphs were referred to as *colors*. The process of labeling the graph is called **coloring the graph**. The minimum number of labels, or colors, that can be used is known as the **chromatic number** of the graph. The chromatic number for the graph in Figure 4.14 is 3.

Questions of this type first attracted interest in the nineteenth century when mathematicians such as Augustus de Morgan, William Rowan Hamilton, and Arthur Cayley became interested in a problem known as the four-color conjecture. The problem stated that any map that could be drawn on the surface of a sphere could be colored with, at most, four colors.

This problem intrigued mathematicians for over 100 years. During that time, many claimed to have proved the conjecture, but flaws were always found in the proofs. It wasn't until 1976 that Kenneth Appel and Wolfgang Haken of the University of Illinois solved the famous problem, and the four-color conjecture became known as the four-color theorem.

Appel and Haken proved this theorem in a way very different from earlier attempts. They used a high-speed computer in their verification. When the proof was finally complete, they had used over 1,200 hours of computer time and had examined approximately 1,936 basic forms of maps. Because of the unusual method of proof and the inability to emulate it by hand, criticism arose from many in the mathematical community.

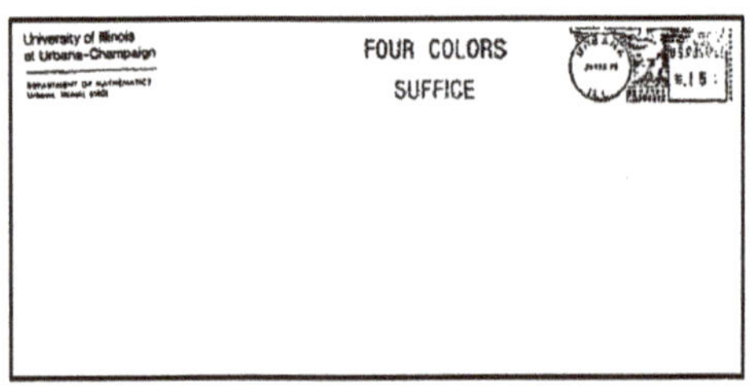

Postage meter stamp used by the University of Illinois to commemorate the proof of the four-color theorem.

In the late 1990s, Neil Robertson, Daniel Sanders, Paul Seymour, and Robin Thomas constructed a new proof. Even though this new proof is computer assisted, it is simpler. It has removed the cloud of doubt hanging over the complex original proof of Appel and Haken. In addition, in 2005, Georges Gonthier used general purpose theorem proving software to prove the theorem.

Are you wondering what scheduling classes and coloring maps have in common? As it turns out, they are both problems that can be solved with the help of graph coloring techniques. One way to approach the problem of coloring a map is to represent each region of the map with a vertex of a graph. Two vertices are then connected by an edge if the regions they represent have a common border. The process of coloring the graph resulting from a map is the same as when you colored the scheduling graph model.

Example

Color the following map using four or fewer colors.

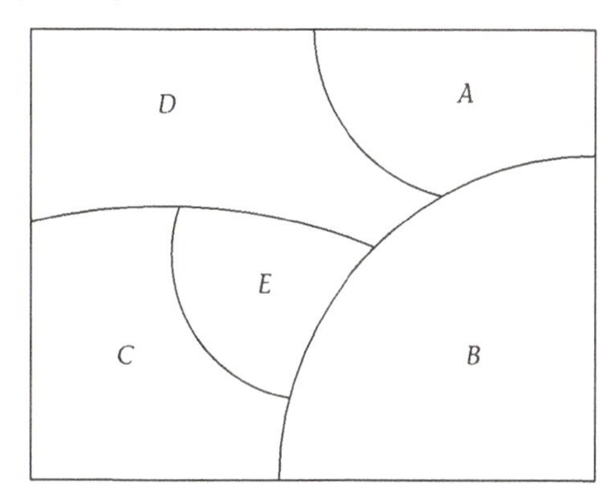

Solution:

First represent the map with a graph in which each vertex represents a region of the map. Then draw edges between vertices if the regions on the map have a common border. Label the graph using a minimum number of colors.

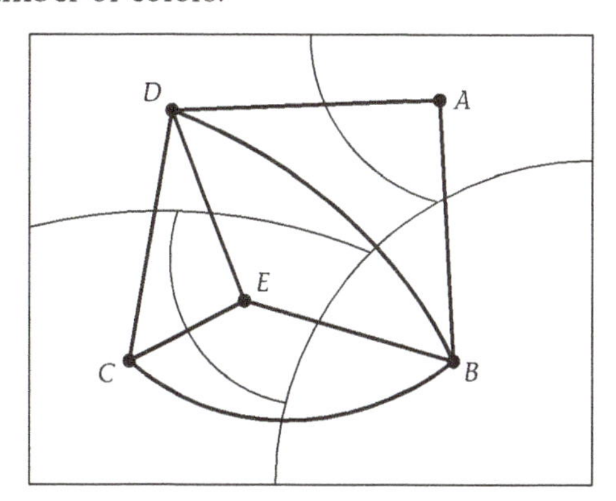

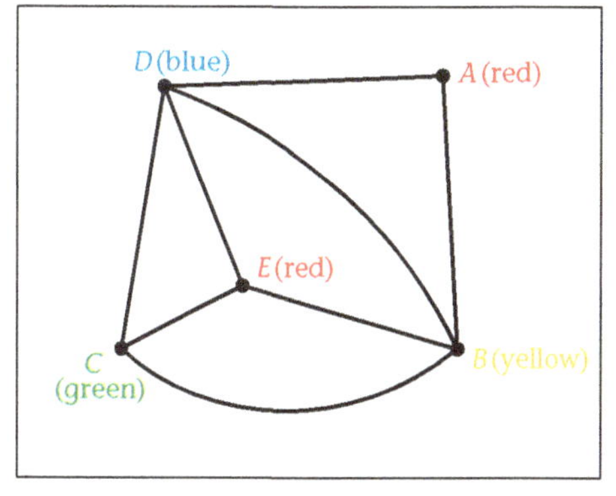

Four colors are necessary to color this map.

Exercises

1. Find the chromatic number for each of the following graphs.

a.

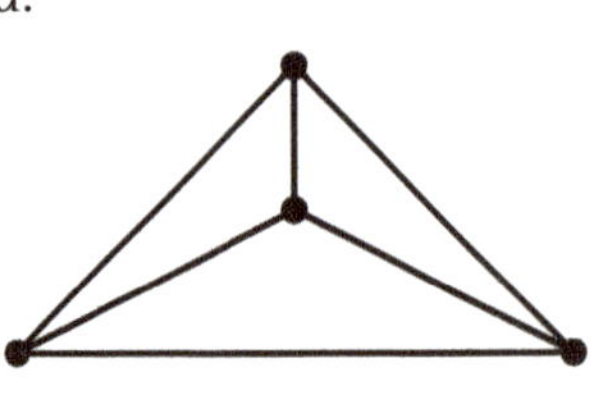

b.

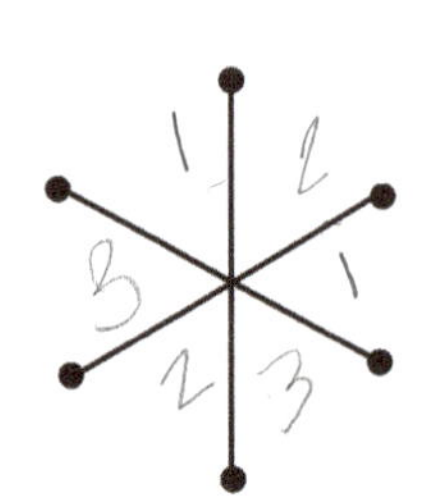

c.

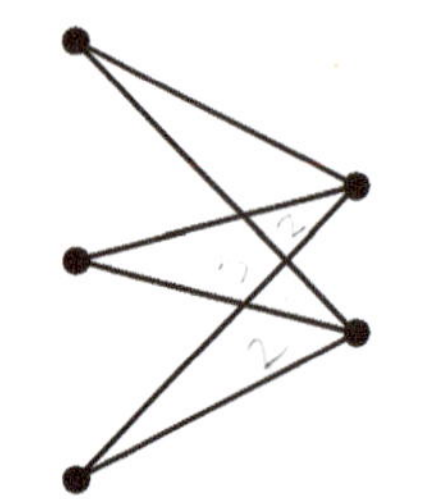

d.

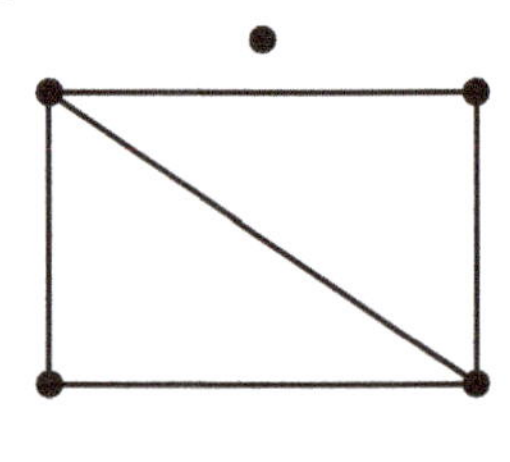

2. a. Draw a graph that has four vertices and a chromatic number of 3.

   b. Draw a graph that has four vertices and a chromatic number of 1.

3. As the number of vertices in a graph increases, a systematic method of labeling (coloring) the vertices becomes necessary. One way to do this is to create a coloring algorithm.

   a. It is possible to begin the coloring process in several different ways, but one way is to color first the vertices with the most conflict. How can the vertices be ranked from those with the most conflict to those with the least?

   b. After having colored the vertex with the most conflict, which other vertices can receive that same color?

   c. Then which vertex would get the second color? Which other vertices could get that same second color?

   d. When would the coloring process be complete?

   e. Refer back to parts a to d of this exercise and create an algorithm that colors a graph.

4. Use the algorithm that you developed in Exercise 3 to color the following graph. What is the chromatic number of the graph? Did your algorithm find the correct chromatic number?

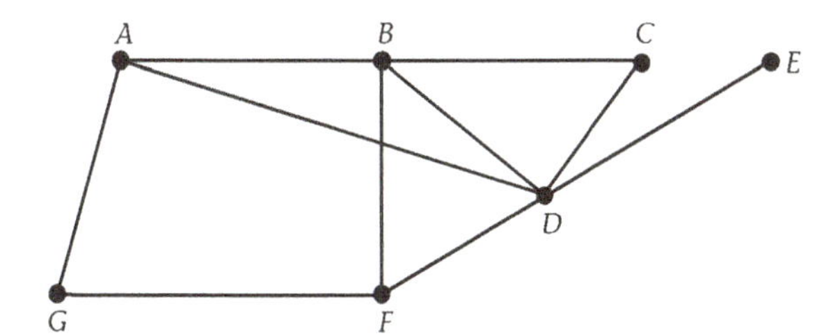

5. If a graph has a chromatic number of 1, what do you know?

6. a. What is the chromatic number of K_2? K_3? K_4? K_5?

   b. What can you say about the number of colors needed to color a complete graph? Explain your reasoning.

   c. Draw a complete graph with four vertices. Use your algorithm from Exercise 3 to color it. Does your algorithm give the correct chromatic number?

7. A **cycle** is a path that begins and ends at the same vertex and does not use any edge or vertex more than once.

   a. If a cycle has an even number of vertices, what is its chromatic number?

   b. What is the chromatic number of a cycle with an odd number of vertices?

   c. Draw a cycle with six vertices. Use your algorithm to color it. Does your algorithm give the correct chromatic number?

8. If your algorithm failed to give a minimal coloring for one of the graphs you tried, do not be too concerned, because it does not mean necessarily that you have a poor algorithm. Mathematicians continue to search for "good" coloring algorithms, but so far they have been unable to find one that colors every graph in the fewest number of colors possible. If you've not found a graph that causes your algorithm to fail, try to draw a graph that will do so.

9. Ms. Suzuki is planning to take her history class to the art museum. Following is a graph that models those students who are not compatible. Assuming that the seating capacity of the cars is not a problem, what is the minimum number of cars necessary to take the students to the museum?

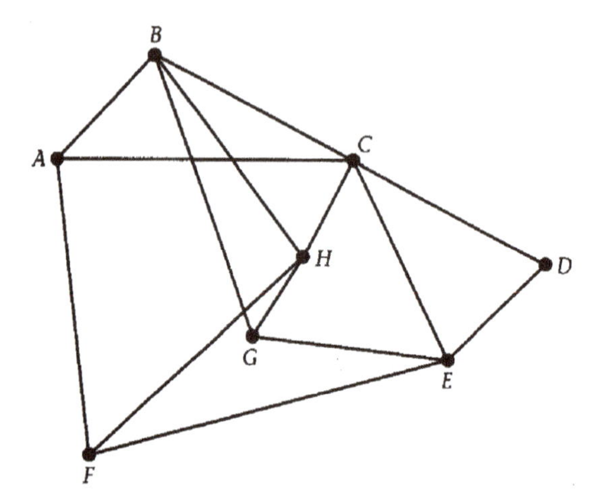

10. Refer to Exercise 1, Lesson 4.3, page 185. What is the minimum number of fish tanks needed to house the fish?

11. Following is a list of chemicals and the chemicals with which each cannot be stored.

Chemicals	Cannot Be Stored With
1	2, 5, 7
2	1, 3, 5
3	2, 4
4	3, 7
5	1, 2, 6, 7
6	5
7	1, 4, 5

How many different storage facilities are necessary in order to keep all seven chemicals?

12. Color the following map using only three colors.

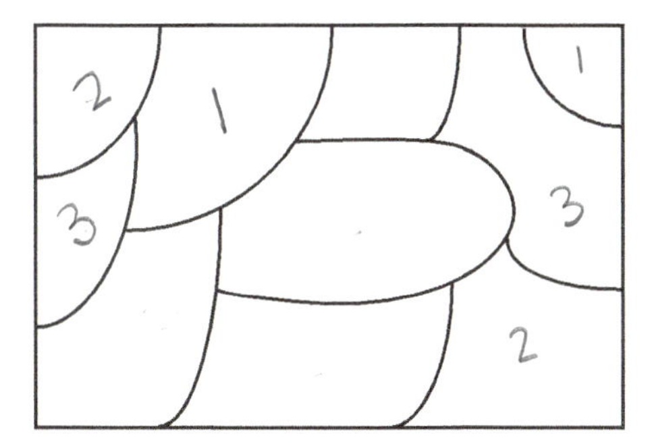

13. Draw graphs to represent the following maps. Color the graphs. What is the minimum number of colors needed to color each map?

a.

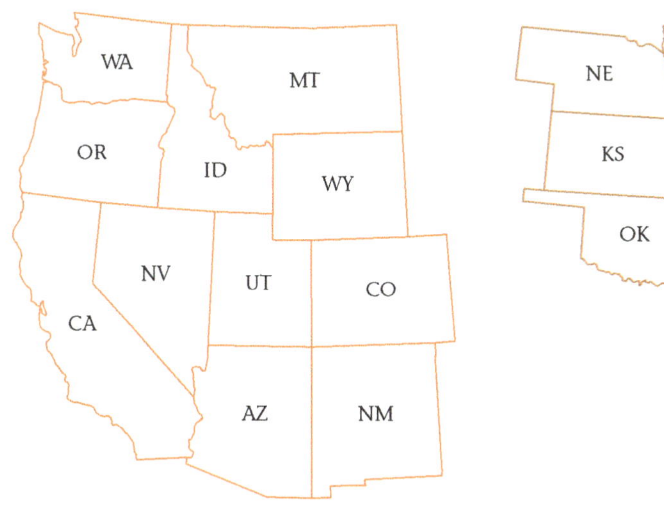

b.

Project

14. Research and report on the mapping industry. Explore questions such as: On the average, how often do maps change? Who are the people/agencies that create maps? How are maps designed and produced? What kind of educational background do cartographers need?

Chapter Extension

Eulerizing Graphs

Now that you've completed Chapter 4, you have the ability to examine a graph, to determine whether it has an Euler circuit, and to find a circuit if one exists. But, unfortunately, most graphs that represent real-world information do not have Euler circuits. So what does the city street inspector, garbage collector, and utility meter reader do when their street-by-street travels must be made in an optimal manner?

In Lesson 4.4, Exercise 8 (page 197), you solved a problem such as this. The street network graph for the exercise was small. By careful examination, you were able to find a way to inspect all streets and repeat only a minimal number of them.

This type of problem is often referred to as the Chinese Postman Problem. It was first studied by the Chinese mathematician Meigu Guan in 1962. The Chinese Postman Problem differs slightly from the situation that you solved in Lesson 4.4. In the postman problem, actual street lengths are examined and attempts are made to minimize the total retraced lengths. In Exercise 8, the assumption was made that all the streets were of equal length, so minimizing total length was equivalent to minimizing the number of reused edges.

Consider the following representation of a street network.

Since this graph has eight vertices of odd degree, there is no way to begin a circuit at *A*, trace each of the edges, and return to *A* without retracing some of the edges. One way to find a circuit that allows for the necessary retracing is to find the degree for each of the vertices. Connecting each odd vertex with an additional edge can then eliminate the vertices of odd degree. This process is called **eulerizing the graph**. Once the graph is eulerized, an Euler circuit can be found. The duplicate edges can then be viewed as streets that must be traveled more than once.

What is the "best" way to add the edges in the eulerization process? When adding edges, it is desirable to add ones that duplicate the fewest number of edges in the original graph. If an edge is added that spans more than one existing edge, then possibly a better eulerization can be found. Since only duplicates of edges in the original graph can be added, be careful not to add "new edges." See the following graphs.

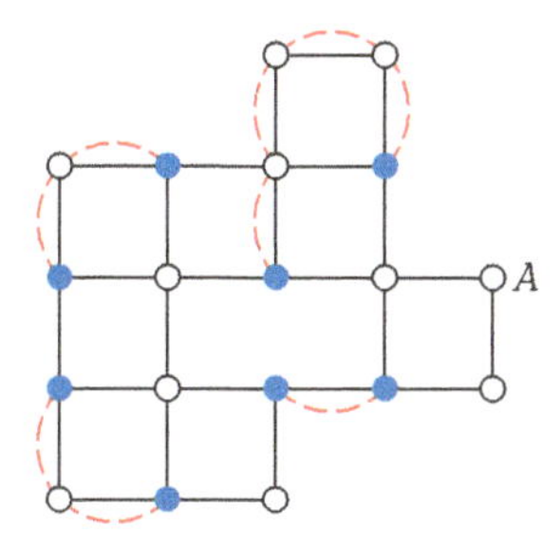

One possible eulerization with nine duplicate edges.

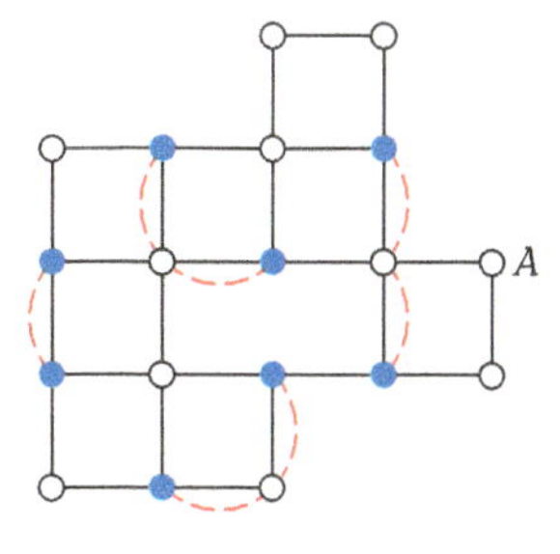

A better eulerization with seven duplicate edges.

As you might have guessed, this problem becomes much more difficult as the size of the graph and the number of vertices with odd degree increase. The following theorem gives a lower bound for the number of duplications that are needed when eulerizing a graph with n vertices.

If a graph has n vertices of odd degree, any circuit in the graph that covers every edge at least once must have at least $n/2$ duplications.

Be aware that this theorem gives only the lower bound and not the exact number of duplications. In other words, there can be no fewer than $n/2$ duplications. For example, in the preceding graph there are eight odd vertices, so at the very least there must be $8/2 = 4$ duplications. In the preceding figure an eulerization was found that has seven retraced edges. The theorem says that it might be possible to do better. Can you find a better eulerization?

If you are interested in knowing more about eulerizing graphs, additional information can be obtained by doing an Internet search or referring to the bibliography at the end of this chapter.

Chapter 4 Review

1. Write a summary of what you think are the important points of this chapter.

2. Draw a graph for the following task table.

Task	Time	Prerequisites
Start	0	—
A	2	None
B	4	*A*
C	4	*A*
D	3	*B*
E	2	*C*
F	1	*C*
G	2	*D*
H	5	*D, E, F*
I	3	*G, H*
Finish		

3. a. List the vertices of the following graph and give their earliest-start time.

   b. Determine the minimum project time.

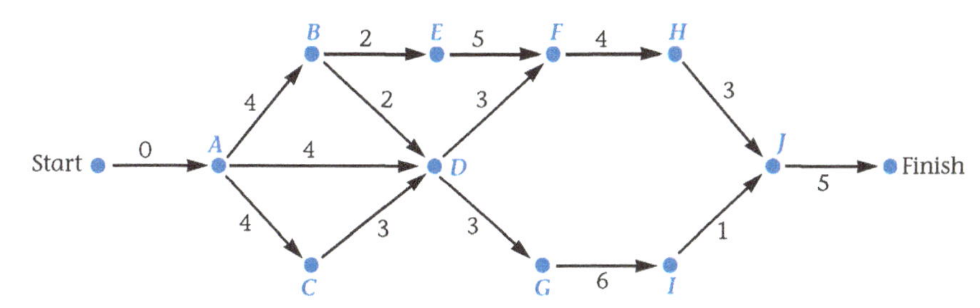

4. Complete the task table for the following graph.

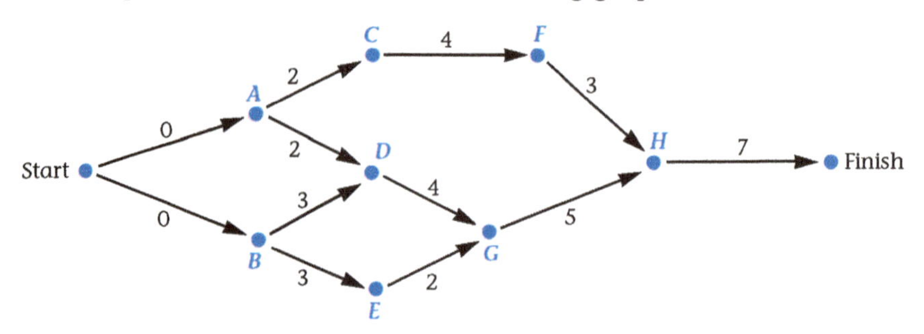

Task	Time	Prerequisites
Start	0	—
A		
B		
C		
D		
E		
F		
G		
H		
Finish		

5. Use your graph from Exercise 2.

   a. Recopy it and label each vertex with its EST.

   b. Determine the critical path and the minimum project time.

6.

Task	Time	Prerequisites
Start	0	—
A	6	None
B	9	None
C	8	None
D	5	*A*
E	4	B, *C*
F	10	*C*
G	5	*D, E*
H	7	*D*
I	8	*F*
Finish		

a. Draw a graph using the information in the table.

b. Label each vertex with its earliest-start time.

c. Determine the minimum project time.

d. Determine the critical path(s).

7.

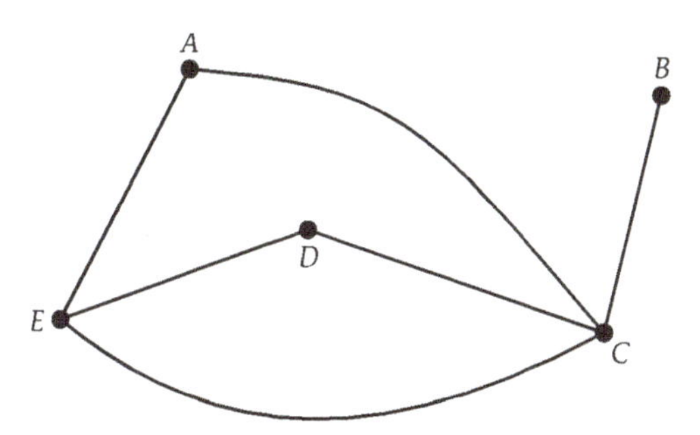

a. Is this graph connected? Explain why or why not.

b. Is this graph complete? Explain why or why not.

c. Name two vertices that are adjacent to vertex *E*.

d. Name a path from *B* to *E* of length 3.

e. What is the degree of vertex *C*?

f. Determine an adjacency matrix for the graph.

8. a. Draw a K_5 graph. Label the vertices E, F, G, H, and I.

   b. Deg(G) = ___?___.

   c. Does the graph have an Euler circuit? Explain.

9. Try to construct a graph with five vertices that is not connected so that two of the vertices have degree 3 and two of the vertices have degree 2. No loops or multiple edges may be used.

10. Tell whether the following graphs have an Euler circuit, an Euler path, or neither. Explain your answers.

    a.

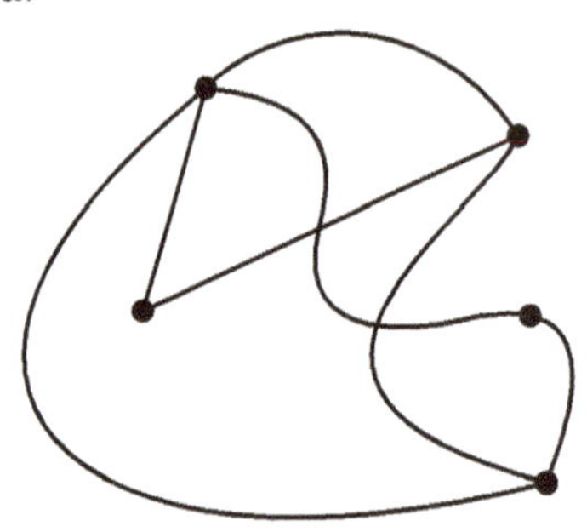

    b.

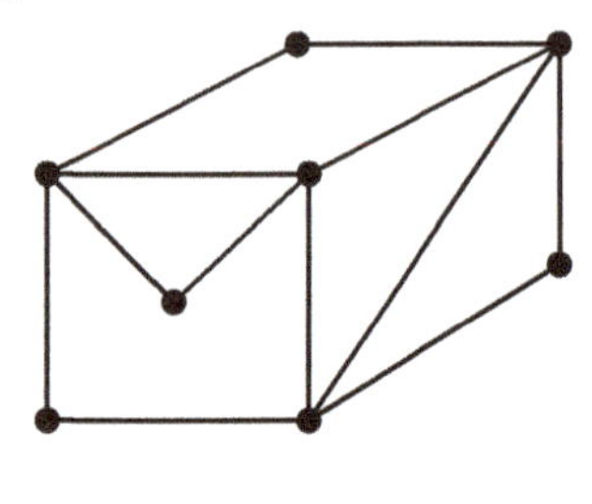

11. Construct a graph for each of the following.

    a. $V = \{A, B, C, D, E\}$

       $E = \{AE, AB, CD, BC, DE\}$

    b.

	A	B	C	D
A	0	0	1	1
B	0	0	1	1
C	1	1	0	1
D	1	1	1	0

12. Following is a multigraph that represents the downtown area of a small city. The local post office has decided that the mail drop boxes, which are located at the intersection of each street, must be monitored twice daily.

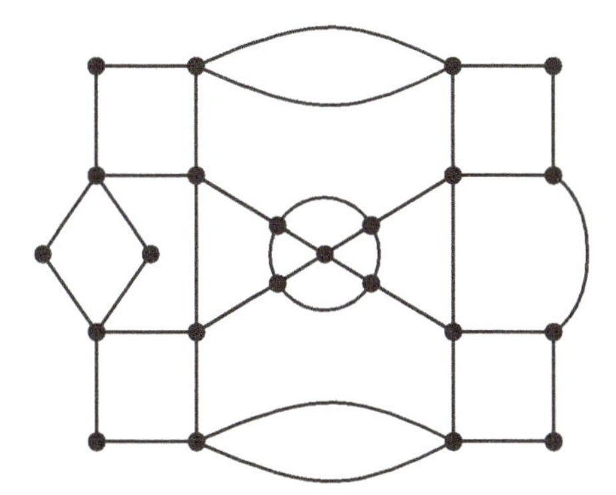

a. Is it possible to find a circuit that begins and ends at the same intersection and visits each drop box exactly once?

b. If not, is there a path that begins at one drop box, visits each drop box exactly once, and ends at a different drop box?

c. If either route exists, copy the figure onto your paper and darken the edges of your proposed route.

13. Use the graph in Exercise 12.

a. Is it possible for the local street inspector to begin at an intersection and inspect each street exactly once?

b. Is it possible for the inspector to finish her route at the same intersection from which she began? Explain why or why not.

14. Find an adjacency matrix for the following digraph.

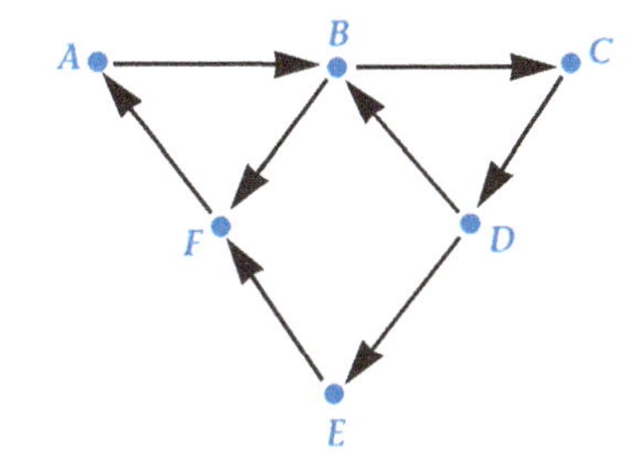

15. Consider the following set of preference schedules.

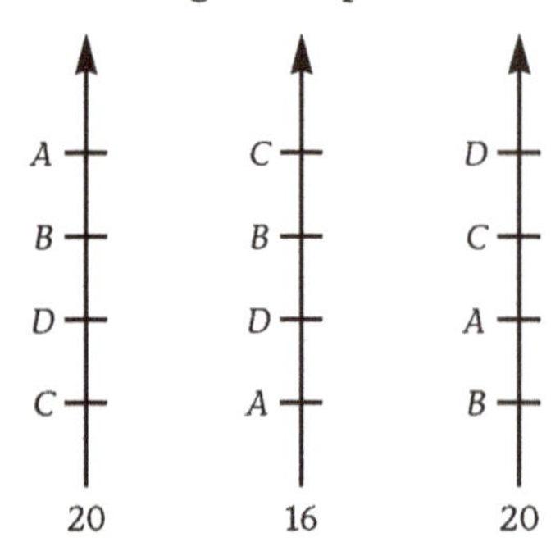

a. Represent this election with a cumulative preference tournament.

b. Is there a Condorcet winner? Explain why or why not.

c. Find several Hamiltonian paths for your graph.

d. Show how to use a Hamiltonian path to construct a pairwise voting scheme (see Exercise 3, Lesson 1.3, page 20) that results in B winning the election.

16. Draw a connected graph with four vertices that has a Hamiltonian circuit but no Euler circuits or paths.

17. In scheduling the final exam for summer school at Central High, six different tests must be scheduled. The following table shows the exams that are needed for seven different students.

	Students						
Exams	1	2	3	4	5	6	7
(*M*) Math	X	—	X	—	X	—	X
(*A*) Art	—	X	—	X	—	X	—
(*S*) Science	X	X	—	—	—	—	X
(*H*) History	—	—	X	—	—	X	—
(*F*) French	—	—	—	X	X	—	—
(*R*) Reading	X	X	—	X	X	—	X

a. Draw a graph that models which exams have students in common with other exams.

b. What is the minimum number of time slots needed to schedule the six exams?

18. The Federal Communications Commission (FCC) is in charge of assigning frequencies to radio stations so that broadcasts from one station do not interfere with broadcasts from other stations. Suppose the FCC needs to assign frequencies to eight stations. The following table shows which stations cannot share frequencies.

Station	Cannot Share with
A	*B*, *F*, *H*
B	*A*, *C*, *F*, *H*
C	*B*, *D*, *G*
D	*C*, *E*, *G*
E	*D*
F	*A*, *B*, *H*
G	*C*, *D*
H	*A*, *B*, *F*

a. Model this situation with a graph.

b. Find the minimum number of frequencies needed by the FCC.

19. Consider the following digraph.

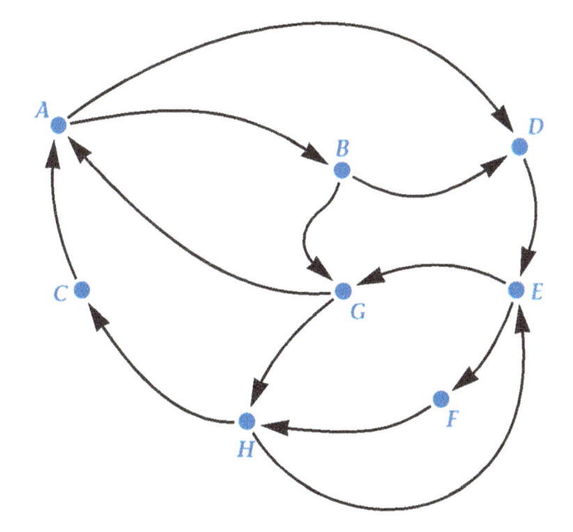

a. Does it have a directed Euler circuit? Explain why or why not. If it does, list one.

b. Does it have a directed Euler path? Explain why or why not. If it does, list one.

20. a. Represent the following map with a graph.

    b. Color your graph.

    c. What is the minimum number of colors needed to color the map?

Bibliography

Biggs, N. L., E. K. Lloyd, and R. J. Wilson. 1998. *Graph Theory 1736–1936.* Oxford: Clarendon Press.

Busch, D. 1991. *The New Critical Path Method.* Chicago: Probus.

Chavey, Darrah. 1992. *Drawing Pictures with One Line.* (HistoMAP Module 21). Lexington, MA: COMAP, Inc.

COMAP, Inc. 2013. *For All Practical Purposes: Mathematical Literacy in Today's World.* 9th ed. New York: W. H. Freeman.

Copes, W., C. Sloyer, R. Stark, and W. Sacco. 1987. *Graph Theory: Euler's Rich Legacy.* Providence, RI: Janson.

Cozzens, Margaret B., and R. Porter. 1987. *Mathematics and Its Applications.* Lexington, MA: D. C. Heath and Company.

Cozzens, Margaret B., and R. Porter. 1987. *Problem Solving Using Graphs.* (HiMAP Module 6). Lexington, MA: COMAP, Inc.

Dossey, John, A. Otto, L. Spence, and C. Vanden Eynden. 2006. *Discrete Mathematics.* 5th ed. Upper Saddle River, NJ: Pearson.

Francis, Richard L. 1989. *The Mathematicians' Coloring Book.* (HiMAP Module #10). Lexington, MA: COMAP, Inc.

Kenney, Margaret J., ed. 1991. "Discrete Mathematics across the Curriculum, K–12." *1991 Yearbook of the National Council of Teachers of Mathematics.* Reston, VA: NCTM.

Malkevitch, J., and W. Meyer. 1974. *Graphs, Models, and Finite Mathematics.* Englewood Cliffs, NJ: Prentice Hall.

Tannenbaum, P., and R. Arnold. 2014. *Excursions in Modern Mathematics.* 8th ed. Upper Saddle River, NJ: Pearson.

Wilson, Robin. 2003. *Four Colors Suffice.* Princeton, NJ: Princeton University Press.

CHAPTER 5

Modeling with Subgraphs and Trees

Elaborate communication networks span the country and most of the earth. The ability to transmit information quickly and easily through these networks affects many aspects of our lives—the way we work, the way we learn, and the way in which we are entertained.

- How are communication networks that link several locations together constructed at the least possible cost?
- How is the most efficient route between two locations in a network found?
- Can the models used to find the best route between points in a communication network also be used to plan the best route for an automobile or plane trip?

The mathematics of graph theory plays an important role in solving these and many other problems that are important in our ever-changing world.

Lesson 5.1

Planarity and Coloring

In Lesson 4.6, problems involving conflict were solved by modeling them with graphs and then coloring the graphs. The four-color theorem states that any map that can be drawn on the surface of a sphere can be colored with four colors or fewer. If this is true, then why does it take more than four colors to color some graphs?

Explore This

Try to redraw the graphs in Figures 5.1 and 5.2 so that their edges intersect only at the vertices. Try to think of the edges of the graph as rubber bands.

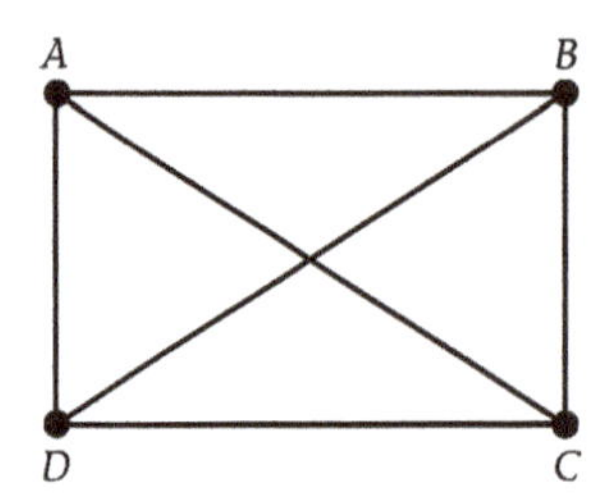

Figure 5.1. K_4 graph.

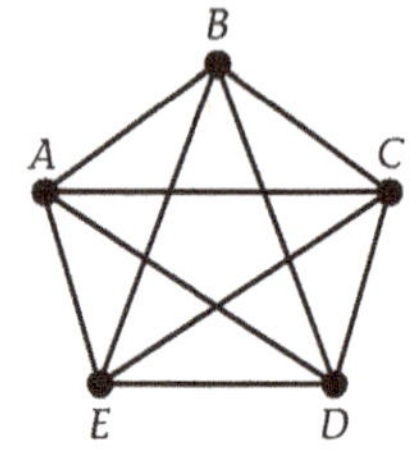

Figure 5.2. K_5 graph.

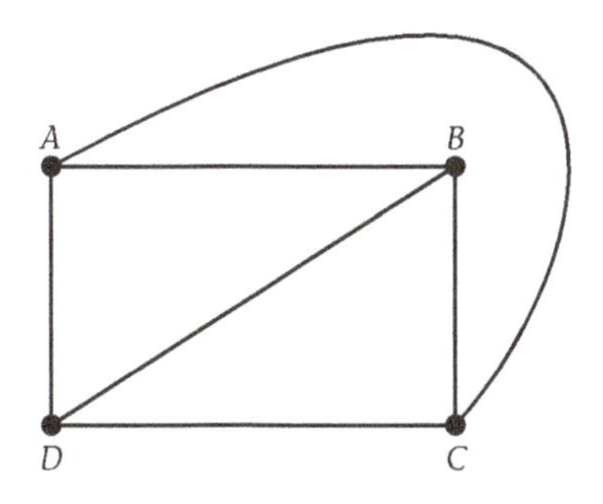

Figure 5.3. Graph in Figure 5.1 redrawn with edges not intersecting.

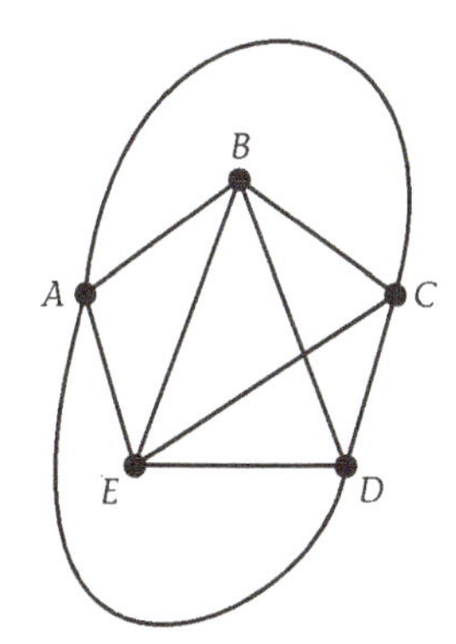

Figure 5.4. An attempt to redraw Figure 5.2 with edges not intersecting.

It is relatively easy to redraw Figure 5.1 so that the edges do not cross (see Figure 5.3). But no matter how hard you try, at least two edges of Figure 5.2 will always intersect (see Figure 5.4). A graph that can be drawn so that no two edges intersect except at a vertex is called a **planar graph**. Figure 5.1 shows a planar graph. Figure 5.2 shows a graph that is not planar.

If regions of a map are represented by vertices of a graph and edges are drawn between vertices when boundaries exist between regions, the resulting graph is planar. In other words, when a map in a plane or on a sphere is modeled by a graph, the resulting graph is always planar.

Hence, the four-color theorem can be stated in a different way:

Every planar graph has a chromatic number that is less than or equal to four.

The question asked earlier about why some graphs require more than four colors can now be answered. Planarity is the key. If a graph is not planar, we do not know how many colors it will take to color it.

One type of graph that is not planar, a K_5, is shown in Figure 5.2. Another nonplanar graph about which many problems have been written is shown in Figure 5.5. Try to redraw it without the edges crossing. Once again you will discover that this is not possible.

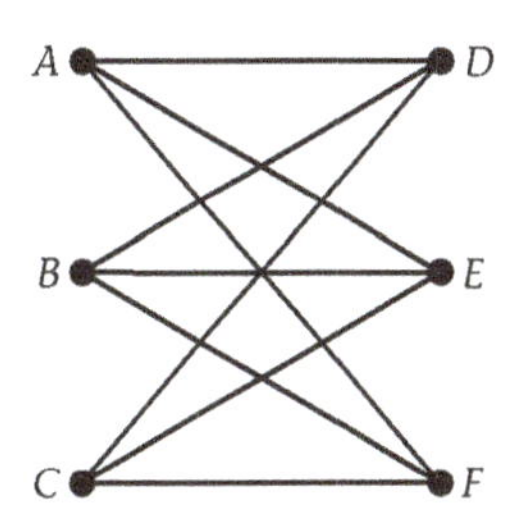

Figure 5.5 $K_{3,3}$ graph.

Bipartite Graphs

The graph in Figure 5.5 has interesting characteristics other than the fact that it is not planar. It is one example of a group of graphs known as **bipartite** graphs. A graph is bipartite if its vertices can be divided into two distinct sets so that each edge of the graph has one vertex in each set.

A bipartite graph is said to be complete if it contains all possible edges between the pairs of vertices in the two distinct sets. Complete bipartite graphs can be denoted by $K_{m,n}$, where m and n are the number of vertices in the two distinct sets. Hence, Figure 5.5 is a $K_{3,3}$ graph.

Figure 5.6 is an example of a complete bipartite graph $K_{3,2}$, since its vertices can be separated into two distinct sets $\{A, B, C\}$ and $\{X, Y\}$, every edge has one vertex in each set, and all possible edges from one set to the other are drawn.

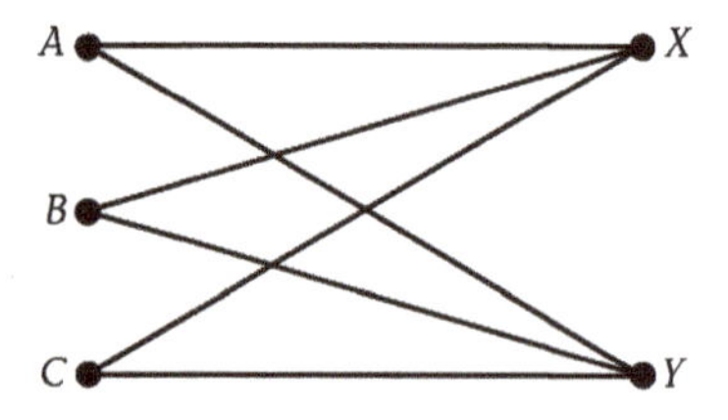

Figure 5.6. $K_{3,2}$ graph.

One way to determine whether a given graph is planar is to try to redraw the graph without edges crossing. For a very large graph, this could prove to be both difficult and time-consuming. In 1930 Kazimierz Kuratowski, a Polish mathematician, provided a partial resolution to this problem of determining the planarity of a graph. He proved that if a graph has a K_5 or $K_{3,3}$ subgraph, it is not planar. A graph G' is said to be a **subgraph** of graph G if all of the vertices and edges of G' are contained in G

Point of Interest

In practice, approximately 99% of all nonplanar graphs of modest size can be shown to be nonplanar because of a $K_{3,3}$ subgraph rather than a K_5 subgraph.

In addition to proving that graphs with K_5 or $K_{3,3}$ subgraphs are not planar, Kuratowski proved that the inverse of his theorem is not true. That is, the lack of a K_5 or $K_{3,3}$ subgraph does not guarantee that the graph is planar (see Exercises 23 and 24 on pages 239 and 240).

Example

Determine whether the following graph is planar or nonplanar.

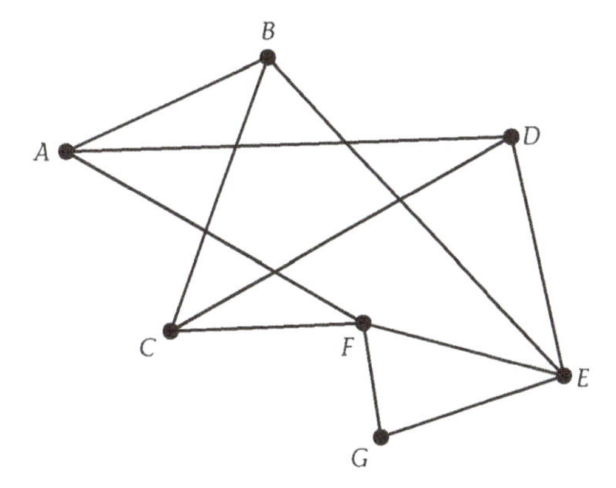

Solution:

On close inspection of the graph and vertices, *A, B, C, D, E,* and *F,* a $K_{3,3}$ subgraph can be found. Therefore, according to Kuratowski's theorem, the graph is nonplanar.

Exercises

In Exercises 1 through 4, decide whether the graph is planar or nonplanar. If the graph is planar, redraw it without edge crossings.

1.

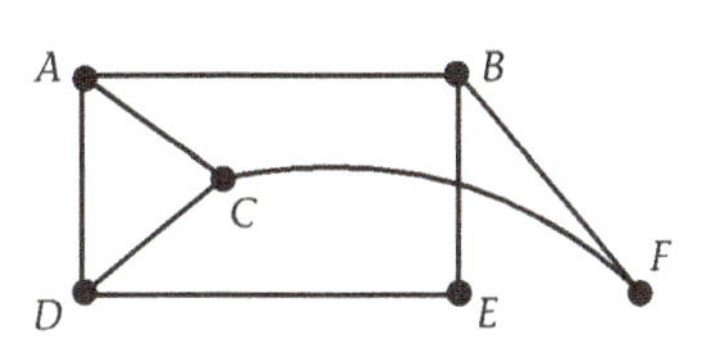

2.

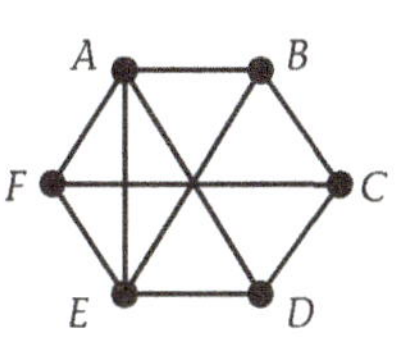

3.

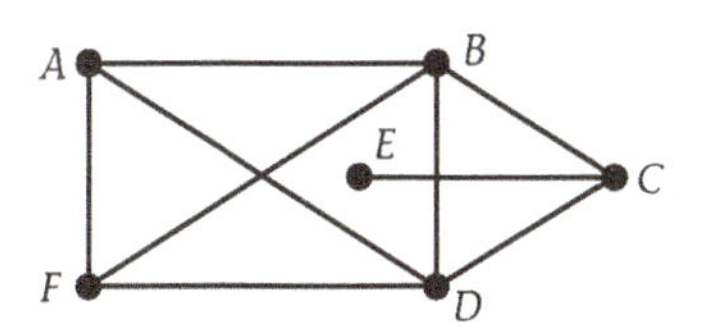

4.

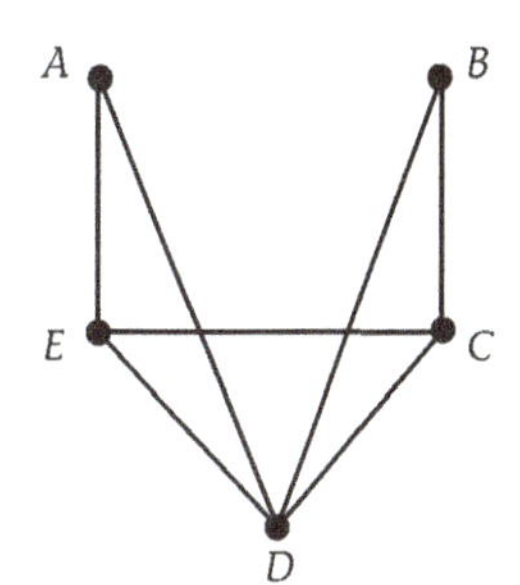

5. The following graph is planar. Draw it without edge crossings.

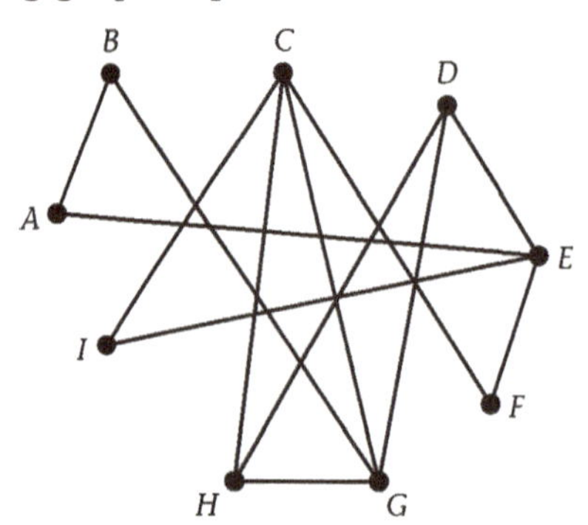

6. By looking at the graph in Exercise 5, how can you tell that it does not contain a K_5 subgraph?

7. Devise a systematic method of searching a graph for a K_5 subgraph. Describe your method in a short paragraph and try it on the following graph. Does the graph contain a K_5 subgraph?

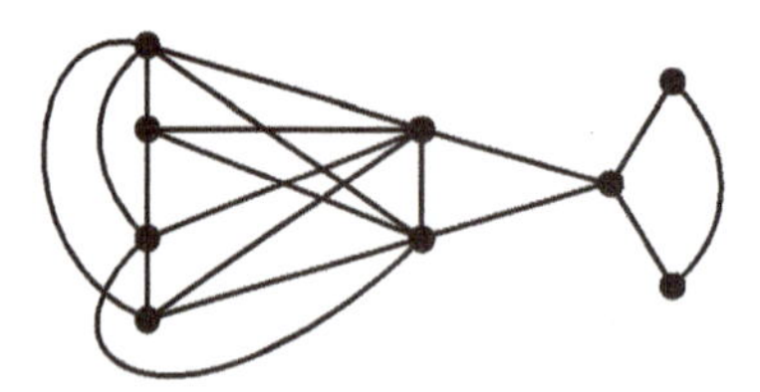

8. The **complement** of a graph G is customarily denoted by $\bar{G}$. The complement $\bar{G}$ has the same vertices as G, but its edges are those not in G. The edges of G and $\bar{G}$ along with vertices from either set would make a complete graph. Draw the complement of the following graph.

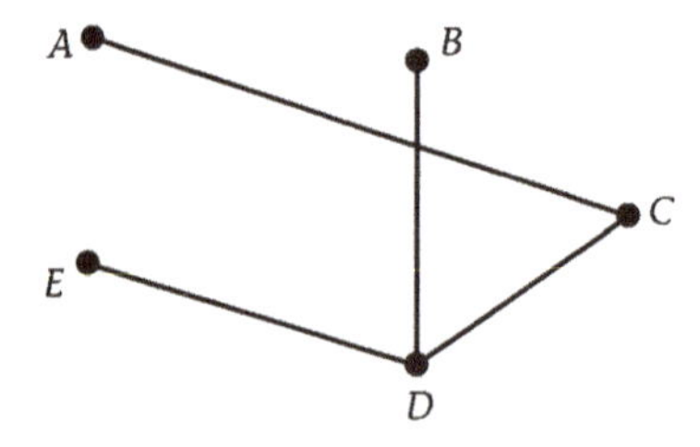

9. Every planar graph with nine vertices has a nonplanar complement. Verify this statement for one case by drawing a planar graph with nine vertices and then drawing its complement. For your case, is the complement nonplanar?

10. The concept of planarity is extremely important to printing circuit boards for the electronics industry. Explain why.

11. Construct the following bipartite graphs.

 a. $K_{2,3}$ b. $K_{2,4}$

12. For each of the following bipartite graphs, list the two distinct sets into which the vertices can be divided.

 a.

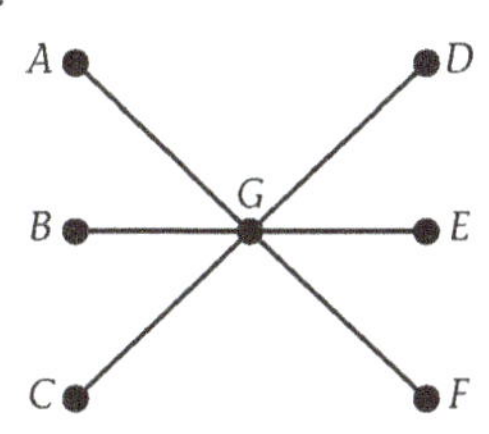

 b.

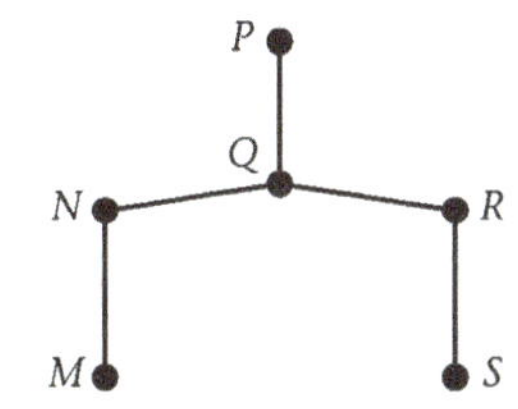

 c.

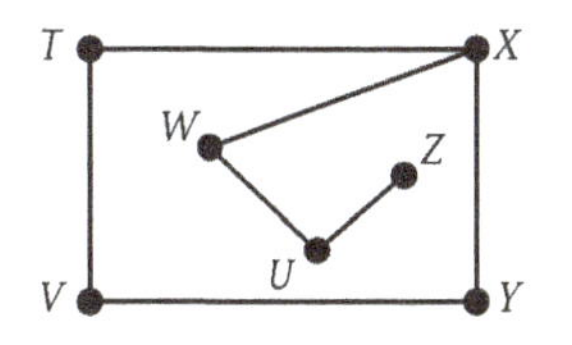

 d.

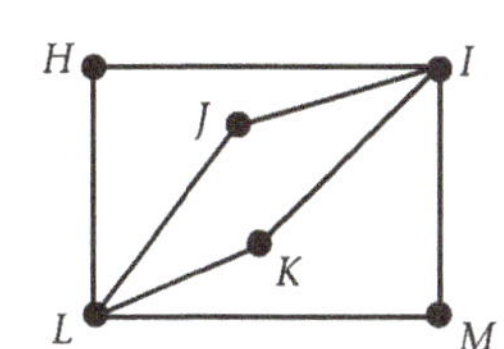

13. State whether the following graphs are bipartite. Explain why or why not.

 a.

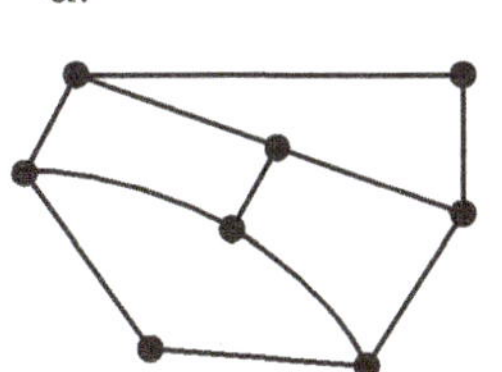

 b.

 c.

14. Devise a method of telling whether a graph is bipartite. Write a short algorithm for your method.

15. How many edges are in a $K_{2,3}$ graph? A $K_{4,3}$ graph? A $K_{m,n}$ graph?

16. When does a bipartite graph $K_{m,n}$ have an Euler circuit?

17. What is the chromatic number of a $K_{m,n}$ graph?

18. At Ms. Johnson's party, six men and five women walk into the dining room. If each man shakes hands with each woman, how many handshakes will occur? Represent this situation with a graph. What kind of a graph is it?

19. Describe a situation that can be represented by a bipartite graph that is not complete.

20. The following puzzle is often referred to as the *Wells and Houses problem* or the *Utilities problem*.

    Three houses and three wells are built on a piece of land in an arid country. Because it seldom rains, the wells often run dry, and so each house must have access to each well. Unfortunately, the occupants of the three houses dislike one another and want to construct paths to the wells so that no two paths cross.

    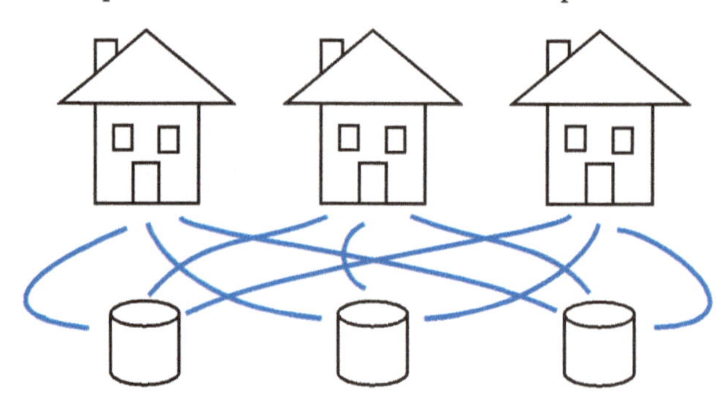

    Draw a graph to model this problem. Is it possible to satisfy the wishes of the feuding families? Explain why or why not.

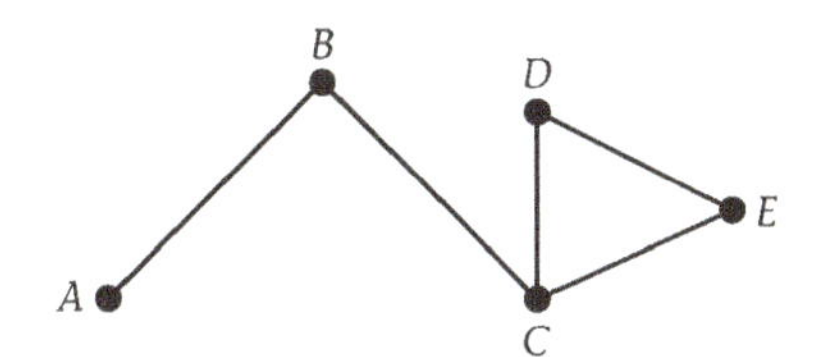

21. The preceding graph is a subgraph of which following graph(s)? Explain how you know.

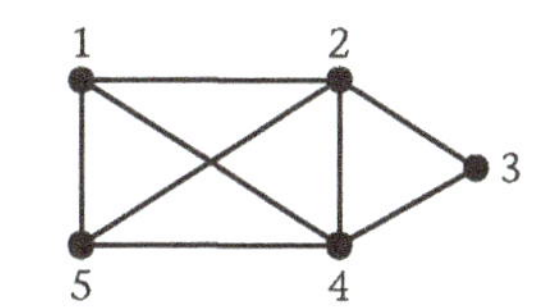

a.

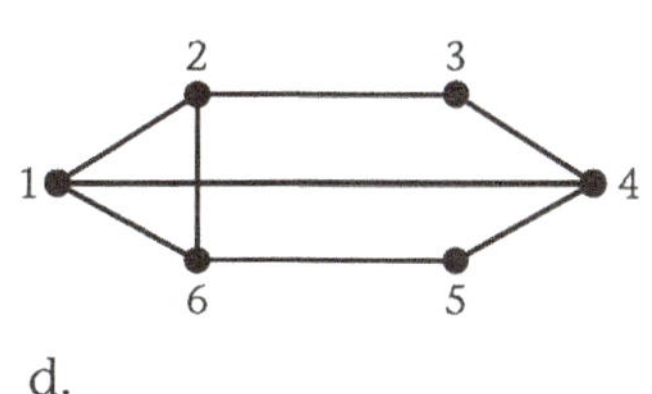

b.

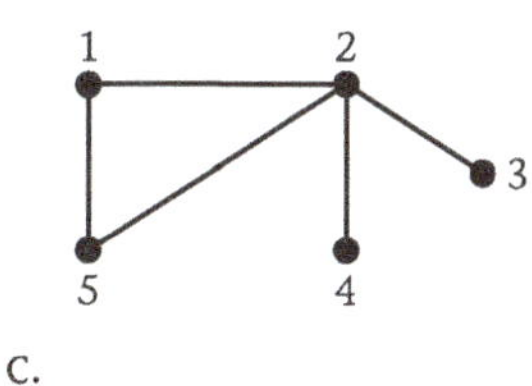

c.

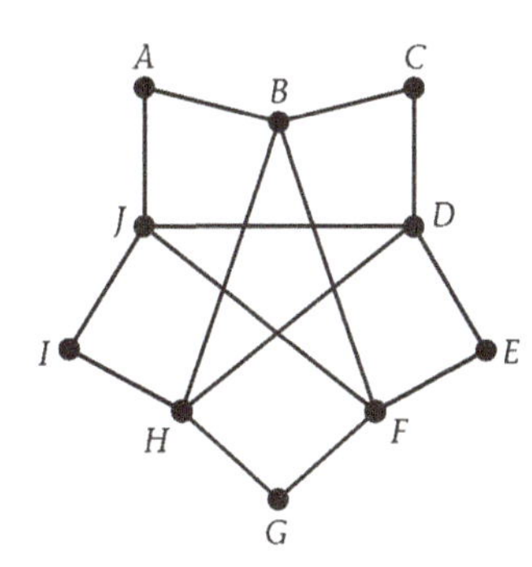

d.

22. Kuratowski proved the following conditional statement.

    If a graph contains a K_5 or a $K_{3,3}$ subgraph, then it is not planar.

    He also proved that the inverse of the statement is false:

    If a graph does not contain a K_5 or a $K_{3,3}$ subgraph, then it is planar.

    State the converse of the conditional statement. Do you think that it is true or false?

23. Is the following graph planar? Does it contain a K_5 or a $K_{3,3}$ subgraph?

24. For many years mathematicians thought that all nonplanar graphs had either a K_5 or a $K_{3,3}$ subgraph. Then nonplanar graphs such as the one in Exercise 23 were found. The graph in Exercise 23 is said to be an **extension** of a K_5 because it was formed by adding a vertex or vertices to the edges of the K_5 graph. Extensions of K_5 and $K_{3,3}$ graphs are nonplanar. This discovery shows that the converse of Kuratowski's theorem is false (Exercise 22).

 Redraw the graph in Exercise 23 to show that it is an extension of a K_5.

Lesson 5.2

The Traveling Salesperson Problem

In Lesson 4.5, you explored circuits that visited each vertex of your graph exactly once (Hamiltonian circuits). In this lesson, you will extend your thinking on this original problem and examine a type of problem known as a **traveling salesperson problem** (TSP). These TSPs involve finding a Hamiltonian circuit of minimum value such as time, distance, or cost. Optimization problems of this type are becoming increasingly important in the world of communications, warehousing, airlines networking, delivery truck routing, and building wiring.

Explore This

Suppose you are a salesperson who lives in St. Louis. Once a week you have to travel to Minneapolis, Chicago, and New Orleans and then return home to St. Louis. The graph in Figure 5.7 represents the trips that are available to you. The edges of the graph are labeled with the cost of each possible trip. For instance, the cost of making a trip from Chicago to New Orleans is $910. When each edge of a graph is assigned a number (weight), the graph is called a **weighted graph**. The numbers associated with the edges often represent such things as distance, cost, or time.

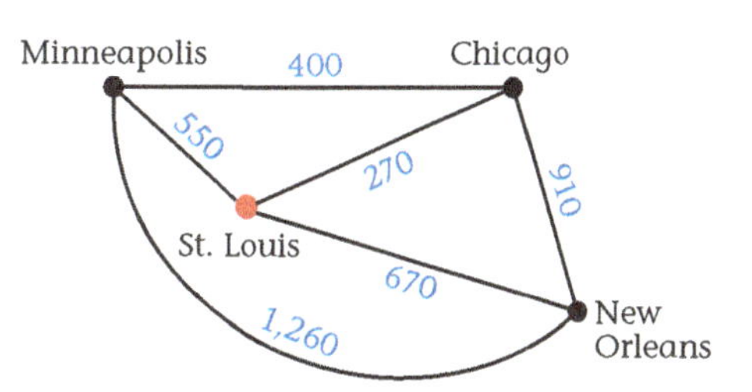

Figure 5.7. Graph of four cities with the costs of traveling between them.

Since you own your own business, it is important that you minimize travel costs. To help save money, find the least expensive route that begins in St. Louis, visits each of the other cities exactly once, and returns to St. Louis.

One way to solve the problem of finding the least expensive route in Figure 5.7 is to use the **brute force method** of listing every possible circuit, along with its cost. A tree diagram like the one in Figure 5.8 can help you list the possible routes. Inspection of the tree diagram shows that the optimal solution is the circuit *S, M, C, N, S* or the circuit in reverse order *S, N, C, M, S*.

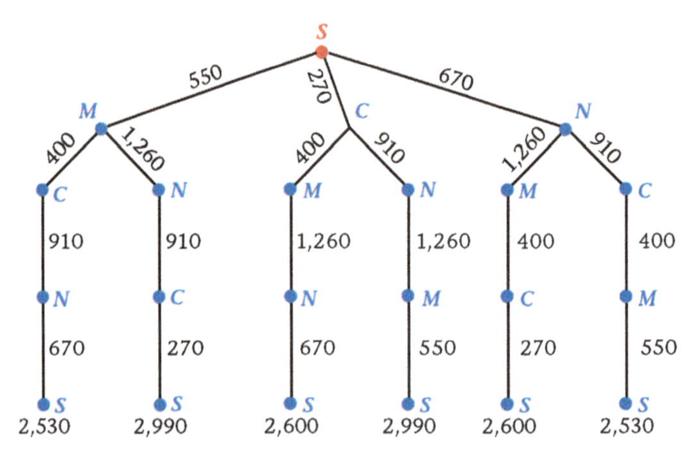

Figure 5.8. Tree diagram of every possible circuit from St. Louis to each of the other cities and back to St. Louis.

Solving the TSP for four vertices is not too difficult or time-consuming because only six possibilities need to be considered. But as the number of vertices increases, so does the number of possible circuits. Hence, checking all possibilities soon becomes impractical, if not impossible. Even with the help of a computer that can do computations at the rate of 1 billion per second, it would take more than 19 million years for the computer to find the weights of every circuit for a graph with 25 vertices!

More Efficient Algorithms

Is it possible that a faster, more efficient algorithm exists? You might try, for example, to begin at a vertex, look for the vertex nearest to your starting vertex, move to it, and continue until you complete the circuit.

To explore this method for the graph in Figure 5.7 on page 241, begin at St. Louis, move to the nearest neighboring vertex (Chicago), then to the nearest vertex not yet visited, and return to St. Louis when you have visited all of the other cities (270 + 400 + 1,260 + 670 = 2,600). This procedure is known as the **nearest-neighbor algorithm**. Although the solution was reached quickly, notice that in this case, it is not the best possible one. A method such as this, which produces a quick and reasonably close-to-optimal solution, is known as a **heuristic method**.

The choice of method now becomes a trade-off. The brute force method of inspecting every possible path guarantees the best route but is prohibitively slow. The nearest-neighbor method is quick but does not necessarily produce the optimal solution. There is no known computationally efficient method of solving all traveling salesperson problems. But with the discovery of more and more efficient algorithms, hope has increased that better solutions will be forthcoming. These solutions to TSPs are of great interest because they translate into savings of millions of dollars for certain areas of the economy.

Point of Interest

The current record-setting solution for a TSP was set in 2006 when an optimal 85,900-city tour was found. In this problem, the cities represented locations on a customized computer chip, and finding the shortest possible movement of the laser equated to minimizing the cost production cost of the chip.

The table below shows several TSP milestones.

TSP MILESTONES

Year	Number of Cities
1954	49
1971	64
1975	80
1977	318
1987	2,392
1992	3,038
1998	13,509
2004	24,978
2006	85,900

According to William J. Cook, the ultimate travel challenge is currently a 1,904,771-city problem that consists of every town, city, and village in the world.

Exercises

In Exercises 1 through 4,

a. Construct a tree diagram showing all possible circuits that begin at *S*, visit each vertex of the graph exactly once, and end at *S*.

b. Find the total weight of each route.

c. Identify the shortest circuit.

d. Use the nearest-neighbor algorithm to find the shortest circuit.

e. Does the nearest-neighbor algorithm produce the optimal solution?

1.

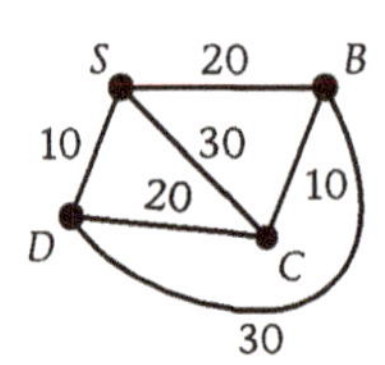

2.

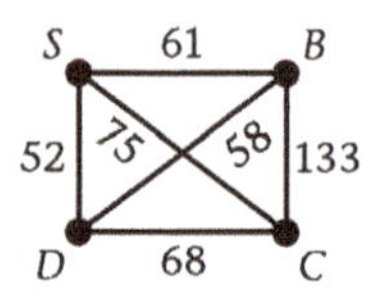

3.

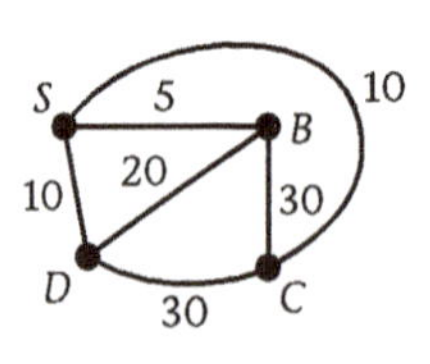

4.

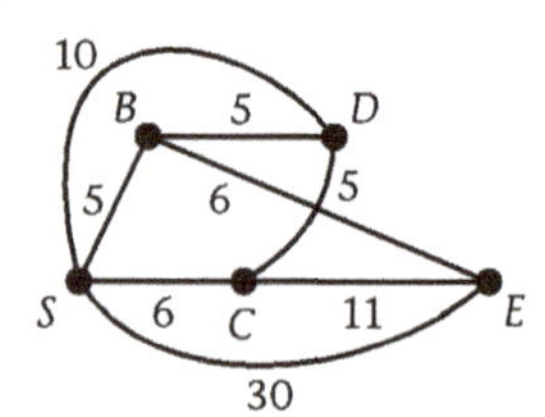

5. In a graph with 10 vertices, 9!, or $9 \cdot 8 \cdot 7 \cdot 6 \cdot 5 \cdot 4 \cdot 3 \cdot 2 \cdot 1$, possible Hamiltonian circuits exist if the beginning vertex is known.

   a. Assume that a computer can perform calculations at the rate of 1 million per second. About how long will it take the computer to check 9! possibilities? What if the graph had 15 vertices (14! possible circuits)?

   b. According to a September 29, 2004 news report, the IBM BlueGene/L computer system can sustain speeds of 36 teraflops—36 trillion computations per second! How long will it take this new system to check a graph with 9! possible circuits? With 14! possible circuits? With 20! possible circuits?

6. Give two examples of a situation in which a solution to the traveling salesperson problem would be beneficial.

7. The following figure shows a circuit board and the distances in millimeters between holes that must be drilled by a drilling machine. Since it is advantageous in terms of time to minimize the distance traveled, find the shortest possible circuit for the machine to travel and the total distance for that circuit. (Assume the machine has to begin and end at point *S*.)

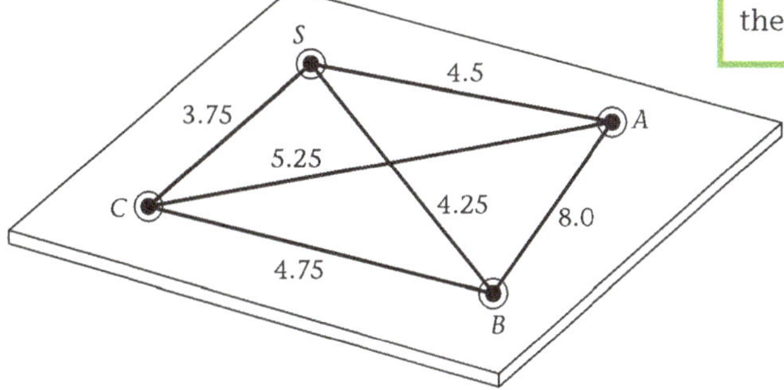

Point of Interest

Mathematicians in IBM's Tokyo Research Laboratory have developed a drill route optimization system that has a TSP approximation algorithm as its core. The new system reduces the drill route length on a circuit board by an average of 80% and the operation time by 15% on the average.

8. A delivery person must visit each of his warehouses daily. His delivery route begins and ends at his garage (G). The table below shows the approximate travel time (in minutes) between stops.

	G	A	B	C	D
G	—	25.0	16.5	18.0	43.0
A	25.0	—	21.0	17.5	15.0
B	16.5	21.0	—	23.5	19.5
C	18.0	17.5	23.5	—	21.0
D	43.0	15.0	19.5	21.0	—

a. Draw a weighted complete graph with 5 vertices to represent the information in the table.

b. Use the nearest-neighbor algorithm to find the quickest route starting at G, visiting each of the vertices of the graph exactly once, and ending at G.

c. What is the travel time of the route in part b?

Projects

9. An algorithm, known by such names as the **sorted-edges algorithm** and the **cheapest-link algorithm**, provides another quick and simple way to find an approximate solution to the TSP. Investigate this algorithm. Write a brief report on the algorithm and give an example to show how it works.

10. Create a TSP poster that illustrates the use of the nearest-neighbor algorithm. At a minimum, your poster should include the following elements.

    - A map of your neighborhood, town, or state with five or six selected locations highlighted.
    - A table that shows the distances between each pair of selected locations.
    - A weighted graph that models the situation.
    - An explanation of how the algorithm is used to find the minimum distance from one selected starting location to each of the others and to return to the original starting point.

11. Select five colleges or universities that you might be interested in attending. After researching the schools of choice, write a paper that incorporates the following.

    - A short paragraph on each school explaining your interest.
    - A weighted, complete graph that shows the five schools and your home.
    - An optimal route based on the shortest path that starts at your home, visits each of the five schools, and returns to your home.

Lesson 5.3

Finding the Shortest Route

The traveling salesperson problem asks that a Hamiltonian circuit of least total weight be found for a graph. What if you didn't need to visit every vertex in the graph and return back to the starting point, but instead you needed only to find the shortest path from one vertex in the graph to another?

Does an efficient method of solving this type of problem exist? The answer is yes, and one algorithm used in finding the shortest path from a given vertex of a graph to any other vertex in that graph is attributed to E. W. Dijkstra.

Mathematician of Note

Edsger W. Dijkstra (1930–2002)

Professor Dijkstra is considered one of the original theorists of modern computer science. He is well known for his amazingly efficient shortest path algorithm that was first published in 1959. Throughout his career, Dijkstra received many honors and awards. He held the Schlumberger Centennial Chair in Computing Sciences at the University of Texas at Austin, 1984–1999, and retired as Professor Emeritus in 1999.

The following algorithm is a modification of Dijkstra's algorithm.

Shortest Path Algorithm

1. Label the starting vertex S and circle it. Examine all edges that have S as an endpoint. Darken the edge with the shortest length and circle the vertex at the other endpoint of the darkened edge.
2. Examine all uncircled vertices that are adjacent to the circled vertices in the graph.
3. Using only circled vertices and darkened edges between the vertices that are circled, find the lengths of all paths from S to each vertex being examined. Choose the vertex and the edge that yield the shortest path. Circle this vertex and darken this edge. Ties are broken arbitrarily.
4. Repeat steps 2 and 3 until all vertices are circled. The darkened edges of the graph form the shortest routes from S to every other vertex in the graph.

Example

Use the shortest path algorithm to find the shortest path from A to F in the graph.

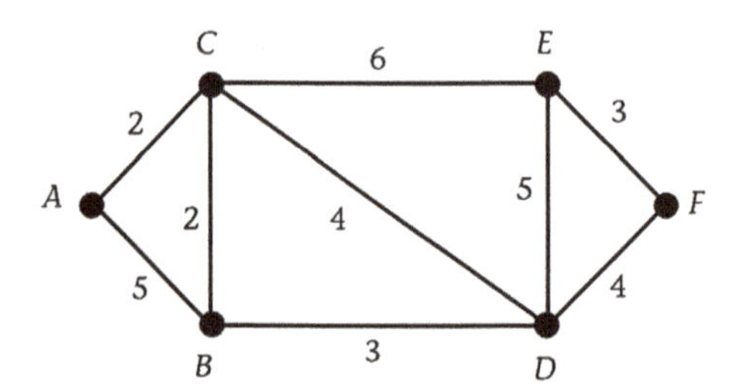

Solution:

Begin by labeling vertex A with an S. Circle the S. Then examine all vertices that are adjacent to S.

	Adjacent Vertices	Path from S to Vertex	Length of Path
Adjacent to S	B	SB	5
	C	SC	2
1. Circle C, darken edge SC.			
Adjacent to S	B	SB	5
Adjacent to C	B	SCB	4
	E	SCE	8
	D	SCD	6
2. Circle B, darken edge CB.			
Adjacent to C	E	SCE	8
	D	SCD	6
Adjacent to B	D	$SCBD$	7
3. Circle D, darken edge CD.			
Adjacent to C	E	SCE	8
Adjacent to D	E	$SCDE$	11
	F	$SCDF$	10
4. Circle E, darken edge CE.			
Adjacent to E	F	$SCEF$	11
Adjacent to D	F	$SCDF$	10
5. Circle F, darken edge DF.			

The shortest route from A to F is A,C,D,F, and the length is 10. The darkened edges also show the shortest routes from A to the other vertices in the graph.

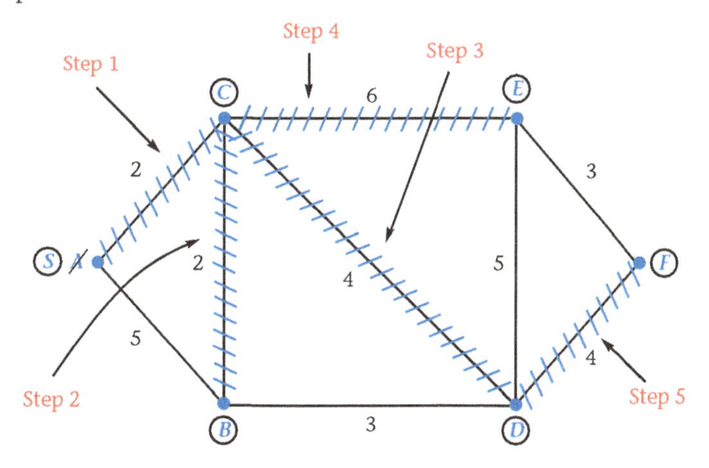

Exercises

1.

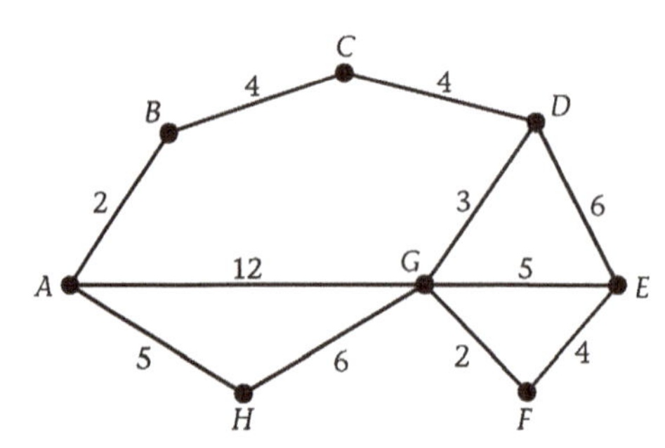

Julian began using the shortest path algorithm to find the shortest route from *A* to *E* for the preceding graph. The work that he was able to complete before he had to stop is shown here.

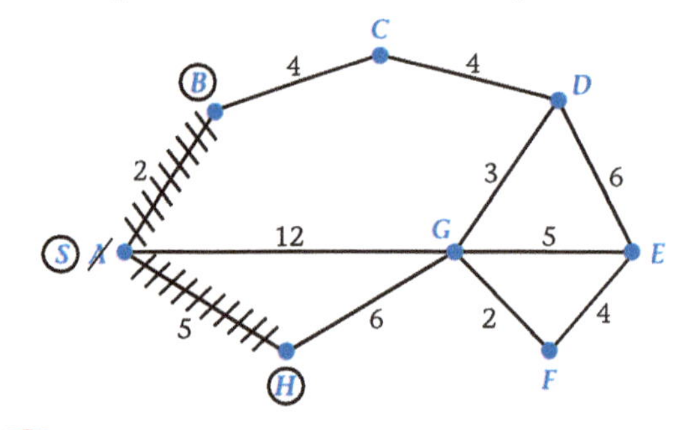

1. *SB* – 2
   SC – 5   Circle *B*, darken *SB*.
   SG – 12
2. *SBC* – 6
   SG – 12   Circle *H*, darken *SH*.
   SH – 5
3. *SBC* – ?
   SG – ?   Circle ?, darken ?
   SHG – ?

Fill in the missing distances, vertex, and edge in step 3. Then complete Julian's problem of using the shortest path algorithm to find the shortest path from *A* to *E*.

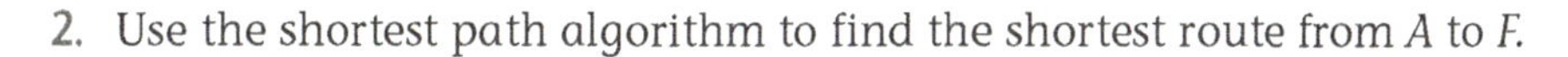

2. Use the shortest path algorithm to find the shortest route from A to F.

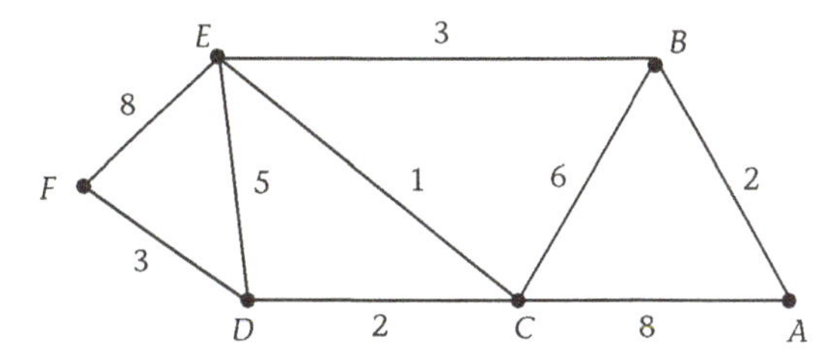

3. When might it not be necessary to repeat the procedure in the algorithm until all of the vertices are circled?

4. Use the shortest path algorithm to determine the shortest distance from S to each of the other vertices in the following graph.

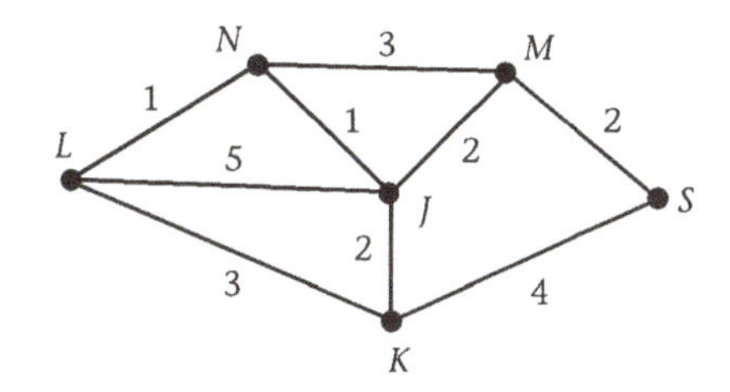

5.

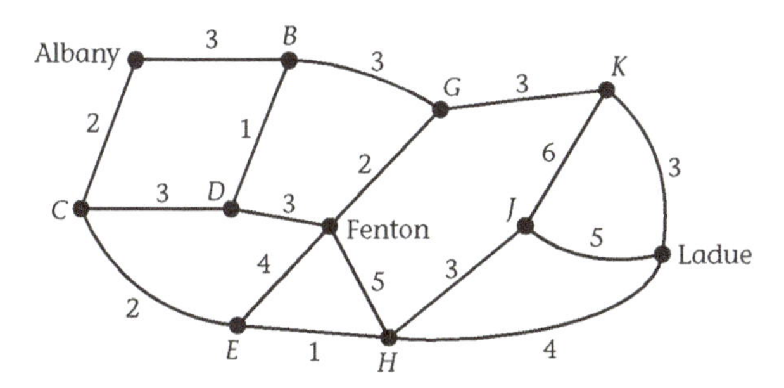

a. Use the shortest path algorithm to find the shortest route from Albany to Ladue in the preceding graph.

b. Assume that it is necessary to travel from Albany to Fenton to deliver a package and then to continue from there to Ladue. Find the shortest route for this trip. Explain why the solution to this question might be different than the shortest route from Albany to Ladue.

6. In the shortest path algorithm, each time you examine the uncircled vertices that are adjacent to the circled ones, you have to recalculate the lengths of the paths from the starting vertex. Explain how the efficiency of the algorithm might be improved by modifying it to avoid such recalculation.

7. What is the shortest distance from S to X in the following graph?

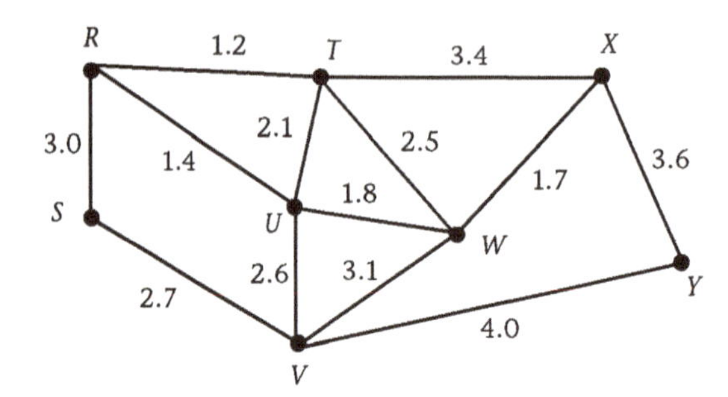

a. 6.7

b. 7.5

c. 7.6

d. 10.9

8. The shortest path algorithm can be applied to digraphs if slight modifications are made. Make the appropriate changes, and try your revised algorithm on the following digraph to find the shortest route from A to F.

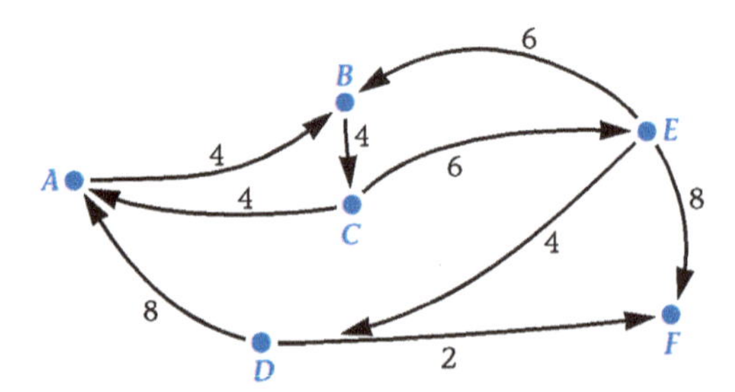

9. Mail Packages, Inc., ships from certain cities in the United States to others. A table of the company's shipping costs follows.

From \ To	Albany	Biloxi	Center	Denver	Evert	Fargo	Gale
Albany	—	7	—	—	4	—	—
Biloxi	—	—	—	—	—	—	6
Center	2	—	—	—	2	—	—
Denver	—	—	1	—	—	—	—
Evert	—	—	—	—	—	—	4
Fargo	—	—	—	—	3	—	2
Gale	1	6	—	—	—	1	—

Since a package can't be shipped directly from Denver to Biloxi, construct a digraph to represent the cost table and apply the shortest path algorithm to find the least charge for shipping the package.

Project

10. Interview several firefighters, ambulance drivers, or paramedics in your community to find out how they determine the shortest route from their facility to an emergency situation. Write a short report on your findings.

How Bumblebees Tackle the Traveling Salesman Problem

Science Daily
February 8, 2013

It is a mathematical puzzle which has vexed academics and traveling salesmen alike, but new research from Queen Mary's School of Biological and Chemical Sciences can reveal how bumblebees effectively plan their route between the most rewarding flowers while traveling the shortest distances.

The research, led by Dr. Mathieu Lihoreau and published in the British Ecological Society's journal *Functional Ecology*, explored the movement of bumblebees as they collected nectar from five artificial flowers varying in reward value.

According to Dr. Lihoreau, "Animals which forage on resources that are fixed in space and replenish over time, such as flowers which refill with nectar, often visit these resources in repeatable sequences called trap-lines. While trap-lining is a common foraging strategy found in bees, birds, and primates, we still know very little about how animals attempt to optimize the routes they travel."

Research into optimizing routes based on distance and the size of potential rewards is reminiscent of the well known Traveling Salesman problem in mathematics, which was first formulated in 1930, but remains one of the most intensively studied problems in optimization.

Co-author Dr. Nigel Raine explained, "Computers solve the problem by comparing the length of all possible routes and choosing the shortest. However, bees solve simple versions of it without computer assistance using a brain the size of grass seed."

The team set up a bee nest-box, marking each bumblebee with numbered tags to follow their behavior when allowed to visit five artificial flowers which were arranged in a regular pentagon. "When the flowers all contain the same amount of nectar, bees learned to fly the shortest route to visit them all," said Dr. Lihoreau. However, by making one flower much more rewarding than the rest, we forced the bees to decide between following the shortest route or visiting the most rewarding flower first."

In a feat of spatial judgment the bees decided that if visiting the high reward flower added only a small increase in travel distance, they switched to visiting it first. However, when visiting the high reward added a substantial increase in travel distance, they did not.

The results revealed a trade-off between either prioritizing visits to high reward flowers or flying the shortest possible route. Individual bees attempted to optimize both travel distance and nectar intake as they gained experience of the flowers.

"We have demonstrated that bumblebees make a clear trade-off between minimizing travel distance and prioritizing high rewards when considering routes with multiple locations," concluded co-author Lars Chittka. "These results provide the first evidence that animals use a combined memory of both the location and profitability of locations when making complex routing decisions, giving us a new insight into the spatial strategies of trap-lining animals."

Lesson 5.4

Trees and Their Properties

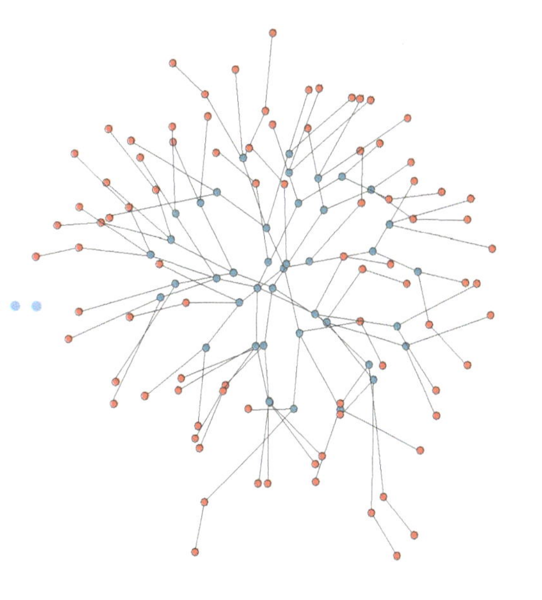

In Lesson 5.2, a special type of graph called a tree was used to organize information and list all possible routes for a traveling salesperson problem.

Tree diagrams have been used since ancient times. But it wasn't until the nineteenth century that their properties were studied in detail. In 1847, Gustav Kirchoff used trees in his study of electrical networks. Ten years later, Arthur Cayley used them in his investigations of certain chemical compounds. Today, trees are one of the most useful structures in discrete mathematics. They are invaluable to computer scientists. Many computer programmers depend on trees when creating, sorting, and searching programs.

Before exploring some of the properties and applications of this type of graph, it is necessary to define a tree. Recall that a **cycle** in a graph is any path that begins and ends at the same vertex and no other vertex is repeated. A **tree** is then defined as a connected graph with no cycles.

Example

Which of the following graphs are trees? Why?

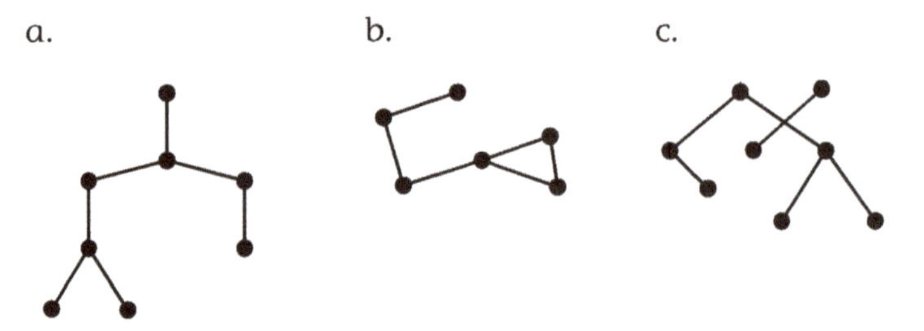

Figure 5.9. Possible trees.

Solution:

1. Figure 5.9a is a tree because it is a connected graph with no cycles.
2. Figure 5.9b is not a tree because it has a cycle.
3. Figure 5.9c is not a tree because it is not connected.

Trees have many applications in the real world. They can be used to list and count possibilities, as was done in the traveling salesperson problem. They can also be used to model family genealogical histories (Figure 5.10), to structure decision-making processes (Figure 5.11), to represent chemical compounds (Figure 5.12), and to help organize probabilities (Lesson 6.4).

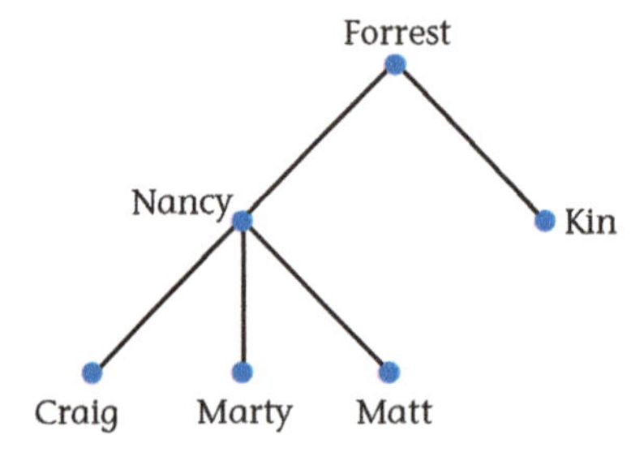

Figure 5.10. Family tree.

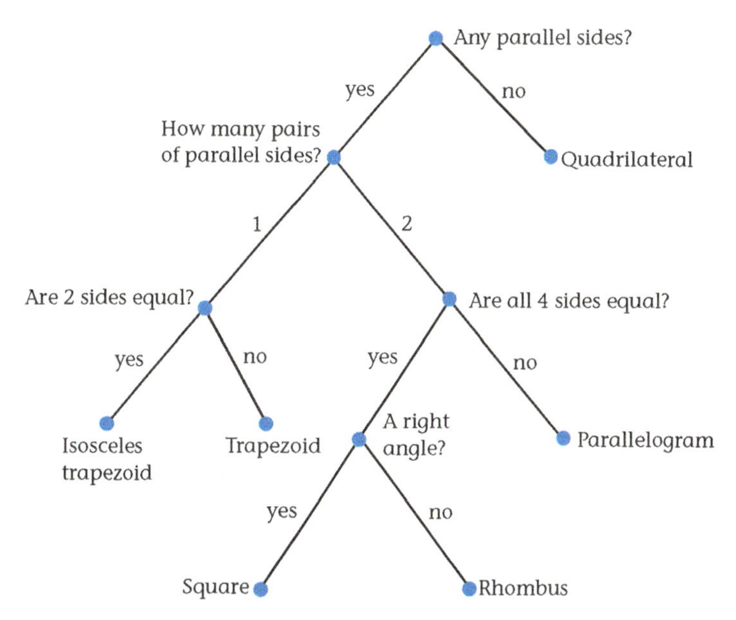

Figure 5.11. Sorting quadrilaterals.

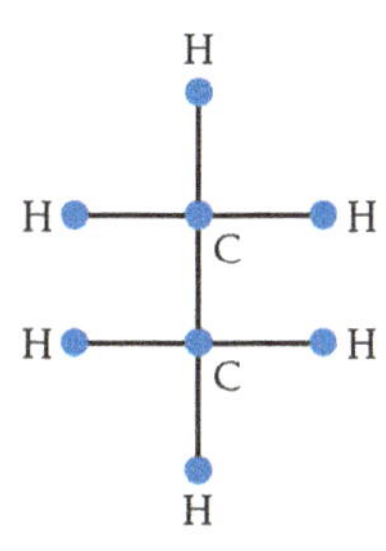

Figure 5.12. C_2H_6 (ethane).

Exercises

1. Examine the following graph for cycles. List as many cycles as you can find.

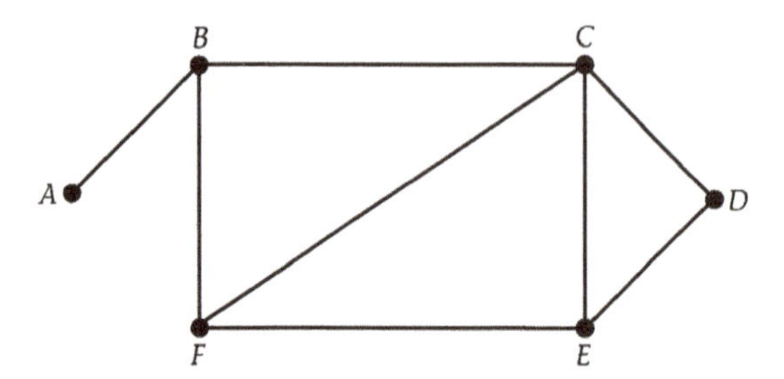

2. Determine whether the following graphs are trees. If the figure is not a tree, explain why.

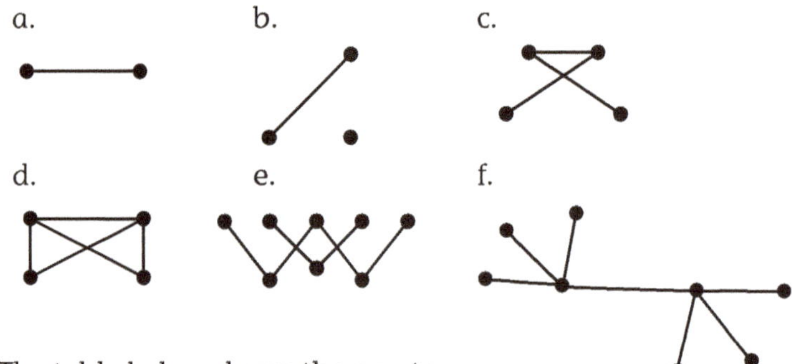

3. The table below shows the one tree that can be drawn with two vertices, the one tree that can be drawn with three vertices, and the two distinct trees that can be formed from four vertices.

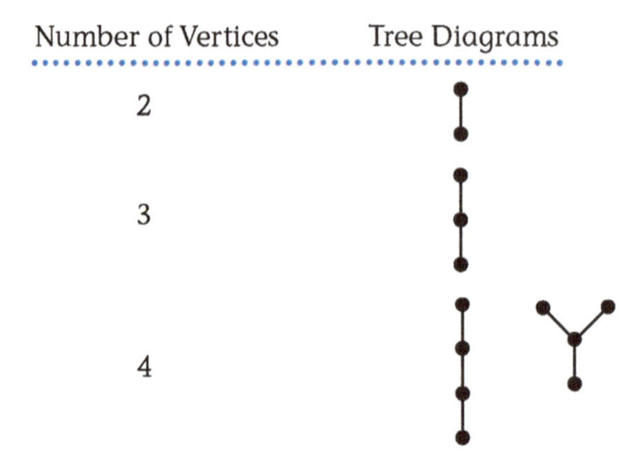

a. Draw all of the trees that are possible for five vertices.

b. Draw all of the trees that are possible for six vertices.

4. Complete the following table for trees with the indicated numbers of vertices.

Number of Vertices	Number of Edges
1	0
2	1
3	
4	
n	

a. How many edges does a tree with 19 vertices have?

b. How many vertices does a tree with 15 edges have?

c. What is the relationship between the number of vertices of a tree and the number of edges?

5. What happens to a tree if an edge is removed from it?

6. Draw a tree with six vertices that has exactly three vertices of degree 1.

7. a. Complete the following table for trees with the indicated numbers of vertices.

Number of Vertices	Sum of the Degrees of the Vertices	Recurrence Relation
1	0	$S_1 = 0$
2	2	$S_2 = S_1 + 2$
3	4	$S_3 =$ ______
4	______	______
5	______	______
6	______	______

b. Write a recurrence relation that expresses the relationship between the sum of the degrees of the vertices of a tree with n vertices and the sum of the degrees of the vertices of a tree with $n - 1$ vertices.

8. Explain why the sum of the degrees of the vertices of a tree with n vertices is equal to twice the number of vertices minus 2. (Hint: Draw several different trees.)

9. Much of the terminology connected with trees is botanical in nature. For instance, a graph that consists of a set of trees is called a **forest**, and a vertex of degree 1 in a tree is called a **leaf**. Draw a forest of three trees. Circle the leaves of your graph.

10. Suppose that a forest has 7 vertices and 2 trees. How many edges does it have?

11. In a hierarchical tree as in Figures 5.10 and 5.11 on pages 256 and 257, it is natural to speak of the **root** of the tree. A tree is rooted when all of the edges are directed away from the chosen vertex (root). In Figures 5.10 and 5.11, the edges are directed downward. Draw a family tree for your family beginning with one of your grandfathers as the root of the tree. What do the leaves of your tree have in common?

12. Refer to the tree in Figure 5.11 on page 257. This graph is called a **decision tree**, and the leaves represent the final outcomes of the different decisions. Using this tree, what is the name of a quadrilateral with two pairs of parallel sides, four sides equal, and no right angles?

13. For the following graph, find two different subgraphs that are trees.

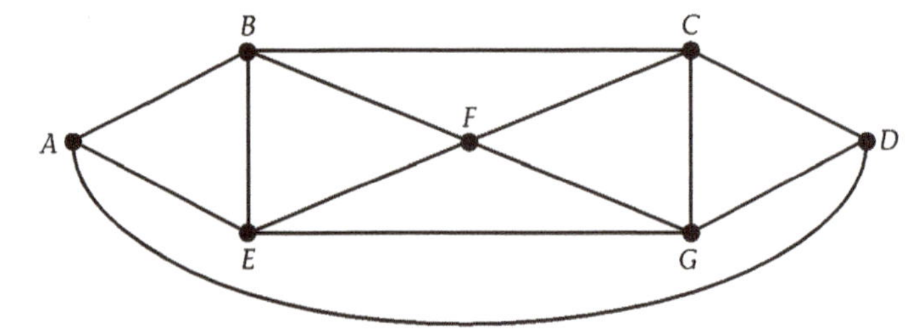

14. Suppose the children in a certain neighborhood want to communicate with one another via their very own communication network. To avoid the expense of connecting each house with every other house, a system needs to be devised that uses as few lines as possible yet allows messages to get to each person. Create such a network for the following houses.

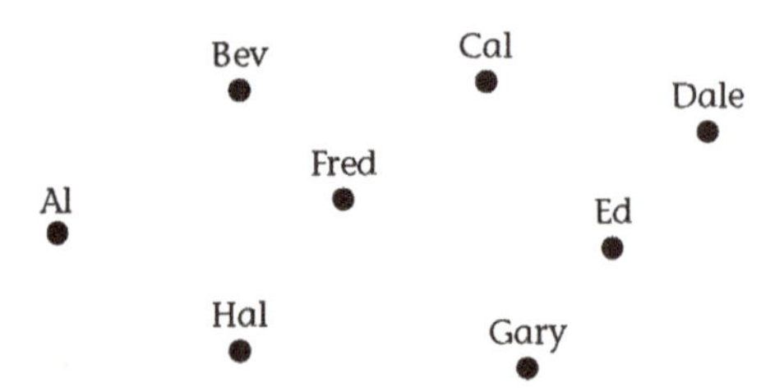

15. Draw a tree with 8 vertices and 10 edges. If it is not possible, explain why.

16. Clock solitaire is a card game in which the 52 cards are dealt face down into 13 piles that correspond to the 12 numbers and the center on the face of a clock. In Figure 5.13, clock positions 11 and 12 are identified with the jack and queen, and the thirteenth, or center, pile is identified with the king.

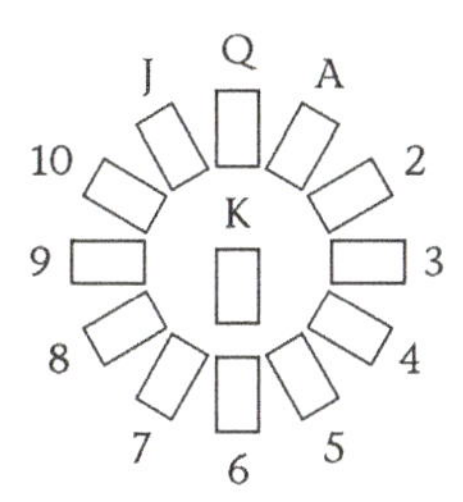

Figure 5.13. Diagram of the 13 piles for a game of clock solitaire.

The game is played by first turning over the top card of the king pile and putting it face up under the pile that corresponds to its value on the clock. Now turn up the top card of the pile under which you just put the card. Continue in this manner. The game is won when you have turned up all 52 cards. If a fourth king is turned up before this happens, play cannot continue, and the game is lost.

In his book *Fundamental Algorithms,* mathematician Donald E. Knuth noted two very interesting things about this game. One is that the probability of winning is 1/13. The other is that by checking the bottom card of the 12 clock piles, you can determine whether you will win the game.

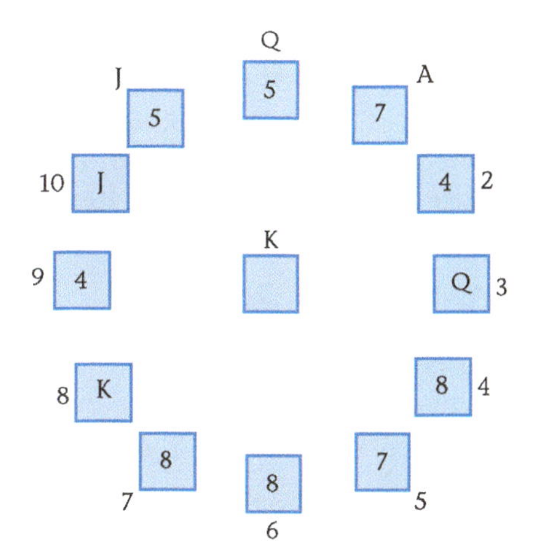

Figure 5.14. One possible configuration of the bottom cards in the game.

To determine whether you can win a game, turn over the bottom card of each pile except for the king pile (see Figure 5.14). Draw an edge from each of the 12 cards to the clock position that corresponds to the card's numeric value. For example, draw an edge from A to 7 (see Figure 5.15). Now redraw the graph with the vertices labeled A, 2, 3, . . . J, Q, and K. You will win the game if the resulting graph is a tree that includes all 13 piles (see Figure 5.16).

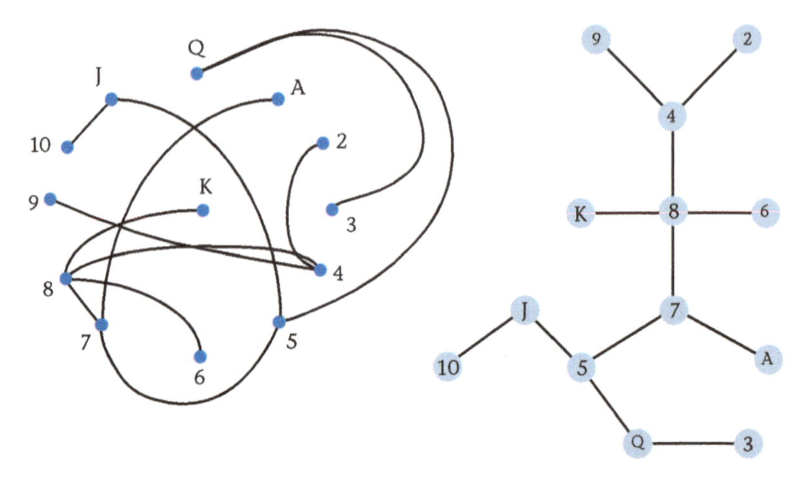

Figure 5.15. Graph showing the edges from the cards to the clock position.

Figure 5.16. Figure showing that the graph in Figure 5.15 is a tree.

Give the game a try, but before you do, record the bottom 12 cards. Predict whether you will win or lose the game by drawing a graph. Notice that only the bottom 12 cards determine your success. The arrangement of the other 40 cards makes no difference.

Modeling Project

17. Research your family tree and report on your discoveries. Be sure to include a graph (tree) in your report to illustrate your findings.

Lesson 5.5

Minimum Spanning Trees

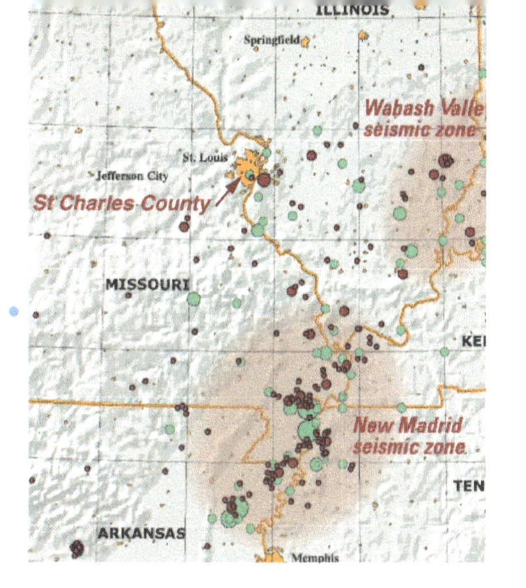

As in many of the previous lessons, this lesson focuses on optimization. Problems and applications here center on two types of problems: finding ways of connecting the vertices of a graph with the least number of edges and finding ways of connecting them with the least number of edges that have the smallest total weight.

Explore This

In making earthquake preparedness plans, the St. Charles County government needs a design for repairing the county roads in case of an emergency. Figure 5.17 is a map of the towns in the county and the existing major roads between them. Devise a plan that repairs the least number of roads but keeps a route open between each pair of towns.

Figure 5.17. The towns in St. Charles County.

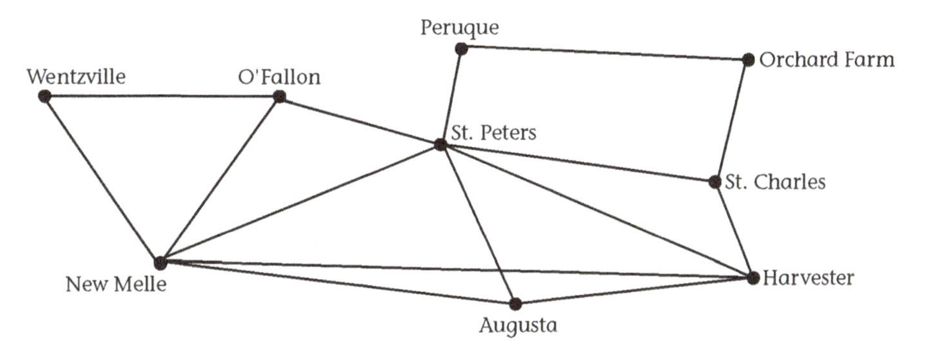

Examine your graph. If it connects each of the towns (vertices) and has no cycles, you've found a spanning tree. A **spanning tree** of a connected graph G is a tree that is a subgraph of G and contains every vertex of G. A spanning tree of the graph in Figure 5.17 would model the minimum number of roads (edges) needed to connect each town in case of an emergency.

Compare your plan with the other plans developed in your class. They should all contain the *same number* of edges but not necessarily the same edges. It is possible for a graph to have many different spanning trees. And as you may have guessed, for a graph that is not connected, no spanning tree is possible.

One systematic way to find a spanning tree for a graph is to delete an edge from each cycle in the graph. Unfortunately, this is not an easy procedure for a very large graph. But there are other ways of finding a spanning tree for a graph if one exists. One such method that can be easily adapted to computers is called the *breadth-first search algorithm.*

Breadth-First Search Algorithm for Finding Spanning Trees

1. Pick a starting vertex, S, and label it with a 0.
2. Find all vertices that are adjacent to S and label them with a 1.
3. For each vertex labeled with a 1, find an edge that connects it with the vertex labeled 0. Darken those edges.
4. Look for unlabeled vertices adjacent to those with the label 1 and label them 2. For each vertex labeled 2, find an edge that connects it with a vertex labeled 1. Darken that edge. If more than one edge exists, choose one arbitrarily.
5. Continue this process until there are no more unlabeled vertices adjacent to labeled ones. If not all vertices of the graph are labeled, then a spanning tree for the graph does not exist. If all vertices are labeled, the vertices and darkened edges are a spanning tree of the graph.

Example

Use the breadth-first search algorithm to find a spanning tree for the following graph.

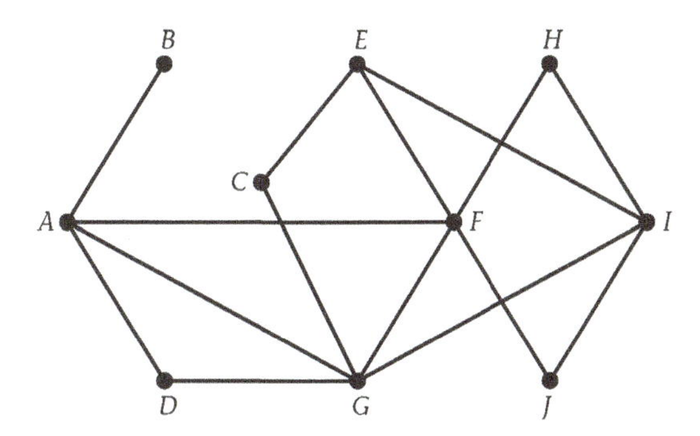

Solution:

As shown in the following figure, the algorithm begins by picking a starting vertex, calling it *S*, and labeling it with a 0. The labeling and darkening of edges then proceed according to steps 2 through 5 of the algorithm. As you probably noticed, this is not a unique solution. It is just one of the graph's many spanning trees.

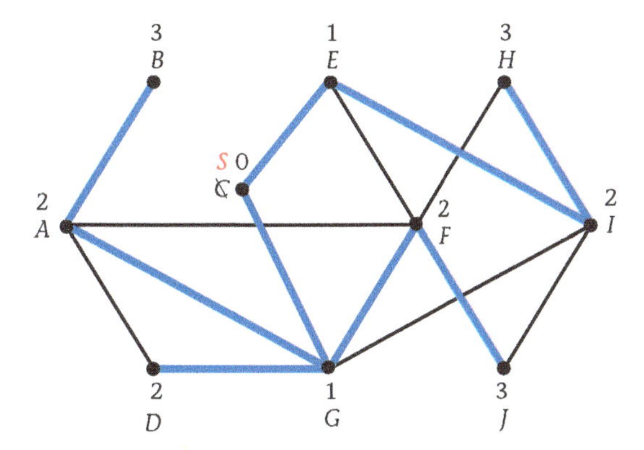

Many applications are best modeled with weighted graphs. When this is the case, it is often not sufficient to find just any spanning tree, but to find one with minimal or maximal weight.

Return to the earthquake preparedness situation and reconsider the problem when distances between towns are added to the graph (see Figure 5.18).

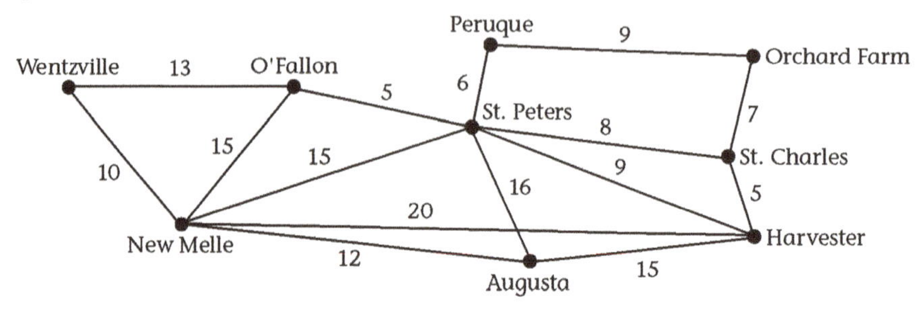

Figure 5.18. Map of St. Charles County with mileage shown.

Refer back to your solution of the original problem and find the total number of miles of road that would need to be repaired if your plan were implemented. Compare your plan with others in your class. Which plan or plans yield the minimum number of miles?

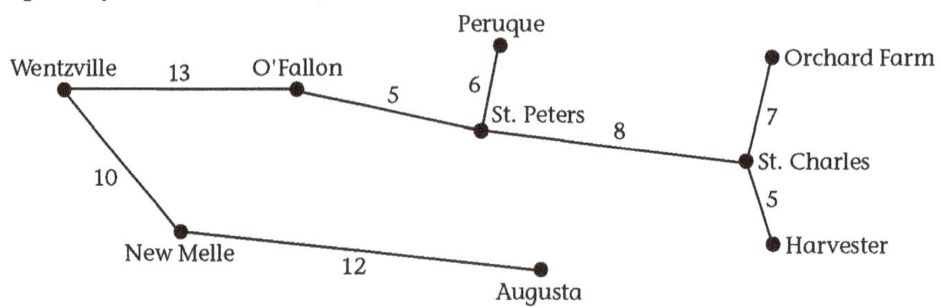

Figure 5.19. Spanning tree of minimum weight for the towns in St. Charles County.

For this particular problem, the minimum possible number of miles of road is 66. A spanning tree with that total weight is shown in Figure 5.19.

A spanning tree of minimal weight is called a **minimum spanning tree**. One algorithm for finding a minimum spanning tree for a graph is known as Kruskal's algorithm. It was developed in 1956 and named after its designer, Joseph B. Kruskal, a leading mathematician at Bell Laboratories.

Example

Use Kruskal's algorithm to find a minimum spanning tree for the following graph.

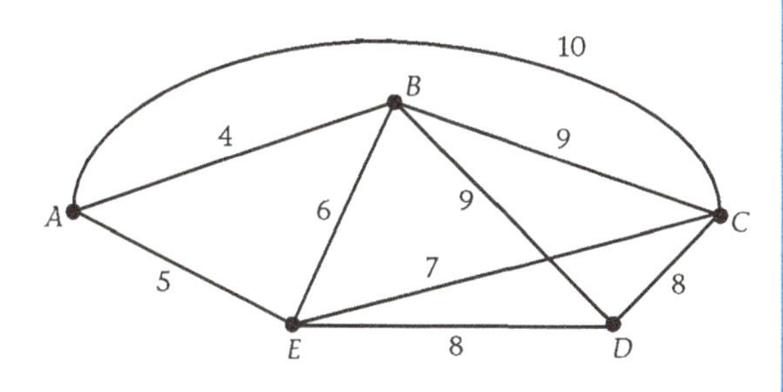

Solution:

There are five vertices in the graph, so four edges must be chosen. List the edges from shortest to longest. First on the list is *AB* (4). Darken it. Then darken *AE* (5). The next shortest edge is *BE*, but if picked, it will form a cycle. So pick *EC* (7). For the last edge there are two edges of length 8. Either *CD* or *ED* can be darkened. The darkened edges of the following graph form one of the minimum spanning trees of the graph. It has a minimal weight of $4 + 5 + 7 + 8 = 24$.

Kruskal's Minimum Spanning Tree Algorithm

1. Examine the graph. If it is not connected, there will be no minimum spanning tree.
2. List the edges in order from shortest to longest. Ties are broken arbitrarily.
3. Darken the first edge on the list.
4. Select the next edge on the list. If it does not form a cycle with the darkened edges, darken it.
5. For a graph with n vertices, continue step 4 until $n - 1$ edges of the graph have been darkened. The vertices and the darkened edges are a minimum spanning tree for the graph.

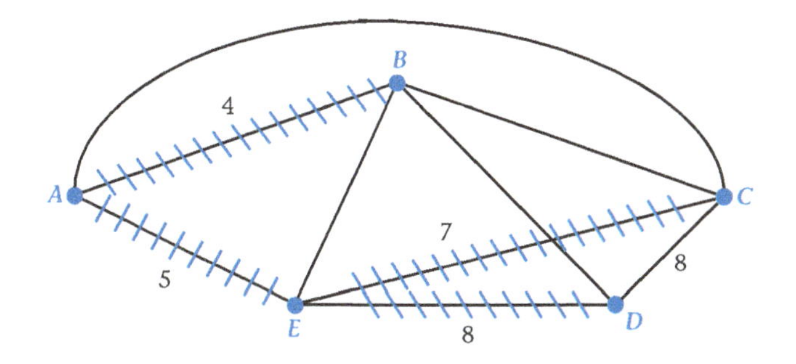

List of Edges from Shortest to Longest

Edge	Length
AB	4
AE	5
BE	6
EC	7
CD	8
ED	8
BD	9
BC	9
AC	10

Notice that both Kruskal's and Dijkstra's (Lesson 5.3) algorithms produce spanning trees. But unlike Dijkstra's shortest path algorithm, which gives you a spanning tree of shortest paths, Kruskal's algorithm yields a spanning tree of minimal total weight.

Exercises

In Exercises 1 through 5, find a spanning tree for each graph if one exists.

1.

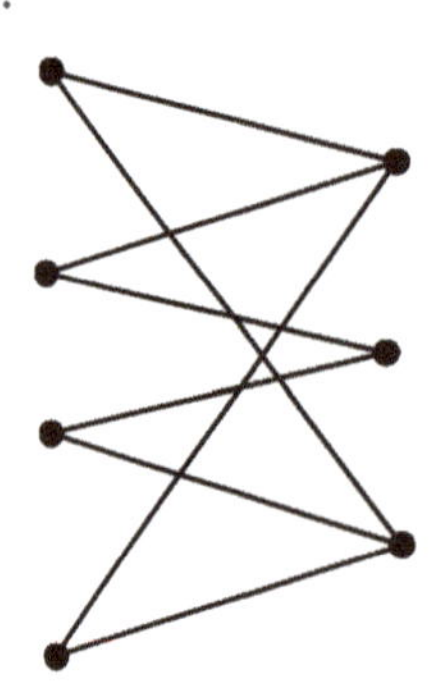

2.

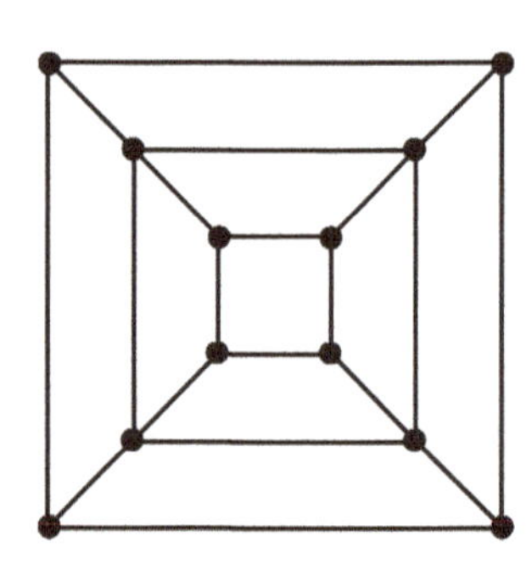

3.

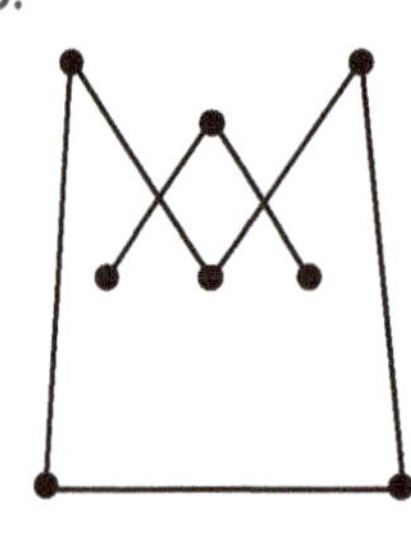

4.

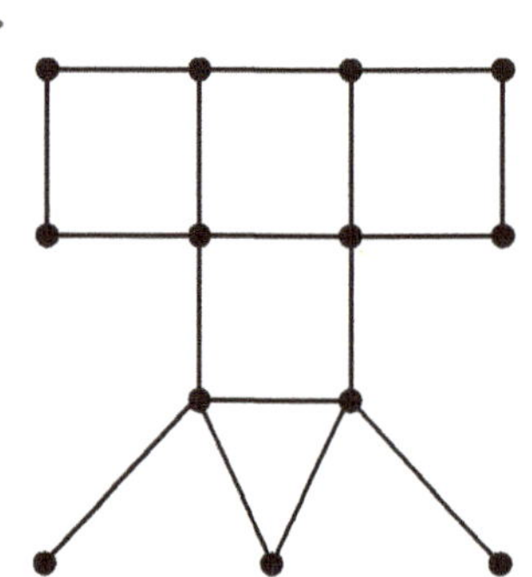

5.

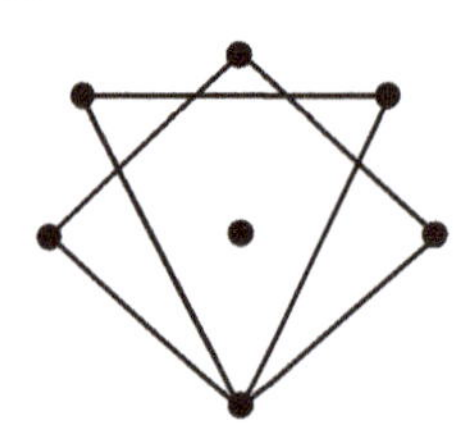

6. Draw a spanning tree for a K_4 graph.

7. 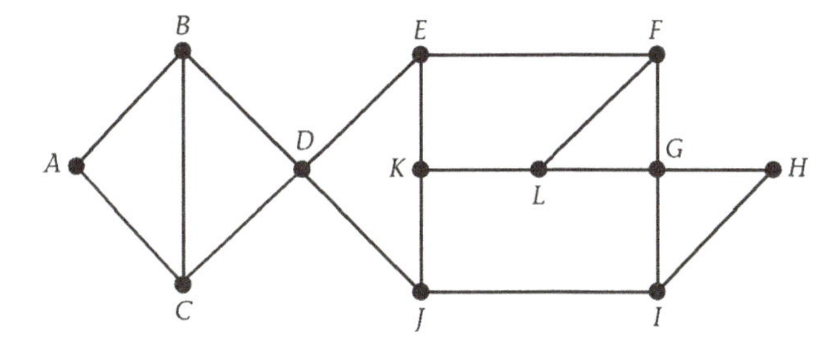

Sid began using the breadth-first search algorithm to try to find a spanning tree for the preceding graph. He began with vertex *A*, labeled it with a 0, and labeled *B* and *C* with 1s. He then darkened edges *AB* and *AC*, looked for vertices adjacent to the 1s, and selected vertex *D*. He labeled it with a 2 and darkened edge *BD*.

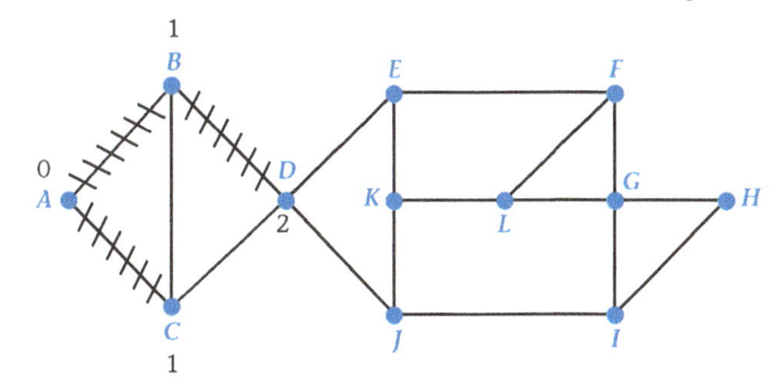

a. Could Sid have darkened *CD* instead of *BD*?

Copy Sid's graph and complete the search for Sid by answering the following questions.

b. Which vertices receive 3s for labels? Label these vertices.

c. Which edges subsequently are darkened? Darken these edges.

d. Three vertices should be labeled 4. Which ones? Label these vertices and darken the appropriate edges. Your graph could look like the one at right.

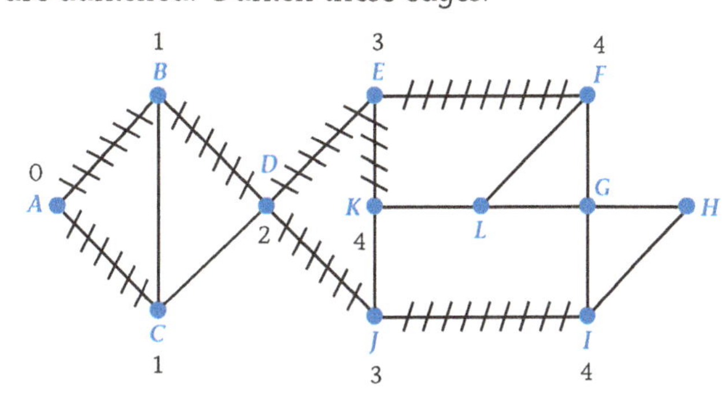

Continue the algorithm until all vertices are labeled. Check your darkened edges to make sure they form a spanning tree.

8. Use the breadth-first search algorithm to find a spanning tree for this graph. Begin at vertex A.

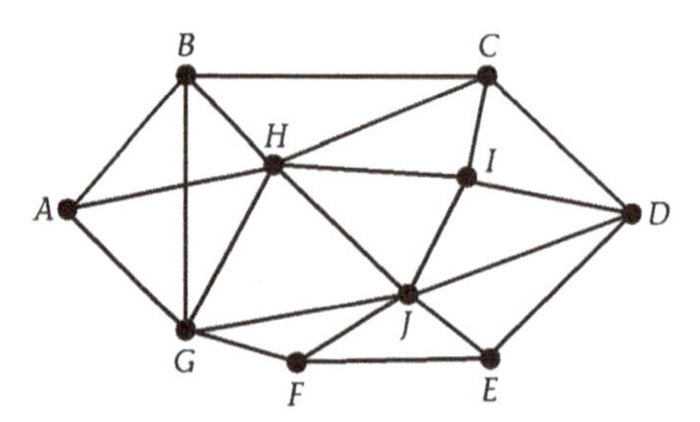

9. Use mathematical induction on the number of edges to prove that every connected graph has a spanning tree.

10. The breadth-first search algorithm can be applied to digraphs if slight changes are made. Modify the algorithm on page 264 so that it can be used with digraphs. Apply your modified breadth-first search algorithm to the following digraph.

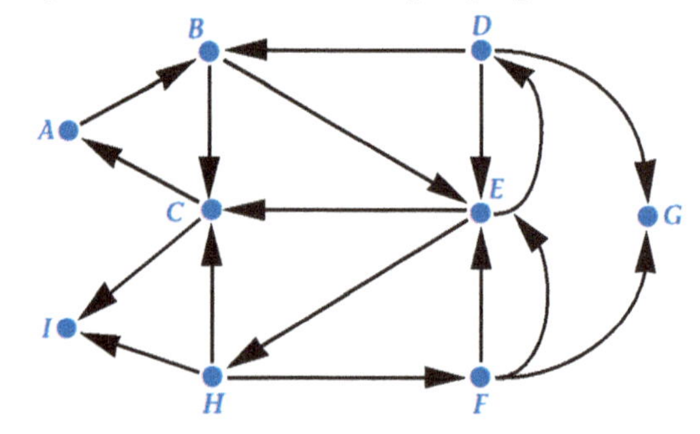

Use Kruskal's algorithm to find a minimum spanning tree for the graphs in Exercises 11 through 14. What is the minimal weight in each case?

11.

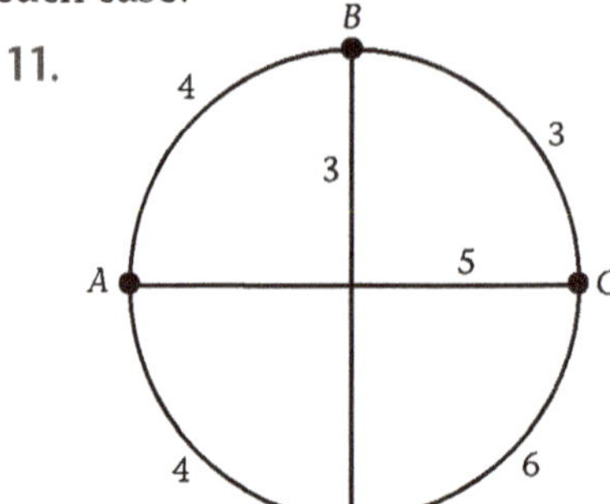

12.

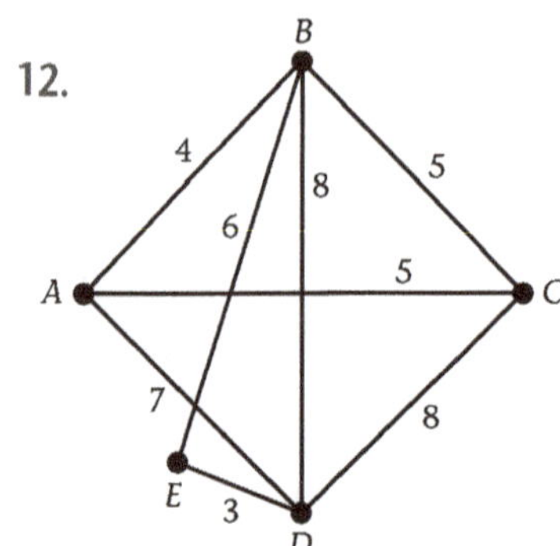

13.

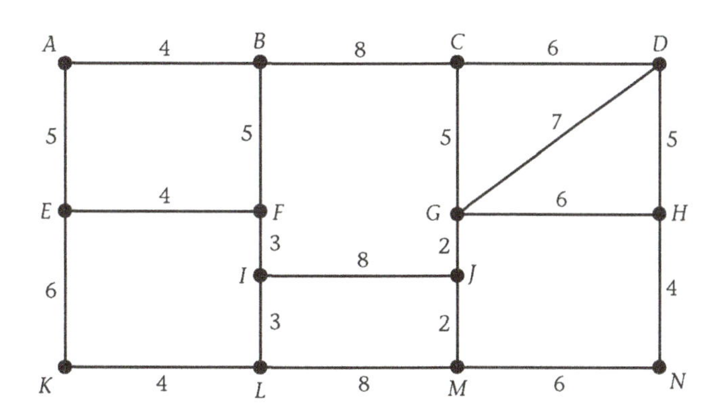

14.

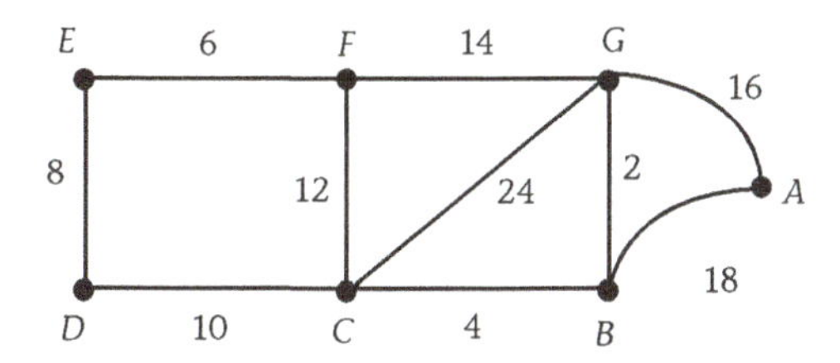

15. The computers in each of the offices at Pattonville High School need to be linked by cable. The following map shows the cost of each link in hundreds of dollars. What is the minimum cost of linking the five offices?

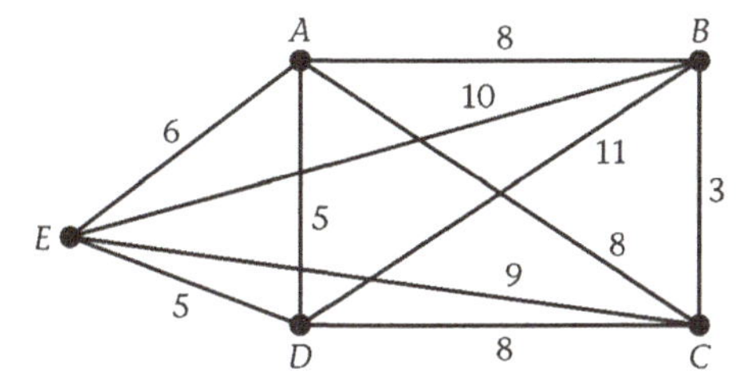

16. Suppose that the cable in Exercise 15 was installed by a disreputable firm that used only the most expensive links. What would be the maximum cost for the four links?

17. How might Kruskal's minimum spanning tree algorithm be modified to make it a maximum spanning tree algorithm?

Another algorithm that can be used to find a minimum spanning tree is attributed to R. C. Prim, a mathematician at the Mathematics Center at Bell Labs.

Prim's Minimum Spanning Tree Algorithm

1. Find the shortest edge of the graph. Darken it and circle its two vertices. Ties are broken arbitrarily.
2. Find the shortest remaining undarkened edge having one circled vertex and one uncircled vertex. Darken this edge and circle its uncircled vertex.
3. Repeat step 2 until all vertices are circled.

18. Use Prim's algorithm to find a minimum spanning tree for the following graph. What is the minimal weight?

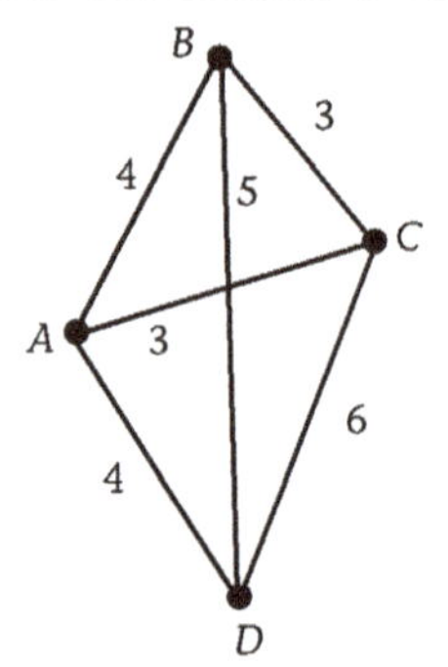

19. Use Prim's algorithm to find a minimum spanning tree for the following graph. What is the minimal weight?

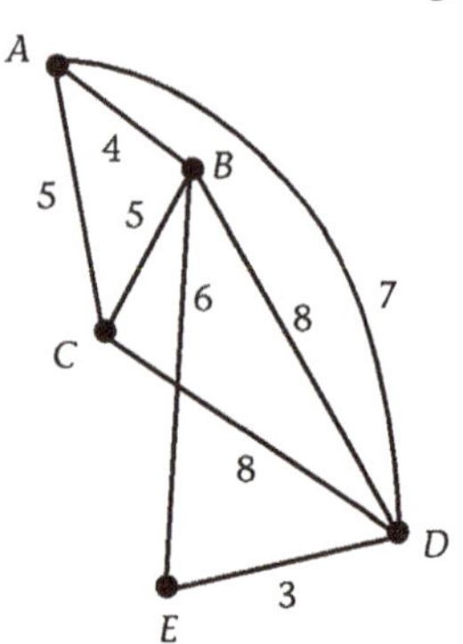

20. When the shortest path algorithm from Lesson 5.3 is applied until all vertices of a graph are used, it yields a spanning tree of the graph. Is it always a minimum spanning tree?

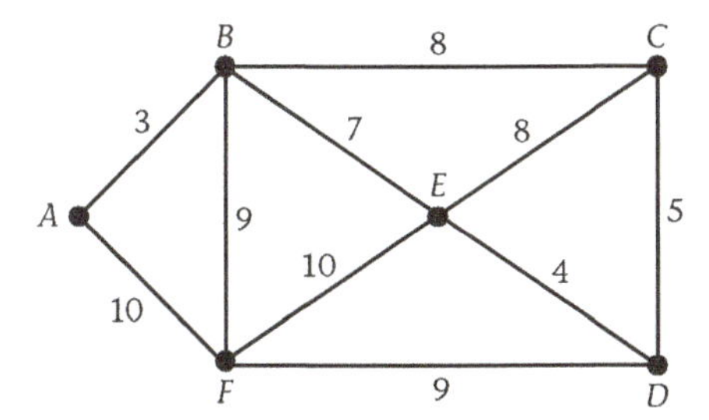

Check your answer to this question by doing the following.

a. Find a minimum spanning tree of the graph above using either Kruskal's or Prim's algorithm. What is the total weight of the minimum spanning tree?

b. Find the shortest route from *A* to each of the other vertices using the shortest path algorithm (page 248). Give the lengths of each of these routes.

c. Is the shortest route tree from *A* to each of the other vertices a minimum spanning tree of this graph? Explain why or why not.

21. Traveling salesperson problems, shortest route problems, and minimum spanning tree problems are often confused because each type of problem can be solved by finding a subgraph that includes all of the vertices of the graph. Compare and contrast what each type of problem asks and when each type of problem is used.

Project

22. In this lesson, you have applied two of the three classical minimum spanning tree algorithms, Kruskal's and Prim's. The third algorithm of this group was designed by O. Borůvka. Investigate Borůvka's algorithm, learn to apply it, and report on how it differs from Kruskal's and Prim's algorithms.

Splitting Terrorist Cells

Science News Online

How can you tell if enough members of a terrorist cell have been captured or killed so there's a high probability that the cell can no longer carry out an attack? A mathematical model of terrorist organizations might provide some clues. The question is what sort of mathematical model would work best.

One way to describe a terrorist organization is in terms of a graph. In this model, each node would represent an individual member of a given cell, and a line linking two nodes would indicate direct communication between those two members.

In this hypothetical four-member cell, represented by a graph, Boromir, Celeborn, and Denethor communicate directly with each other, but Aragorn communicates only with Boromir.

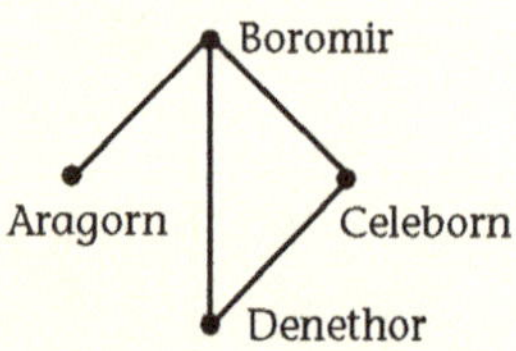

When a cell member is removed, the corresponding node is deleted from the graph. Deleting enough nodes leads to disruption. Mathematically, you could ask the question: How many nodes must you remove from a given graph before it splits into two or more separate pieces?

In this seven-member cell, removing *A* would have little effect on the organization. Removing *E* and *G* instead would split the cell into two units that presumably would be less effective on their own.

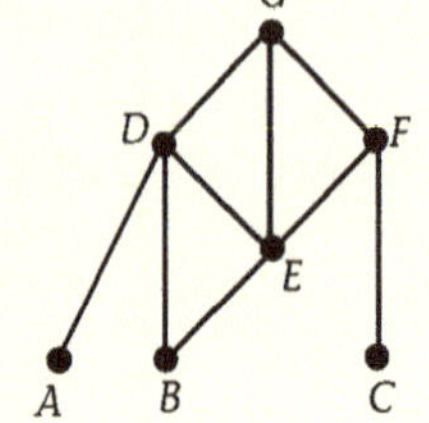

A graph model, however, may not be the best one available for representing a typical terrorist organization, mathematician Jonathan D. Farley of the Massachusetts Institute of Technology contends.

Farley has proposed an alternative approach that reflects an organization's hierarchy. "My method uses order theory to quantify the degree to which a terrorist network is still able to function," he says.

In this case, the relationship of one individual to another in a cell becomes important. Leaders are represented by the topmost nodes in a diagram of the ordered set representing a cell and foot soldiers are nodes at the bottom. Disrupting the organization would be equivalent to disrupting the chain of command, which allows orders to pass from leaders to foot soldiers.

In this ordered-set representation of a terrorist cell, points represent individual members and lines show communication links. Members *A*, *B*, and *C* are leaders and rank higher than all other members. *I*, *J*, and *K* have the lowest rank.

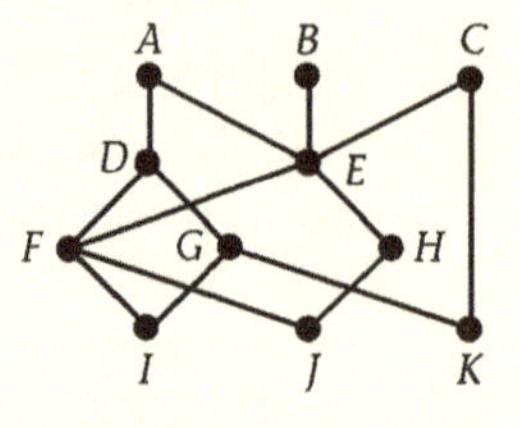

A given ordered set may have several such chains of command that link a leader with a foot soldier. All of these chains must be broken for a cell (or remnant) to be considered ineffective.

Farley's model has several shortcomings. Law enforcement often doesn't know how a terrorist cell is organized or even its full membership. Nonetheless, Farley contends, "this model enables law enforcement to plan its operations in less of an ad hoc fashion than they might be able to do otherwise."

Lesson 5.6

Binary Trees, Expression Trees, and Traversals

Decision trees and family trees are two examples of a special kind of tree known as a **rooted tree**. A rooted tree is a directed tree in which every vertex except the root has an indegree of 1, while the root has an indegree of 0. Figure 5.20 shows an example of a rooted tree in which vertex R is the root.

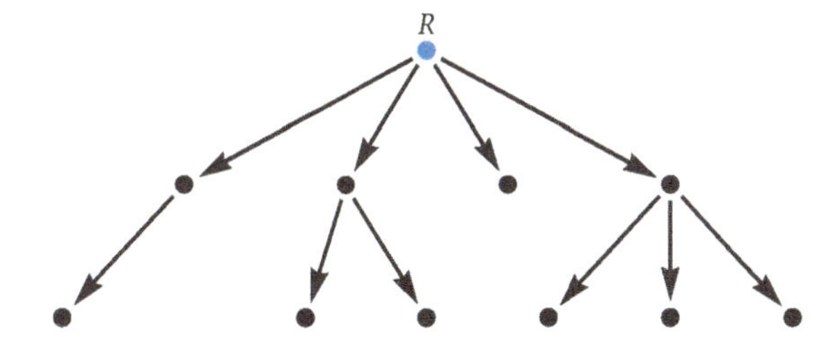

Figure 5.20. Rooted tree.

Since all edges are directed away from the root, it is not necessary to draw the arrowheads on the ends of the edges (see Figure 5.21).

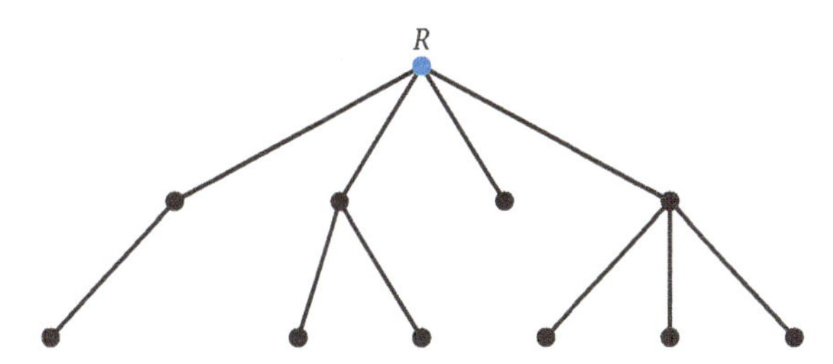

Figure 5.21. Rooted tree without arrowheads.

Rooted trees are used to model situations that are multistaged or hierarchical in structure.

Example

A couple decides to have three children. What are the possible outcomes?

Solution:

Draw and label a rooted tree.

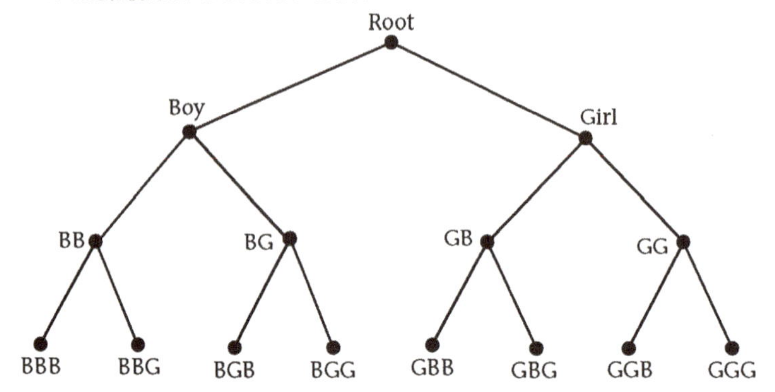

The eight possible outcomes are shown on the vertices of the tree that have outdegrees of 0.

In a rooted tree, a vertex V is said to be at level K if there are K edges on the path from the root to V. The root is at level 0, and the vertices adjacent to the root are at level 1. If a vertex V is at level 4, then any vertex adjacent to V at level 3 is called the **parent** of V, and any adjacent vertex of level 5 is called a **child** of V. A rooted tree in which each vertex has at most two children is called a **binary tree**.

Example

Which trees are binary trees? For those that are binary trees, name the parent of V and the children of V.

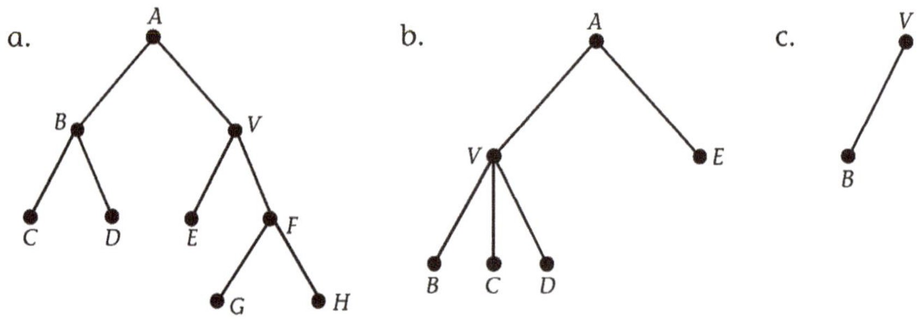

Solution:

All three of these graphs are trees, but only Figures a and c are binary trees. In a, the parent of V is A and the children are E and F. In c, V has no parent and B is the only child.

In computer science applications, binary trees are used to evaluate arithmetic expressions. When you write the expression (4 + 6) * 8 - 4/2, you understand how to find its value because you are familiar with the order of operations for expressions. Unfortunately, a computer cannot efficiently imitate your methods. However, if an expression is represented as a binary tree, the computer can quickly and efficiently evaluate it.

To represent the expression (4 + 6) * 8 – 4/2 as a binary tree, first find the operation in the expression that is performed last. Make that operation the root of the tree. The right and left sides of this operation become the children of the root (see Figure 5.22).

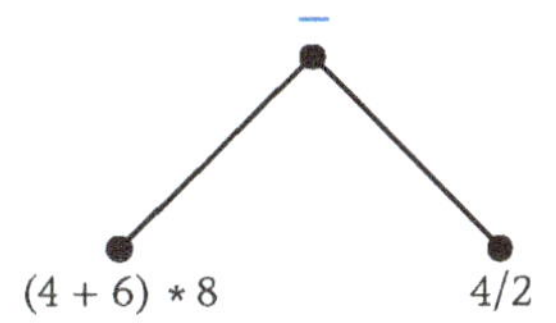

Figure 5.22. First step of representing an expression as a binary tree.

Continue this recursive process of placing operations at each internal vertex (the children) and putting operands on the leaves until no expression that contains operations appears on the leaves (see Figure 5.23).

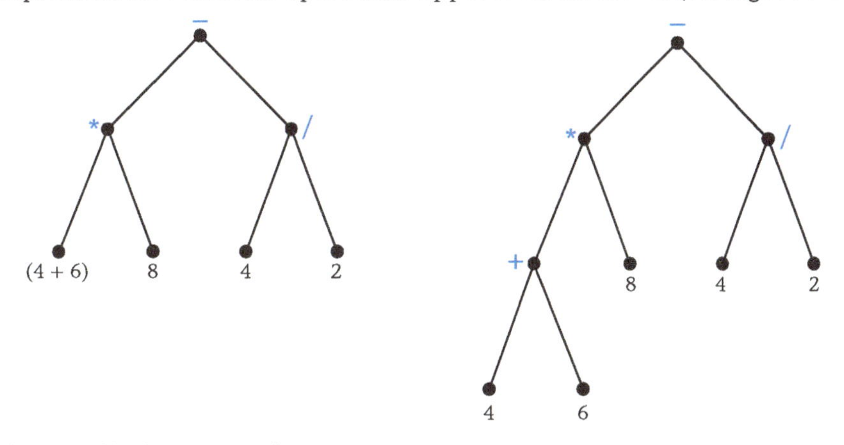

Figure 5.23. Continuing the recursive process.

The final binary tree in Figure 5.23 is called an **expression tree.**

Example

Represent $A/B + C * (D - E)$ as an expression tree.

Solution:

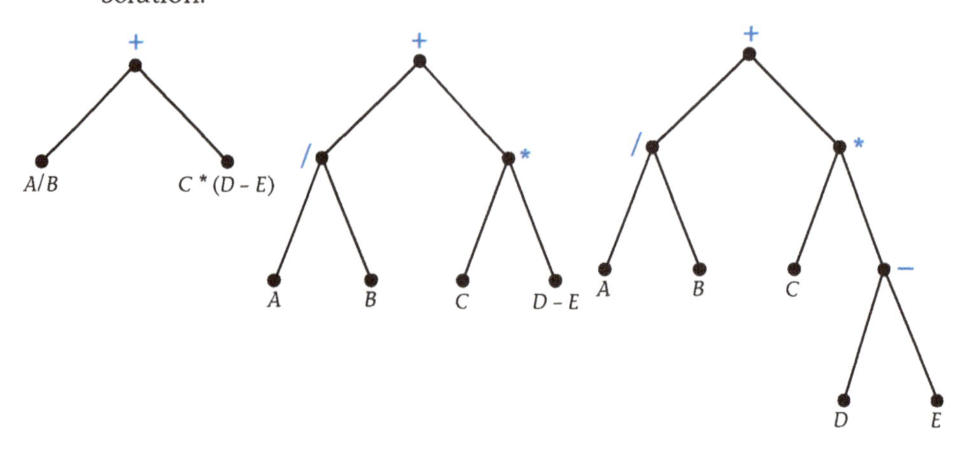

Tree Traversals

Once an expression is represented by a binary tree, the computer must have some systematic way of "looking at" the tree in order to find the value of the original expression. This organized procedure for obtaining information by visiting each vertex of the tree exactly once is called a **traversal** of the graph.

There are many different types of traversals, including one called a **postorder traversal**. This traversal differs from other traversals in that it visits the left child of the tree first, then the right child, and finally the parent or root (see Figure 5.24).

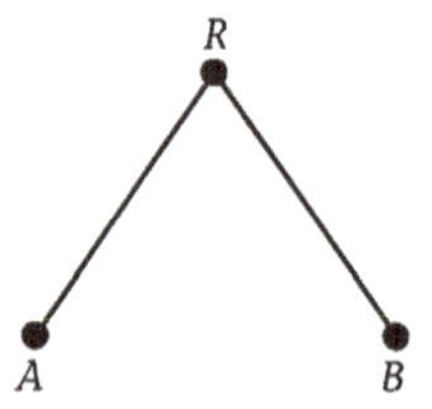

Figure 5.24. Postorder traversal A, B, R.

To find the postorder listing of the vertices of the tree in Figure 5.25, begin by moving to the left subtree of *A* and doing a postorder traversal on that subtree, which has *B* as its root. This requires you to branch to the left subtree of *B* and do a postorder traversal. Since the left subtree of *B* consists of only the vertex *D*, visit that vertex by numbering it with a 1. Now go to the right subtree of *B* and do a postorder traversal. Again, this subtree consists of only one vertex. Visit *E* and number it with a 2. Since the left and right children of *B* have been traversed, visit *B* (the root of that subtree). Number it 3.

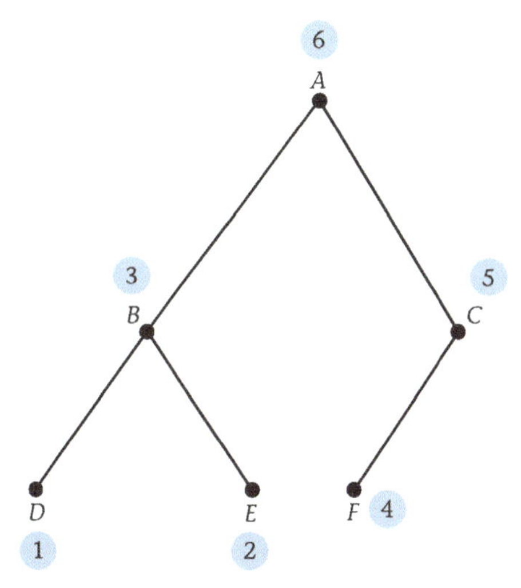

Figure 5.25. Postorder traversal of a binary tree.

You have traversed the left subtree of *A* and must now traverse the right subtree. To do this, you begin with a postorder traversal on the subtree that has *C* as its root. Move to the left subtree of *C*, visit *F*, and number it 4. Since there is no right subtree of *C*, visit the root *C*, and number it 5. Since both the left and right subtrees of root *A* have been visited, *A* can now be visited and numbered with a 6. The postorder traversal is complete and the postorder listing is *DEBFCA*.

Example

Give a postorder listing for the following expression tree.

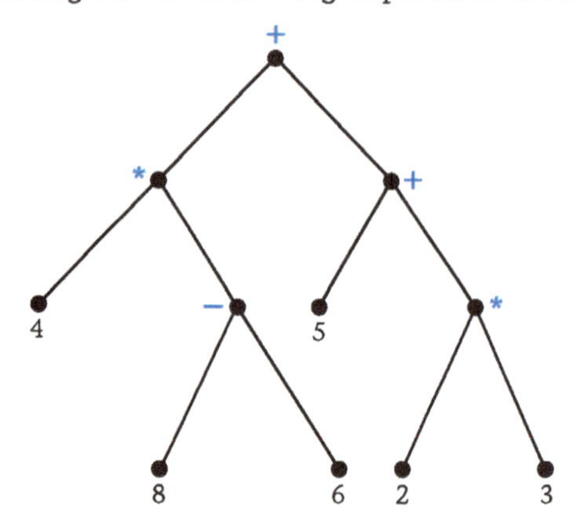

Solution:

Left subtree of A

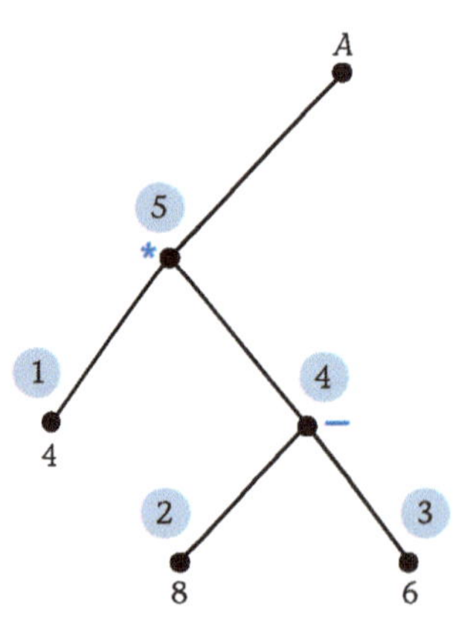

Right subtree of A

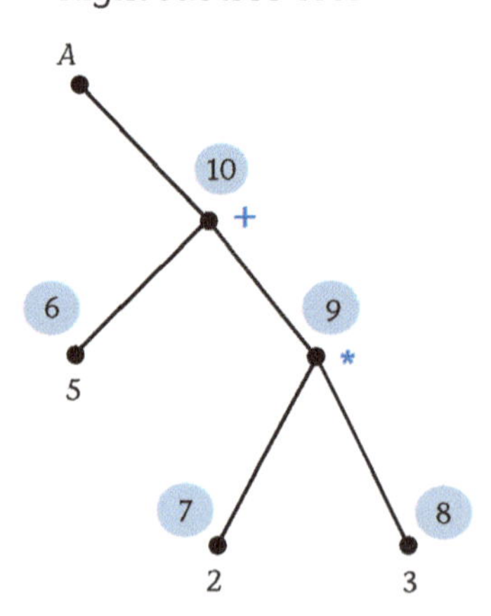

The root is visited last and 11 is assigned to A. The postorder listing is 4 8 6 - * 5 2 3 * + +.

The notation obtained by doing a postorder traversal is known as **reverse Polish notation (RPN)**. This notation with operations next to each other and no parentheses may look strange to you, but to owners of certain calculators, this notation is familiar and easy to use. RPN works well with calculators and computers because no parentheses are ever needed to indicate the desired order of operations.

So how do you find the value of the expression 4 8 6 − * 5 2 3 * + +? To evaluate RPN, scan the expression from the left until you find two numbers followed by an operation sign, in this case, 8 6 −. This says to you to take the 8 and 6 and subtract. Substitute the result, 2, back in the expression and repeat the process. This continues until you have evaluated the expression.

```
4 (8 6 −) * 5 2 3 * + +
(4 2 *) 5 2 3 * + +
8 5 (2 3 *) + +
8 (5 6 +) +
(8 11 +)
19
```

People who become accustomed to using this type of notation find it very quick and convenient to use because there is never a question about the order in which to perform the operations.

Point of Interest

Reverse Polish Notation was proposed in the 1950s by the Australian philosopher Charles L. Hamblin. It provided a way to write a mathematical expression without using parentheses and brackets.

Hewlett-Packard Company adapted RPN for its first hand-held scientific calculator. Currently, calculators such as the HP 50g support RPN as one of their entry modes.

Exercises

1. Tuleh wants to buy a car. She has the options of two different brands of radios and four different exterior colors. Draw a tree diagram to show all possible outcomes of choosing a radio and a color for the car.

2. A coin is tossed three times. Draw a tree diagram to show the possible outcomes.

In Exercises 3 through 6, examine each tree. If the tree is a binary tree, (a) give the level of vertex V, (b) name the parent of V, and (c) name the children of V.

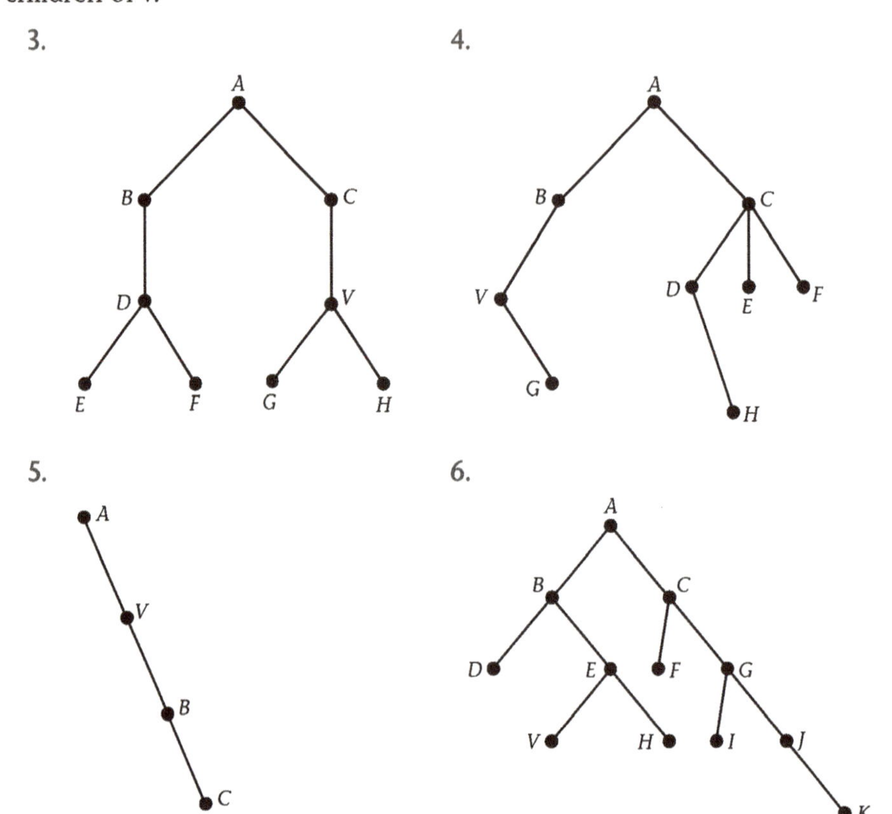

7. Jeff's brother Tom has a first-grade spelling book that contains five chapters. Each odd-numbered chapter has two lessons and each even-numbered chapter has three lessons. The second lesson of each chapter has two questions whereas all others have one. Draw a rooted tree that models Tom's book. How many questions are in the book?

In Exercises 8 through 11, represent the expression as a binary expression tree.

8. (2 − 5) * (4 + 7)
9. (2 + 3) * 4
10. 2 + 3 * 4 − 6/2
11. $A * B + (C - D/E)$

In Exercises 12 through 15, find the postorder listings for each binary tree.

12.

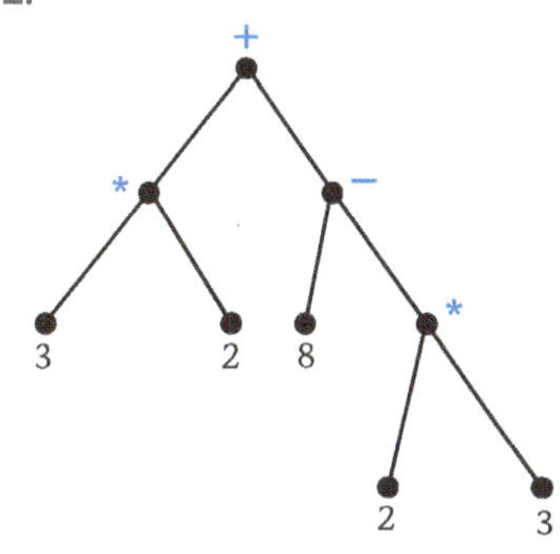

13.

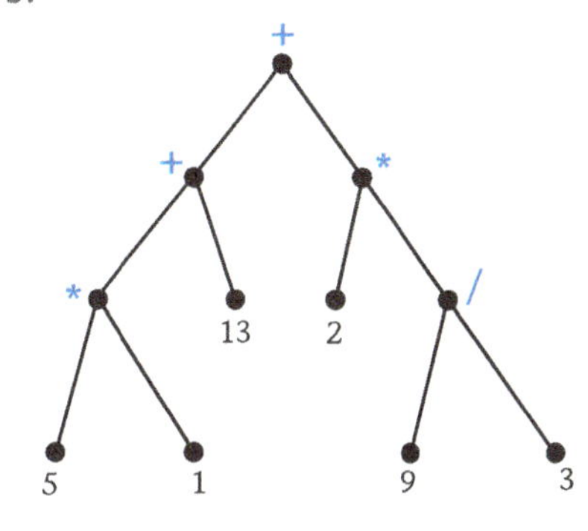

14.

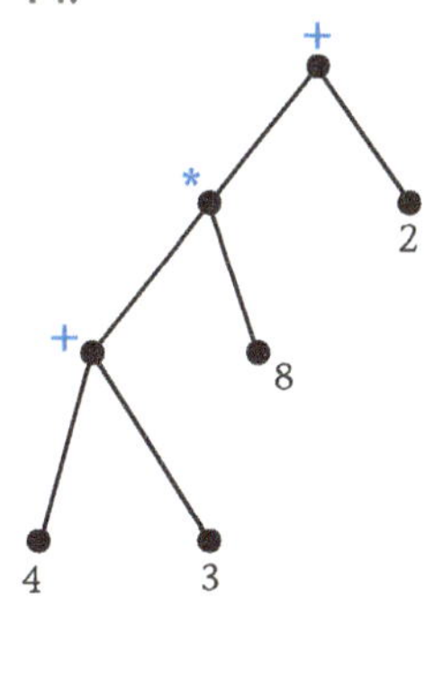

15.

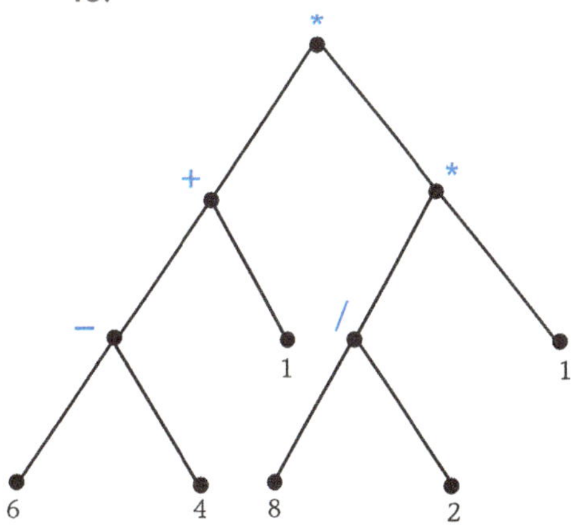

16. Evaluate the following reverse Polish notations.

a. 6 2 − 7 * 3 2 + +
b. 6 5 4 3 2 − + / +
c. 1 2 + 4 3 − + 6 2 / 2 + +
d. 4 3 + 8 2 − + 4 + 3 −
e. 7 2 5 + / 6 1 − *
f. 8 2 * 3 5 + /

17. Give the reverse Polish notation for each of the following expressions.

    a. $2 + 3 * 6 - (4 + 1)$ b. $(5 - 3) * 2 + (7 - 6/2)$

18. Construct an expression tree that would have the following reverse Polish notation.

    $A\ B\ *\ C\ D\ +\ E\ -\ +$

19. Construct a binary tree that has *ABC* as its postorder listing. Is your answer unique? If not, construct an additional tree(s).

20. A traversal that visits first the parent or root of the tree, then the left child, and finally the right child is called a **preorder traversal.**

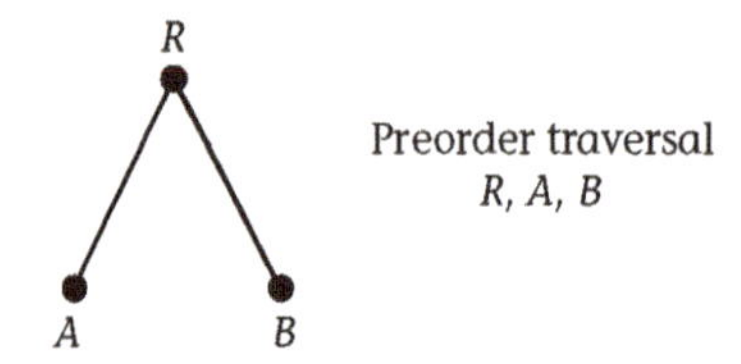

The preorder listing for the following binary tree is *ABDECFG*.

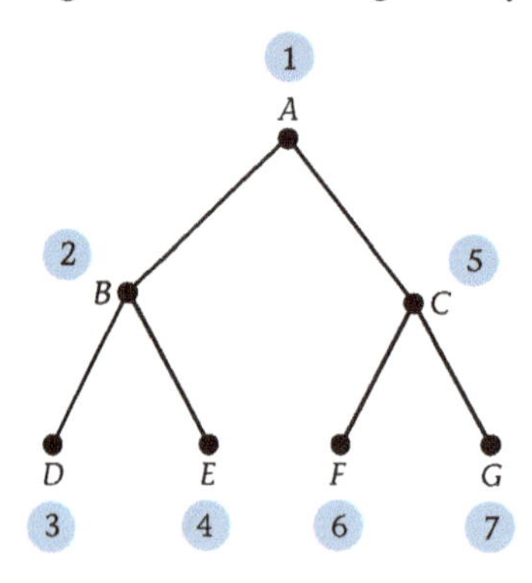

Find the preorder listing for the following binary tree.

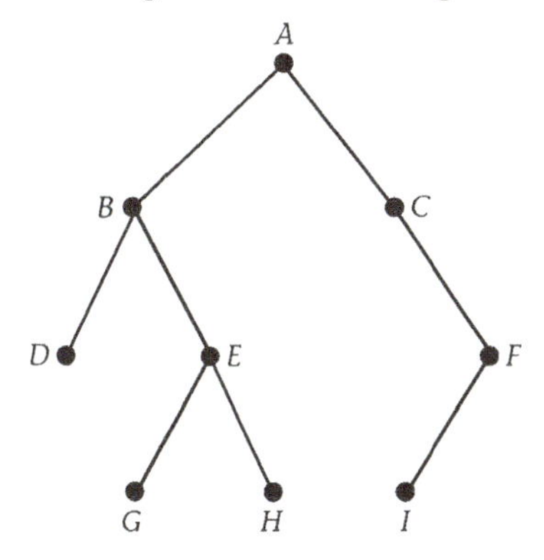

21. Find the preorder listings for the binary trees in Exercises 12 through 15.

22. The notation obtained from a preorder traversal is called **Polish notation.** To evaluate the expression, scan it from the left until you come to an operation followed by two numbers. Perform that operation, place the result back in the expression and continue. For example, * 2 3 is evaluated as 6. Complete the following evaluation.

+ + 4 (* 2 3) + 5 / 6 3
+ (+ 4 6) + 5 / 6 3
+ 10 + 5 / 6 3

23. Evaluate the following Polish notations.

a. + * 3 2 − 8 * 2 3
b. + / 6 3 + 4 3
c. + / 12 * 3 4 + 3 1
d. / * 8 2 + 3 5

Project

24. A postorder traversal of an expression tree yields reverse Polish notation, in which the operations follow the operands. A preorder traversal yields Polish notation, in which the operations precede the operands. Create a traversal rule that yields a notation in which the operations are between the operands. In a report, describe your procedure and show how it works on an expression tree. Also find a rule that evaluates your listings. Try your rule on several different expressions. Does it work for all of them? If not, explain why you think it's flawed.

Chapter Extension

Steiner Trees

Dr. Terry has three computers in her office that need to be networked. If the following graph shows the shortest distance between each computer, what is the minimum amount of cable required?

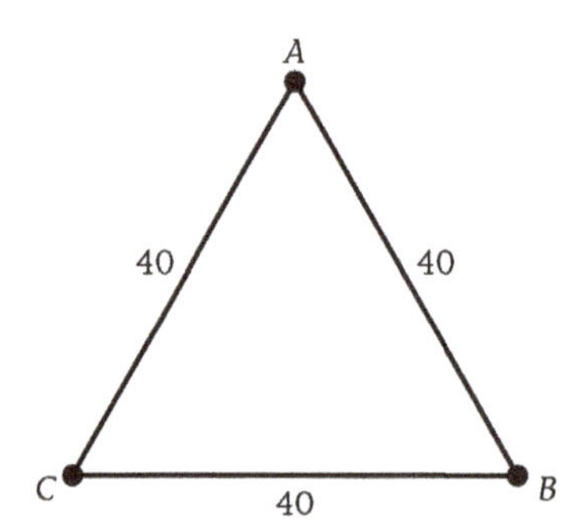

Finding a minimum spanning tree to link the three computers yields a total length of 80 feet of cable, but is there a better way?

The answer to this question is yes, if you do not have to follow the edges in the graph, and creating a junction someplace other than at one of the computers is not a problem. For example, if the cables for these computers can be placed and joined anywhere in the room, you can use less than the 80 feet required for the minimum spanning tree solution (see Figures a and b).

a.

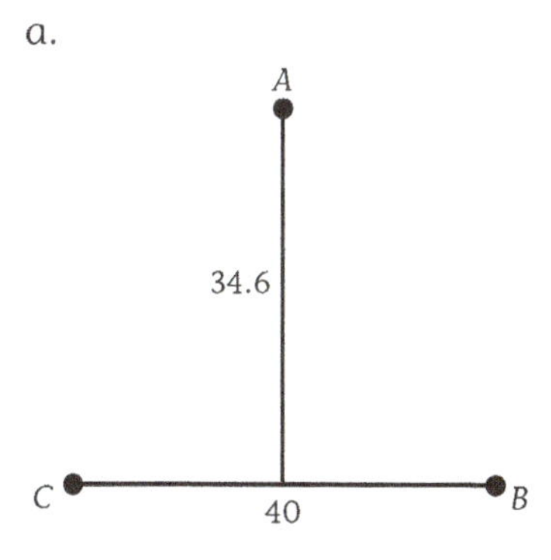

Total length of cables: 74.6 ft.

b.

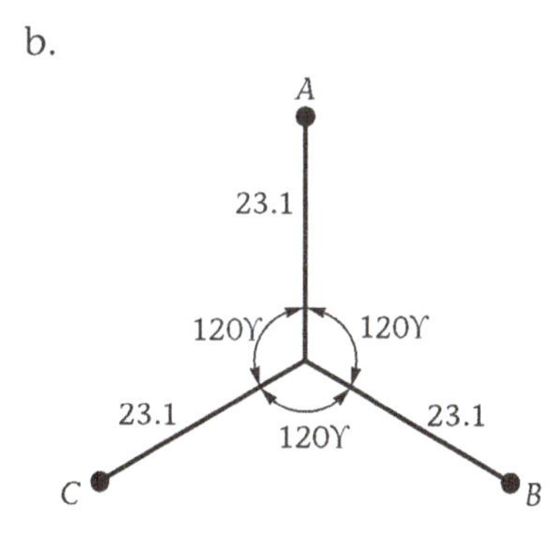

Total length of cables: 69.3 ft.

A tree using the newly created junction point, such as the one shown in Figure b, is known as a **Steiner tree**. The junction point where the three edges meet at 120° angles is called a **Steiner point**.

A Steiner point inside a triangle can be found using the following procedure, known as the Torricelli procedure.

Step 1 Assume that the largest angle of ΔABC is less than 120° and that BC is the longest side of the triangle. On the longest side of the triangle BC construct an equilateral triangle BCP.

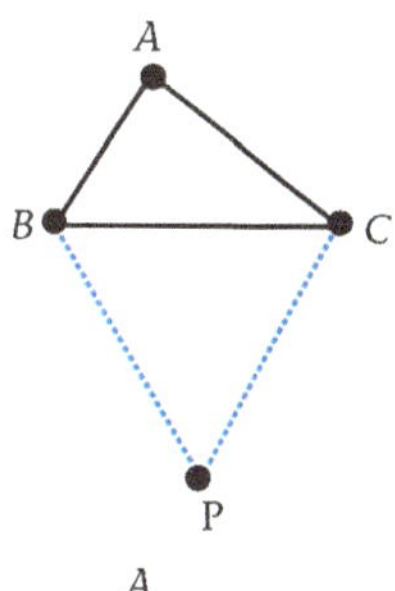

Step 2 Circumscribe a circle around ΔBCP.

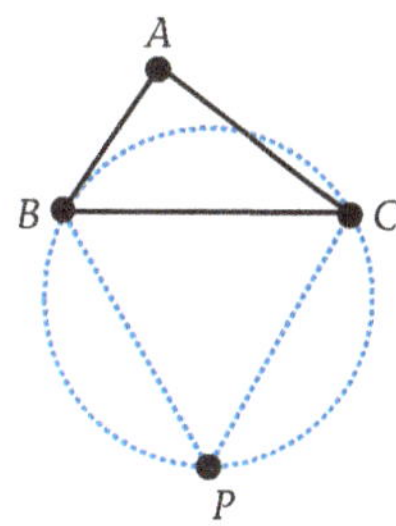

Step 3 Join P to A. The point S where the line segment PA intersects the circle is the desired Steiner point.

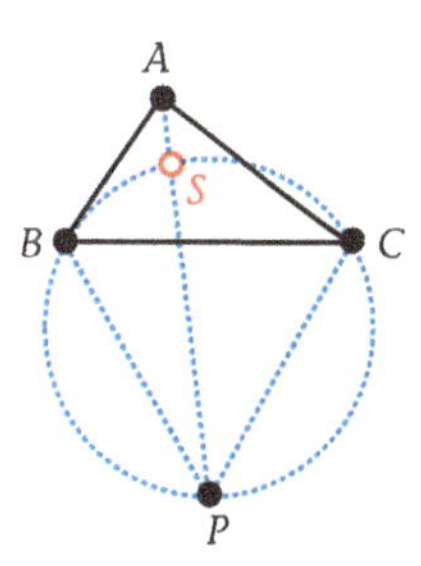

Creating Steiner trees brings up many questions that can be explored using a drawing utility such as the Geometer's Sketchpad or Cabri:

- If the computers are positioned on the vertices of a triangle that is not equilateral, will there be a Steiner tree?
- In some three-point cases, the minimum spanning tree solution is the best networking solution. When will this happen?
- If there were four computers to network instead of three, how many Steiner points would there be?
- Will there ever be more than one Steiner tree? If so, will they always be equal in length?

Of course, the questions continue to expand as the number of vertices increases. As you may have discovered already, the optimal way of connecting the vertices of a graph (or computers, in this case) is either to find the minimum spanning tree if no interior junctions are allowed or to construct an optimal Steiner tree.

As the number of vertices in a graph increases, the number of Steiner trees increases very rapidly. For example, the number of Steiner trees in a graph with ten vertices totals in the millions. Therefore, finding the Steiner tree with the least weight becomes very complicated.

Since their discovery, mathematicians have made significant findings about Steiner trees and have even created efficient algorithms that approximate optimal solutions. But unfortunately, as with the traveling salesperson problem, there is no efficient algorithm that always finds the desired minimum distance.

Chapter 5 Review

1. Write a summary of what you think are the important points of this chapter.
2. Is the following graph planar or nonplanar? If it is planar, redraw it without edge crossings.

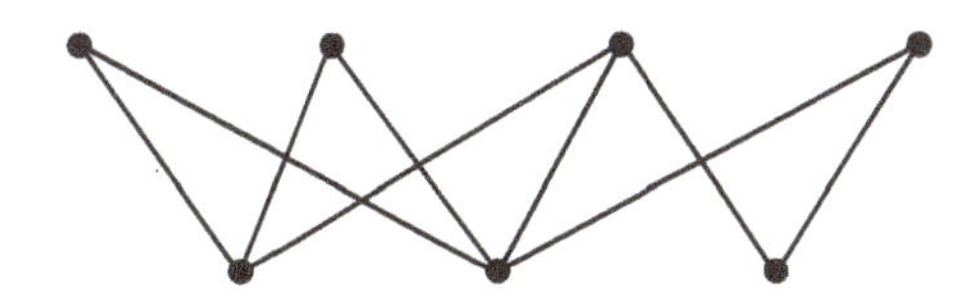

3. Show that the graph described by the following adjacency matrix is planar.

$$\begin{array}{c} \\ A \\ B \\ C \\ D \\ E \end{array} \begin{array}{c} \begin{array}{ccccc} A & B & C & D & E \end{array} \\ \begin{bmatrix} 0 & 1 & 1 & 1 & 0 \\ 1 & 0 & 0 & 1 & 1 \\ 1 & 0 & 0 & 1 & 1 \\ 1 & 1 & 1 & 0 & 1 \\ 0 & 1 & 1 & 1 & 0 \end{bmatrix} \end{array}$$

4. What is the chromatic number for each of the following?
   a. any tree with five vertices
   b. any tree with an odd number of vertices
   c. any tree with an even number of vertices
   d. any tree with two or more vertices

5.

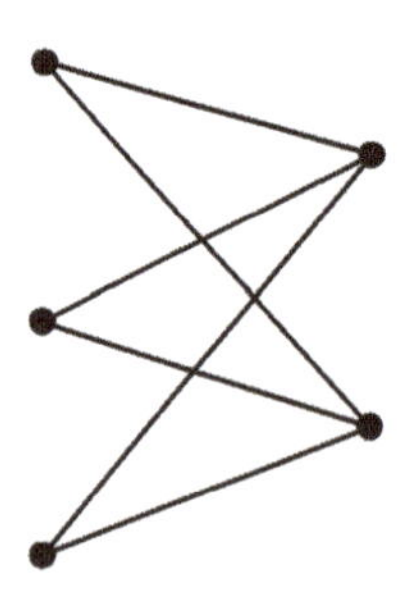

a. Explain why this graph is a bipartite graph.

b. Is this graph a complete bipartite graph? Explain why or why not.

c. Is this graph planar? If so, find a planar drawing for the graph.

d. What is the chromatic number for this graph?

6. For the following graph, draw a subgraph that has 4 vertices and 4 edges.

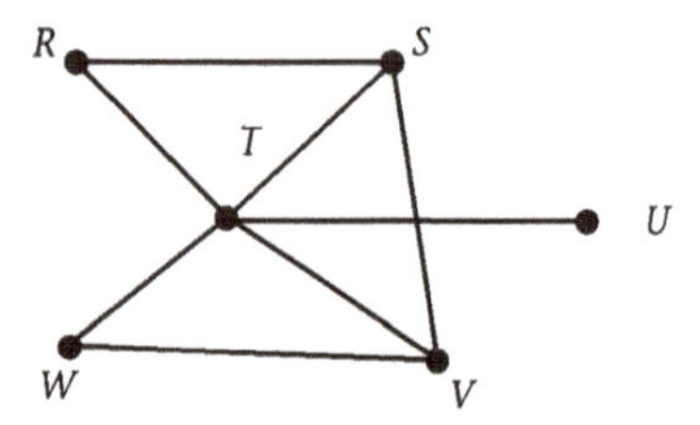

7. Mr. Gonzalez, the principal at Central High School, leaves his office once an hour to visit the math, science, and civics classrooms, and then returns to his office. The distances between rooms are shown on the following graph.

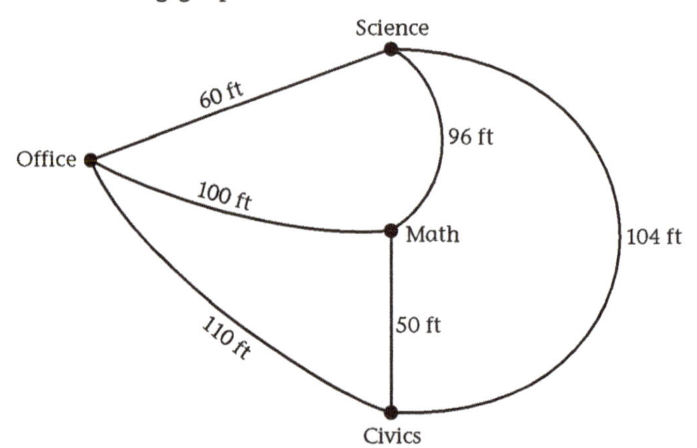

a. Find the shortest route possible for Mr. Gonzalez.

b. What is the total distance of the shortest route in part a?

c. What kind of circuit does Mr. Gonzalez make?

8. For the following complete graph, find the total weight of the nearest-neighbor route starting at A.

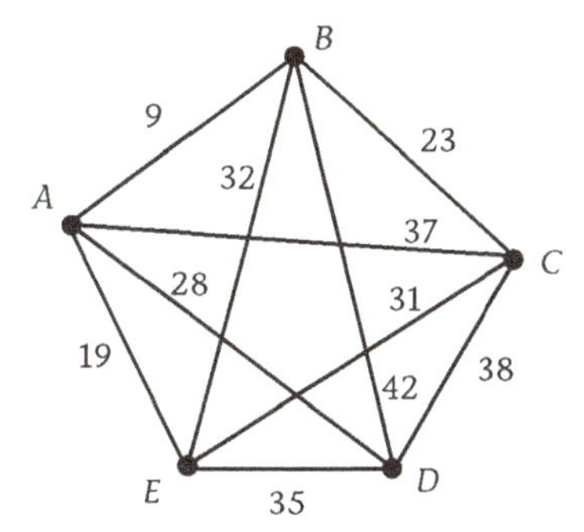

9. Find a spanning tree for the following graph if one exists.

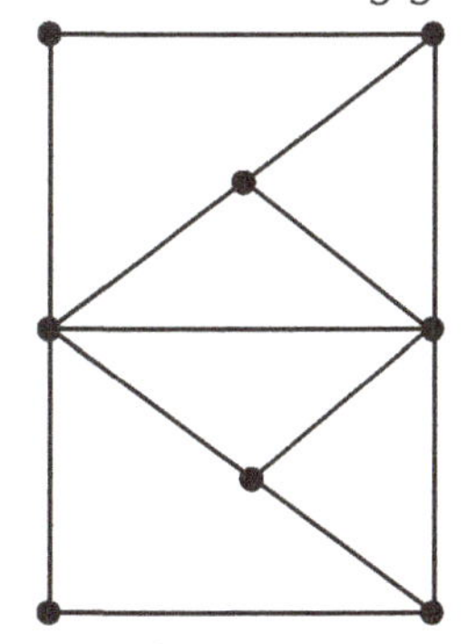

10. How many different spanning trees are there for a cycle with 3 vertices? With 4 vertices? With 5 vertices? With n vertices?

11. For the following graph, explain why the darkened edges are not a spanning tree.

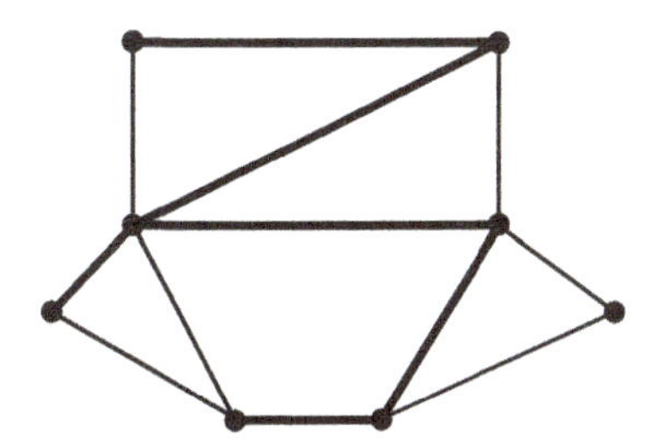

12. When given the position where it is currently located and its destination, a certain robot car is programmed to find the shortest path for the trip. The routes that the car can travel are shown on the following graph.

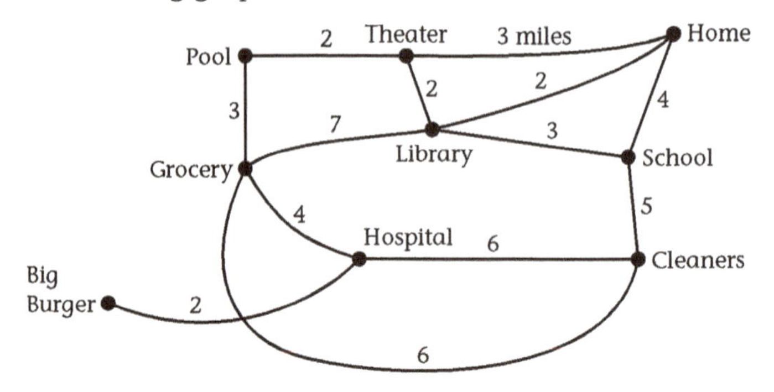

a. Use inspection to find the shortest path from Home to Big Burger.

b. What is the minimum distance from Home to Big Burger?

c. Use the shortest path algorithm (page 248) to find the shortest paths from Home to each of the other locations on the graph.

13. Assume that all locations represented by the graph in Exercise 12 need to be connected by cable. Find the minimum amount of cable needed to link the nine locations.

14. Are the following graphs trees? Explain why or why not.

a.

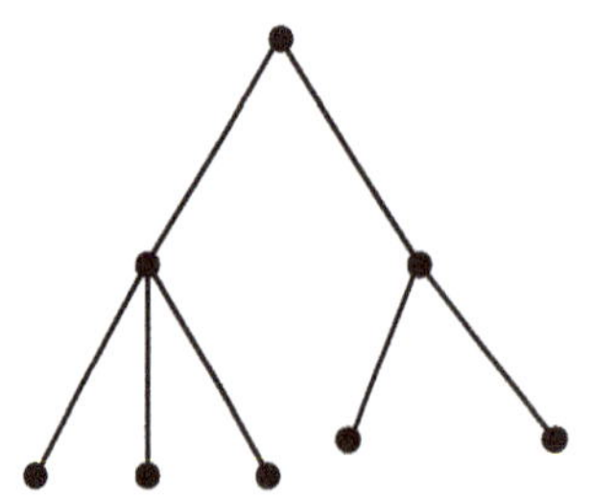

b.

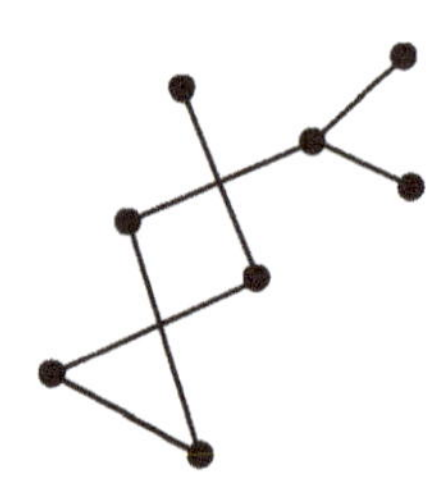

c. d.

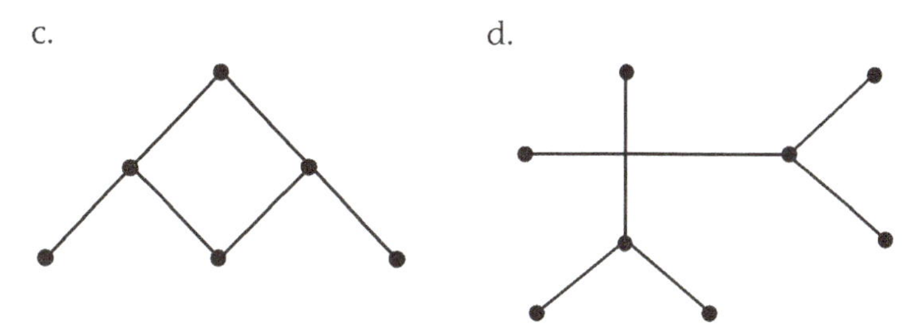

15. What happens to a tree if the number of edges is increased by one?

16. Use the breadth-first search algorithm from page 264 to find a spanning tree for the following graph. Begin the algorithm at the vertex labeled *S*.

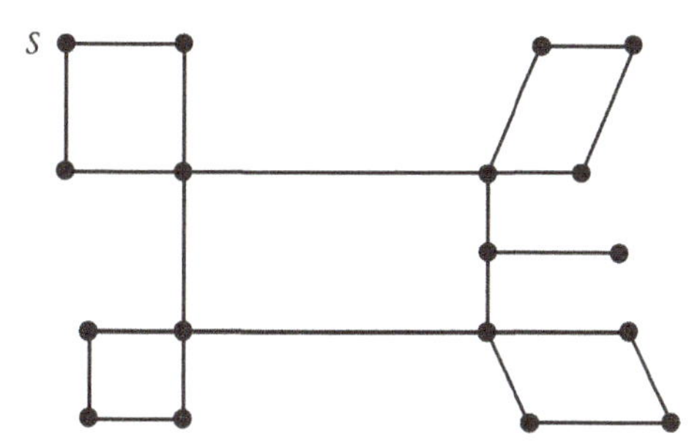

17. The vertices of the following graph represent buildings on a small college campus. Administrators at the campus want to connect the buildings with fiber-optic cable and are interested in finding the least expensive way of doing so. The costs of connecting buildings (in thousands of dollars) are shown as weighted edges of the graph.

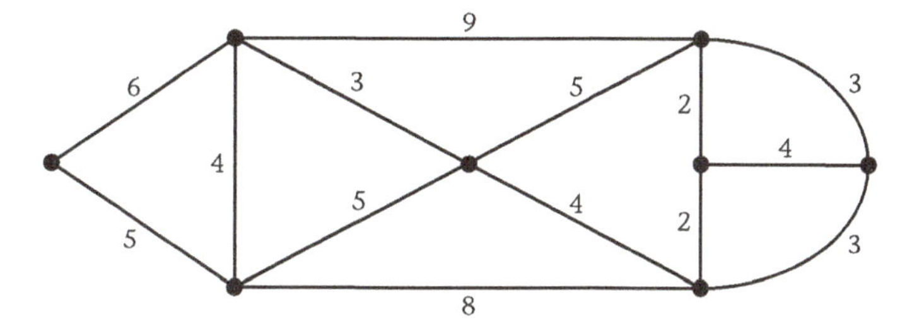

a. Use one of the spanning tree algorithms to find a minimum spanning tree for the graph.

b. What is the total cost of connecting the buildings?

18. Draw a tree with eight vertices that has exactly four vertices of degree 1.
19. Create a problem that can be solved by using one of the minimum spanning tree algorithms and find the solution to your problem. Then give your problem to a classmate and ask him or her to solve it. If the answer differs from yours, determine the correct solution.
20. You roll a die with faces numbered 1–6. Then you flip a coin. Draw a tree diagram to show the possible outcomes.
21. Represent (4 – 3) * 8 + 5 as a binary expression tree.
22. Find a postorder listing for the following binary tree.

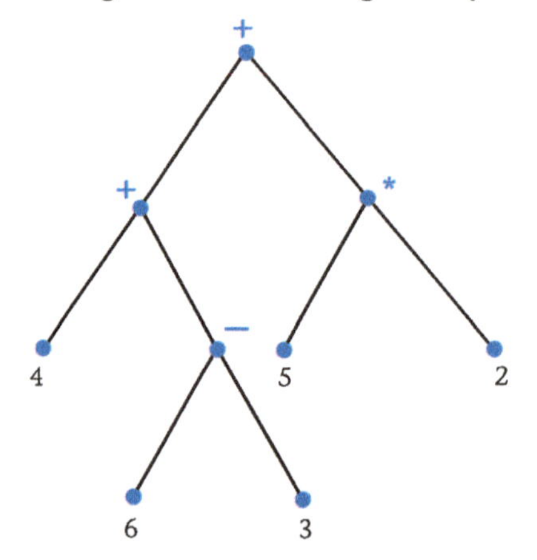

23. Evaluate the following reverse Polish notation.

    7 1 + 3 * 2 4 + -

24. Create a mathematical expression of your own, represent it with an expression tree, and find the postorder listing for the tree. Give your listing to another student in your class and have him or her evaluate the reverse Polish notation. Check the value of the notation with the value of your original mathematical expression.

Bibliography

Chartrand, Gary. 1985. *Introductory Graph Theory.* Mineola, NY: Dover Publications, Inc.

Chavey, Darrah. 1992. *Drawing Pictures with One Line.* (HistoMAP Module 21). Lexington, MA: COMAP, Inc.

COMAP. 2013. *For All Practical Purposes: Mathematical Literacy in Today's World.* 9th ed. New York: W. H. Freeman.

Cook, William J. 2012. *In Pursuit of the Traveling Salesman: Mathematics at the Limits of Computation.* Princeton, NJ: Princeton University Press.

Copes, W., C. Sloyer, R. Stark, and W. Sacco. 1987. *Graph Theory: Euler's Rich Legacy.* Providence, RI: Janson.

Cozzens, Margaret B., and R. Porter. 1987. *Mathematics and Its Applications.* Lexington, MA: D. C. Heath and Company.

Cozzens, Margaret B., and R. Porter. 1987. *Problem Solving Using Graphs.* Lexington, MA: COMAP, Inc.

Crisler, Nancy, and Walter Meyer. 1993. *Shortest Paths.* Lexington, MA: COMAP, Inc.

Dossey, John, A. Otto, L. Spence, and C. Vanden Eynden. 2006. *Discrete Mathematics.* 5th ed. Upper Saddle River, NJ: Pearson.

Francis, Richard L. 1989. *The Mathematician's Coloring Book.* (HiMAP Module 10). Lexington, MA: COMAP, Inc.

Ore, O. 1990. *Graphs and Their Uses.* Washington, DC: Mathematical Association of America.

Queen Mary, University of London. "How bumblebees tackle the traveling salesman problem." *ScienceDaily,* 29 Jun. 2011. Web. 8 Feb. 2013.

CHAPTER

6

Counting and Probability

Chance and risk play a role in everyone's life. No doubt you have often heard questions like "What are the chances?" Some risks are avoidable, but others are not. For example, everyone lives with the risk of catching a cold, but people do not have to play gambling games such as lotteries. The popularity of lotteries attests to the fascination that people have with risk: In 2011, Americans spent $251 per person on lotteries.

Often questions of probability are associated with questions about counting. In order to determine the probability of winning a lottery jackpot, a mathematician must calculate the number of ways of selecting numbers from those that are available.

- In how many ways can a lottery participant choose several numbers from those on a lottery ticket?
- What is the probability of winning a lottery jackpot?
- What is the probability that a medical test's results are correct?
- How has an understanding of probability helped improve the reliability of U.S. space shuttle launches?

In this chapter, you will examine a variety of questions about counting and probability that are fundamental to modeling random events.

Lesson 6.1

A Counting Activity

Probability calculations are important in many modeling situations. A meteorologist, for example, must calculate the probability of rain, a lottery commission must calculate the probability a player will win, and a medical researcher must calculate the probability that the results of tests are correct.

Many probability calculations require knowing the number of ways in which an event can happen, such as the number of ways a lottery player can fill out a lottery card. Often the numbers involved are quite large, and careful methods must be used to be sure counting is done properly. However, the best way to begin your work is by considering some situations involving relatively small numbers.

The Wizard of Id by Brant Parker and Johnny Hart. Reprinted by permission of John L. Hart FLP, and Creators Syndicate, Inc.

Explore This

The Central High School student council is discussing three fundraising proposals. Pierre suggests that the council operate a game at the annual school fair. His idea is to write each of the letters of the school's team name, *Lions*, on a Ping-Pong ball and have participants draw two of the balls from an opaque container. If the letters spell (in the order drawn) a legal word, the participant wins a prize. Some council members are critical of the idea because they think the game is too easy to win.

Hilary proposes printing cards with the numbers 1 through 9 displayed in a square matrix and having participants mark two of the numbers (see Figure 6.1). A winning pair is generated at random, and a prize is given to any participant who matches both winning numbers. Her scheme leaves council members uncertain about how many winners can be expected in a school of 1,000 students.

1	2	3
4	5	6
7	8	9

Figure 6.1. A card for Hilary's game.

Chuck also wants to operate a game at the school fair. His game involves a board with the numbers 1 through 6 displayed (see Figure 6.2). A participant places $1 on any of the numbers, rolls two dice, and wins a dollar for each time the chosen number appears. Several council members feel that the organization would lose money on this game.

1	2	3	4	5	6

Figure 6.2. The board for Chuck's game.

Following are three sets of questions related to the games suggested by Pierre, Hilary, and Chuck. As time permits, discuss one or more of the three sets with a few other people.

Here is one way to divide the sets of questions among small groups in your class. At the direction of your instructor, divide your class into groups of three people. Write the numbers 1, 2, or 3 on each of several slips of paper. Have each group draw one of the slips from a bag or box. Each group should consider the set of questions whose number corresponds to the number drawn.

5,631 People Win N.Y. Lotto with 911 on 9/11

wnbc.com
September 14, 2002

On the first anniversary of the terrorist attacks on New York City, a date often referred to as simply 9/11, the evening numbers drawn in the New York Lottery were 911.

Lottery officials said Thursday that 5,631 people selected the winning sequence. They will each win $500. Others won with box bets.

The liability limit for the midday and evening draws is $10,000 in sales to ensure a maximum payout. The maximum for a "straight" bet, or matching the exact sequence of winning numbers, is $500 for a $1 ticket. Bettors can also win $250 on a 50-cent ticket.

On any given day, 7–10 sets of numbers are "closed out." By 5:09 p.m. Tuesday, the 9-1-1 combination for both midday and evening draws Wednesday had reached its limit.

According to past winning numbers listed on the Lottery Website, Wednesday was the first time in more than a year that the 911 combination came up.

A similar coincidence occurred Nov. 12 when the numbers 5-8-7 came up in the New Jersey Lottery the day American Airlines Flight 587 crashed on the New York coast.

After all groups have finished their discussions, a spokesperson for each group should present the results of the group's discussion to the class. The groups that discussed set 1 should report first, and so forth.

1. Analyze Pierre's proposal. How many different two-letter "words" are there? How many of them are real words? If each of the school's 1,000 students enters exactly once and pays a $1 entry fee, how many winners might there be? How much should each winner receive if the council hopes to raise $500?
2. Analyze Hilary's proposal. In how many ways can a student fill in the entry form? If each of the school's 1,000 students enters exactly once and pays a $1 entry fee, how many winners might there be? How much should each winner receive if the council hopes to raise $500?
3. Analyze Chuck's proposal. In how many ways can the two dice fall? How often would the council pay the participant $1? $2? How often would the council make $1? Do you think that the council can raise $500 if that is the goal? (If you want to try the game, check with your teacher to see if dice are available in your classroom.)

Exercises

1. One of the goals of this chapter is to develop a few techniques that can be used to determine the number of ways in which an event can happen. The most basic such technique is making a list of all possible ways. This is a reasonable method as long as the number of items in the list is not too large. Make a list of all possible "words" that can be made by using two letters of the word *Lions*.

2. Suppose the Ping-Pong balls in Pierre's game are drawn one at a time, and the first is kept out of the container while the second is drawn. How many different letters could appear on the first Ping-Pong ball? The second? What is the connection between these numbers and the number of "words" in the list you made in Exercise 1? If the school's team name were *Tigers*, how many "words" of two letters would there be?

3. Make a list of all possible ways of choosing a pair of numbers from the nine available on one of Hilary's cards. How many are there?

4. If you are filling in one of Hilary's cards, in how many ways can you select your first number? After you've picked your first number, in how many ways can you pick your second number? How are these two numbers related to the number of pairs you listed in Exercise 3?

5. Since Chuck's game involves two dice, it is important to be able to distinguish them. Therefore, imagine the dice are different colors, say red and green. One way the dice can fall is the red die a 3 and the green die a 4. This can be written in shorthand as the pair (3, 4). This outcome is different from the red die a 4 and the green die a 3, which can be written as (4, 3). Make a list of all possible ways the red die and the green die can fall together. How many pairs are in your list?

6. The red die can land in six different ways, as can the green. How are these two sixes related to the number of things in the list you made in Exercise 5?

A second counting technique is known in mathematics as the *fundamental multiplication principle*. It says that if events A and B can occur in a and b ways separately, then there are $a \times b$ ways that the events can occur together.

To use the principle, make a blank for each event and write the number of ways each event can occur in a blank. Then multiply these numbers. For example, to determine the number of ways that a die and a coin can fall together, make two blanks: ___ ___. Then write the number of possibilities for the die and the coin in the blanks: 6 2, and multiply to get 12.

If a full list of the 12 is needed, a systematic way to ensure that all items are listed is to make a tree diagram like that shown here.

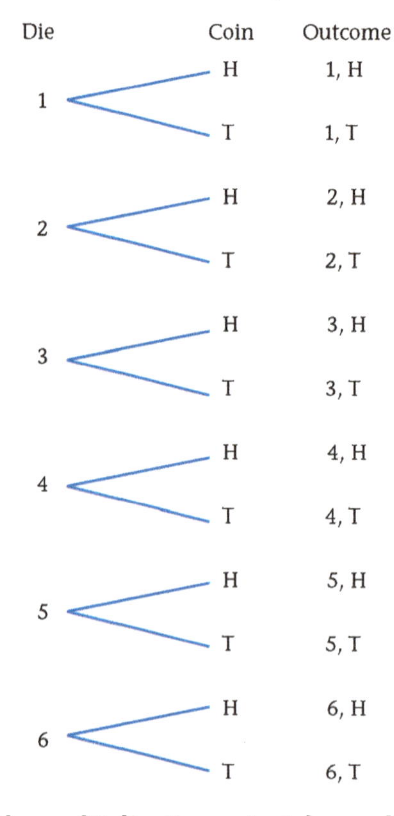

7. Explain how the multiplication principle can be applied to find the number of different "words" of two letters that can be made from the letters of *Lions*.

8. A utility company in North Dakota once sponsored a contest to promote energy conservation. The contest was to find all legal words that can be made from the letters of *insulate* without using any letter more than once.

   a. Use the multiplication principle to determine the number of "words" of two letters that are possible.

   b. The multiplication principle can be extended to three or more events. Show how to apply the principle to determine the number of "words" that can be made from three letters of *insulate.*

9. It is possible to modify the multiplication principle to find the number of ways of selecting two numbers on one of Hilary's cards. Explain how this can be done.

10. Why is it necessary to modify the multiplication principle in Exercise 9?

11. Lotteries often require the participant to select several numbers from a collection of numbers printed on a card. If a state lottery has the numbers 1 through 25 printed in a square matrix, in how many ways can a participant select two of them? Explain.

12. Explain how the multiplication principle can be used to determine the number of ways in which two dice can fall.

13. Make a tree diagram to show all the outcomes when a red die and a green die are tossed together.

14. Make a tree diagram to show all the possibilities when filling out one of Hilary's cards.

15. You are playing Chuck's game and decide to bet on the number 5.

   a. Use the tree diagram you made in Exercise 13 to count the number of ways in which you can win or lose. In how many ways can you win $1? $2? In how many ways can you lose $1?

   b. If you played many times, do you think you would win or lose money in the long run? Explain.

16. In a common carnival dice game, three dice are rolled. Use the multiplication principle to determine the number of ways in which three dice can fall.

17. Read the news article on page 300. Use the counting techniques you learned in this lesson to find the number of ways that a player can place a bet in the New York lottery game described in the article.

18. Counting techniques are useful in many modeling situations other than the analysis of games. An example is genetics. As you may know from your study of biology, a female inherits an X chromosome from her mother and another X chromosome from her father. A male inherits an X chromosome from his mother and a Y chromosome from his father. Use the counting techniques you learned in this lesson to explain the different ways in which chromosomes can be passed from parents to offspring.

Computer/Calculator Explorations

19. Many calculators have built-in random-number generators that can be modified to simulate random situations. Adapt the random-number generator of a calculator to simulate the games proposed by Pierre, Hilary, and Chuck. Present your work to the members of your class.

Projects

20. Research and report on the impact of lotteries in the United States. What are the benefits and problems associated with lotteries?

Lesson 6.2

Counting Techniques, Part 1

In Lesson 6.1 you examined several situations in which the answer to the question, "In how many ways can this be done?" is important. Two techniques that can be used to answer this question are the multiplication principle and making a complete list. The latter can often be done with the aid of a tree diagram or some other systematic procedure such as an algorithm.

In this and the next lesson, you will consider other counting techniques, beginning with the addition principle.

The Addition Principle

Recall that the multiplication principle says that if events A and B can occur in a and b ways, respectively, then events A and B can occur together in $a \times b$ ways. The **addition principle** says that if A and B can occur in a and b ways, respectively, then either event A or event B can occur in $a + b$ ways.

For example, if the student council at Central High consists of 17 members, of which 9 are girls and 8 are boys, and if one girl *and* one boy are to be selected to hold two different offices on the council, then there are $9 \times 8 = 72$ ways of filling the two offices. If a single student, who may be either a boy *or* a girl, is to be selected to hold a single office, then there are $9 + 8 = 17$ ways of making the selection.

The word *and* in the description of an event often indicates that the multiplication principle should be used, and the word *or* often indicates that the addition principle should be used.

The events "selecting a boy" and "selecting a girl" are called **mutually exclusive** or **disjoint** because a person cannot be both a boy and a girl. On the other hand, events such as "selecting a member of your school's football team" and "selecting a member of your school's basketball team" are not mutually exclusive if there is a person who is a member of both teams. When events are not mutually exclusive, the addition principle requires a modification that you will consider in this lesson's exercises.

Using the Multiplication and Addition Principles Together

The multiplication and addition principles are often used together, as the following example shows. The Central High council members are considering a contest in which words of any length are made from the team name *Lions.*

A word of one letter may be composed in only five ways: *l, i, o, n,* and *s*. A word of two letters requires a first letter *and* a second letter, which must be different from the first letter. Thus, there are $5 \times 4 = 20$ ways of composing a word of two letters. A word of three letters requires a first letter and a second letter and a third letter, so there are $5 \times 4 \times 3 = 60$ ways of composing a word of three letters. Similarly, there are $5 \times 4 \times 3 \times 2 = 120$ ways of composing a word of four letters and $5 \times 4 \times 3 \times 2 \times 1 = 120$ ways of composing a word of five letters.

A word may be composed by using one letter *or* by using two letters *or* by using three letters *or* by using four letters *or* by using five letters. Therefore, the total number of words is $5 + 20 + 60 + 120 + 120 = 325$. (Note that the events "composing a word by using one letter," "composing a word by using two letters," "composing a word by using three letters," and "composing a word by using four letters" are mutually exclusive.)

Factorials, Permutations, and Probability

The calculation of the number of words of five letters that can be made from the letters of *Lions* requires multiplying all integers from 1 through 5. This product is an extension of the multiplication principle and is known as the **factorial** of 5, or just 5 factorial. A factorial is symbolized by an exclamation mark: 5!. Most calculators have a factorial key or function. If you have never done a factorial on your calculator, try doing so now.

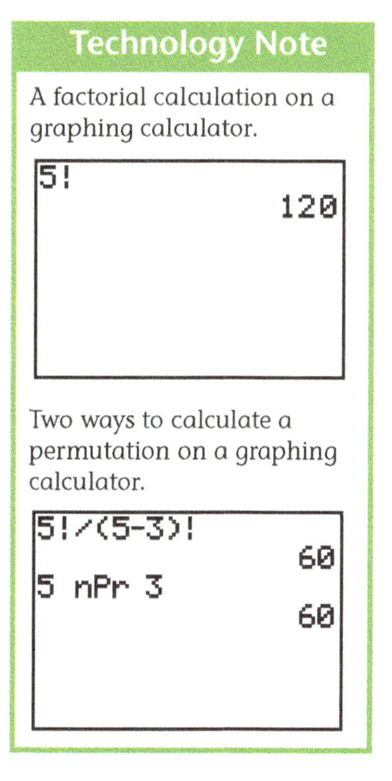

The term **permutation** is often used to describe an ordering (or arrangement) of several objects. For example, the game proposed by Pierre is one in which order matters: the words *is* and *si* are not the same. (Situations in which order does not matter are considered in Lesson 6.3.)

The number of permutations in a situation can be computed by using your calculator's factorial key. For example, to find the number of "words" of three letters that can be formed from the letters of *lions,* divide 5! by 2!. Note that this calculation produces the correct result because

$$5 \times 4 \times 3 = \frac{5 \times 4 \times 3 \times 2 \times 1}{2 \times 1}.$$

There are two commonly used symbols for the number of permutations of three things from a group of five: $P(5, 3)$ or ${}_5P_3$. Either is calculated by evaluating the expression $5!/(5 - 3)!$. (Many calculators have a special permutation function.)

In general, $P(n, m)$ is calculated by evaluating the expression $\frac{n!}{(n-m)!}$

Lesson 6.1 began with several questions about the frequency with which certain events can occur. An event's **probability** is the ratio of the number of ways the event can occur to the total number of possibilities in that situation. For example, there are 325 "words" that can be formed from the letters of *Lions,* but only 20 of them have two letters. Thus, the probability of forming a two-letter "word" from the letters of *Lions* is 20/325. Probabilities can be expressed as fractions, decimals, or percentages. As a decimal, the probability of forming a two-letter "word" from *Lions* is .0615, so you would expect a two-letter word about 6 times out of 100.

Because the numerator of a probability is never smaller than 0 and never larger than the denominator, probabilities always range between 0 and 1, inclusive. An event that cannot happen has probability 0; an event that is certain to happen has probability 1.

You now have several counting techniques at your disposal:

1. Making a list of all possibilities, for which tree diagrams are often helpful
2. The multiplication principle and the related factorial and permutation formulas
3. The addition principle

Skill at using these techniques develops with practice. The following exercises help develop that skill and also demonstrate some refinements of the three techniques.

Exercises

1. Which is equivalent to $P(10, 4)$, $10!/4!$ or $10!/6!$? Find the value of $P(10, 4)$.
2. At right are the final *USA Today* 2012–2013 season rankings of high school girls basketball teams. If you are a sportswriter voting for the top teams and you can rank only your top 5, in how many ways can you form your ranking from the 25 teams shown?
3. A multi-speed bicycle has a chain that is moved to change the bicycle's speed. The rider uses the bicycle's front and rear shift mechanisms to move the chain from one front or rear sprocket to another.

   a. If a bicycle has three front sprockets and five rear sprockets, how many speeds does it have?
   b. Is it correct to say that a particular speed requires a particular front sprocket *and* a particular rear sprocket, or is it correct to say that a particular speed requires a particular front sprocket *or* a particular rear sprocket?

Super 25 Girls Basketball Rankings

USA Today
April 9, 2013

1. Duncanville High School, Duncanville TX
2. St. Mary's High School, Phoenix AZ
3. Riverdale High School, Murfreesboro TN
4. Marion County High School, Lebanon KY
5. Dutch Fork High School, Irmo, SC
6. St. John's College High School, Washington DC
7. Bishop O Dowd High School, Oakland CA
8. Hopkins High School, Hopkins MN
9. Malcom X. Shabazz High School, Newark NJ
10. Windward School, Los Angeles CA
11. Incarnate Word Academy, St. Louis MO
12. Mater Del High School, Santa Ana CA
13. Southwood High School, Shreveport LA
14. Millbrook High School, Raleigh NC
15. Dr. Phillips High School, Orlando FL
16. Hoover High School, Birmingham AL
17. St. Mary's High School, Stockton CA
18. Marian Catholic High School, Chicago Heights IL
19. Bedford North Lawrence High School, Bedford IN
20. Norcross High School, Norcross GA
21. Fairmont High School, Kettering OH
22. Spring-Ford Senior High School, Royersford PA
23. North Point High School, Waldorf MD
24. Grand Haven High School, Grand Haven MI
25. Long Island Lutheran High School, Glen Head NY

4. (See Exercise 8 in Lesson 6.1, page 303.)

   a. How many different "words" of any length can be made from the letters of *insulate*? (Hint: You can make a word of one letter *or* a word of two letters *or* a word of three letters . . . *or* a word of eight letters.)

   b. A group of students is considering entering the contest by programming a computer to print all possible "words" that can be made from the letters of *insulate* and then checking the list against an unabridged dictionary. If the computer prints the words in four columns of 50 words each on a page of paper, how many pages are needed?

5. Some states have vehicle license "numbers" with three letters followed by three digits. Often the letters *I, O,* and *Z* are not used because they can be confused with the numerals 1, 0, and 2, respectively.

   a. If these restrictions apply and if characters may be repeated, how many different license plates are possible?

   b. What is the probability that a vehicle selected at random has a license number that begins with *CAT*?

6. a. In how many ways can the coach of a baseball team arrange the batting order of nine starting players?

   b. A sportscaster once suggested that a baseball team try every possible batting order for its nine starters in order to determine which one worked best. If a team decides to do so and plays one game each day of the week with a different batting order in each game, how long will it take to complete the experiment?

7. Three math students and three science students are taking final exams. They must be seated at six desks so that no two math students are next to each other and no two science students are next to each other.

   a. In how many ways can the students be seated if the desks are in a single row? (Hint: Draw six blanks and use the multiplication principle.)

   b. What is the probability that a math student occupies the first seat in the row?

   c. What is the probability that math students occupy the first seat and the last seat?

   d. What is the probability that a math student occupies either the first seat or the last seat?

8. The multiplication principle states that the number of permutations of the letters of the word *math* is 4!. The permutation formula says that $P(4, 4)$ is $\frac{4!}{(4-4)!}$. The denominator of this expression is 0!, which is meaningless. However, for 4! and $\frac{4!}{(4-4)!}$ to give the same result, what value must 0! have? Explain.

9. The U.S. Postal Service began using five-digit zip codes in 1963. Every post office was given its own zip code, which ranged from 00601 in Adjuntas, Puerto Rico, to 99950 in Ketchikan, Alaska.

   a. If the only five-digit zip code that could not be used was 00000, how many zip codes were possible in 1963?

   b. Some five-digit zip codes are prone to errors because they are still legal five-digit zip codes when read upside down. When this happens, a letter goes to the wrong post office and must be returned. How many zip codes are legal when read upside down? (Hint: Draw five blanks, think carefully about which digits can go in each blank, and apply the multiplication principle.)

   c. How many of the zip codes you counted in part b are not prone to errors because they read the same when turned upside down?

10. a. In how many ways can a person draw two cards from a standard 52-card deck if the first card is returned to the deck before the second card is drawn?

    b. In how many ways can two cards be drawn if the first card is not returned?

11. The addition principle cannot be used as stated in this lesson when two events are not mutually exclusive. For example, if there are 15 people on your school's basketball team and 40 people on your school's football team, then there are 55 ways of choosing one person from either team only if there are no people on both teams.

    a. If there are eight people who play both football and basketball, in how many ways can a person be selected from either team?

    b. Write the appropriate number of people in each of the three regions of the following Venn diagram.

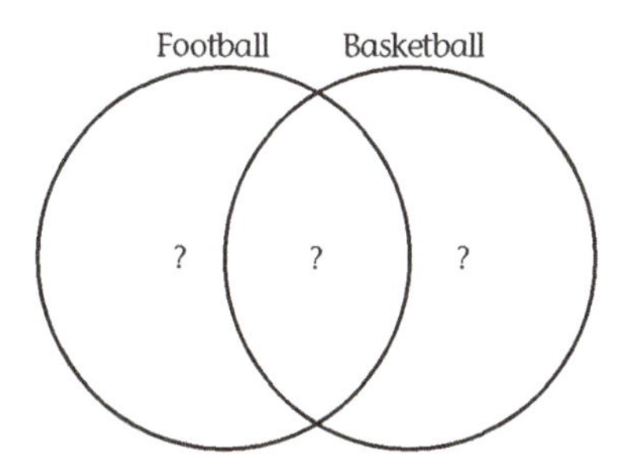

c. Describe how the addition principle is applied when two events are mutually exclusive and how it is applied when two events are not mutually exclusive.

d. Event A and event B can occur in a and b ways, respectively, and events A and B have c items in common. Write an algebraic expression for the number of ways in which event A or event B can occur.

e. Central High's soccer team has 37 members, and its basketball team has 14 members. If there are a total of 43 students involved, how many are on both teams? Explain.

12. Before the 1992 major league baseball season began, Joe Torre, who then managed the St. Louis Cardinals, said he had picked his starting lineup. He also said he had determined his first three batters but not the order in which they would bat. In how many ways could Joe arrange his batting order if the pitcher bats last? (Hint: Draw nine blanks and apply the multiplication principle.)

13. a. In how many different ways can a teacher arrange 30 students in a classroom with 30 desks?

b. The radius of the earth is approximately 6,370 kilometers. A standard medical drop is $\frac{1}{10}$ cubic centimeter. Use the formula for the volume of a sphere, $V = \frac{4}{3}\pi r^3$, to find the volume of the earth in drops of water. Compare this with the number of seating arrangements in part a.

14. There are three highways from Claremont to Upland and two highways from Upland to Pasadena.

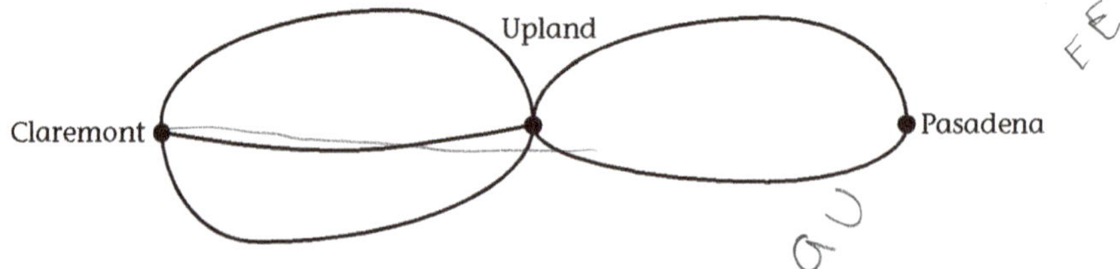

a. In how many ways can a driver select a route from Claremont to Pasadena?

b. Is it correct to say that a trip from Claremont to Pasadena requires a road from Claremont to Upland *and* a road from Upland to Pasadena, or is it correct to say that the trip requires a road from Claremont to Upland *or* a road from Upland to Pasadena?

c. In how many ways can a driver plan a round-trip from Claremont to Upland and back?

15. Radio station call letters in the United States consist of three or four letters, of which the first must be either a *K* or a *W*. Assuming that letters may be repeated, determine the number of radio stations that can be assigned call letters.

The logo for radio station KURE at Iowa State University.

16. Six different prizes are given by drawing names from the 68 Central High orchestra members attending the orchestra's annual picnic. In how many ways can the prizes be given if no one can receive more than one prize?

17. Three-digit telephone area codes were introduced in 1947. At that time, the first digit could not be a 0 or a 1, the second digit could be only a 0 or a 1, and the third could be any digit except 0.

    a. How many area codes were possible?

    b. Because of a shortage of area codes, beginning in 1995 any digit became a legal second digit. How many area codes were possible after 1995?

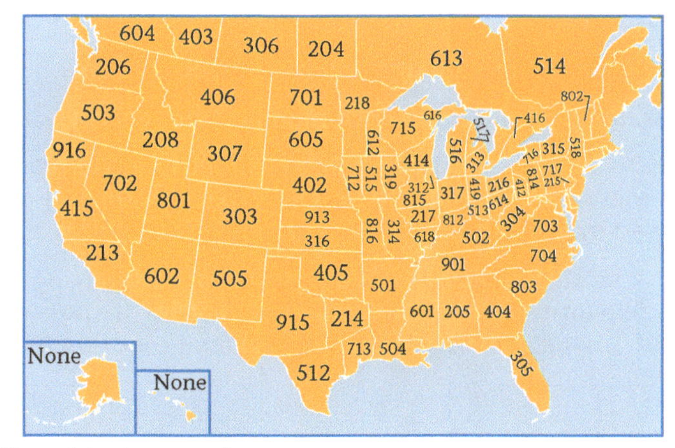

The original area codes.

New Area Code in Southern Nevada Means 10-digit Calls

Las Vegas Review-Journal
November 16, 2012

The 411 is that 702 is about to get cozy with 725. And that means we're all going to have to dial three more digits when making local calls starting in 05 of 14.

In conventional speak, new residents of Southern Nevada, from Indian Springs in the northwest to Mesquite in the northeast to Laughlin at the very bottom of the state, will be issued a 725 area code in an overlaying setup with the existing 702 area code, beginning in June 2014.

The Public Utilities Commission of Nevada on Thursday approved a second area code because available prefixes in the existing 702 area code will be exhausted in 2014.

Because an overlaying area code will mean a new next-door neighbor will have a 725 area code while yours remains 702, local calls throughout the valley will require using an area code.

Overlaying area codes are becoming more common across the United States. Other cities about to implement overlaying area codes are Boston and San Jose.

18. The UPC bar codes consist of two groups of five digits each. One group, as assigned by the Uniform Code Council in Dayton, Ohio, represents the manufacturer, and the other group represents the products of that manufacturer.

    a. How many different manufacturers can be encoded?

    b. How many products can each manufacturer encode?

19. Various word puzzles with cash prizes can be found in places such as Internet sites and newspapers. One type of puzzle is a modified crossword in which a clue is given and only two choices are offered.

    a. Consider such a puzzle with 20 questions, each having two possible answers. How many different entries are possible? (Hint: Imagine 20 blanks and use the multiplication principle.)

b. Someone embarks on the ambitious project of submitting every possible entry. Suppose that it takes 5 minutes to do one entry and that the piece of paper on which an entry is written is 0.003 inch thick. How long would it take to prepare the entries, and how thick a stack of paper would result?

20. A carnival game called Chuk-a-Luk is similar to the one proposed by Chuck in Lesson 6.1, except that three dice are used.

    a. In how many ways can three dice fall? Explain.

    b. Determine the number of ways you can win $1, win $2, win $3, or lose $1 in the game of Chuk-a-Luk if you win $1 for each die showing your number. (Hint: You win $2 if the first *and* second dice show your number *and* the third die doesn't, *or* if the first *and* third dice show your number *and* the second die doesn't, *or* if the second *and* third dice show your number *and* the first die doesn't. Draw several sets of three blanks and then use the multiplication and addition principles.)

    c. In the long run, do you think a player should expect to win or lose money in the game of Chuk-a-Luk? Explain.

21. The news article on page 300 discusses two types of bets that can be made by selecting three digits from 0 through 9, "straight" bets and "box" bets. Players who made a box bet won with 911, 191, or 119. Discuss the difference in the two types of bets and determine the probability of winning for each type.

22. Factorials can be described recursively. Let $f(n)$ represent $n!$. Write a recurrence relation that expresses the relationship between $f(n)$ and $f(n-1)$.

Projects

23. Research the history of the study of probability. How did it begin? What roles did Jerome Cardan, Blaise Pascal, and Pierre Fermat play? What problems interested them?

24. Investigate the number of permutations of several objects, of which some look alike. For example, the letters of *math* can be arranged in $4! = 24$ ways; how many different permutations are there of the letters of *look?* What if there are several sets of identical letters, such as in *Mississippi?* Write a summary that includes a general principle for handling such situations and several examples.

25. Investigate the number of permutations of several objects arranged in circular fashion. For example, Ann, Sean, Juanita, and Herb can be seated along one side of a rectangular table in $4! = 24$ ways. In how many different ways can they be seated around a circular table? Write a summary that includes a general principle for handling such situations and several examples. Explain your interpretation of the meaning of the word *different.*

26. Investigate the use of the addition principle with three events that are not mutually exclusive. Suppose, for example, that the football, basketball, and track teams of Central High have 41, 15, and 34 members, respectively. If 6 people play both football and basketball, 7 are on both the basketball and track teams, 15 are on both the football and track teams, and 4 play all three sports, how many people are involved in one sport or another? Develop a general principle for modeling situations of this type and draw a Venn diagram to represent it. Can the principle be extended to four or more events? How?

Lesson 6.3

Counting Techniques, Part 2

The counting techniques developed in Lessons 6.1 and 6.2 are only one method short of forming a fairly complete tool kit for modeling a variety of probability problems. In this lesson, you consider a technique for counting in situations in which the order of occurrence is unimportant.

Combinations

The game proposed by Hilary in Lesson 6.1 is a simple lottery in which participants select two numbers from nine printed on a card. Participation requires a selection, not an ordering. That is, if the winning numbers are 2 and 6, it does not matter whether the participant selects 2 or 6 first.

The term **combination** is used to describe a selection of several objects. The number of combinations in a situation can be counted by modifying the technique used to count permutations. For example, if Hilary's game requires an ordering of two numbers instead of a selection, then the number of ways of filling out a ticket is counted as a permutation: $P(9, 2) = \frac{9!}{(9-2)!} = 72$. Because this permutation counts a pair such as 2 and 6 as different from the pair 6 and 2, every possible pair is counted twice. Thus, the number of combinations of two things selected from a group of 9 is $72/2 = 36$. Two commonly used symbols for the number of combinations of two things from a group of nine are $C(9, 2)$ or ${}_9C_2$.

If Hilary's game requires picking three numbers in the proper order, then the number of ways of filling out a card is $P(9, 3) = 504$. If order does not matter, then 504 is too large. For example, if the winning numbers are 2, 5, and 8, then 504 counts any arrangement of 2, 5, and 8 as different. The number of ways of arranging 2, 5, and 8 is $3 \times 2 \times 1 = 6$. Therefore, 504 is six times too large, and $C(9, 3) = \frac{P(9, 3)}{3!} = 504/6 = 84$.

In general, $C(n, m)$ is calculated by evaluating the expression $\frac{P(n, m)}{m!}$. But, $P(n, m) = \frac{n!}{(n-m)!}$ so $C(n, m) = \frac{n!}{(n-m)!m!}$.

Technology Note

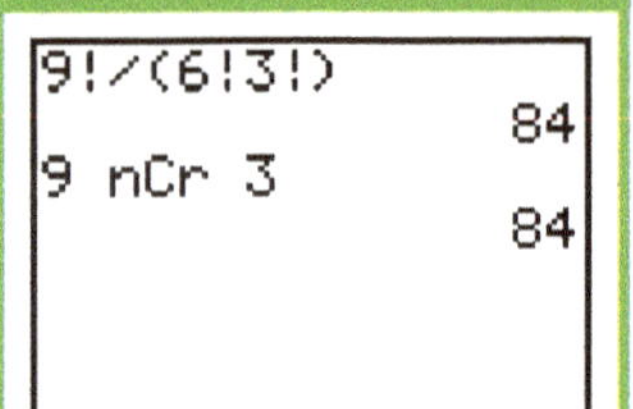

Combinations can be calculated by using a calculator's factorial function or, on some calculators, by using a combination function.

Since there are 36 ways of filling in one of Hilary's lottery tickets, the probability that any one ticket wins is 1/36, or about .028. If 1,000 tickets are sold, Hilary can expect about $1{,}000 \times .028 = 28$ winners. If the game requires the selection of three numbers, the probability a single ticket wins is 1/84, or about .012. If 1,000 tickets are sold, about $1{,}000 \times .012 = 12$ winners can be expected.

Using Combinations with Other Counting Techniques

Combinations are often used along with other counting techniques. For example, the 17-member student council at Central High consists of 9 girls and 8 boys. A committee of 4 council members is being selected. If the committee members are not arranged in a particular way (i.e., chair, secretary, and so forth), then the order of selection is unimportant. The number of ways the committee can be selected is $C(17, 4) = \frac{17!}{13!4!} = 2{,}380$.

If the committee must have two girls and two boys, there are $C(9, 2) = \frac{9!}{7!2!} = 36$ ways of selecting the 2 girls and $C(8, 2) = \frac{8!}{6!2!} = 28$ ways of selecting the two boys. Because the committee must consist of 2 girls and 2 boys, apply the multiplication principle to conclude that there are $36 \times 28 = 1{,}008$ ways of forming the committee. If the 4 committee members are selected at random, the probability that the committee consists of 2 girls and 2 boys is $\frac{1{,}008}{2{,}380}$, or about .424.

Now suppose that the committee must consist of either all boys or all girls. There are $C(9, 4) = \frac{9!}{5!4!} = 126$ ways of selecting 4 girls and $C(8, 4) = \frac{8!}{4!4!} = 70$ ways of selecting 4 boys. Because the committee must consist of either 4 girls or 4 boys and because all-boy and all-girl committees are mutually exclusive, apply the addition principle to conclude that there are 126 + 70 = 196 ways of forming the committee. Again, if the 4 committee members are selected at random, the probability the committee consists of either all boys or all girls is $\frac{196}{2,380}$, or about .082.

Exercises

1. Which is larger: $C(10, 2)$ or $C(10, 8)$?

2. Find the sum of all possible combinations of four things. That is, find $C(4, 0) + C(4, 1) + C(4, 2) + C(4, 3) + C(4, 4)$. Do the same for all possible combinations of three things and all possible combinations of five things. On the basis of your results, make a guess about the sum of all possible combinations of six things. Describe any pattern you notice.

3. In this lesson the number of all-boy four-person committees on the Central High student council is calculated as $C(8, 4) = 70$, the number of all-girl four-person committees is calculated as $C(9, 4) = 126$, and the number of four-person committees that are half boys and half girls is calculated as $C(8, 2) \times C(9, 2) = 1,008$.

   a. How many four-person committees consist of three girls and one boy?

   b. How many committees consist of one girl and three boys?

   c. Find the sum of the numbers of committees that consist of four boys, no boys, two boys, three boys, and one boy. Compare this sum with the total number of four-person committees calculated by $C(17, 4)$ in this lesson (see page 318).

4. Darrell Dewey has just left his Central High social studies class and bumped into his friend Carla Cheetham. Darrell informs Carla that Ms. Howe gave a ten-question true/false quiz today. When Carla asks about the quiz, Darrell says he found it easy and thinks that four of the answers are false.

   a. When Carla takes the quiz, in how many ways can she select four questions to mark false?

   b. In how many ways can Carla select six questions to mark true?

   c. In how many ways can Carla fill in the quiz if she ignores Darrell's hint?

5. A standard deck of cards contains 13 different cards from each of four suits: spades and clubs, which are black in color, and diamonds and hearts, which are red in color.

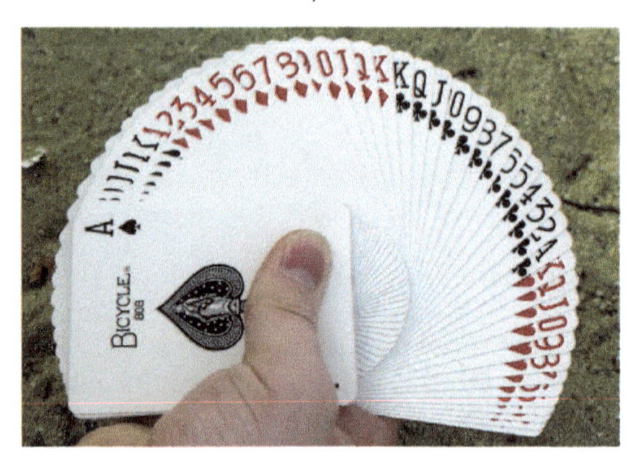

   a. In how many ways can 2 cards be dealt from a standard 52-card deck?

   b. In how many ways can 2 red cards be dealt from a standard 52-card deck?

   c. What is the probability that 2 cards dealt from a standard 52-card deck are both red?

6. Maria has a part-time summer job selling ice cream from a small vehicle she drives through residential areas of her community. She carries eight different flavors and sells a two-scoop cone for $1.80.

   a. How many two-scoop cones are possible if both scoops are the same flavor?

   b. How many two-scoop cones are possible if each scoop is a different flavor?

   c. All together, how many two-scoop cones are possible?

7. Hedy Foans, who writes a music column in the Central High *Scribbler*, decides to poll students on their favorite songs. She prepares a list of ten current favorites, from which students are asked to rank their top three. In how many ways can a student pick a first, second, and third choice from Hedy's ten?

8. Ms. Howe has a planter in one of her classroom windows that is divided into five sections. She purchases two geraniums and three marigolds to plant in the five spaces.

   a. In how many ways can Ms. Howe select the two sections in which to plant the geraniums?

   b. In how many ways can Ms. Howe select the three sections in which to plant the marigolds?

9. In February 1992, an Australian company sent representatives to Virginia in an attempt to purchase one ticket for every possible selection in the state's lottery. The representatives spread their purchases among eight retail chains that had a total of 125 outlets. One representative bought a total of 2.4 million tickets at a single retail chain headquarters. When time ran out, the group had purchased 5 million tickets, or about 70% of all possible selections.

   One of the tickets purchased by the group matched the winning numbers: 8, 11, 13, 15, 19, 20. After a controversy over the legality of the purchase, the lottery decided to award the $27 million jackpot to the Australian group, which represented about 2,500 investors who paid an average of $3,000 each. Each investor stood to receive an average of $10,800, at the rate of $540 a year over the 20-year payment period.

   a. At the time of the Australian purchase, a Virginia lottery ticket contained the numbers 1 through 44, from which a participant selected six. In how many ways can a selection be made?

   b. If it takes 5 seconds to purchase a Virginia lottery ticket, how long would it take one person working 40 hours a week to purchase a ticket for every possible selection?

   c. If each Virginia lottery form has space for five entries and if each form has a thickness of 0.003 inch, how thick is a stack of forms of all possible selections?

   d. Until October of 1999, a Florida lottery ticket contained the numbers 1 through 49, from which a participant selected six. In how many ways can this be done?

   e. An individual once bought 80,000 tickets in the Florida lottery. What was this person's probability of winning a share of the $94 million jackpot that had accrued at that time?

f. After October of 1999, the Florida lottery required the selection of six numbers from 53. How does the probability of winning the jackpot in the Florida lottery today compare with the probability of winning the jackpot prior to the change?

Frank and Ernest © reprinted by permission of Newspaper Enterprise Association, Inc.

10. Most lotteries include several prizes besides the jackpot. For example, the Florida lottery gives second prizes to tickets that match 5 of the 6 winning numbers, third prizes to those that match 4 of the 6, and fourth prizes to those that match 3 of the 6.

    a. How many different ways are there to receive a second prize? (Hint: The ticket must match 5 of the 6 winning numbers *and* 1 of the 47 non-winning numbers.)

    b. How many different ways are there to receive a third prize?

    c. How many different ways are there to receive a fourth prize?

11. Chapter 1 discusses various voting models. Suppose there are seven choices on a ballot.

    a. In how many ways can a voter rank the seven choices?

    b. Recall that when approval voting is used, the choices are not ranked. In how many ways can you select three choices of which to approve?

12. Dee Noat, the director of Central High's music department, is holding tryouts for the school's jazz band. There are 7 students competing for three saxophone positions, 8 for two piano spots, 5 for two percussion spots, and 12 for three places as guitarists. In how many ways can Dee select her band?

13. The figure below was drawn by marking nine equally spaced points on the circumference of a circle and connecting every pair of points.

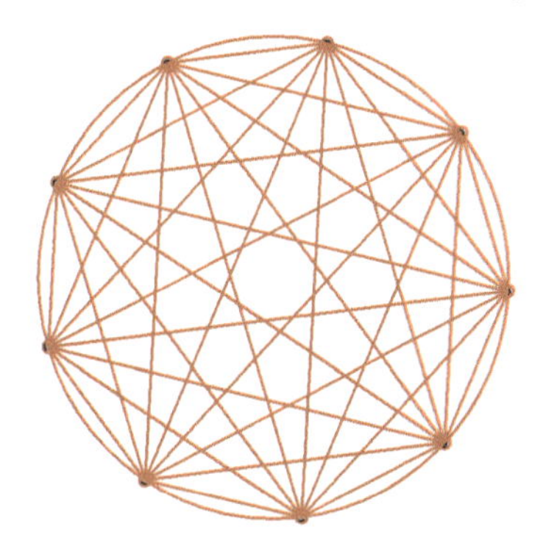

    a. How many chords are there?

    b. Number the points from 1 through 9, and explain why drawing the chords is analogous to filling out every possible selection in Hilary's lottery.

    c. Recall that a complete graph is one in which every pair of vertices is connected with an edge. How many edges are there in a complete graph with ten vertices?

14. Emily's Pizza Emporium can prepare a pizza with any one or more of nine ingredients. In how many different ways can a pizza be ordered at Emily's? (Hint: A pizza can be ordered with one ingredient, or two ingredients, or three ingredients, ...)

15. College Inn Pizza claims that it offers 105 different two-topping pizzas. How many different toppings do you think College Inn Pizza uses? Explain.

16. Carl Burns, coach of the Central High Lions basketball team, has 12 players on his squad. Of these, 3 are centers, 4 are forwards, and 5 are guards.

    a. Is it correct to say that a team requires a center *and* 2 forwards *and* 2 guards, or is it correct to say that a team requires a center *or* 2 forwards *or* 2 guards?

    b. In how many ways can Coach Burns select his starting team?

17. A telephone exchange consists of all seven-digit phone numbers with the same three-digit prefix.

    a. How many different phone numbers are possible in a given exchange?

    b. If a community has 95,000 telephone subscribers, what is the minimum number of exchanges needed?

    c. How many phone numbers are possible in a given three-digit area code? (Assume all possible exchanges are permitted.)

18. Allison Gerber, a math teacher at Central High, gives prizes to students in her class who improve their average grade. At the end of each term she places the names of all qualifying students in a container and draws three.

    a. If there are 21 qualifying students and the prizes are three Central High Lions T-shirts, in how many ways can the prizes be awarded?

    b. If there are 21 qualifying students and the prizes are a new calculator, a Lions T-shirt, and a discrete mathematics book, in how many ways can the prizes be awarded?

19. Electronic data encryption is important to most people because their bank accounts and other financial data are accessible over the Internet. For many years, the United States government certified a 56-bit encryption system. Each bit can be either a 0 or a 1.

    a. Read the news article on page 325. Explain how the number of unique DES keys is calculated.

    b. The article reports that in 2004, a key could be broken in 1/64th of the time it took to break it in 1998. Does this claim seem reasonable if the article's claim that computer speeds double every 18 months is accurate?

    c. The AES encryption system uses 128 bits. Assuming that computer speeds double every 18 months, in what year would you expect a 128-bit key to be breakable in one week using a system comparable to the one reported in the article?

    d. Does the article's use of the word "combination" seem appropriate to you? Explain.

DES Encryption is Inadequate says NIST

Computer Weekly
July 30, 2004

The National Institute of Standards and Technology (NIST) is proposing that the Data Encryption Standard (DES) lose its certification for use in software products sold to the government.

The algorithm uses a 56-bit key to encrypt blocks of data, and can produce up to 72,000,000,000,000,000 unique keys.

While that number of unique combinations was formidable in the 1970s and 1980s, given the power of computers at that time, experts were aware that the growth of computing power would, in time, render the algorithm breakable, and that DES had at most a 15-year life span, according to NIST.

By the 1990s, computers had become powerful enough that breaking the DES algorithm was achievable, even for groups with limited resources. In a 1998 experiment funded by the non-profit civil liberties group the Electronic Frontier Foundation, Paul Kocher and his colleagues designed a machine for about $250,000 that could break one DES key a week.

With computers doubling in speed every 18 months, a similar system designed with 2004 technology could presumably break a key in 1/64th of that time using so-called "brute force" methods, which essentially try every possible key combination until the correct combination is guessed.

20. Dominoes come in different-sized sets. A double-six set is the most common. In a double-six set, each half of a domino may have any number of spots from 0 through 6. The two halves of a given domino in the set pair a number of spots with the same number of spots or a different number of spots.

0-0
1-1
2-2
⋮

a. How many dominoes with the same number of spots on each half are there in a double-six set? 7

b. If every possible pairing is included in the set, how many dominoes with a different number of spots on each half are there in a double-six set? $C(7,2) = 21$

c. What is the total number of dominoes in a double-six set?

$7 + C(7,2) = 21 + 7 = 28$

d. Write a description of the way a domino in a double-six set is formed. Explain how the words *and* or *or* in your description reflect the calculations you made to obtain your answer to part c.

e. If you select a domino at random from a double-six set, what is the probability that it has the same number of spots on each half?

f. How many dominoes are there in a double-twelve set?

21. How many different sums of money can be made from a $1 bill, a $5 bill, a $10 bill, and a $20 bill? (Hint: You can use one bill at a time or two bills at a time or three bills at a time or four bills at a time.)

22. Many card games involve 5-card hands. (See the description of a standard deck of cards in Exercise 5.)

 a. How many different 5-card hands can be dealt from a standard 52-card deck?

 b. In how many ways can a selection of 3 aces be made from the 4 aces that are found in a standard deck?

 c. In how many ways can 3 cards of the same kind (aces, twos, threes, and so forth) be dealt from a standard deck? (Hint: You can deal 3 aces or 3 twos or 3 threes or . . . .)

 d. Repeat part c for 2 cards of the same kind.

 e. In how many ways can a hand consisting of 3 of one kind and 2 of another (a full house) be dealt from a standard deck?

23. To win the jackpot in the California Fantasy 5 lottery game, a participant must match 5 numbers from 39 that are available. If you buy ten tickets per week in the California lottery, about how often could you expect to win the jackpot? Explain.

24. Some bike locks allow the user to set a four-digit code that opens the lock. These locks are convenient because the user can select a familiar number and is thereby less likely to forget the code.

 a. How many different codes are possible with such a lock?

 b. Locks like these are often called combination locks. Do you think this is an appropriate name? Explain.

25. The term *odds* is often used in the media. Although related, odds are different from probabilities, and the difference can cause confusion. The definition of the odds in favor of an event is the ratio of the number of ways the event can occur to the number of ways it can fail. For example, consider a simple game in which you roll a single die and win if one or two spots show and lose otherwise. The odds in favor of your winning are 2:4 or 1:2. The probability of your winning is $\frac{2}{6}$ or $\frac{1}{3}$. The odds against your winning are 4:2 or 2:1.

   a. Based on the news article below about the 2013 Kentucky Derby, what is the probability that Revolutionary will win?

   b. Revolutionary lost the race. According to the article, what probability was assigned to this event?

   c. Odds are expressed in various ways. Colons and dashes are commonly used. But odds can also be written as fractions and converted to decimals. In the news article on the 2013 Kentucky Derby, use decimals to compare the odds in favor of horses with the probability of their winning. When are the odds and the probability nearly equal?

Kentucky Derby: Revolutionary Remains Pre-Race Favorite

SB Nation
May 4, 2013

With post time now less than an hour and a half away, the odds for the 2013 Kentucky Derby remain steady. Revolutionary, who became the favorite as odds moved due to wet conditions, is still favored at 5-1.

Goldencents remains at 7-1 odds, but has now been joined in a tie for second place by Orb, whose odds recently moved from 8-1 to 7-1.

Normandy Invasion at 8-1, Itsmyluckyday at 9-1 and Verrazano at 9-1 round out the top contenders. Falling Sky continues to have the lowest odds although the horse has moved up from 37-1 to 36-1. Vyjack remains with long odds at 29-1.

Projects

26. Research one or more of the lotteries in your area. How large are the jackpots? How many tickets are usually sold? What portion of the proceeds goes to the players? What happens to the rest of the money? Are there any rules to prevent the kind of purchase made by the Australian group in the Virginia lottery (see Exercise 9)? What kinds of strategies are known to be used by players?
27. Investigate probabilities of common card hands. Show how to calculate as many as possible.

Whist Players Astonished After Each Receives Full Suit in One Hand

The Daily Mail
November 24, 2011

It is an occurrence that comes with mind-boggling odds of a thousand quadrillion - or a thousand million million million million - to one.

But a group of whist-playing pensioners say they were stunned when each player was dealt a complete suit in an opening hand.

Wenda Douthwaite, 77, and her three friends were left 'gobsmacked' during the game in their village hall last week.

Mathematicians say the odds of this happening are a jaw-dropping 2,235,197,406,895,366, 368,301,559,999 to one.

The 28-digit figure is the equivalent odds of a person finding a specific drop of water in the Pacific Ocean.

Mrs Douthwaite, from Kineton, Warkickshire, who has attended whist drives for 50 years, said: 'We've never seen anything like it before. Everything was done as usual.

'The cards were shuffled, cut and dealt as normal but that was the only thing that was normal. And it was the first game of the night as well. As soon as I picked up my cards I saw I had a complete set of spades.

'Suddenly someone around the table said they'd got a complete suit too. We compared cards and were totally shocked when one of us had all the hearts, another had the diamonds, another had the clubs and I had the spades. I was shaking when we laid the cards down on the table.

Lesson 6.4

Probability, Part 1

The counting techniques discussed in the first three lessons of this chapter can be used to find probabilities of simple events. However, many probability models involve compound events that are composed of two or more simple events. This lesson is concerned with rules that govern operations on two or more probabilities.

Recall that the probability of an event is the ratio of the number of ways the event can occur to the total number of possibilities. For example, the probability that a die falls with an even number showing is 3/6 because three of the six possibilities are even. For convenience, the statement "the probability that a die falls even" is abbreviated *p*(a die falls even).

The Addition Principle for Probabilities

The addition and multiplication principles that you previously used to find the number of ways in which events can occur have probability counterparts.

The addition principle for probabilities states that:

$p(A \text{ or } B) = p(A) + p(B) - p(A \text{ and } B)$, for any two events A and B;

$p(A \text{ or } B) = p(A) + p(B)$, for mutually exclusive (disjoint) events A and B.

The subtraction of $p(A \text{ and } B)$ is unnecessary when events are mutually exclusive because $p(A \text{ and } B) = 0$. When applying the addition principle, do not assume events are mutually exclusive unless you are certain.

Consider Exercise 11 from Lesson 6.2 (page 311). In that exercise, 40 people are on a football team, 15 are on the basketball team, and 8 people are on both teams. To determine the total number of people involved, you add 40 and 15, then subtract 8 to get 47.

If a person is chosen at random from this group, the probability of selecting a football player is $\frac{40}{47}$, the probability of selecting a basketball player is $\frac{15}{47}$; and the probability of selecting someone who is both a football and basketball player is $\frac{8}{47}$. To determine the probability of selecting either a football or basketball player, perform a calculation similar to the one in the previous paragraph: $\frac{40}{47} + \frac{15}{47} - \frac{8}{47} = \frac{47}{47}$, or 1 (see Figure 6.3). Note that without a subtraction of $\frac{8}{47}$, the answer exceeds 1, which is impossible.

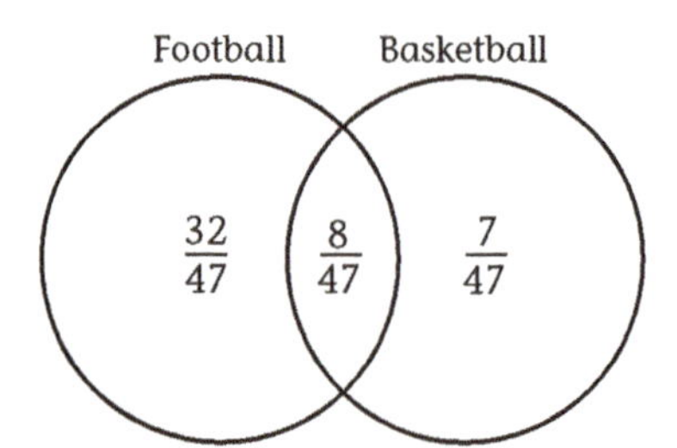

Figure 6.3. The addition principle for probabilities.

The Multiplication Principle and Conditional Probability

The multiplication principle is similar to the addition principle in that it is applied differently depending on the type of events. The addition principle can be shortened if the events are mutually exclusive; the multiplication principle can be shortened if the events are independent.

As an example, consider the following data on the student population at Central High, which has exactly 1,000 students.

	Male	Female	Total
Seniors	156	144	300 (30%)
Juniors	168	172	340 (34%)
Sophomores	196	164	360 (36%)
Total	520 (52%)	480 (48%)	

Many probabilities can be found from these data. For example, the probability of selecting a junior is .34, and the probability of selecting a girl from the junior class is $\frac{172}{340} = .5059$.

A tree diagram with appropriate probabilities written along the branches is a convenient way to organize these probabilities (see Figure 6.4).

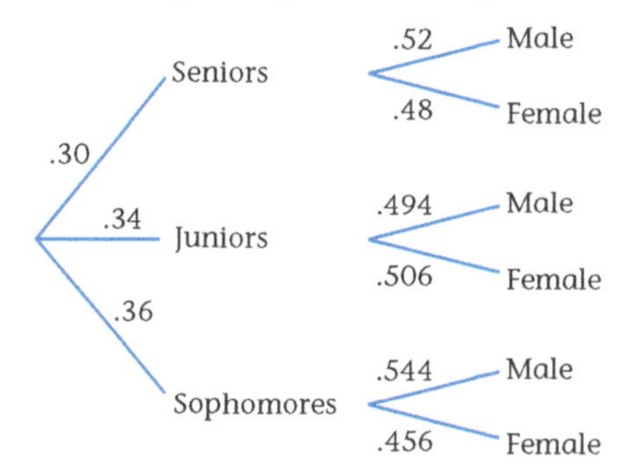

Figure 6.4. Organizing probabilities with a tree diagram.

The probability of selecting a girl from the junior class is called a **conditional probability** because the event describes the selection of a girl under the condition that the person selected is a junior. (Verbal descriptions of conditional events often use the word *from*. Other commonly used words are *if, when,* and *given that*.) The probability of *A* from *B* is sometimes written symbolically as $p(A/B)$.

Note that conditional probabilities change if the order of events is reversed. For example, the probability of selecting a girl from the juniors $\left(\frac{172}{340} = .5059\right)$ is different from the probability of selecting a junior from the girls $\left(\frac{172}{480} = .3583\right)$.

Since 172 of the 1,000 students are junior girls, the probability of selecting a student who is both a junior and a girl is .172. Refer to the tree diagram in Figure 6.4 and note that the probabilities written along the junior branch and the female branch that follows it are .34 and .506, respectively. The product of .34 and .506 is approximately .172. Thus, the probability of selecting a student who is a junior and a girl equals the product of the probability of selecting a junior and the probability of selecting a girl from the juniors.

The **multiplication principle** for probabilities states that for two events A and B, $p(A \text{ and } B) = p(A) \times p(B \text{ from } A)$. The products that result from applying the multiplication principle can be written to the right of the tree diagram, as shown in Figure 6.5.

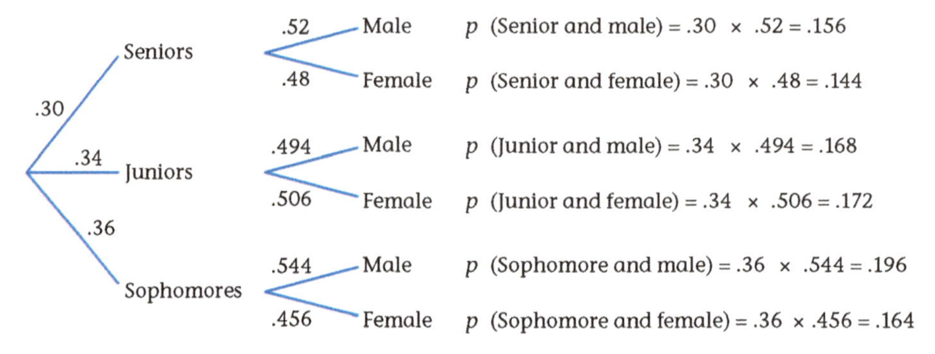

Figure 6.5. The multiplication principle for probabilities.

Refer back to the table and note that 48% of the students are girls. Refer to the tree diagram in Figure 6.4; you'll note something unusual about the senior class: the percentage of seniors who are girls is exactly the same as the percentage of girls in the entire school. For this reason, the events selecting a girl and selecting a senior are called independent.

When events are independent, the probability that one occurs is not affected by the occurrence of the other: If you want to know the probability of selecting a girl, it makes no difference if the selection is made from the entire school or from the senior class. However, note that the probability of selecting a girl from the entire school (.48) is not the same as the probability of selecting a girl from the junior class $\left(\frac{172}{340} = .5059\right)$. Therefore, the events selecting a girl and selecting a junior are not independent.

Two events A and B are **independent** if $p(B \text{ from } A) = p(B)$ or if $p(A \text{ from } B) = p(A)$. Thus, for independent events, the multiplication principle states that $p(A \text{ and } B) = p(A) \times p(B)$.

When using the multiplication principle, be careful not to assume that events are independent unless you are certain. In some cases, independence is quite obvious. For example, a toss of a coin has nothing to do with the outcomes of previous tosses: the probability that a particular toss is heads is $\frac{1}{2}$ regardless of the previous toss. This independence means that it is correct to calculate the probability of obtaining two heads in a row by multiplying $\frac{1}{2}$ by $\frac{1}{2}$ to obtain $\frac{1}{4}$.

Just as independence is obvious in some cases, so is the lack thereof. For example, a man who has a beard is more likely to have a mustache than are men in general. Therefore, having a beard and having a mustache are not independent. It is incorrect to calculate the probability that a man has both a beard and a mustache by multiplying the probability that a man has a beard by the probability that a man has a mustache.

In other cases, it can be difficult to determine whether two events are independent without inspecting data or probabilities. For example, are people who own poodles more or less likely to own pink cars than are people in general? If pink car ownership is either more or less common among poodle owners than among the general public, then the events "owning a poodle" and "owning a pink car" are not independent.

There are two ways to determine whether events A and B are independent:

1. Compare $p(A)$ with $p(A \text{ from } B)$. If they are the same, A and B are independent. (You can also compare $p(B)$ with $p(B \text{ from } A)$, but it is not necessary to make both comparisons.)
2. Multiply the probability of A and the probability of B. If the result is the same as the probability of both A and B occurring, then A and B are independent.

Example: Checking for Independence

Consider the following data on ownership of pink cars and poodles in a community. Are owning a pink car and owning a poodle independent in this community?

	Own Poodles	Don't Own Poodles
Own Pink Cars	250	450
Don't Own Pink Cars	1,250	18,350
Totals	1,500	18,800

To apply the first method of checking for independence, note that the probability of selecting someone who owns a pink car is $\frac{700}{20{,}300} = .0345$, and the probability of selecting someone who owns a pink car from the poodle owners is $\frac{250}{1{,}500} = .167$. Because the two probabilities are not equal, owning a pink car and owning a poodle are not independent.

To apply the second method of checking for independence, note the following probabilities: $p(\text{owning a poodle}) = \frac{1{,}500}{20{,}300} = .0739$,

owning a pink car) $= \frac{250}{20{,}300} = .0123$. Calculate the product of the first two probabilities and compare it with the third: $.0739 \times .0345 = .00255 \neq .0123$.

Attention to detail is important when either the addition or multiplication principles are applied. In this lesson's exercises you will use both principles in a variety of settings and consider the consequences of improper application.

Exercises

1. Here again is the table of Central High student population data.

	Male	Female	Total
Seniors	156	144	300 (30%)
Juniors	168	172	340 (34%)
Sophomores	196	164	360 (36%)
Total	520 (52%)	480 (48%)	

a. What is the probability of selecting a male?

b. What is the probability of selecting a male from the sophomore class?

c. Use your answers to parts a and b to determine whether the events "selecting a male" and "selecting a sophomore" are independent.

d. What is the probability of selecting a sophomore?

e. What is the probability of selecting someone who is both a male and a sophomore? Compare this probability with the product of the probability of selecting a male and the probability of selecting a sophomore.

f. Multiply the probability of selecting a male from the sophomore class by the probability of selecting a sophomore. Compare this with the probability of selecting someone who is both a male and a sophomore.

g. Is the probability of selecting a male from the sophomore class the same as the probability of selecting a sophomore from the males?

2. Consider events that describe the performance of your school's football team.

   a. Do you think the events "the team wins" and "the team wins at home" are independent? Explain.

   b. Describe the conditions that would have to exist if the events "the team wins" and "the team wins in bad weather" are independent.

3. Use the table of Central High student population data in Exercise 1.

   a. What is the probability of selecting someone who is either a male or a sophomore?

   b. Add the probability of selecting a sophomore and the probability of selecting a male. Compare this with the probability of selecting someone who is either a sophomore or a male.

   c. Are the events selecting a male and selecting a sophomore mutually exclusive? Explain.

4. Read the *Florida Today* editorial below about opposition to the Cassini launch.

   a. How was the probability $\frac{1}{350}$ obtained?

   b. What assumptions were made in the calculation of this probability?

Cassini Foes Exaggerate Risk of Saturn Mission

Florida Today
October 9, 1997

Washington—A small but vociferous group of anti-nuclear activists is fighting against Monday's launch of the international Cassini mission, destined for Saturn, because they fear a potential release of plutonium from the on-board power supply. While their concern is understandable, an examination of the issue shows the safety and environmental risks to be very small and the knowledge to be gained very large.

The Cassini spacecraft is designed with a power system that has been employed on 23 planetary missions over the past three decades. It uses plutonium to generate heat, which is converted to electricity to operate the probe. To protect against an accident, the plutonium is encased in special containers that can withstand high impact and temperatures.

To be sure, plutonium is radioactive and toxic. There is a measurable but small danger from Cassini's plutonium. The probability of an accident during initial launch in which there could be a release of plutonium is 1 in 1,500. The chance that there could be a release in the final launch phase is about 1 in 450. When the probe swings by Earth in a gravity assist two years after launch, the likelihood there could be a release from an accidental re-entry is less than 1 in 1 million.

Put together, the total probability of plutonium release is estimated at about 1 in 350.

5. The following tree diagram represents the Central High data from Exercise 1, with the events "selecting a male" and "selecting a female" occurring first.

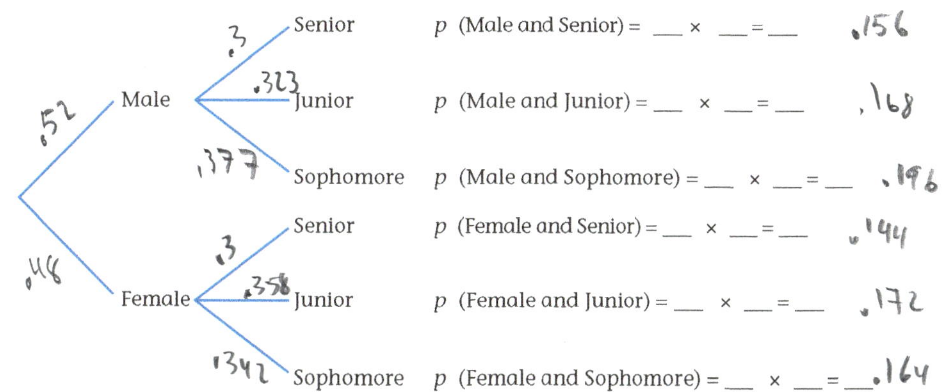

a. Write the correct probabilities along each branch and complete the calculations at the right.

b. Find the sum of the probabilities you calculated on the right.

6. A card is drawn at random from a standard 52-card deck.

a. What is the probability of drawing an ace?

b. What is the probability of drawing an ace from the diamonds only?

c. Are the events "drawing an ace" and "drawing a diamond" independent? Explain.

d. What is the probability of drawing a diamond?

e. What is the probability of drawing a card that is both a diamond and an ace? Compare this probability with the product of the probability of drawing an ace and the probability of drawing a diamond.

f. What is the probability of drawing a card that is either an ace or a diamond?

g. Are the events "drawing an ace" and "drawing a diamond" mutually exclusive?

h. Analyze the events "drawing a king" and "drawing a face card" (jack, queen, or king). Are they independent? Are they mutually exclusive? Explain your answers.

7. Two cards are drawn separately from a standard deck.

   a. What is the probability that the first card is red?

   b. What is the probability that the second card is red if the first card is red and is not put back in the deck before the second card is drawn?

   c. What is the probability that the first card is red and the second card is red if the first card is not put back in the deck before the second card is drawn?

   d. Compare your answer for part c with that for part c of Exercise 5 from Lesson 6.3 (page 320).

   e. The following tree diagram represents the colors of cards when two cards are drawn in succession from a standard deck without the first card being put back. Write the correct probabilities along each branch and complete the calculations at the right.

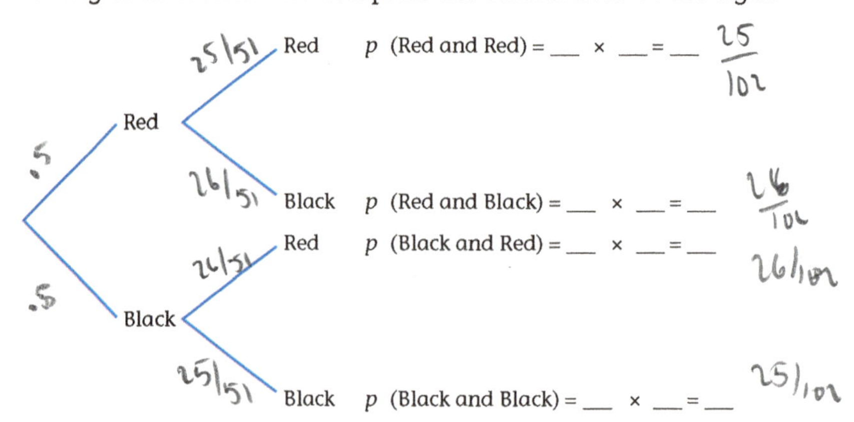

   f. What is the probability that exactly one of the two cards is red? (This occurs if the first card is red and the second card is black or if the first card is black and the second card is red. Find this probability by adding two of the probabilities you calculated and wrote to the right of the tree diagram.)

8. Again, two cards are drawn separately from a standard deck.

   a. What is the probability that the first card is red?

b. What is the probability that the second card is red if the first card is red and is put back before the second card is drawn? 1/2

c. What is the probability that the first card is red and the second card is red if the first card is put back in the deck? 1/4

d. The following tree diagram represents the colors of cards when two cards are drawn in succession from a standard deck with the first card replaced before the second is drawn. Write the correct probabilities along each branch and complete the calculations at the right.

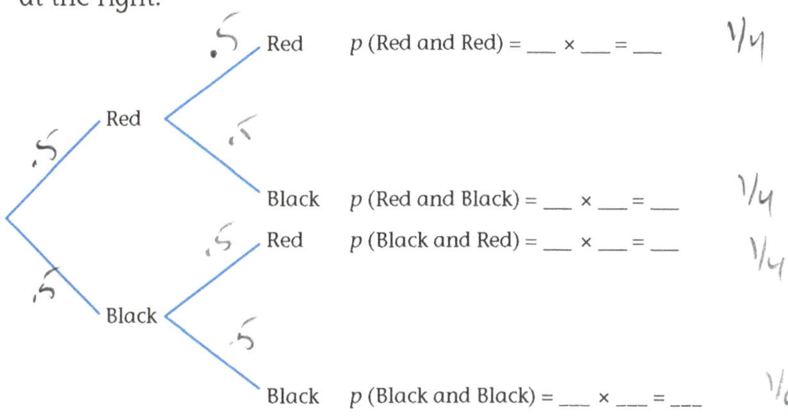

e. What is the probability that exactly one of the two cards is red? 1/2

f. Is the second draw independent of the first in the situation of Exercise 7 or in the situation of this exercise?

9. The probability that Coach Burns's Central High basketball team wins its first game of the season is .9, and the probability the team wins its second game of the season is .7.

   a. What is the probability the team wins both its first and second games? .9×.7= .63

   b. What assumption did you make about the outcomes of the first and second games in answering the previous question? Do you think that this is a reasonable assumption? Explain.

   Independent

10. The probability of rain today is .3. Also, 40% of all rainy days are followed by rainy days and 20% of all days without rain are followed by rainy days. The following tree diagram represents the weather for today and tomorrow.

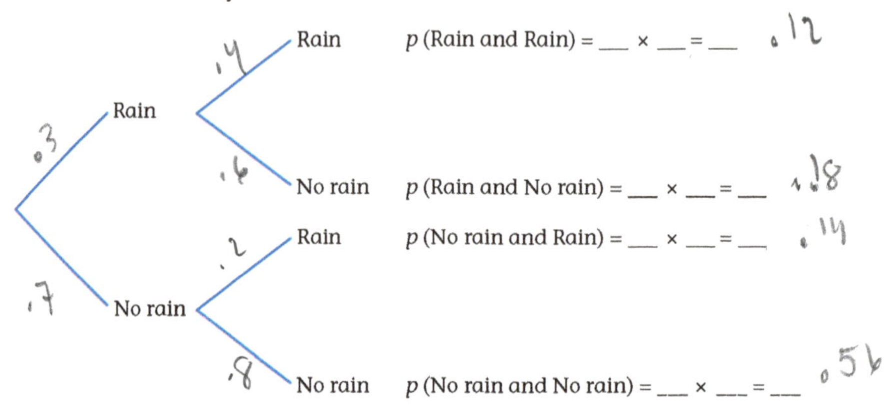

a. Write the correct probabilities along each branch and complete the calculations at the right.

b. What is the probability that it rains on both days?

c. What is the probability that it rains on one of the two days?

d. What is the probability that it does not rain on either day?

e. Is tomorrow's weather independent of today's? Explain.

11. Some Americans favor mandatory HIV screening for workers in certain professions such as health care. However, medical tests are seldom perfect. When imperfect medical tests are used on people who lack symptoms, the results must be carefully interpreted because false positives are common. (A false positive is a positive test result for someone who does not have the condition that the test is supposed to detect.) For example, consider a test that is 98% accurate. That is, it fails to report the existence of the disease in only 2% of those who have it, and it incorrectly reports the existence of the disease in 2% of those who do not have it. About 2 people out of 1,000 have the disease.

a. Write the appropriate probabilities along each branch of the tree diagram and complete the calculations at the right.

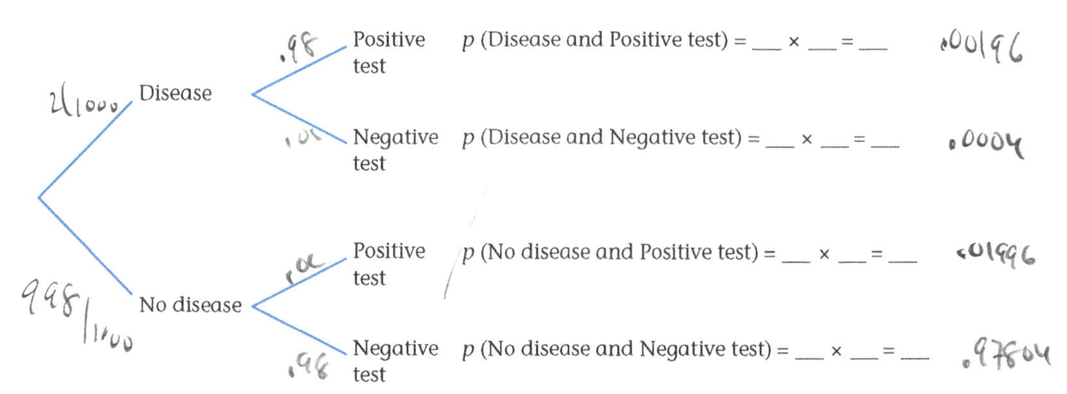

b. If 100,000 people are screened for the disease, about how many can be expected to test positive?

c. How many of those who test positive actually have the disease?

d. What is the probability that a person who tests positive for the disease actually has it?

e. Why is the tree diagram in part a ordered the way it is? That is, why does it not make sense to show test results along the first branch and the presence of disease or lack thereof second?

12. Consider the dice game proposed by Chuck in Lesson 6.1 and suppose that you bet on the number 5.

    a. Are the outcomes of the two dice independent of each other? Explain.

    b. What is the probability that a 5 appears on the first die and on the second?

c. The following tree diagram represents the outcomes of the two dice. Write the correct probabilities along each branch and complete the calculations at the right.

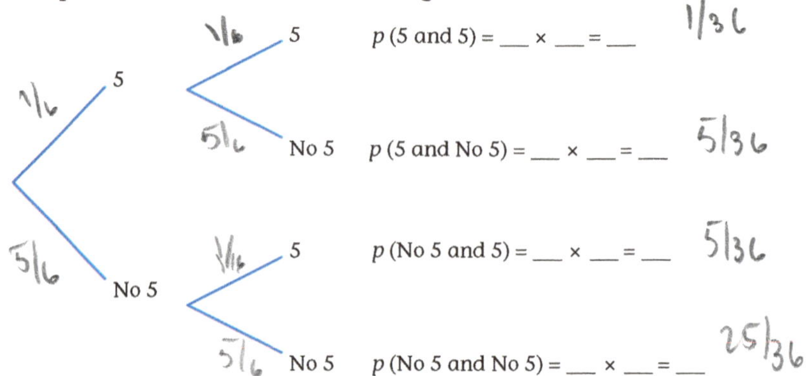

d. A single 5 can appear if either the first die shows 5 or the second die shows 5. What is the probability that exactly one 5 appears?

e. What is the probability that no 5s appear? How is this probability related to the probabilities you found in parts b and d?

f. Do you think you can expect to win or lose money in the long run if you played this game? Explain.

13. The multiplication principle for independent events can be used for more than two such events. The carnival game Chuk-a-Luk, for example, is similar to the game proposed by Chuck in Lesson 6.1, except that three dice are rolled. You can calculate the probability that all three dice show a 5 by cubing the probability that a single die shows a 5. What is the probability that all three dice show a 5?

14. The probability that one of Ms. Howe's plants blooms is .9. What is the probability that all five bloom?

15. European roulette wheels contain the numbers 0 through 36. The ball spun around the perimeter of the wheel has an equal chance of landing on any of these numbers. The Scottish actor Sean Connery once bet on the number 17 three times in a row, and the ball landed on 17 all three times. What is the probability that the ball lands on the same number three times in a row?

Unlike the European roulette wheel, an American wheel has a 00.

16. On January 28, 1986, the space shuttle *Challenger* exploded over Florida, killing astronauts Greg Jarvis, Christa McAuliffe, Ron McNair, Ellison Onizuka, Judy Resnik, Dick Scobee, and Mike Smith. Many people feel this accident could have been prevented if closer attention had been paid to the laws of probability. The rocket that carried the shuttle aloft had several sections that were sealed by large rubber O-rings. A presidential commission found that the accident was caused by leakage of burning gases from these rings. Studies found that the probability of a single O-ring's working properly was about .977.

    a. If the *Challenger's* six O-rings are truly independent of one another, what is the probability that all six function properly?

    b. After the commission issued its report, some people likened the probabilities in the *Challenger* accident to those in the game of Russian roulette. Compare the probability that all six O-rings function properly with the probability that a six-chamber revolver fires if only one chamber contains a bullet and a chamber is selected at random.

This photo was taken just seconds after the space shuttle *Challenger* exploded.

    c. Each joint was sealed by a system of two O-rings that were supposed to be independent of each other but were not. If the probability that one O-ring fails is 1 – .977 = .023, what is the probability that both O-rings in a system of two fail if the two are independent of each other?

    d. Subtract your previous answer from 1 to determine the probability that a single joint's seal works and then find the probability that all six function properly. (Keep in mind that your answer is based on a faulty assumption of independence of the two O-rings in a single joint.)

Cracks in NASA's Orion Model Spur Lockheed Martin to Revisit Design

Denver Post
November 28, 2012

Lockheed Martin Space Systems is revisiting the crew-capsule design of NASA's Orion spacecraft after three structural cracks were found during proof pressure testing of the first flight model at Kennedy Space Center.

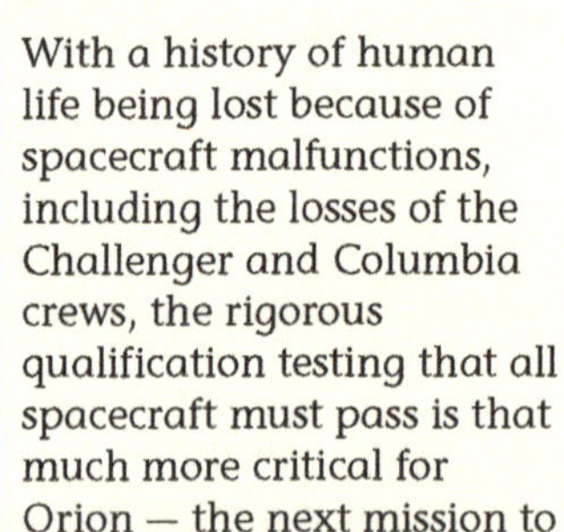

"It shouldn't have cracked," said Mark Geyer, Orion program manager at NASA. "We predicted that it was going to be high stress, but we didn't think it would crack."

Geyer and Cleon Lacefield, Orion program manager at Lockheed Martin, were at the company's Jefferson County facility Tuesday for progress reports and to discuss how the cracks happened and what they are doing to fix the problem.

The capsule was designed in Colorado, although much of it was manufactured elsewhere.

"The team has a repair in place," Geyer said. "It'll basically distribute the stress across that location. It's a simple repair."

With a history of human life being lost because of spacecraft malfunctions, including the losses of the Challenger and Columbia crews, the rigorous qualification testing that all spacecraft must pass is that much more critical for Orion – the next mission to carry humans into space following the now-retired shuttle program. But project leaders noted that the testing system seems to be working since the flaws were found long before flight.

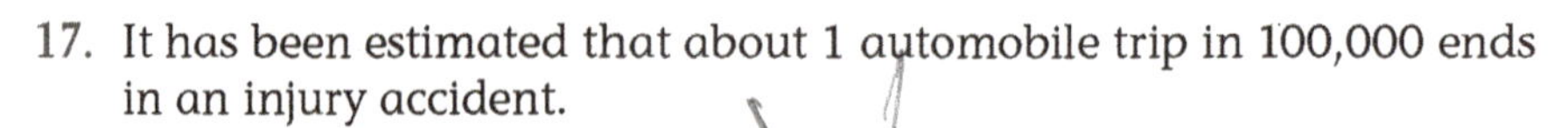

17. It has been estimated that about 1 automobile trip in 100,000 ends in an injury accident.

    a. What is the probability that a given automobile trip does not end in an injury accident?

    b. If you make approximately 3 automobile trips a day, about how many do you make over a 30-year period?

    c. If the outcome of a particular automobile trip is independent of the previous one, what is the probability none of the trips you make over a 30-year period end in an injury accident?

18. Of the inhabitants of Wilderland, 40% are Hobbits and 60% are humans. Furthermore, 20% of all Hobbits wear shoes and 90% of all humans wear shoes.

    a. Make a tree diagram to show the breakdown of residents into Hobbits who either do or do not wear shoes and humans who either do or do not wear shoes. Write the appropriate probabilities along the branches and write the appropriate events and their calculated probabilities to the right of the diagram.

    b. Suppose 10,000 residents are selected at random. About how many would you expect to be Hobbits who wear shoes?

    c. About how many of the 10,000 would you expect to be shoeless Hobbits?

    d. About how many of the 10,000 would you expect to be humans who wear shoes?

    e. About how many of the 10,000 would you expect to be shoeless humans?

    f. What percentage of the inhabitants who wear shoes are Hobbits?

    g. If an inhabitant is selected at random, are any of the events selecting a Hobbit, selecting a human, selecting someone who wears shoes, or selecting someone who doesn't wear shoes independent? Are any of them mutually exclusive? Explain.

19. A witness to a crime sees a man with red hair fleeing the scene and escaping in a blue car. Suppose that one man in ten has red hair and that one man in eight owns a blue car.

    a. What is the probability that a man selected at random has red hair and owns a blue car?

    b. What assumption did you make when you answered part a?

    c. The following table represents the men in the community in question. Use the data to present an argument either in favor of or against the assumption.

	Blue Car	No Blue Car
Red hair	1,990	14,010
Nonred hair	18,015	125,985

20. Many games involve rolling a pair of dice.

    a. In how many ways can a pair of dice fall?

    b. What is the probability of rolling a pair of 6s?

    c. What is the probability of rolling a pair of 6s twice in succession?

    d. What is the probability of not rolling a pair of 6s?

    e. What is the probability of not rolling a pair of 6s twice in succession?

    f. A New York gambler nicknamed "Fat the Butch" once bet that he could roll at least one pair of 6s in 21 chances. Find the probability of not rolling a pair of 6s in 21 successive rolls of a pair of dice. (By the way, he lost about $50,000 in the course of several bets.)

21. The circles in the following Venn diagram represent events A and B; the variables x, y, and z represent the number of things in each region. Use the diagram to explain why the multiplication principle for probabilities, $p(A \text{ and } B) = p(A) \times p(B \text{ from } A)$, is true.

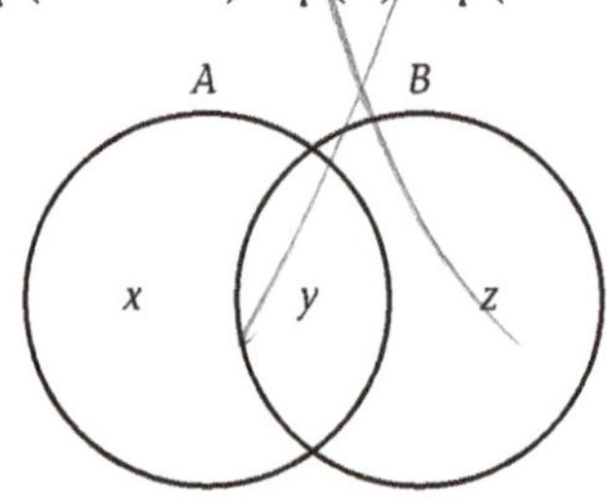

22. Edward Kaplan, a professor at Yale University, once wrote an article in which he complained that the U.S. State Department had overstated the risk of traveling to Israel. He said that in the prior 442 days terrorists had killed only 120 Israelis within "Israel proper." Based on a population of 6.3 million, he said this implies a risk of 19 in one million. Moreover, Kaplan noted that Israel had 461 traffic deaths in 2001, which is a risk of 73 in a million; while the United States has 145 traffic deaths per million in a year. Kaplan calculated that his risk of dying from terrorism or a car crash during a 1-week visit to Israel was 1.7 in a million and concluded that the riskiest part of his trip was the drive from Yale to Kennedy airport in New York. Explain the probability calculations that Kaplan used to arrive at 1.7 in a million.

Projects

23. Gather data on your school's student population broken down into the categories of the Central High population in this lesson. Determine whether the data exhibit any events that approximate independence.

24. Research the probabilities of about a dozen real-world events, such as being killed in an automobile accident, winning a lottery in your area, and being killed in a plane accident. Rank them according to their probability of occurring. Do you think the attention given to events by our culture is proportionate to their rate of occurrence? Explain.

25. Real-world data can exhibit events that come close to passing a test for independence. How close is close enough? If you have taken or are taking a statistics course, prepare a report to share with other members of your class that discusses how statisticians decide whether a difference is statistically significant.

26. Research and report on misuses of probability in the courtroom. For example, in a 1968 California case, People v. Collins (68 Cal 2d319), probabilities were erroneously multiplied. What precedent was set by this case? What are other probability-related precedents that have been established?

27. Bayes' Theorem (or rule/law) is a formula describing probability relationships that you have used tree diagrams to analyze in this lesson. Research and report on some of the uses of Bayesian analysis in mathematical modeling of real-world events.

Mathematician of Note

Thomas Bayes (1701-1761)

An English mathematician and minister, Thomas Bayes is remembered for discovering a formula relating conditional probabilities. He never published his discovery, which became known to the world after his death.

Lesson 6.5

Probability, Part 2

This chapter began with an examination of several games proposed for a school fundraiser. The counting and probability techniques you learned in the preceding sections are important tools in modeling these games and many real-world situations that involve probability. However, the questions of whether the school can expect to make money on these games and how much it can expect to make (or lose) have not been answered completely. This lesson considers two important ideas—probability distributions and expected value—that will enable you to complete your analysis of the games proposed by Pierre, Hilary, and Chuck.

Binomial Probability Distributions

As a first example, consider a brief quiz of three questions in which the answers are either true or false. Because there are only two possible outcomes for any one question, the process of answering a single question is called **binomial**. When the process is repeated several times, the multiplication principle implies that there are more outcomes than the two that are possible in a single trial. For example, with three trials there are $2 \times 2 \times 2 = 8$ outcomes. A tree diagram can be used as an aid to listing all eight outcomes (see Figure 6.6).

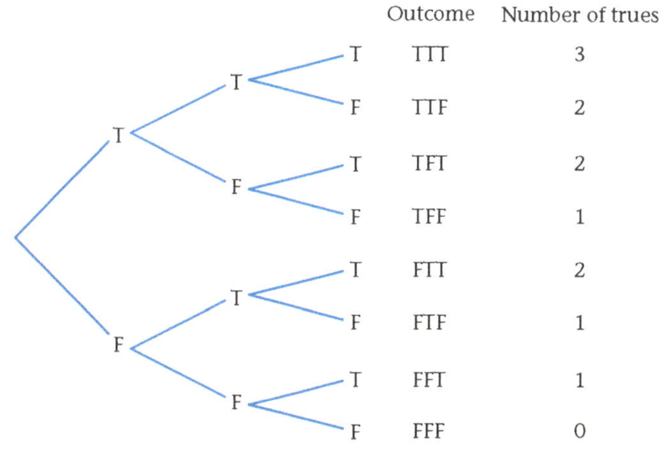

Figure 6.6. A three-question true/false quiz.

Because each outcome is as likely as any other, the probability associated with each of the eight is $\frac{1}{8}$. The probabilities can also be calculated by first writing $\frac{1}{2}$ on each branch of the tree diagram, and then multiplying (see Figure 6.7).

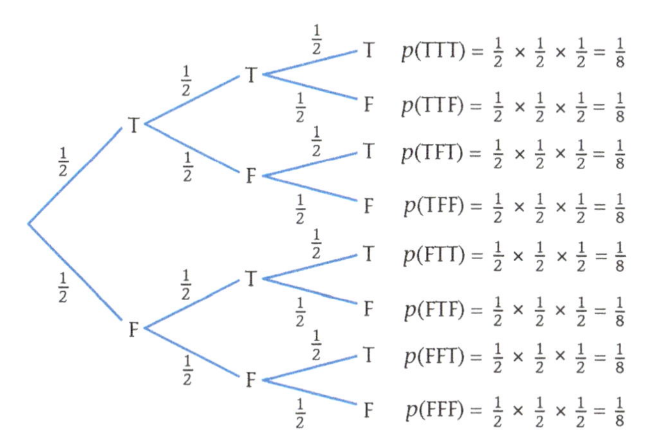

Figure 6.7 Probabilities in a three-question true/false quiz.

The results can be collected into a table called a **probability distribution** table. The way in which the table is constructed depends on whether you are interested in the number of true answers or the number of false answers. If you are interested in the number of true answers, then the table looks like this:

Number of True	0	1	2	3
Probability	$\frac{1}{8}$	$\frac{3}{8}$	$\frac{3}{8}$	$\frac{1}{8}$

A Binomial Probability Shortcut

The tree diagram approach is tedious if the number of outcomes is large, which is often the case. The counting techniques and the multiplication principle you learned in previous lessons of this chapter can be used to develop an alternative to the tree diagram approach.

For example, consider the probability that exactly two of the three answers are true. The blanks shown below represent the three questions. There are three ways you can select the two to mark true.

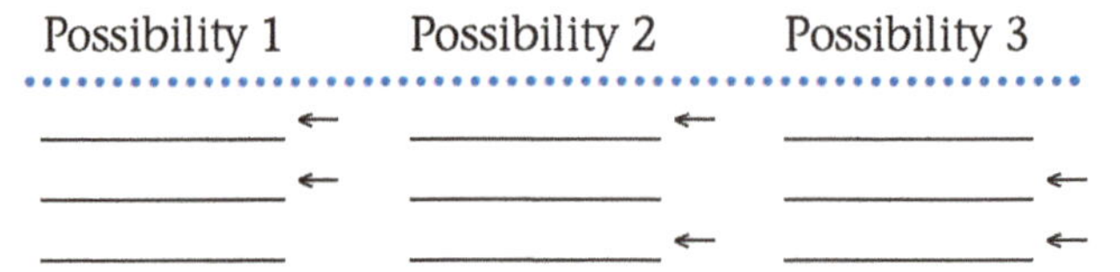

Because the order of selection of the two questions to mark true does not matter, the number of ways of selecting them can be counted as $C(3, 2) = \frac{3!}{2!1!} = 3$. The probability associated with each true answer is $\frac{1}{2}$, and the probability associated with each false answer is also $\frac{1}{2}$, so the probability of two true answers followed by one false answer is $\left(\frac{1}{2}\right)^2\left(\frac{1}{2}\right)$. This probability is multiplied by 3 because there are three ways that the two true answers can occur.

As a second example, consider a quiz of ten questions and the probability that exactly four of them are true. If there are four true, there must be six false, so the probabilities that must be multiplied are four $\frac{1}{2}$s for the true answers and six $\frac{1}{2}$s for the false, which gives $\left(\frac{1}{2}\right)^4\left(\frac{1}{2}\right)^6$. The number of ways that the four true questions can be selected is $C(10, 4)$. Therefore, the probability is $C(10, 4)\left(\frac{1}{2}\right)^4\left(\frac{1}{2}\right)^6 = 210\left(\frac{1}{16}\right)\left(\frac{1}{64}\right)$, or about .205.

Note that the denominators of the combination formula match the powers of the probabilities.

$$C(10,4)\left(\frac{1}{2}\right)^4\left(\frac{1}{2}\right)^6 = \frac{10!}{4!6!}\left(\frac{1}{2}\right)^4\left(\frac{1}{2}\right)^6$$

When using this method of calculating binomial probabilities, the probability of a single question being true is multiplied by itself several times. It is essential, therefore, that the individual outcomes be independent of one another. If, for example, the answer to one question depends on an answer to another, then this technique should not be used.

The probability associated with a given outcome is often different from $\frac{1}{2}$. Consider, for example, the dice game proposed by Chuck in Lesson 6.1. Because this game involves only two dice, it is not difficult to model it with a tree diagram. However, to serve as an example, the following analysis uses the counting technique just discussed.

If you bet on the number 5 in Chuck's game, then you win \$2 if two 5s appear, win \$1 if a single 5 appears, or lose \$1 if no 5s appear. Consider the possibility that one 5 appears. There are $C(2, 1)$ ways to select the die that shows 5. The probability of one 5 is $\left(\frac{1}{6}\right)\left(\frac{5}{6}\right)$ because one of the dice must show 5 and the other must show anything but 5. The correct probability is therefore $C(2, 1)\left(\frac{1}{6}\right)^1\left(\frac{5}{6}\right)^1$, or about .278.

Technology Note

```
binompdf(10,.5,4
)
              .205078125
```

The first screen shows a binomial probability calculation on a calculator with a binomial probability function. The number of trials is the first number, the probability of a single true is the second, and the number of trues is the third. The second screen shows a binomial probability calculation using a calculator's combination function.

```
10 nCr 4*.5^4*.5
^6
              .205078125
```

Similarly, the probabilities of no 5s and two 5s can be calculated as $C(2, 0)\left(\frac{1}{6}\right)^0\left(\frac{5}{6}\right)^2$ and $C(2, 2)\left(\frac{1}{6}\right)^2\left(\frac{5}{6}\right)^0$, respectively. The probabilities are summarized in the following distribution table.

Amount Won	–1	1	2
Probability	.694	.278	.028

In general, if p is the probability associated with a single binomial outcome, the probability of n successes in m attempts is $C(m, n)(p)^n (1 - p)^{m-n}$, provided that individual trials are independent.

Expectation

The question that remains to be answered is how a player could expect to do in Chuck's game in the long run. The player's **expectation** (also known as the expected value of the player's probability distribution) is used to answer this question.

The calculation of expectation for a player in Chuck's game weights each amount that Chuck could win (or lose) according to its probability: $-1(.694) + 1(.278) + 2(.028) = -.36$. The expectation can be interpreted as the average amount the player can expect to win per play of the game. If, for example, the game is played 100 times, the player can expect to lose about $100(.36) = \$36$.

If the Central High council decides to use this game as a fundraiser, the council's viewpoint is the opposite of that of the player: It loses $2 when two 5s appear, loses $1 when one 5 appears, and makes $1 when no 5s appear. Therefore, the council's expectation is $0.36. The council can expect to make about $36 for each 100 times the game is played.

A Binomial Probability/Expectation Example

A quality control engineer at the manufacturing plant of an electronics company randomly selects five compact disc players from the assembly line and tests them for defects. If a problem on the assembly line causes the factory to produce 10% defective, what is the probability that the problem will be detected by the engineer's test?

The following table shows the probability distribution for the number of defective players that the engineer might find in a sample of 5. The probabilities have been rounded to three decimal places.

Number Defective	Probability Calculation	Probability
0	$C(5, 0)(.1)^0(.9)^5$	.590
1	$C(5, 1)(.1)^1(.9)^4$	.328
2	$C(5, 2)(.1)^2(.9)^3$	.073
3	$C(5, 3)(.1)^3(.9)^2$	.008
4	$C(5, 4)(.1)^4(.9)^1$	.000
5	$C(5, 5)(.1)^5(.9)^0$	.000

There is about a 59% chance that the engineer will fail to find a defective player in a sample of 5 if the defective level is 10%. The probability that the engineer will find at least one defective player is about 41%, which can be found by subtracting 59% from 100% or by adding the last five probabilities in the table.

The engineer's expectation is calculated as 0(.590) + 1(.328) + 2(.073) + 3(.008) + 4(.000) + 5(.000), or about .498, which can be interpreted as the average number of defective CD players the engineer detects in samples of 5. To put it another way, if the engineer goes through this routine once each day when the defective level is at 10%, the engineer can expect, on average, to detect about one-half a defective CD player a day, or one every two days.

The following exercises treat modeling questions that involve binomial probability and expectation.

Technology Note

```
binomcdf(10,.4,3
)
        .3822806016
```

Some calculators have a function that finds the sum of probabilities in a binomial distribution. This screen shows such a calculation in a situation with a .4 probability of success. The calculated probability is the sum of the probabilities of 0, 1, 2, or 3 successes in 10 trials.

Sherlock Holmes Nemesis Helped by Oxford Mathematicians

BBC
January 11, 2012

A pair of Oxford mathematicians have contributed to the latest Sherlock Holmes film by supplying mathematical formulae, codes and lecture notes for the detective's arch-enemy.

Alain Goriely and Derek Moulton were originally asked to invent some 19th Century equations to write on Professor Moriarty's blackboard. The mathematicians from Oxford's Mathematical Institute went on to devise codes and ciphers for the film Sherlock Holmes: A Game of Shadows.

Dr Goriely, a professor of mathematical modelling, said the work was "fun".

Their equations, transcribed by a calligrapher, are visible in the background in a scene in which Holmes confronts Moriarty in his office.

"On the board you have different aspects of his work, works on the binomial theorem, one on the dynamics of asteroids," said Dr Goriely. "Hopefully they are all mathematically correct. "You can see all his equations and also you should see hints of the code he uses to encode his message to his crew."

The mathematicians were careful to make their equations historically accurate in terms of annotations and symbols typical of the 19th Century period, when the film is set.

Sherlock Holmes enthusiasts have speculated that the author Sir Arthur Conan Doyle based the fictional character Moriarty on the American astronomer Simon Newcomb and mathematicians such as Carl Friedrich Guass and Srinivasa Ramanujan. Guass and Ramanujan were known for their papers on the dynamics of asteroids and generalizations

Exercises

1. Hale Ault, a student at Central High, is known for occasionally neglecting his studies. When he finds a question on an exam that he cannot answer, he uses one of several random processes as an aid. Examine each of Hale's schemes and discuss first whether each outcome has the same probability of occurring as does each of the others and, second, whether several successive applications of the scheme are independent of one another.

   a. On a true/false question, flip a coin and answer true if the result is heads and false if it is tails.

   b. On a three-choice multiple-choice question, flip two coins. Mark the first answer if both coins are heads, the second if both are tails, and the third if the result is one head and one tail.

   c. On a four-choice multiple-choice question, associate each of your fingers on one hand with one of the choices, slap your fingers against the desk, and select the one that stings the most.

   d. On a four-choice multiple-choice test that allows the use of scientific calculators, use the calculator's random-number generator to display a random number between 0 and 1. Mark the first answer if the number is below 0.25, the second if it is between 0.25 and 0.5, the third if it is between 0.5 and 0.75, and the fourth if it is between 0.75 and 1.

2. Ms. Howe is giving a five-question true/false quiz.

   a. In how many ways can a student select three of the questions to mark true?

   b. Show how to calculate the probability that exactly three of the answers are true.

   c. Complete the probability distribution for the number of true answers:

Number of True	0	1	2	3	4	5
Probability						

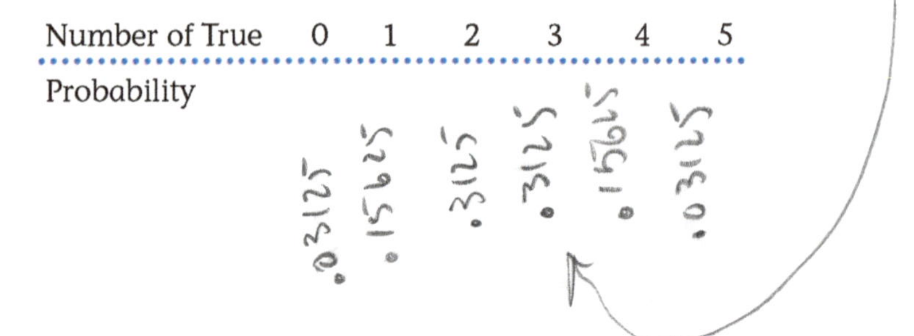

d. Ms. Howe has a bias toward true answers, and so her questions have true answers about 70% of the time. Recalculate the probability distribution:

Number of True	0	1	2	3	4	5
Probability						

3. Hale Ault is taking a ten-question true/false quiz on which the answers have equal chances of being true or false. Hale is doing the quiz by guessing and needs at least six correct in order to pass.

   a. Find the probability of exactly six correct answers.

   b. Find the probability of exactly seven correct answers.

   c. Find the probability of exactly eight correct answers.

   d. Find the probability of exactly nine correct answers.

   e. Find the probability of exactly ten correct answers.

   f. Hale passes if he gets six right or if he gets seven right or if he gets eight right or if he gets nine right or if he gets ten right. What is the probability that Hale passes the quiz?

4. Recall that the game of Chuk-a-Luk is played with three dice. You win $1 for each time your number shows but lose $1 if your number does not show.

   a. Suppose you bet on the number 5. Show how to calculate the probability that 5 shows exactly once.

   b. Complete the probability distribution for the number of times that 5 shows:

Number of 5s	0	1	2	3
Probability				

   c. Complete the following calculation of your expected winnings: (Note that the distribution of winnings is different from the distribution of the number of 5s because zero 5s results in a loss of $1.)

   −1(______) + 1(______) + 2(______) + 3(______) = ______

   d. If you play the game 100 times, about how much money should you expect to win or lose? Explain.

5. A list of people eligible for jury duty contains about 40% women. A judge is responsible for selecting six jurors from this list.
   a. If the judge's selection is made at random, what is the probability that three of the six jurors are women?
   b. Prepare a probability distribution table for the number of women among the six jurors.
   c. Suppose that the judge's selection includes only one woman. Do you think this is sufficient reason to suspect the judge of discrimination? Explain.

North Carolina Judge Vacates Death Penalty Under Racial Justice Law

Chicago Tribune
April 21, 2012

In a landmark ruling, a North Carolina judge on Friday vacated the death penalty of a black man convicted of murder, saying prosecutors across the state had engaged in deliberate and systematic racial discrimination when striking black potential jurors in death penalty cases.

The ruling was the first under North Carolina's Racial Justice Act, passed in 2009, which allows judges to reduce death sentences to life in prison without parole when defendants can prove racial bias in jury selection.

The decision could have an effect on death penalty cases nationwide; for years, such cases have included arguments by black defendants and civil rights lawyers that prosecutors keep blacks off juries for racial reasons.

6. Sickle cell anemia is a genetic disease that strikes an estimated 1 in 400 African-American children in the United States. The disease causes red blood cells to have a crescent shape rather than the normal round shape, which inhibits their ability to carry oxygen. Victims suffer from severe pain and are susceptible to pneumonia and organ failure. Children of parents who are both carriers of the sickle cell gene are frequently stricken.

   a. Healthy parents have two normal A genes, and carrier parents have one normal A gene and one sickle S gene. A victim of the disease has two S genes. A child inherits one gene independently from each parent. Complete the probability calculations in the tree diagram representing parents who are both carriers.

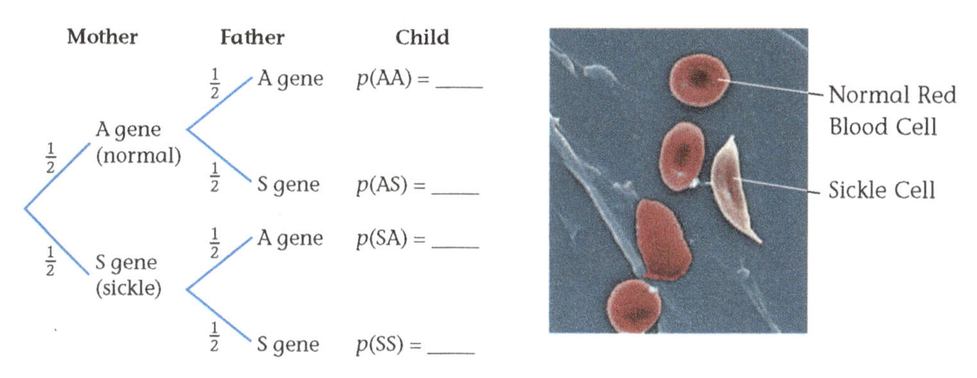

   b. What percentage of children of two carrier parents have sickle cell anemia? What percentage are carriers? What percentage are healthy?

   c. A couple who are both carriers have five children. Complete the following probability distribution for the number of children who have the disease.

Number of Children	Probability
0	
1	
2	
3	
4	
5	

   d. Calculate the expected value for the distribution in part c. Interpret it in this case.

7. A quality control engineer at a widget factory randomly selects three widgets each day for a thorough inspection. Suppose the assembly process begins producing 20% defective.

    a. Prepare a probability distribution for the number of defective widgets the engineer will find in the sample of three.

    b. Do you think the engineer's quality control model is a good one? If not, suggest a way to improve it.

8. A lottery ticket costs $1 and requires a player to select 6 numbers from the 44 available.

    a. If a $27 million jackpot is the only prize and you do not have to share it, complete the following probability distribution for your expected winnings (see Exercise 9 in Lesson 6.3, page 321).

Amount Won	$27 million	–$1
Probability		

    b. Calculate the expected value for this distribution.

    c. Recall that an Australian group purchased 5 million tickets in the Virginia lottery. Assume that the jackpot is the only prize, revise the distribution, and recalculate the expectation for the Australian group's winnings.

9. A country has a series of three radar defense systems that are independent of one another. An enemy plane has a 15% chance of escaping any one system.

    a. Prepare a probability distribution for the number of radar systems that a plane escapes.

    b. What is the probability that an enemy plane penetrates the country's radar defenses?

10. Recall the lottery proposed by Hilary in Lesson 6.1. It requires selecting two numbers from the nine available. The proposed price of a ticket is $1. The proposed prizes are $20 for matching both numbers and $1 for matching one of the two winning numbers.

    a. Prepare a probability distribution for the amount a player can expect to win.

    b. Calculate the player's expectation.

c. Can the Central High council expect to make money on this game? Explain. If you think not, suggest a revision of Hilary's plan for awarding prizes so that the council can expect to make money.

11. A fair coin is tossed several times.

a. Find the probability of obtaining exactly 5 heads in 10 tosses. (Do not do the entire probability distribution.)

b. Find the probability of obtaining exactly 10 heads in 20 tosses. Compare this with the previous answer.

c. Prepare a table showing the probabilities of obtaining 4, 5, or 6 heads in 10 tosses.

d. Prepare a table showing the probabilities of obtaining 8, 9, 10, 11, or 12 heads in 20 tosses.

e. Are you more likely to obtain between 40% and 60% heads in 10 tosses or in 20? Explain.

12. One variety of Extrasensory Perception (ESP) is the ability to communicate with another person without speaking. One common test for ESP has one person concentrate on a card selected at random from a special deck and another person record the image that is perceived.

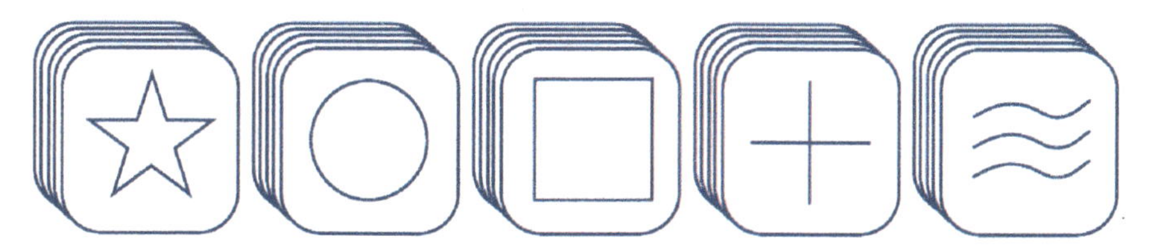

a. If the deck consists of five of each of the cards shown here, what is the probability of guessing any one card correctly?

b. Consider an experiment in which one person selects a card and concentrates on it and another person records his or her impression. The card is placed back in the deck, the deck is shuffled, and the experiment is repeated a total of five times. Prepare a probability distribution for the number of cards the receiver can guess correctly.

c. Suppose the receiver gets more than three correct. What is the probability of this happening by chance?

13. Recall the word game proposed by Pierre in Lesson 6.1. Suppose that the only two-letter words made from the letters of *Lions* that are considered legal are *in, is, on, no,* and *so.*

    a. What is the probability that a player draws a legal word from the letters recorded on the Ping-Pong balls?

    b. Pierre proposes that the charge for playing the game be $0.50 and the prize for selecting a legal word be $1. Prepare a probability distribution for a player's winnings.

    c. Calculate the player's expectation.

    d. Should the council expect to make money on the game? If you think not, suggest a revision of Pierre's plan so that the council can expect to make money.

14. Sara Swisher, a Central High Lions' star basketball player, has a field goal percentage of 62.

    a. Sara attempts seven field goals in the first quarter of tonight's game. Prepare a probability distribution for the number of field goals that Sara makes.

    b. What assumption have you made? Do you think this is a realistic assumption?

    c. Calculate the expectation. What does it mean in this case?

15. A distribution's expected value can be used to make decisions. For example, a person might decide to make an investment if the expectation has a positive dollar value but not to do so if it has a negative one.

    a. The price of one share of a stock is $35. You estimate that the probability the price will fall to $30 is .3, the probability the price will fall to $25 is .1, the probability the price will increase to $38 is .4, and the probability the price will increase to $42 is .2. Should you buy the stock? Explain.

    b. A lottery ticket costs one dollar. To win the jackpot, a participant must match 6 numbers from 42 available. Assuming that the participant does not share the jackpot with anyone else and that the jackpot is the only prize, how large must the jackpot be for the player's expectation to be positive?

c. Is a positive expected value a good reason to play a lottery? Explain.

16. Two probability students who are also baseball fans are discussing the number of games the World Series should last if the teams are evenly matched and the games are independent of each other.

    a. The students agree that the probability a series lasts four games is $2 \times \left(\frac{1}{2}\right)^4 = .125$. Explain this calculation. Why did they multiply by 2?

    b. One student says that the probability a series lasts 5 games is $2 \times C(5, 4) \times \left(\frac{1}{2}\right)^4\left(\frac{1}{2}\right)^1$ or $2 \times C(5, 4) \times \left(\frac{1}{2}\right)^5$. The other student says that this probability is too large because $C(5, 4)$ over-counts the number of ways one team can win a 5-game series. Which student is correct, and what is the probability?

    c. In how many ways can a given team win a 6-game series? Use your answer to find the probability a series ends in 6 games.

    d. What is the probability a series ends in 7 games?

17. *Pascal's triangle* is an array of numbers that you may have seen in an algebra class. To construct the triangle, begin with a row of two 1s: 1 1. Each new row starts and ends with a 1, and the other numbers are found by adding the numbers above and on either side of them in this way:

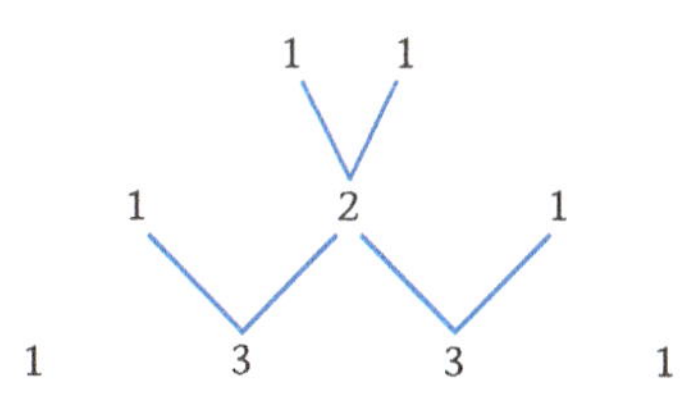

    a. Continue the triangle for three additional rows.

    b. Calculate all possible combinations of five things: $C(5, 0)$, $C(5, 1)$, $C(5, 2)$, $C(5, 3)$, $C(5, 4)$, and $C(5, 5)$.

    c. Where do your answers in part b occur in the triangle?

d. The following tree diagram shows all possible ways of answering true/false quizzes with up to four questions. Fill in the distributions of the number of true answers for quizzes of one, two, three, and four questions by tracing the paths of the diagram. Compare these numbers to those in Pascal's triangle.

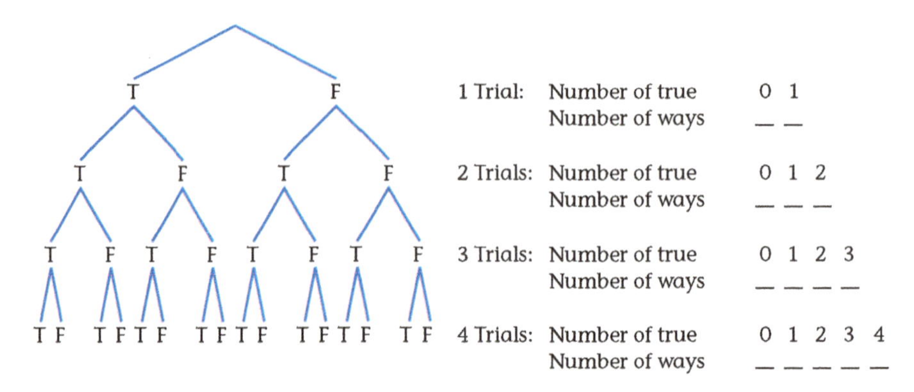

Computer/Calculator Explorations

18. Write a program for your computer or programmable calculator that does probability distributions of the type discussed in this lesson. The program should accept as input the probability of a single success and the number of trials. It should calculate and display each number of successes and the related probability.

Projects

19. Pascal's triangle contains many patterns other than those in Exercise 17. Investigate and report on some of them.

20. Research and report on the use of expected value as a decision-making tool. How, for example, is it used in business?

21. This lesson discusses a combinatorial technique for calculating binomial probabilities. Although useful, this technique has limitations, particularly if the number of trials is very large. Mathematicians sometimes use the Poisson distribution to approximate binomial probabilities. Research and report on the use of the Poisson distribution to approximate binomial probabilities.

22. On September 9, 1990, the Sunday newspaper supplement *Parade* carried a column by Marilyn vos Savant in which she responded to a problem posed by a reader about the television game show *Let's Make a Deal*. The show features three "mystery" doors from which a contestant picks; there could be anything from an expensive prize to worthless junk behind each door. The problem asked whether a contestant should switch doors after the contestant's selection of one of three available doors prompted the show's host to open a door containing a worthless prize. Marilyn's response that the contestant should switch brought a flood of mail, most of which disagreed with her. Research the controversy and prepare a report on the arguments on both sides. Select the answer with which you agree and defend it.

What Looks Like Play May Really Be a Science Experiment

Health Day
September 27, 2012

You may think a toddler is just playing in the sand box, but she may really be conducting a sophisticated scientific experiment and learning something new every time she pours out another scoop of sand, new research suggests.

"Children have the same brains we do. Everyone can learn from data and know if a hypothesis is good or not," explained Alison Gopnik, a professor of psychology at the University of California, Berkeley.

She said that in the past people thought preschoolers were irrational and illogical. However, during the 1980s and 1990s, researchers realized that young children actually had coherent, structured thoughts and could make causal inferences about the world around them.

One experiment reviewed by Gopnik illustrated how even babies can act like mini-scientists and use a probability model. In this experiment, a researcher showed babies a box full of red and white balls. Then the researcher closed her eyes and randomly removed some balls from the box and placed them in another small bin. If the sample was truly random, the distribution of the balls should be close to that of the original container.

But, sometimes the researchers switched the samples, giving the babies an unexpected result. When the sample of balls didn't match the expected distribution of balls, the babies stared at the non-matching sample longer.

Chapter Extension

Monte Carlo Models

Mathematician of Note

Stanislaw Ulam (1909–1984)

He came to the United States from Poland in 1935 at the invitation of fellow mathematician John von Neumann. He worked first at the Institute for Advanced Study at Princeton University, then at Los Alamos, where he was instrumental in the development of the hydrogen bomb.

Direct calculation of probabilities of real-world events is sometimes difficult, even for professional mathematicians. Such was the experience of Stanislaw Ulam when he worked in a laboratory in Los Alamos, New Mexico, in the 1940s. Ulam used a computer to simulate the occurrence of events for which he could not calculate probabilities directly. He chose the name "Monte Carlo" to describe his approach.

Today's computers are faster and more readily available than those Ulam used at Los Alamos. Even hand-held calculators can do many of the tasks only computers could perform a half-century ago.

```
0 → T: 0 → F
For (N,1,100)
int(2 * rand) → R
If R = 0
1 + F → F
If R = 1
1 + T → T
End
Disp F,T
```

Consider, for example, the answers to true/false questions. A random selection of a single answer can be simulated by a coin flip, but it can also be simulated by generating a random number. Many calculators have a random-number function that generates random decimals between 0 and 1. The instruction int(2 * rand) doubles this range, then drops the decimal; and thus generates random 0s and 1s. (Some calculators have a second random-number function that generates random integers over a specified range.) A simple program that simulates the random selection of 100 answers to true/false questions is shown to the left. It runs on most Texas Instruments graphing calculators.

Monte Carlo's Role in Retirement Planning

Morningstar
April 23, 2013

A Monte Carlo simulation might sound like a ride you'd find at Disneyworld, but it's actually a statistical method used to determine probability and assess risk. At its most basic, a Monte Carlo simulation allows the user to determine the likelihood of different outcomes based on a set of assumptions and how those assumptions respond to random variables.

The name Monte Carlo naturally brings to mind the gambling mecca located in Monaco, and its use in this case is no coincidence. Scientists working on the Manhattan Project during World War II applied the name to the simulation method used to develop the atomic bomb because of the method's use of random chance.

Since then the Monte Carlo method has been applied to everything from economics to traffic patterns. What these applications all have in common is that they apply random variables (also known as a stochastic technique) to test how a given piece of information will react under various conditions. For example, an oil company might use a Monte Carlo simulation to determine a range of potential outcomes in helping it decide whether to drill at a specific location.

Monte Carlo simulations are particularly useful in the realm of finance. Given the unpredictable nature of the stock market, the Monte Carlo method can help financial planners model how a particular portfolio will perform under various market conditions, thus helping them make more informed investment decisions. This approach is especially useful in retirement planning, in which investors try to figure out which savings rates, allocations, market returns and spending patterns will allow them to make their nest eggs last a lifetime.

The accuracy of answers obtained with Monte Carlo models depends on the number of trials that are run: A larger number of trials is more likely to produce an answer close to the correct one than is a smaller number of trials. Thus, you can expect the percentage of trues given by this simulation to be closer to 50 if 1,000 trials are used instead of 100. Computers, it should be noted, can perform a large number of trials much more quickly than calculators.

The following is an example of a difficult problem that can be modeled and solved by Monte Carlo methods.

Engineers sometimes need to determine the temperature of a point on a rectangular metal plate from the temperatures at the four edges.

Figure 6.8 represents a 10 centimeter × 6 centimeter plate. Coordinates have been assigned to facilitate the simulation. The point in question is at (6, 4). An edge temperature is written along each edge.

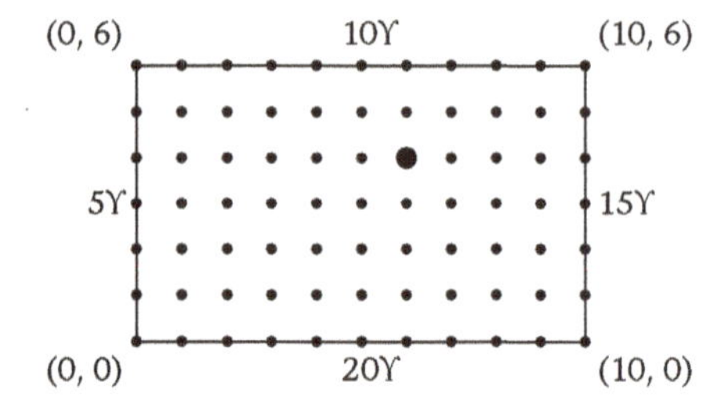

Figure 6.8. A 10 centimeter × 6 centimeter plate.

A Monte Carlo model for this problem envisions a random walk starting at (6, 4). A random selection determines a movement of either one unit to the right, one unit to the left, one unit upward, or one unit downward. The selection is repeated until the walk terminates at one of the edges. After many trials, the temperature at each edge is weighted according to the portion of trials that terminated at the edge. The weighted average of the four edge temperatures gives an estimate of the temperature at the selected point.

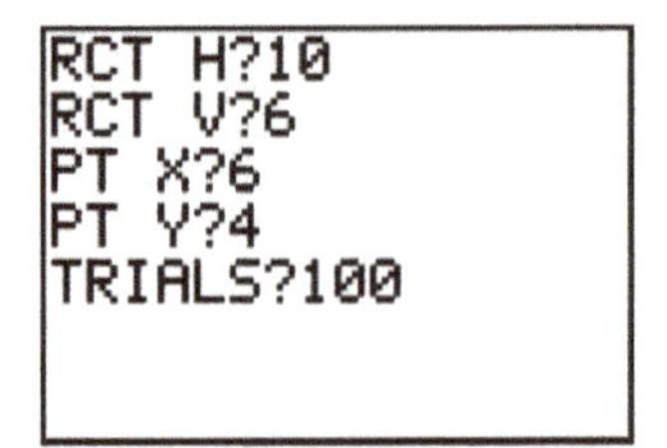

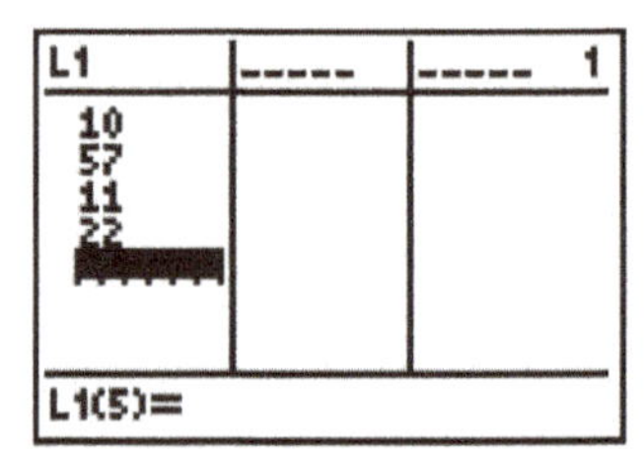

The above screens show the input (left) and results (right) of the TI-83 program RCTRNDWK, which accompanies this book. The user is prompted for the rectangle's horizontal and vertical dimensions, the coordinates of the point, and the number of trials. The numbers of times the walk ended at the left, top, right, and bottom edges are stored in that order in the calculator's list L1. Based on these 100 trials, an estimate of the temperature at the point (6, 4) is 5(.10) + 10(.57) + 15(.11) + 20(.22), or about 12°. A more reliable estimate can be made from 1,000 trials.

Chapter 6 Review

1. Write a summary of what you think are the important points of this chapter.
2. The following table represents ownership of cats among professionals and non-professionals in a community. A person is selected at random from this group.

	Own a Cat	Don't Own a Cat
Professionals	2,300	11,400
Non-professionals	5,600	27,600
Totals	7,900	39,000

   a. What is the probability of selecting someone who owns a cat?

   b. What is the probability of selecting a cat owner from the professionals?

   c. What is the probability of selecting a professional?

   d. What is the probability of selecting a person who owns a cat and is a professional?

   e. What is the probability of selecting someone who owns a cat or is a professional?

   f. Are the events selecting a cat owner and selecting a professional independent? Explain.

   g. Are the events selecting a cat owner and selecting a professional mutually exclusive? Explain.

3. In the summer of 2004, New South Wales (Australia) introduced new license plates. Each plate has two letters, followed by two numbers and two more letters. The first plate issued was AA00AA.

   a. How many plates are possible if all letters and digits are permitted?

   b. In a report released on July 21, 2004, the Australian Broadcasting Company quoted the New South Wales Roads Minister as saying there are over 31 million new combinations. Comment.

4. Lesson 1.5 examined voting situations in which some voters receive more votes than others. Recall that a coalition is a collection of voters. In how many ways can you form a coalition of one, two, three, four, or five voters from a group of five voters?

5. Teams A and B are playing a 5-game series that ends when one team wins three games. Each team has a 50% chance of winning any game.

   a. What is the probability that team A wins the first three games?

   b. What is the probability that team B wins the first three games?

   c. The series ends in three games if either team A or team B wins the first three games. What is the probability that the series ends in three games?

6. a. In how many ways can six books be arranged on a shelf?

Peanuts © reprinted by permission of United Feature Syndicate, Inc.

   b. If two of the books are math books, how many arrangements have the math books in the first two positions? (Hint: Draw six blanks and use the multiplication principle.)

   c. What is the probability that the math books are in the first two positions?

   d. In how many arrangements are the math books next to each other?

e. What is the probability that the math books are next to each other?

7. You are playing a game in which you flip two coins. If both show heads, you win $2; if both show tails, you win $1; but if the coins do not match, you lose $1.

   a. Construct a probability distribution for the amount won on a single play of the game.

   b. Calculate your expectation.

   c. If you play the game 100 times, about how much would you expect to win or lose?

8. In 2002, McDonald's restaurants in London were criticized for signs that claimed there were 40,312 possible combinations from eight items offered on the menu.

   a. How many different ways can you combine 8 items, assuming that you must use at least one of them?

   b. How might McDonald's have arrived at 40,312?

9. Being listed first on an election ballot is known to improve a candidate's chances. In order to minimize this effect, ballots could be constructed in such a way that each candidate is first on some ballots but not on all. There are three candidates for mayor and five candidates for city council in a local election.

   a. How many different orderings are possible if the mayoral candidates must be listed before the council candidates?

   b. If 20,000 people vote, about how many would see each ballot?

10. Are you more likely to win the jackpot in a lottery that requires the selection of 5 numbers from 39 or in one that requires the selection of 6 numbers from 36? If 3,000,000 tickets are sold in each lottery, about how many winning tickets would you expect in each?

11. Two cards are drawn from a standard deck. The first card is not put back before the second card is drawn.

    a. What is the probability that the first card is a heart?

    b. What is the probability that the second card is a heart if the first card is a heart?

c. What is the probability the first card is a heart and the second card is a heart?

d. The following tree diagram represents the two draws. Write the correct probabilities along the branches and calculate the probabilities at the right.

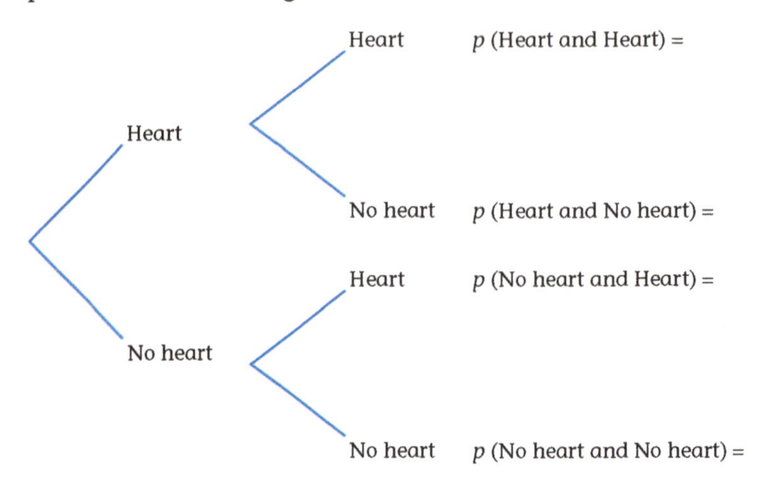

e. Is the second draw independent of the first? Explain.

12. Read the news article below about the Pennsylvania Cash lottery.

a. How many numbers are available for a player to select? Explain how you arrived at your answer.

b. Comment on the use of the word "odds" in the article.

Pennsylvania Cash 5 Lottery Has Three Winners

Yahoo News
April 30, 2013

Jackpot-winning Pennsylvania Lottery Cash 5 tickets from the April 29 drawing – worth a combined total of more than $1.3 million – were sold in Blair, Bucks and Schuylkill counties.

Three tickets matched the winning Cash 5 numbers drawn, 05-11-16-31-38. Each jackpot-winning ticket is worth $440,667, less 25 percent federal withholding.

Odds of winning the jackpot prize are 1-in-962,598.

13. There are five boys and six girls in a group, and a committee of three is being selected.

    a. In how many ways can the committee be formed?

    b. How many committees consist of one boy and two girls?

    c. What is the probability that the committee has exactly one boy?

    d. How many committees consist of one boy or one girl?

    e. What is the probability that the committee has one boy or one girl?

14. Suppose 90% of all drivers are good, that 5% of all good drivers get tickets, and that 70% of all bad drivers get tickets.

    a. Write the appropriate probabilities along each branch of the following tree diagram and calculate the probabilities shown at the right.

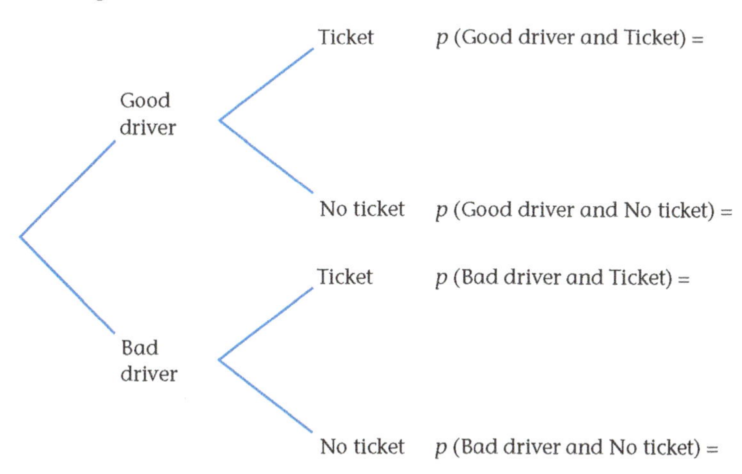

    b. If a community has 50,000 drivers, about how many can be expected to get tickets?

    c. How many of the people who get tickets are bad drivers?

    d. What is the probability that a person who gets a ticket is a bad driver?

15. Suppose that you and three friends each choose a number between 1 and 10.

    a. What is the probability that all three of your friends pick the same number that you pick?

    b. What is the probability that all three of your friends pick a number different from yours?

    c. What is the probability that each of the four picks a number that is different from the number picked by each of the others?

16. Two different prizes are being awarded in a group of ten people.

    a. In how many ways can this be done if the same person can win both prizes?

    b. In how many ways can this be done if each person can win no more than one prize?

    c. In how many ways can this be done if each person can win no more than one prize and both prizes are the same?

17. A fair die is rolled five times.

    a. What is the probability that a 6 appears exactly twice?

    b. What is the probability that a 6 appears two times or fewer?

    c. What is the probability that a 6 never appears?

18. By some estimates, 60% of all email is spam. Various kinds of filters are available that try to identify and eliminate spam. However, they are not 100% accurate. Suppose a given filter correctly identifies spam as spam 90% of the time, but it also incorrectly identifies legitimate email as spam 20% of the time.

    a. What percentage of email is identified as spam by this filter?

    b. If this filter identifies an email as spam, what is the probability that it really is?

    c. What percentage of email that passes the filter is spam?

19. Egbert fixes an omelet for breakfast every morning. Depending on what is in his refrigerator, he adds one or more of the following ingredients to the eggs: mushrooms, green peppers, cheddar cheese, and ham. How many different omelets can Egbert make?

20. This Venn diagram shows the probabilities associated with events A and B.

    a. What is $p(A)$?

    b. What is $p(A \text{ and } B)$?

    c. What is $p(A \text{ or } B)$?

    d. What is $p(A \text{ from } B)$?

    e. Are A and B mutually exclusive? Explain.

    f. Are A and B independent? Explain.

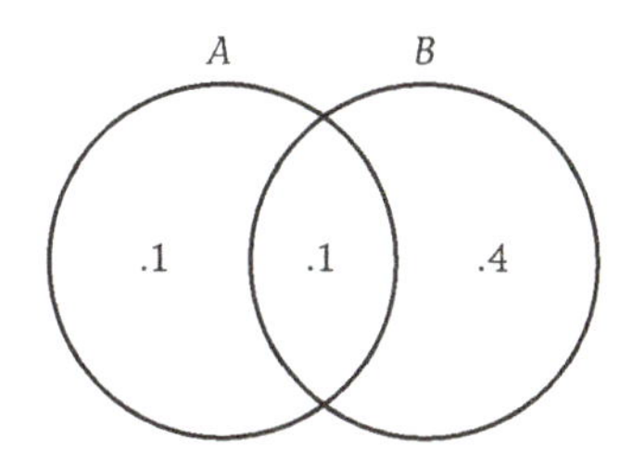

21. In the novel *Debt of Honor* by Tom Clancy, the character Colonel Zacharias plans to bomb ten missile silos. The colonel targets each silo with two bombs, each of which is 95% accurate.

    a. The novel states that there is "less than half a percent chance" that a given silo will not be destroyed. Assuming independence, do you agree?

    b. Referring to the half percent chance, the novel concludes "that number times ten targets meant a 5% chance that one missile would survive, and that could not be tolerated." Again, assume independence; also assume that "one" means at least one. Do you agree?

22. The sensitivity (the ability to detect a disease in a person who has it) of a medical test is 90%. The specificity (the ability to detect the absence of a disease in a person who doesn't have it) is 95%. About 400 people in 100,000 have the disease.

    a. If the test is used to screen a large group of people who lack symptoms of the disease, what percentage tests positive?

    b. What is the probability that someone who tests positive does not have the disease (a false positive)?

    c. If 20 people are screened, estimate the number that have false positive tests.

23. A die is rolled. Consider the events "the number rolled is divisible by 2," "the number rolled is divisible by 3," and "the number rolled is divisible by 5." Which pairs of events are mutually exclusive? Which pairs are independent?

24. Read the news article below about fizzy drinks.

    a) Based on the study reported in the article, are the events "drinking one sugar-sweetened drink a day" and "getting kidney stones" mutually exclusive? Explain.

    b) Based on the study reported in the article, are the events "drinking one sugar-sweetened drink a day" and "getting kidney stones" independent? Explain.

One Fizzy Drink a Day Increases Risk of Kidney Stones

Daily Mail
May 15, 2013

Drinking a can of fizzy drink a day could increase the risk of kidney stones by almost a quarter, new research shows.

The study found that drinking sugar-sweetened drinks makes the painful stones more likely to develop.

Other drinks - such as coffee, tea and orange juice - reduced the risk, the research by Brigham and Women's Hospital, in Boston, found.

25. In his book *Chance Rules*, Brian Everitt compares the probability of winning a lottery jackpot in the United Kingdom with the probability of dying before the lottery is drawn. One of the largest lotteries in the United States is the multi-state Powerball lottery. To win the Powerball jackpot, a player must match five of 59 numbered white balls and one of 35 numbered red balls.

    a. What is the probability that a single entry wins the Powerball jackpot?

    b. The CIA *World Factbook* estimates the annual death rate in the United States in 2013 at 8.39 per 1,000. Compare the probability you found in part a to the probability that an American who buys one Powerball ticket one hour before the drawing dies before the drawing is held.

    c. If a player who buys one Powerball ticket wants the probability of winning the jackpot to be higher than the probability of dying before the drawing, when should the player buy a ticket?

Bibliography

Bernstein, Peter L. 1996. *Against the Gods: The Remarkable Story of Risk.* New York: John Wiley & Sons.

Campbell, Stephen K. 2004. *Flaws and Fallacies in Statistical Thinking.* New York: Dover.

Campbell, Stephen K. 1999. *Statistics You Can't Trust: A Friendly Guide to Clear Thinking About Statistics in Everyday Life*. Parker, CO: Think Twice Publications.

Casti, John L. 1998. *Would-Be Worlds: How Simulation is Changing the Frontiers of Science.* New York: John Wiley & Sons.

Charpak, Georges, and Henri Broch. 2004. *Debunked!: ESP, Telekinesis, and Other Pseudoscience*. Baltimore: Johns Hopkins University Press.

David, F. N. 1998. *Games, Gods, and Gambling.* Mineola, NY: Dover Publications.

Dewdney, A. K. 1996. *200% of Nothing.* New York: John Wiley & Sons.

Everitt, Brian. 2008. *Chance Rules: An Informal Guide to Probability, Risk, and Statistics,* 2nd ed. New York: Springer.

Freedman, David, et. al. 2007. *Statistics.* 4th ed. New York: W. W. Norton & Company.

Gigerenzer, Gerd. 2002. *Calculated Risks*. New York: Simon & Schuster, Inc.

Green, Thomas A., and Charles L. Hamburg. 1986. *Pascal's Triangle.* Palo Alto, CA: Dale Seymour.

Haigh, John. 2003. *Taking Chances: Winning with Probability*. New York: Oxford University Press.

Holland, Bart. 2002. *What Are the Chances? : Voodoo Deaths, Office Gossip, and Other Adventures in Probability*. Baltimore: Johns Hopkins University Press.

Huff, Darrell. 1964. *How to Take a Chance.* New York: W. W. Norton & Company.

Kaplan, Michael and Ellen Kaplan. 2007. *Chances Are: Adventures in Probability.* New York: Penguin Books.

McGrayne, Sharon Bertsch. 2012. *The Theory That Would Not Die: How Bayes' Rule Cracked the Enigma Code, Hunted Down Russian Submarines, and Emerged Triumphant from Two Centuries of Controversy.* New Haven: Yale University Press.

Orkin, Mike. 1991. *Can You Win? The Real Odds for Casino Gambling, Sports Betting, and Lotteries.* New York: W. H. Freeman.

Packel, Edward W. 2006. *The Mathematics of Games and Gambling,* 2nd ed. Washington, DC: Mathematical Association of America.

Paulos, John Allen. 2001. *Innumeracy: Mathematical Illiteracy and Its Consequences.* New York: Hill and Wang.

Peterson, Ivars. 1997. *The Jungles of Randomness.* New York: John Wiley & Sons.

Ropeik, David, and George Gray. 2002. *Risk: A Practical Guide for Deciding What's Really Safe and What's Dangerous in the World Around You.* New York: Mariner.

Ropeik, David. 2010. *How Risky Is It, Really?: Why Our Fears Don't Always Match the Facts.* New York: McGraw-Hill.

Rosenhouse, Jason. 2009. *The Monty Hall Problem: The Remarkable Story of Math's Most Contentious Brain Teaser.* New York: Oxford University Press.

Rosenthal, Jeffrey S. 2006. *Struck by Lightning: The Curious World of Probabilities.* Washington: Joseph Henry Press.

Weaver, Warren. 1982. *Lady Luck: The Theory of Probability.* Mineola, NY: Dover Publications.

Ulam, Stanislaw M. 1991. *Adventures of a Mathematician.* Berkeley: University of California Press.

vos Savant, Marilyn. 1997. *The Power of Logical Thinking.* New York: St. Martin's Press.

CHAPTER

7

Matrix Models

The daily business activity that supplies us with the products and services we need generates large quantities of data. To help understand this information, these data often need to be organized into matrices. Proper organization of information is necessary not only for understanding but also for effective planning.

- How can a company that provides batteries for another company's compact disc players be sure that it will have enough batteries on hand to fill orders?
- How does a fast-food chain determine prices that will allow it to do as well as possible against a competitor?
- How can a meteorologist use data about recent weather activity to predict the weather for tomorrow or a week from now?
- How can a park service use birth rates and survival rates of deer in managing herd populations?

Matrix models demonstrate remarkable versatility in helping to solve these and other real-world problems.

Lesson 7.1

The Leontief Input-Output Model, Part 1

In Chapter 3 you explored the Leslie model, a mathematical model that can be used to determine the growth of a population. In this lesson, you will examine a new model, one that can be used to analyze the flow of goods among sectors in an economy.

This model, known as the Leontief input-output model, was developed by Wassily Leontief of Harvard University in the 1960s. It can be applied to complex economies with hundreds of production sectors, such as a country, or to a situation as small as a single company that produces only one product.

Mathematician of Note

Wassily Leontief (1906–1999)

Leontief began his study of input-output analysis in the early 1930s by constructing input-output tables that described the flow of goods and services among various sectors of the economy in the United States. He was awarded the Nobel Prize in economics in 1973 for his work in this area.

To explore this model, begin by looking at a simple case.

Suppose a company manufactures a particular type of battery that is used to power various kinds of electric motors. However, not all the batteries produced by the company are available for sale outside the company. For every 100 batteries produced, 3 (3%) are used within the company. Thus, if the company produces 500 batteries during a week's time, 15 will be used within the company and 485 will be available for external sales.

In general, for a total production of P batteries by this company, $0.03P$ batteries will be used internally and $P - 0.03P$ will be available for external sales to customers. If D represents the number of batteries available for external sales **demand**, then the number of batteries available for external sales equals the total production of batteries less 3% of that total production.

The following linear equation can be used to model this situation.

$$D = P - 0.03P$$

Example

Suppose the company receives an order for 5,000 batteries. What must the total production be to satisfy this external demand for batteries?

Solution:

To find the total production necessary, substitute 5,000 for D in the previous equation and solve for P.

$$D = P - 0.03P$$

$$5{,}000 = P - 0.03P$$

$$5{,}000 = P(1 - 0.03)$$

$$P = 5{,}155 \text{ batteries}$$

Hence, the company must produce a total of 5,155 batteries to satisfy an external demand for 5,000 batteries.

Notice that the total production of batteries equals the number of batteries used within the company during production plus the number of batteries necessary to fill the external demand.

Now take a look at a simple two-sector economy. Suppose the battery company buys an electric motor company and begins producing motors as well as batteries. The company's primary reason for this merger is that electric motors are used to produce batteries. Batteries are also used to manufacture motors. The expanded company has two divisions, the battery division and the motor division.

The production requirements of the newly expanded company's two divisions are:

Battery Division

1. For the battery division to produce 100 batteries, it must use 3 (3%) of its own batteries.
2. For every 100 batteries produced, 1 motor is required from the motor division. (Notice that the number of motors required is 1% of the total number of batteries produced.)

Motor Division

1. For the motor division to produce 100 motors, it must use 4 (4%) of its own motors.
2. For every 100 motors produced, 8 batteries are required from the battery division. (Notice that the number of batteries required is 8% of the total number of motors produced.)

The production needs within this two-sector economy can be represented visually by using a weighted digraph as shown in Figure 7.1. This digraph shows that if the company needs to produce a total of b batteries and m motors, then:

1. The battery division will require $0.03b$ batteries from its own division and $0.01b$ motors from the motor division.
2. The motor division will require $0.04m$ motors from its own division and $0.08m$ batteries from the battery division.

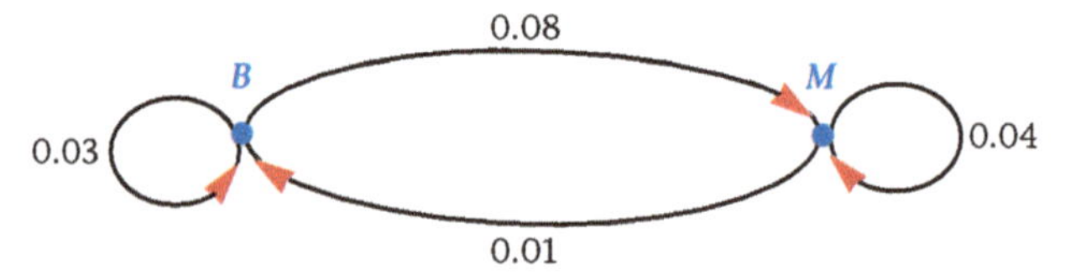

Figure 7.1. Weighted digraph showing the input required by each division.

A third way to present the information regarding the company's production needs is with a matrix. This matrix shows the required input from each sector of the economy. It is called a **consumption matrix** for the economy.

		To	
		Battery	Motor
From	Battery	0.03	0.08
	Motor	0.01	0.04

As an example, to produce 200 batteries, the battery division will need $(0.03)(200) = 6$ batteries from the battery division and $(0.01)(200) = 2$ motors from the motor division. And to produce 50 motors, the motor division will need $(0.08)(50) = 4$ batteries from the battery division and $(0.04)(50) = 2$ motors from the motor division.

Explore This

At the direction of your teacher, divide your class into groups of three or four people. Each group is to represent a production management team for the newly merged battery and motor company just described.

The company has given the battery division a daily total production quota of 1,000 batteries. The daily quota for the motor division is 200 motors. The production management team's problem is to determine how many batteries and motors each of the company's divisions will need during production and how many batteries and motors will be available for sales.

To solve the problem, each management team must complete the following tasks:

1. Find the number of batteries and motors needed by the battery division to meet its quota. Repeat for the motor division.
2. Find the number of batteries available for external sales if both divisions meet their daily quotas. Repeat for the number of motors available for sales outside the company.
3. Write an equation that represents the number of batteries that the company will use internally to meet its production quotas. Let b represent the total number of batteries produced, and m represent the total number of motors produced by the company. Repeat for the number of motors used internally.

4. Explore this problem from another point of view. Suppose the company needs to fill an order of 400 batteries and 100 motors. Estimate the total production needed from each division to fill these orders. Check your estimate and revise it if necessary. Hint: Recall that the amount of a product available to fill an external order (outside demand) equals the total production minus the amount of that product consumed internally by the company.

5. If time permits, find a system of two equations in two unknowns that could be used to find the total production for each division required in task 4.

After all groups have finished tasks 1 through 4, a spokesperson for each production management team should present the results of the team's discussion to the class.

Exercises

1. A utility company produces electric energy. Suppose that 5% of the total production of electricity is used up within the company to operate equipment needed to produce the electricity. Complete the following production table for this one-sector economy.

Total Production Units	Units Used Internally	Units for External Sales
500	0.05(500) = ___	500 – 0.05(500) = ___
900	___	___
___	100	___
___	250	___
___	___	2,375
___	___	7,125
P	___	___

2. Suppose that for every dollar's worth of computer chips produced by a high-tech company, 2 cents' worth is used by the company in the manufacturing process.

   a. What percentage of the company's total production of computer chips is used up within the company?

   b. What would the weighted digraph look like for this situation?

c. Write an equation that represents the dollars' worth of computer chips available for external demands (D) in terms of the total production (P) of chips by the company.

d. What must the total production of computer chips be in order for the company to meet an external demand for \$20,000 worth of computer chips?

3. The high-tech company described in Exercise 2 adds another division that produces computers. Each division within the expanded company uses some of the other division's product:

   Computer Chip Division: Every dollar's worth of computer chips produced requires an input of 2 cents' worth of computer chips and 1 cent's worth of computers.

   Computer Division: Every dollar's worth of computers produced requires an input of 20 cents' worth of computer chips and 3 cents' worth of computers.

   a. Draw a weighted digraph that summarizes the production needs for this two-sector economy.

   b. Construct a consumption matrix for this economy.

   c. Suppose the computer chip division produces \$1,000 worth of chips. How much input does it need from itself and from the computer division?

   d. Suppose the computer division produces \$5,000 worth of computers. How much input does it need from itself and from the computer chip division?

For Exercises 4–6, use the following consumption matrix for a company with two departments: service and production.

$$\begin{array}{l} \\ \text{Service} \\ \text{Production} \end{array} \begin{array}{c} \begin{array}{cc} \text{Service} & \text{Production} \end{array} \\ \begin{bmatrix} 0.05 & 0.20 \\ 0.04 & 0.01 \end{bmatrix} \end{array}$$

4. Draw a weighted digraph to show the flow of goods and services within this company.

5. Complete the following:

   a. For every dollar's worth of output, the service department requires ___ cents' worth of input from its own department and ___ cents' worth of input from the production department.

   b. For every dollar's worth of output, the production department requires ___ cents' worth of input from its own department and ___ cents' worth of input from the service department.

6. Suppose that the total output for the service department is $20 million over a certain period of time and the total output from the production department is $40 million.

   a. How much of the total output for the service department is used within the service department? How much input is required from the production department?

   b. How much of this total output for the production department is used within the production department? How much input is required from the service department?

   c. Combine the information in parts a and b to find how much of the total output from the service and production departments will be available for sales demands outside the company.

For Exercises 7–10, use the following weighted digraph that represents the flow of goods and services in a two-sector economy involving transportation and agriculture. The numbers on the edges represent cents' worth of product (or service) used per dollar's worth of output.

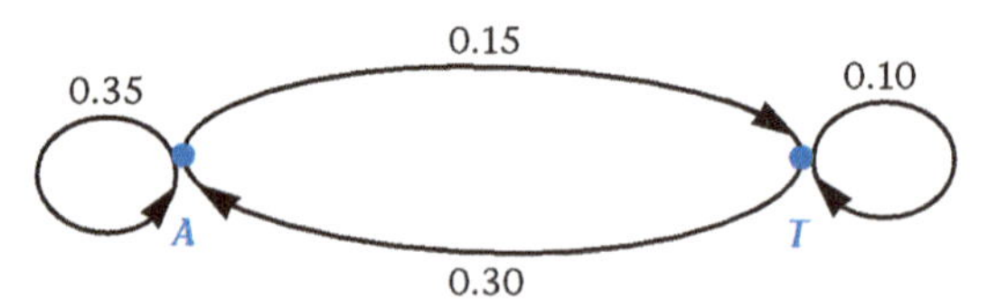

7. Construct a consumption matrix for this situation.

8. Complete the following:

   a. For every dollar's worth of output the agriculture sector requires ___ cents' worth of input from its own sector and ___ cents' worth of input from the transportation sector.

   b. For every dollar's worth of output the transportation sector requires ___ cents' worth of input from its own sector and ___ cents' worth of input from the agriculture sector.

9. If the total output for the agriculture sector is $50 million and the total output for the transportation sector is $100 million over a certain period of time, find the following.

   a. The total amount of agricultural goods used internally by this two-sector economy.

   b. The amount of agricultural goods available for external sales.

   c. The total amount of transportation services used internally by this two-sector economy.

   d. The amount of transportation services available for external sales.

10. a. Write an equation to represent the internal consumption of agricultural products for this economy. Let P_A represent the total production for agriculture and P_T represent the total output for transportation.

    b. Write an equation to represent the internal consumption of transportation services. Use P_A and P_T as in part a.

    c. Use the information from parts a and b to write two equations, one that represents the amount of agricultural products available to fill external consumer demands and the other that represents the amount of transportation products available. Let D_A represent the total available for external demand for agriculture and D_T represent the total available for external demand for transportation. (Hint: Recall that the amount of a product available to fill external demands is equal to the total production less the amount that is used internally.)

    d. Suppose that this economy has an external demand for $10 million worth of agricultural products and $15 million in transportation services. Write a system of equations that could be solved to find the total production of agriculture products and transportation services necessary to satisfy these demands.

11. Solve the system of equations that was found by the production management team in the Explore This (task 5) on page 382.

Project

12. Research and report to the class on the life and work of Wassily Leontief.

Toyota, BMW to Research Lithium-air Battery

Reuters
January 24, 2013

Toyota Motor Corporation and BMW AG will jointly research a lithium-air battery that is expected to be more powerful than the lithium-ion batteries currently used in many hybrid and electric vehicles. The two companies will also work on a fuel cell vehicle system, which includes a hydrogen tank and motor, by 2020. The strengthening of the partnership between the two companies will allow them to cut development costs as competition intensifies globally.

A lithium-air battery has its anode filled with lithium and cathode with air. Theoretically, the battery can generate and store more electricity than the existing lithium-ion battery. The new technology is being studied by researchers, including IBM, who are working to develop a lithium-air battery that will let electric vehicles run 500 miles on one charge.

Lesson 7.2

The Leontief Input-Output Model, Part 2

You may have noticed in Lesson 7.1 that the Leontief model can become complicated very quickly, even when dealing with only one- and two-sector economies. You are probably thinking that there has to be an easier way. You're right. There is an easier way.

Recall the consumption matrix (C) for this economy of the battery and motor company from Lesson 7.1.

In that lesson, you were asked to find a system of two equations in two unknowns that could be solved to find the total production of batteries (b) and motors (m) necessary to meet external sales demands of 400 batteries and 100 motors. To find two such equations, you used the fact that the total production for each product was equal to the amount

$$C = \begin{array}{c} \\ \text{Battery} \\ \text{Motor} \end{array} \begin{array}{c} \begin{array}{cc} \text{Battery} & \text{Motor} \end{array} \\ \begin{bmatrix} 0.03 & 0.08 \\ 0.01 & 0.04 \end{bmatrix} \end{array}$$

of the product used by the two divisions during production plus the external demand for that product (sales outside the company).

For example, the total number of batteries produced (b) equals the number of batteries used within the battery division ($0.03b$) plus the number of batteries sent to the motor division ($0.08m$) plus the number of batteries required for sales outside the company (400). This gives the equation

$$b = 0.03b + 0.08m + 400.$$

Likewise, the total number of motors produced (m) equals the number of motors sent to the battery division ($0.01b$) plus the number of motors used within the motor division ($0.04m$) plus the number of motors required for sales outside the company (100). This gives the equation

$$m = 0.01b + 0.04m + 100.$$

Thus, the system of two equations in two unknowns (unsimplified) looks like

$$b = 0.03b + 0.08m + 400$$
$$m = 0.01b + 0.04m + 100.$$

You can solve this system algebraically, or you can solve it using matrices. If you use matrices, you can use a calculator or computer to help you with the work.

Using Matrices to Represent Systems of Linear Equations

To solve a system of equations using matrices, first explore how you could represent this system of equations with matrices.

A consumption matrix C has already been defined. If you let

$$P = \begin{bmatrix} b \\ m \end{bmatrix}$$

represent a total production matrix (P) and

$$D = \begin{bmatrix} 400 \\ 100 \end{bmatrix}$$

represent a demand matrix (D), then the system of equations can be written as a matrix equation.

$$\underbrace{\begin{bmatrix} b \\ m \end{bmatrix}}_{P} = \underbrace{\begin{bmatrix} 0.03 & 0.08 \\ 0.01 & 0.04 \end{bmatrix}}_{C} \underbrace{\begin{bmatrix} b \\ m \end{bmatrix}}_{P} + \underbrace{\begin{bmatrix} 400 \\ 100 \end{bmatrix}}_{D}$$

This matrix equation can be written more simply as

$$P = CP + D,$$

showing that the total production = internal consumption + external demand.

Notice that this matrix equation resembles the simple linear equation solved in the example in Lesson 7.1 (page 379). Indeed, solving this matrix equation uses the same operations that are used in solving a linear equation. Verify that this is true by tracing the steps for solving a linear equation using ordinary algebra and solving a matrix equation using matrix algebra in the following table.

One-Sector Economy Linear Equation Ordinary Algebra	Two-Sector Economy Matrix Equation Matrix Algebra	Comments
$p = 0.03p + d$	$P = CP + D$	
$p - 0.03p = d$	$P - CP = D$	
$1p - 0.03p = d$	$IP - CP = D$	The identity matrix I times $P = P$.
$(1 - 0.03)p = d$	$(I - C)P = D$	
$\frac{1}{1-0.03}(1 - 0.03)p = \frac{1}{1-0.03}d$	$(I - C)^{-1}(I - C)P = (I - C)^{-1}D$	*
$1p = \frac{1}{1-0.03}d$	$IP = (I - C)^{-1}D$	Recall that $A^{-1}A = I$. (See Exercise 8 in Lesson 3.3, page 144.)
$p = \frac{1}{1-0.03}d$	$P = (I - C)^{-1}D$	

*Up to this point the ordinary algebra operations and the matrix operations have been identical. In solving the linear equation, it would be natural to divide both sides of the equation by (1 – 0.03). But, there is no division operation in matrix algebra. The thing to do, then, is to multiply both sides of the linear equation by the multiplicative inverse of (1 – 0.03). Multiplying both sides of the matrix equation by the multiplicative inverse of $(I - C)$ is a valid matrix operation.

In summary, if the consumption matrix (C) and the external demand matrix (D) are known, then the total production matrix (P) can be found using the matrix equation

$$P = (I - C)^{-1}D.$$

For the battery and motor problem, the solution is

$$\begin{bmatrix} b \\ m \end{bmatrix} = \left(\begin{bmatrix} 1 & 0 \\ 0 & 1 \end{bmatrix} - \begin{bmatrix} 0.03 & 0.08 \\ 0.01 & 0.04 \end{bmatrix}\right)^{-1} \begin{bmatrix} 400 \\ 100 \end{bmatrix}.$$

Using a calculator or computer to do the computations and rounding to the nearest whole number, you will find that

$$\begin{bmatrix} b \\ m \end{bmatrix} = \begin{bmatrix} 421 \\ 109 \end{bmatrix}$$

The results show that to fill an order for 400 batteries and 100 motors, the company must produce 421 batteries and 109 motors.

Solving Systems of Linear Equations Using Matrices

The matrix techniques used for solving systems of equations in this lesson can be used to solve any system of n independent equations in n unknowns. Look, for example, at the following system of two equations in two unknowns.

$$2x_1 + 3x_2 = 23$$
$$5x_1 - 2x_2 = 10$$

This system can be written as a single matrix equation.

If we let

$$A = \begin{bmatrix} 2 & 3 \\ 5 & -2 \end{bmatrix}, X = \begin{bmatrix} x_1 \\ x_2 \end{bmatrix} \text{ and } B = \begin{bmatrix} 23 \\ 10 \end{bmatrix},$$

the matrix equation can be written in the form $AX = B$, which is similar to a simple linear equation such as $ax = b$.

One way to solve this linear equation is to multiply both sides of the equation by the multiplicative inverse of a, $\frac{1}{a}$. The same strategy can be used to solve the matrix equation as shown in the following table.

	Linear Equations Ordinary Algebra	Matrix Equations Matrix Algebra
Step 1	$ax = b$	$AX = B$
Step 2	$\frac{1}{a}ax = \frac{1}{a}b$	$A^{-1}AX = A^{-1}B$
Step 3	$1x = \frac{1}{a}b$	$IX = A^{-1}B$
Step 4	$x = \frac{1}{a}b$	$X = A^{-1}B$

Applying this method to the previous system of equations, we have

$$\begin{bmatrix} x_1 \\ x_2 \end{bmatrix} = \begin{bmatrix} 2 & 3 \\ 5 & -2 \end{bmatrix}^{-1} \begin{bmatrix} 23 \\ 10 \end{bmatrix}.$$

You can use a calculator or computer to do the calculations and verify that the solution for the given system of linear equations is $x_1 = 4$ and $x_2 = 5$.

Exercises

Use either a calculator or computer software to perform matrix operations in the following exercises.

1. Use matrices to solve each of the following systems of equations.

   a. $5x - 3y = 2$

      $x + 2y = 3$

   b. $4x - y = 2$

      $5x + 2y = 9$

2. The consumption matrix C for the high-tech company described in Exercise 3 of Lesson 7.1 is shown below.

$$C = \begin{array}{r} \\ \text{Chips} \\ \text{Computers} \end{array} \begin{array}{c} \begin{array}{cc} \text{Chips} & \text{Computers} \end{array} \\ \begin{bmatrix} 0.02 & 0.20 \\ 0.01 & 0.03 \end{bmatrix} \end{array}$$

   a. Suppose that the total production for the company is \$40,000 worth of computer chips and \$50,000 worth of computers. Write a production matrix, P.

   b. Compute the matrix product CP to find the amount of each product that the company uses internally.

   c. Use the information from parts a and b and the matrix equation $D = P - CP$ to compute the amount of computer chips and computers available for sales outside the company (external demand).

3. The consumption matrix C for the company described in Exercises 4–6 of Lesson 7.1 is shown below.

$$C = \begin{array}{l} \text{Service} \\ \text{Production} \end{array} \overset{\begin{array}{cc} \text{Service} & \text{Production} \end{array}}{\begin{bmatrix} 0.05 & 0.20 \\ 0.04 & 0.01 \end{bmatrix}}$$

   a. Suppose that the total output for the service department of the company is \$20 million and the total output for the production department is \$40 million. Write a production matrix, P.

   b. Compute the matrix product CP to find the output from each department that the company uses internally.

   c. Compute the output that is available for sales demands outside the company (external demand).

   d. The company must meet external demands of \$25 million in service and \$50 million in products over a period of time. What must be the total production in service and products to meet this demand?

4. The consumption matrix C for the two-sector economy described in Exercises 7–10 of Lesson 7.1 is shown below.

$$C = \begin{array}{l} \text{Agriculture} \\ \text{Transportation} \end{array} \overset{\begin{array}{cc} \text{Agriculture} & \text{Transportation} \end{array}}{\begin{bmatrix} 0.35 & 0.15 \\ 0.30 & 0.10 \end{bmatrix}}$$

   Suppose the economy must meet external demands of \$10 million in agricultural products and \$15 million in transportation services. Find the total production of agriculture products and transportation services necessary to satisfy these demands.

5. The techniques developed in this lesson using a two-sector economy can easily be extended to solve problems that involve economies of more than two sectors. For example, examine an economy that has three sectors—transportation, energy, and manufacturing. Each of these sectors uses some of its own products or services as well as some from each of the other sectors, as follows:

   Transportation Sector: Every dollar's worth of transportation provided requires an input of 10 cents' worth of transportation services, 15 cents' worth of energy, and 25 cents' worth of manufactured goods.

Energy Sector: Every dollar's worth of energy produced requires an input of 25 cents' worth of transportation services, 10 cents' worth of energy, and 20 cents' worth of manufactured goods.

Manufacturing Sector: Every dollar's worth of manufactured goods produced requires an input of 20 cents' worth of transportation services, 20 cents' worth of energy, and 15 cents' worth of manufactured goods.

a. Draw a weighted digraph for this three-sector economy.

b. Construct a consumption matrix (C) for this economy. Label the rows and columns of your matrix.

c. The total production over a period of time for this economy is \$150 million in transportation, \$200 million in energy, and \$160 million in manufactured goods. Write a production matrix (P) for this economy. Label the rows and columns of your matrix.

d. Compute the matrix product CP to find the amount of each product that is used internally by the economy. Write your answer as a matrix and label the rows and columns.

e. Use the information from parts c and d and the matrix equation $D = P - CP$ to find the amount of goods available for external demand (sales outside the three sectors described here).

f. Suppose the estimated consumer demand for transportation, energy, and manufactured goods and services in millions of dollars are 100, 95, and 110, respectively. Find the total production necessary to fulfill these demands. Use the matrix equation $P = (I - C)^{-1}D$.

6. An economy consisting of three sectors (services, manufacturing, and agriculture) has the consumption matrix

$$C = \begin{array}{c} \\ \text{Services} \\ \text{Manufacturing} \\ \text{Agriculture} \end{array} \begin{array}{c} \begin{array}{ccc} \text{Services} & \text{Manufacturing} & \text{Agriculture} \end{array} \\ \begin{bmatrix} 0.1 & 0.3 & 0.2 \\ 0.2 & 0.3 & 0.1 \\ 0.2 & 0.1 & 0.2 \end{bmatrix} \end{array}.$$

a. Draw a weighted digraph for this economy.

b. On which sector of the economy is manufacturing the most dependent? The least dependent?

c. If the services sector has an output of $40 million, what is the input in dollars from manufacturing? From agriculture?

d. A production matrix, P, in millions of dollars for this economy is as follows. Use the matrix product CP to find the internal consumption of services and products within this economy. Find the external demand matrix D.

$$P = \begin{matrix} \text{Services} \\ \text{Manufacturing} \\ \text{Agriculture} \end{matrix} \begin{bmatrix} 20 \\ 25 \\ 15 \end{bmatrix}$$

e. An external demand matrix, D, in millions of dollars follows. How much must be produced by each sector to meet this demand?

$$D = \begin{matrix} \text{Services} \\ \text{Manufacturing} \\ \text{Agriculture} \end{matrix} \begin{bmatrix} 4.6 \\ 5.0 \\ 4.0 \end{bmatrix}$$

7. An economy consisting of four sectors (transportation, manufacturing, agriculture, and services) has the consumption matrix (in millions of dollars worth of products)

$$C = \begin{matrix} \\ \text{Transportation} \\ \text{Manufacturing} \\ \text{Agriculture} \\ \text{Services} \end{matrix} \begin{matrix} \text{Transportation} & \text{Manufacturing} & \text{Agriculture} & \text{Services} \\ 0.25 & 0.28 & 0.22 & 0.20 \\ 0.15 & 0.15 & 0.17 & 0.23 \\ 0.19 & 0.20 & 0.21 & 0.15 \\ 0.20 & 0.24 & 0.19 & 0.25 \end{matrix}.$$

a. Draw a weighted digraph for this economy.

b. On which sector of the economy is services the most dependent? The least dependent?

c. If the manufacturing sector has an output of $20 million, what is the input in dollars from services? From transportation?

d. A production matrix, P, in millions of dollars, follows. Use the matrix product CP to find the internal consumption of services and products within this economy. Find the external demand matrix D.

$$P = \begin{array}{l} \text{Transportation} \\ \text{Manufacturing} \\ \text{Agriculture} \\ \text{Services} \end{array} \begin{bmatrix} 50 \\ 40 \\ 45 \\ 50 \end{bmatrix}$$

e. An external demand matrix, D, in millions of dollars, follows. How much must be produced by each sector to meet this demand?

$$D = \begin{array}{l} \text{Transportation} \\ \text{Manufacturing} \\ \text{Agriculture} \\ \text{Services} \end{array} \begin{bmatrix} 10 \\ 12 \\ 10 \\ 15 \end{bmatrix}$$

8. A two-industry system consisting of services and manufacturing has the consumption matrix

$$C = \begin{array}{l} \\ \text{Services} \\ \text{Manufacturing} \end{array} \begin{array}{c} \begin{array}{cc} \text{Services} & \text{Manufacturing} \end{array} \\ \begin{bmatrix} 0.5 & 0.5 \\ 0.2 & 0.3 \end{bmatrix} \end{array}.$$

a. Compute the total production necessary to satisfy a consumer demand for 15 units of services and 25 units of manufacturing.

b. Comment on the productivity of this system. Explain your answer.

c. If the consumer demand is for 30 units of services and 50 units of manufacturing, find the production needed to fill these demands.

d. On the basis of the results in parts a and c above, predict the total production of services and goods for a consumer demand for 45 units of service and 75 units of manufacturing. Check your prediction by computing the production matrix for this case.

9. A company has two divisions: service and production. The flow of goods and services within this company is described by the consumption matrix

$$C = \begin{array}{c} \\ \text{Services} \\ \text{Products} \end{array} \begin{array}{c} \begin{array}{cc} \text{Services} & \text{Products} \end{array} \\ \begin{bmatrix} 0.10 & 0.25 \\ 0.05 & 0.10 \end{bmatrix} \end{array}.$$

a. Draw a weighted digraph for this situation.

b. The total output for the company during one year is $50 million in services and $75 million in products. How much of the total output is used internally by each of the company's divisions?

c. What total output is needed to meet an external consumer demand of $15 million in service and $25 million in products?

d. If the consumer demand increases to $22 million for services and to $30 million for products, what will be the effect on the total production of goods and services?

Projects

10. Research and report to the class on the following.

a. What computer software is available in your school for solving systems of equations?

b. What is the largest number of variables that the software can handle?

c. How long does it take the computer to solve a system with the largest number of variables possible using this software?

d. It took Professor Leontief 56 hours to solve a system of 42 equations in 42 unknowns using the Mark II in the 1940s and 3 minutes to solve a system of 81 equations in 81 unknowns using the IBM 7090 in the 1960s. If the software available in your school can handle systems of 42 and 81 equations, find out:

i. How long it takes to solve a system of 42 independent equations in 42 unknowns.

ii. How long it takes to solve a system of 81 independent equations in 81 unknowns.

e. Investigate parts a and b for computer software such as *Mathematica*, *Maple*, or *MATLAB* that your school may not own.

11. Research and report to the class on parts a to c of Exercise 10 for the graphing calculators that are available in your school. Try, in particular, to find this information for the TI-94 calculator, the TI-Nspire, and the Casio PRIZM.

Computer/Calculator Explorations

12. Write a graphing calculator (or computer) program that uses matrices for solving systems of *n* equations in *n* unknowns.

13. Write a graphing calculator (or computer) program designed to solve the various consumption problems presented in this lesson.

Lesson 7.3

Markov Chains

A **Markov chain** is a process that arises naturally in problems that involve a finite number of events or states that change over time. In this lesson, you will explore a situation that illustrates the significant characteristics of a Markov chain.

Consider a situation where a food service director for a local high school conducted a survey in hopes of predicting the number of students who will eat in the cafeteria in the future. The results of the survey are as follows:

Mathematician of Note

(1856–1922)

Markov was a Russian mathematician whose work influenced many other famous mathematicians and statisticians. His studies of linked chains of events led to the modern study of stochastic processes. As a result of his work, one type of stochastic process is called a Markov chain.

- If a student eats in the cafeteria on a given day, the probability that he or she will eat there again the next day is 70% and the probability that he or she will not eat there is 30%.
- If a student does not eat in the cafeteria on a given day, the probability that he or she will eat in the cafeteria the next day is 40% and the probability that he or she will not eat there is 60%.

Suppose that on Monday, 75% of the students ate in the cafeteria and 25% ate elsewhere. What can be expected to happen on Tuesday?

One way to organize this information is to use a tree diagram like the one in Figure 7.2.

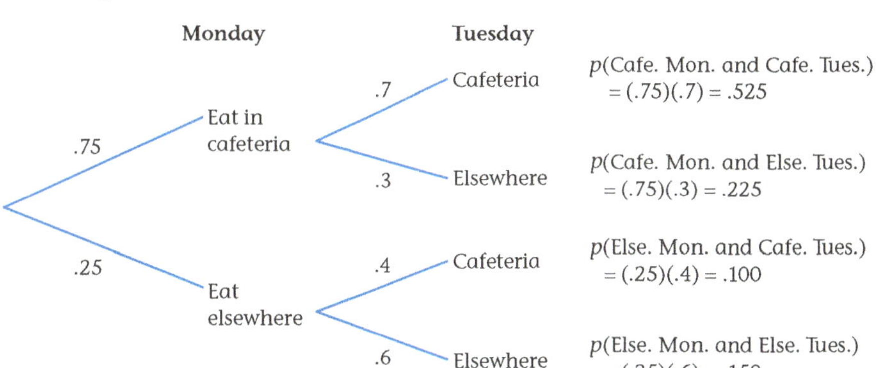

Figure 7.2. Cafeteria statistics organized in a tree diagram.

To determine the percent of the students who will eat in the cafeteria on Tuesday, look at the tree diagram. Notice that this happens if a student eats in the cafeteria on Monday and again on Tuesday. It also happens when a student eats elsewhere on Monday and in the cafeteria on Tuesday. The portion of the student population who will eat in the cafeteria on Tuesday is .525 + .100 = .625, or 62.5%.

Similarly, the portion of students who will eat elsewhere on Tuesday is .225 + .150 = .375, or 37.5%. Note that this could also be calculated by subtracting .625 from 1.

The tree diagram model is fine if only two stages are required to reach a solution. But the number of branches of the tree diagram doubles with each additional day, and the model soon becomes impractical. So an alternative is needed.

Transition Matrices

The Monday student data are called the **initial distribution** of the student body and can be represented by a row (or **initial-state**) matrix, D_0.

$$D_0 = \overset{\begin{matrix}\text{C} & \quad\text{E}\end{matrix}}{\begin{bmatrix} 0.75 & 0.25 \end{bmatrix}},$$

where C = eats in the cafeteria and E = eats elsewhere.

Movement from one state to another is often called a **transition**. So the data about how students choose to eat from one day to the next can be written in a matrix called a **transition matrix**, T.

$$T = \begin{array}{c} \\ C \\ E \end{array} \begin{array}{c} \begin{array}{cc} C & E \end{array} \\ \begin{bmatrix} 0.7 & 0.3 \\ 0.4 & 0.6 \end{bmatrix} \end{array}$$

Notice that the entries of a transition matrix are probabilities, values between 0 and 1 inclusive. Also notice that the transition matrix is a square matrix and the sum of the probabilities in any row is 1.

Now calculate the product of matrix D_0 and matrix T.

$$\begin{aligned} D_0T &= [0.75 \ \ 0.25]\begin{bmatrix} 0.7 & 0.3 \\ 0.4 & 0.6 \end{bmatrix} \\ &= [0.75(0.7) + 0.25(0.4) \quad 0.75(0.3) + 0.25(0.6)] \\ &= [0.625 \quad 0.375] \end{aligned}$$

Compare these calculations with those made in the tree diagram model. The values in the resulting row matrix can be interpreted as the portion of students who eat in the cafeteria and who eat elsewhere on Tuesday. This row matrix is called D_1 to indicate that it occurs one day after the initial day.

To see what happens on Wednesday, it is only necessary to repeat the process using D_1 in place of D_0.

$$\begin{aligned} D_1T &= [0.625 \ \ 0.375]\begin{bmatrix} 0.7 & 0.3 \\ 0.4 & 0.6 \end{bmatrix} \\ &= [0.625(0.7) + 0.375(0.4) \quad 0.625(0.3) + 0.375(0.6)] \\ &= [0.5875 \quad 0.4125] \end{aligned}$$

The resulting row matrix is called D_2 to indicate that it occurs two days after the initial day. Thereafter, D_2 shows that approximately 59% of the students will eat in the cafeteria on Wednesday and 41% will eat elsewhere.

Consider how D_2 was calculated.

$D_2 = D_1T$, but $D_1 = D_0T$, so by substitution, $D_2 = (D_0T)(T)$.

Because matrix multiplication is associative,

$D_2 = (D_0T)(T) = D_0(T^2)$.

This means that the calculation of the distribution of students on Wednesday can be completed by taking the initial-state matrix times the square of the transition matrix.

This observation simplifies additional calculations. For example, if you want to know the distribution on Friday, four days from Monday, calculate $D_4 = D_0(T^4)$ on a calculator that has matrix features or on a computer equipped with matrix software.

$$D_0T^4 = [0.75 \quad 0.25]\begin{bmatrix} 0.7 & 0.3 \\ 0.4 & 0.6 \end{bmatrix}^4 = [0.572875 \quad 0.427125]$$

About 57% of the students can be expected to eat in the cafeteria on Friday.

The movement of students from one state to another can also be shown with a weighted digraph called a **transition digraph** or **state diagram** (see Figure 7.3).

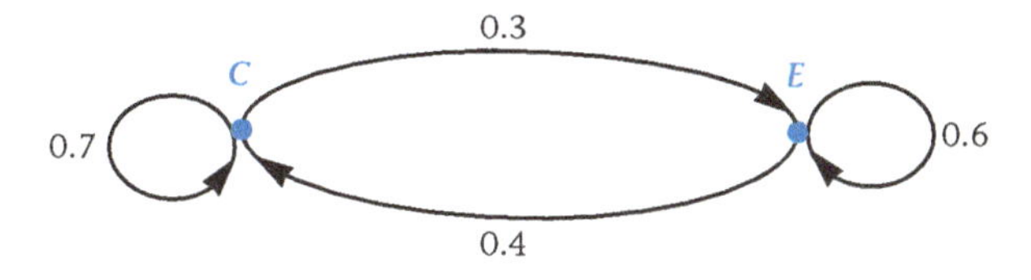

Figure 7.3. Transition digraph for the cafeteria statistics.

Exercises

Use either a calculator or computer software to perform matrix operations in the following exercises.

For Exercises 1–4, use the following transition matrix.

$$T = \begin{array}{c} \\ C \\ E \end{array}\begin{array}{c} \begin{array}{cc} C & E \end{array} \\ \begin{bmatrix} 0.7 & 0.3 \\ 0.4 & 0.6 \end{bmatrix} \end{array}$$

1. For parts a–c, use the initial distribution $D_0 = [0.75 \;\; 0.25]$.
   a. Find the distribution of students eating and not eating in the cafeteria each day for the first week of school.
   b. Find the distribution of students eating and not eating in the cafeteria after 2 weeks (10 school days) have passed. Repeat for 3 weeks (15 days).
   c. Based on your computations in parts a and b, what do you notice?
2. a. Choose any other initial distribution of students and repeat parts a and b of Exercise 1.
   b. Compare the results of part a and Exercise 1, part b. Does the initial distribution appear to make a difference in the long run?
   c. Calculate the 15th power of matrix T. Compare the entries in T^{15} to the distribution after the 15th day.

When successive applications of a Markov process are made and succeeding distributions approach some distribution D, this distribution is called a **steady state distribution** for the Markov chain with a transition matrix T. A sufficient condition, which we will not prove in this text, for a Markov chain to have a steady state distribution is that some power of its transition matrix have only positive entries. Since all the entries in the transition matrix T are nonzero probabilities, this condition is clearly met for the cafeteria Markov chain.

3. What is the steady state distribution for the Markov chain with a transition matrix T?

4. Suppose the entire student body eats in the cafeteria on the first day of school. The initial distribution in this case is $D_0 = [1 \quad 0]$. Repeat parts a and b of Exercise 1 for this distribution. After several weeks, what percentage of students will be eating in the cafeteria?

5. Which of the matrices below could be Markov transition matrices? For the matrices that could not be transition matrices, explain why not.

   a. $\begin{bmatrix} 0.7 & 0.3 \\ 0.6 & 0.6 \end{bmatrix}$ b. $\begin{bmatrix} 0.1 & 0.4 & 0.5 \\ 0.2 & 0.6 & 0.2 \end{bmatrix}$ c. $\begin{bmatrix} 1.2 & -4 \\ 1 & 0 \end{bmatrix}$

   d. $\begin{bmatrix} 0.6 & 0.3 & 0.1 \\ 0.3 & 0.3 & 0.3 \end{bmatrix}$ e. $\begin{bmatrix} 0.75 & 0.25 \\ 1 & 0 \end{bmatrix}$ f. $\begin{bmatrix} 0.45 & 0.55 \\ 0.33 & 0.66 \end{bmatrix}$

6. There is a 60% chance of rain today. It is known that tomorrow's weather depends on today's according to the probabilities shown in the following tree diagram.

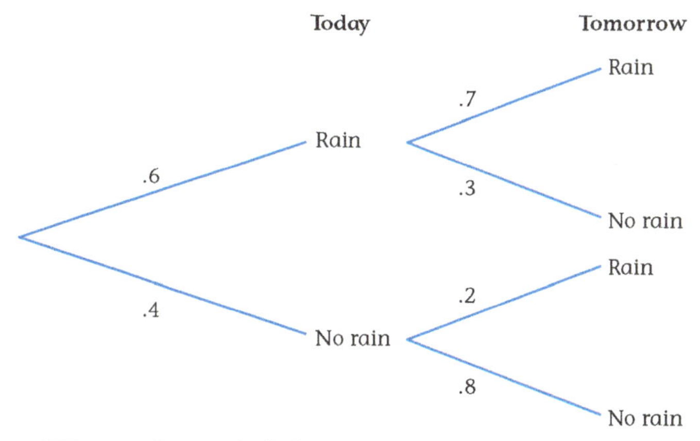

   a. What is the probability it will rain tomorrow if it rains today?

   b. What is the probability it will rain tomorrow if it doesn't rain today?

   c. Write an initial-state matrix that represents the weather forecast for today.

   d. Write a transition matrix that represents the transition probabilities shown in the tree diagram.

   e. Calculate the forecast for 1 week (7 days) from now.

   f. In the long run, for what percentage of days will it rain?

St. Louis Forecast

weather.com
January 27, 2013

Light rain coming to an end around mid-afternoon. Mostly cloudy with temperatures in the mid 40s. Winds SE at 15 mph. Chance of rain 75%. Tomorrow, mostly cloudy, 30% chance of rain. High 67°F.

7. A taxi company has divided the city into three districts—Westmarket, Oldmarket, and Eastmarket. By keeping track of pickups and deliveries, the company found the following:

   - Of the fares picked up in the Westmarket district, only 10% are dropped off in that district, 50% are taken to the Oldmarket district, and 40% go to the Eastmarket district.
   - Of the fares picked up in the Oldmarket district, 20% are taken to the Westmarket district, 30% stay in the Oldmarket district, and 50% are dropped off in the Eastmarket district.
   - Of the fares picked up in the Eastmarket district, 30% are delivered to each of the Westmarket and Oldmarket districts, while 40% stay in the Eastmarket district.

   a. Draw a transition digraph for this Markov chain.

   b. Construct a transition matrix for these data.

   c. Write an initial-state matrix for a taxi that starts off by picking up a fare in the Oldmarket district. What is the probability that it will end up in the Oldmarket district after three additional fares?

   d. Find and interpret the steady state distribution for this Markov process.

8. Emily, Jon, and Gretchen are tossing a football around. Emily always tosses to Jon, and Jon always tosses to Gretchen, but Gretchen is equally likely to toss the ball to either Emily or Jon.

   a. Draw a transition digraph to represent this information.

   b. Represent this information as the transition matrix of a Markov chain.

   c. What is the probability that Emily will have the ball after three tosses if she was the first one to throw it to one of the others?

   d. Find and interpret the steady state distribution for this Markov chain.

   e. Explain why there are no zeros in the steady state distribution even though there were several zeros in the transition matrix.

9. Jim agreed to care for Emily's cat, Ellington, for the weekend. On Friday night Ellington prowled the first floor of Jim's house, randomly moving from room to room, not staying in one room for more than a few minutes. The following floor plan shows the location of the rooms and doorways in Ellington's range. The letters on the floor plan represent Living room, Dining room, Kitchen, Bathroom, and Study.

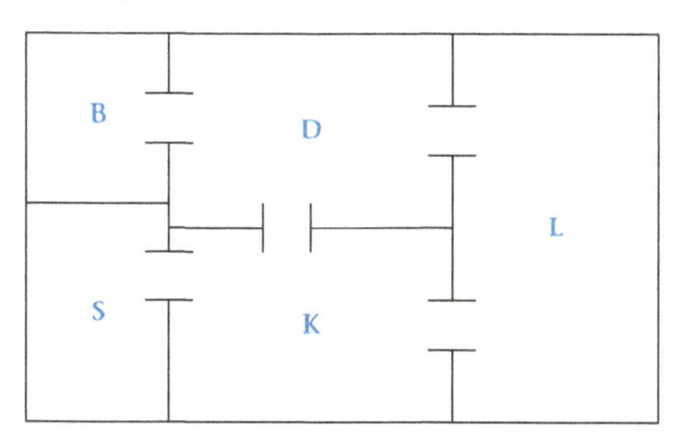

Each of Ellington's movements can be interpreted as a transition in a Markov chain in which a state is identified with the room he is in. The first row of the transition matrix is

$$\begin{array}{c} \\ L \end{array} \begin{array}{c} \begin{array}{ccccc} L & D & K & S & B \end{array} \\ \begin{bmatrix} 0 & \frac{1}{2} & \frac{1}{2} & 0 & 0 \end{bmatrix} \end{array}.$$

a. Complete the following transition matrix for this situation.

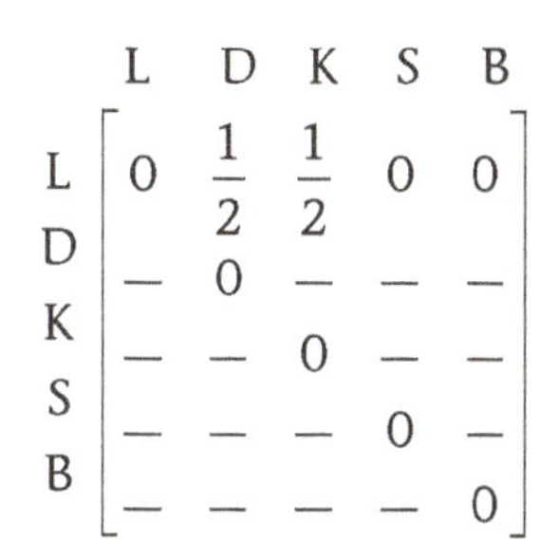

$$\begin{array}{c|ccccc} & L & D & K & S & B \\ \hline L & 0 & \frac{1}{2} & \frac{1}{2} & 0 & 0 \\ D & \_ & 0 & \_ & \_ & \_ \\ K & \_ & \_ & 0 & \_ & \_ \\ S & \_ & \_ & \_ & 0 & \_ \\ B & \_ & \_ & \_ & \_ & 0 \end{array}$$

b. If Ellington starts off in the living room, what is the probability that he will be in the study after two transitions? After three transitions?

c. After a large number of transitions, what is the probability that Ellington will be in the bathroom?

d. In the long run, what percent of the time will Ellington spend in either the kitchen or the dining room?

10. A discrete mathematics student observes a bug crawling from vertex to vertex along the edges of a tetrahedron model on the teacher's desk (see figure below). From any vertex the bug is equally likely to go to any other vertex.

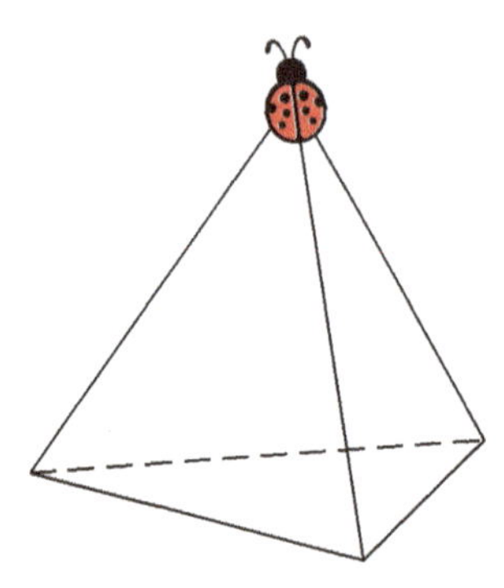

a. Construct a transition matrix for this situation.

b. Determine the probabilities for the location of the bug if it moves to a different vertex about 20 times.

11. Dick's old hound dog, Max, spends much of his time during the day running from corner to corner along the fence surrounding his square-shaped yard. There is a .5 probability that Max will turn in either direction at a corner. The corners of Max's yard point north, east, south, and west.

a. Draw a transition digraph that represents Max's movement.

b. Construct a transition matrix for this situation.

c. Look at the behavior of successive powers of the transition matrix. Notice the oscillation of the transition probabilities between the states represented by the rows of the matrices. Does this system appear to stabilize in some way? Explain your answer.

d. Approximately what percentage of the time will Max spend at each of the corners of his yard? (Note: You need to halve the entries in the matrix to account for the oscillating pattern.)

e. Suppose that Max changes his routine one day, and the pattern of his new movements is represented by the following transition matrix. Approximately what percentage of the time will Max now spend at each of the corners of his yard?

$$\begin{array}{c} \\ N \\ E \\ S \\ W \end{array} \begin{array}{c} \begin{array}{cccc} N & E & S & W \end{array} \\ \begin{bmatrix} 0 & \frac{1}{2} & 0 & \frac{1}{2} \\ \frac{3}{8} & 0 & \frac{5}{8} & 0 \\ 0 & \frac{3}{8} & 0 & \frac{5}{8} \\ \frac{3}{4} & 0 & \frac{1}{4} & 0 \end{bmatrix} \end{array}$$

12. Using mathematical induction, prove that $D_k = D_0T^k$ for any original distribution D_0 and transition matrix T, where k is a natural number.

13. A group of researchers are studying the effect of a potent flu vaccine in healthy (well) and infirm (ill) rats. When the rats are injected with the vaccine, three things may occur. The rat may have no reaction and its health status does not change. The rat may have a mild reaction and become ill, or the rat may have a severe reaction and die. The probabilities of each of these reactions are shown in the following matrix.

$$\begin{array}{c} \\ \text{Well} \\ \text{Ill} \\ \text{Dead} \end{array} \begin{array}{c} \begin{array}{ccc} \text{Well} & \text{Ill} & \text{Dead} \end{array} \\ \begin{bmatrix} 0.8 & 0.2 & 0 \\ 0.1 & 0.6 & 0.3 \\ 0 & 0 & 1 \end{bmatrix} \end{array}$$

a. Write an initial-state matrix for a healthy rat who is injected with the vaccine.

b. In this study the scientists check the status of the rats on a daily basis. Use the transition matrix to predict the health of the rat in part a after 4 days.

c. Use the transition matrix to predict the rat's health in the long run.

d. This Markov chain has a state that is called an **absorbing state**. Which state do you think it is? Why?

14. A hospital categorizes its patients as well (in which case they are discharged), good, critical, and deceased. Data show that the hospital's patients move from one category to another according to the probabilities shown in this transition matrix.

	Well	Good	Critical	Deceased
Well	1	0	0	0
Good	0.5	0.3	0.2	0
Critical	0	0.3	0.6	0.1
Deceased	0	0	0	1

a. Write an initial-state matrix for a patient who enters the hospital in critical condition.

b. If patients are reclassified daily, predict the patient's future after 1 week in the hospital.

c. Predict the future of any patient in the long run.

d. Does this Markov chain have any absorbing states? Which states do you think are absorbing? Why?

Project

15. Research and report to the class on the life and work of A. A. Markov.

Computer/Calculator Exploration

16. Write a graphing calculator (or computer) program that can be used to find the steady state distribution for a Markov chain.

Lesson 7.4

Game Theory, Part 1

When two or more individuals try to control the course of events, conflict often arises. **Game theory** is a branch of mathematics that uses mathematical tools to study such situations. Even though the study of the discipline can be traced back to the seventeenth century, it was not until the 1940s during World War II that game theory was recognized as a legitimate branch of mathematics.

You may tend to think of games as being fun or relaxing ways to spend your time. There are, however, many decision-making situations in fields such as economics and politics that can also be thought of as games. The players in these games may be individuals, teams of people, whole countries, or even forces of nature. Each player (or side) has a set of alternative courses of action called **strategies** that can be used in making decisions. Mathematical game theory deals with selecting the best strategies for a player to follow in order to achieve the most favorable outcomes.

In this lesson, you will explore some examples of games with two players and use matrix models to determine the best strategy for each player to choose.

Mathematician of Note

John von Neumann
(1903–1957)

Von Neumann came to the United States in 1930 from Hungary. He is considered one of the founders of modern game theory. He also is known for his contributions to the development of high-speed computers that enabled the United States to develop and produce the first hydrogen bomb.

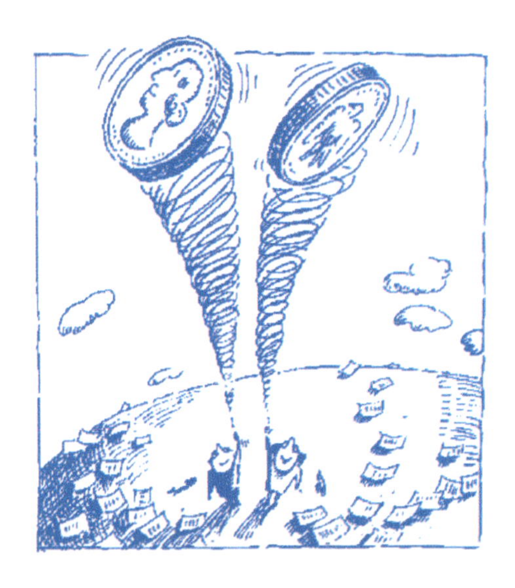

Explore This

Consider a simple coin-matching game that Sol and Tina are playing. Each conceals a penny with either heads or tails turned upward. They display their pennies simultaneously.

- Sol wins three pennies from Tina if both coins are heads.
- Tina wins two pennies from Sol if both are tails, and one penny from Sol if the coins don't match.

Play the game with a partner. What is the best strategy for Tina? What is the best strategy for Sol?

If you think carefully about the game, you will probably decide that it isn't such a good deal for Sol. As long as Tina displays tails, she cannot lose. If Sol knows that Tina is going to play tails, he should display heads because he will lose more if he doesn't.

You probably think this is a rather boring game. In a sense it is, because both players will do the same thing every time.

A game in which the best strategy for both players is to pursue the same strategy every time is called **strictly determined.** Although strictly determined games are fairly boring, there are situations in life in which they cannot be avoided. Knowing how to analyze them properly can be beneficial. Although strictly determined games are often very simple, they can be difficult to analyze without an organizational scheme. Matrices offer a way of doing this.

Mathematician of Note

John Forbes Nash Jr.
(1928–)

Nash, an American mathematician, was born in West Virginia. When he was in his 20s, he did path-breaking research in modern game theory. The "Nash equilibrium" is named after him. In the 1950s, he began to suffer from schizophrenia, and for many years, he was unable to work. Once the disease was in remission, he continued his work in game theory and won the 1994 Nobel Prize in Economics for his efforts. In 1998, Sylvia Nasar wrote a biography of Nash's life entitled, *A Beautiful Mind*. The movie by the same name was adapted from the book and released in 2001.

Matrix Models

The following matrix presents Sol's view of the game. It is customary to write a game matrix from the viewpoint of the player associated with the matrix rows rather than the player associated with the columns. Such a matrix is called a **payoff matrix**. The entries are the payoffs to Sol for each outcome of the game.

$$\begin{array}{cc} & \text{Tina} \\ & \begin{array}{cc} \text{Heads} & \text{Tails} \end{array} \\ \text{Sol} \begin{array}{c} \text{Heads} \\ \text{Tails} \end{array} & \begin{bmatrix} 3 & -1 \\ -1 & -2 \end{bmatrix} \end{array}$$

This matrix is easy to follow if you are Sol, but the entries are just the opposite if you are Tina. If you find it difficult to think of all the numbers as their opposites, you may find it preferable to write a second matrix from the column-player Tina's point of view.

$$\begin{array}{cc} & \text{Tina} \\ & \begin{array}{cc} \text{Heads} & \text{Tails} \end{array} \\ \text{Sol} \begin{array}{c} \text{Heads} \\ \text{Tails} \end{array} & \begin{bmatrix} -3 & 1 \\ 1 & 2 \end{bmatrix} \end{array}$$

Consider the game from Sol's point of view. Sol does not want to lose any more money than necessary, so he analyzes his strategies from the standpoint of his losses. If he displays heads, the worst he can do is to lose 1 cent. If he displays tails, the worst he could do is lose 2 cents. Since it is better to lose 1 cent than lose 2 cents, Sol decides to display heads.

Sol's analysis can be related to the payoff matrix by writing the worst possible outcome of each strategy to the right of the row that represents it. The worst possible outcome of each strategy is the smallest value of each row, often referred to as the **row minimum**.

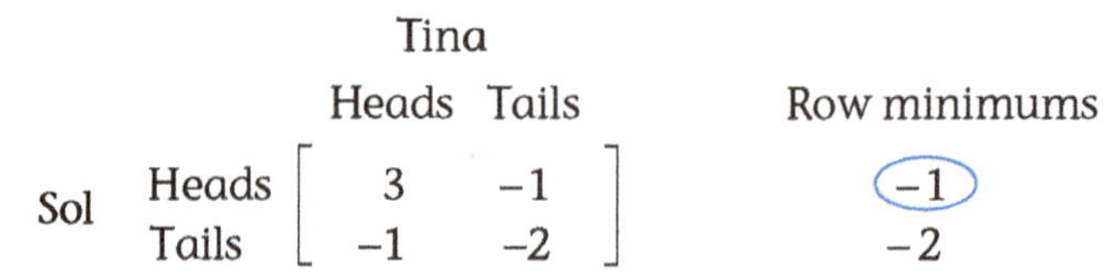

Sol's best strategy is to select the option that produces the largest of these minimums or, in other words, to select the "best of the worst." Because this value is the largest of the smallest row values, it is called the **maximin** (the maximum of the row minimums).

In general, the best strategy for the row player in a strictly determined game is to select the strategy associated with the largest of the row minimums.

Because Tina's point of view is exactly the opposite of Sol's, she views the minimums as maximums and vice-versa. Therefore, her best strategy is the one associated with the smallest of the largest column values, the **minimax** (the minimum of the column maximums).

		Tina Heads	Tina Tails
Sol	Heads	3	−1
	Tails	−1	−2
Column maximums		3	(−1)

In general, the best strategy for the column player in a strictly determined game is to select the strategy associated with the smallest of the column maximums.

In this game, the value selected by both Sol and Tina is the same one, that is, the −1 that appears in the upper right-hand corner of the matrix. This is the identifying characteristic of strictly determined games. If the value selected by the two players is not the same, then the game is not strictly determined. Games that are not strictly determined are considered in the next lesson.

A strictly determined game is one in which the maximin (the maximum of the row minimums) and the minimax (the minimum of the column maximums) are the same value. This value is called the saddle point of the game. The saddle point can be interpreted as the amount won per play by the row player.

Games with More Than Two Strategies

When players have more than two strategies, a game is somewhat harder to analyze. It is often helpful to eliminate strategies that are **dominated** by other strategies. For example, in a competition between two pizza restaurants, Dino's and Al's, both are considering four strategies:

- running no special
- offering a free minipizza with the purchase of a large pizza
- offering a free medium pizza with the purchase of a large one
- offering a free drink with any pizza purchase

A market study estimates the gain in dollars per week to Dino's over Al's according to the following payoff matrix.

		Al's			
		No special	Mini	Medium	Drink
Dino's	No special	200	–400	–300	–600
	Mini	500	100	200	600
	Medium	400	–100	–200	–300
	Drink	300	0	400	–200

What should the managers of Dino's and Al's do?

Suppose you are the manager of Dino's and examine the first two rows carefully. Notice that the first row of the matrix is dominated by the second because each number in row 2 is greater than or equal to its corresponding number in row 1. Row 1 can be eliminated by drawing a line through it. Similarly, the second row dominates the third, and so the third row can be eliminated.

		Al's			
		No special	Mini	Medium	Drink
Dino's	No special	~~200~~	~~–400~~	~~–300~~	~~–600~~
	Mini	500	100	200	600
	Medium	~~400~~	~~–100~~	~~–200~~	~~–300~~
	Drink	300	0	400	–200

Now consider the matrix from Al's point of view. Because all the payoffs to Al's are opposites of the payoffs to Dino's, a column is dominated if all its entries are greater than or equal to, rather than less than, those of another column. Notice that all the values in the first column are larger than the corresponding values in the second column. Because the first column is dominated by the second, it is unwise for Al's to run no special. So this strategy can be eliminated. Similarly, the second column dominates the third, and so the third column can be eliminated.

		Al's			
		No special	Mini	Medium	Drink
Dino's	No special	~~200~~	~~400~~	~~300~~	~~600~~
	Mini	~~500~~	100	~~200~~	600
	Medium	~~400~~	~~100~~	~~200~~	~~300~~
	Drink	~~300~~	0	~~400~~	−200

Once these strategies are eliminated, the game is easier to examine for a minimax and a maximin.

		Al's				
		No special	Mini	Medium	Drink	Row minimums
Dino's	No special	~~200~~	~~400~~	~~300~~	~~600~~	
	Mini	~~500~~	100	~~200~~	600	(100)
	Medium	~~400~~	~~100~~	~~200~~	~~300~~	
	Drink	~~300~~	0	~~400~~	−200	−200
Column maximums			(100)		600	

The game is strictly determined with a saddle point of 100. Dino's best strategy is to offer the free mini, and Al's best strategy is to do the same. By pursuing this strategy, Dino's will gain about $100 a week over Al's.

Exercises

1. Each of the following matrices represents a payoff matrix for a game. Determine the best strategies for the row and column players. If the game is strictly determined, find the saddle point of the game.

a. $\begin{bmatrix} 16 & 8 \\ 12 & 4 \end{bmatrix}$ b. $\begin{bmatrix} 0 & 4 \\ -1 & 2 \end{bmatrix}$ c. $\begin{bmatrix} 2 & -3 \\ -3 & 4 \end{bmatrix}$

d. $\begin{bmatrix} 0 & 1 & 2 \\ 3 & -2 & 0 \end{bmatrix}$ e. $\begin{bmatrix} 0 & -6 & 1 \\ -4 & 8 & 2 \\ 6 & 5 & 4 \end{bmatrix}$ f. $\begin{bmatrix} 0 & 3 & 1 \\ -3 & 0 & 2 \\ -1 & -4 & 0 \end{bmatrix}$

2. For parts a–c, use the following payoff matrix.

$$\begin{bmatrix} -4 & 2 \\ 5 & 3 \end{bmatrix}$$

a. Determine the best strategies for the row and column players and the saddle point of the game.

b. Add 4 to each element in the payoff matrix. How does this affect the best strategies and the saddle point of the game?

c. Multiply each element in the payoff matrix by 2. How does this affect the saddle point of the game and the best strategies?

d. Make a conjecture based on the results of parts b and c.

3. Discuss what would happen in the game given in this lesson if Sol decided to depart from his best strategy. Suppose he switches to displaying tails occasionally. Do you think Tina should still play tails every time? Explain your answer.

4. Use the concept of dominance to solve each of the following games. Give the best row and column strategies and the saddle point of each game.

a.

	E	F	G
A	3	1	7
B	0	1	3
C	4	3	4
D	1	3	6

b.

	E	F	G
A	4	−1	−2
B	0	1	1
C	0	−2	5
D	3	2	4

5. The Democrats and Republicans are engaged in a political campaign for mayor in a small western community. Both parties are planning their strategies for winning votes for their candidate in the final days. The Democrats have settled on two strategies, A and B, and the Republicans plan to counter with strategies C and D. A local newspaper found out about their plans and conducted a survey of eligible voters. The results of the survey are as follows:

   - If the Democrats choose plan A and the Republicans choose plan C, then the Democrats will gain 150 votes.
   - If the Democrats choose A and the Republicans choose D, the Democrats will lose 50 votes.
   - If the Democrats choose B and the Republicans choose C, the Democrats will gain 200 votes.
   - If the Democrats choose B and the Republicans choose D, the Democrats will lose 75 votes.

   Use this information to create a model for the game. Find the best strategies and the saddle point of the game.

6. Two major discount companies, Salemart and Bestdeal, are planning to locate stores in Nebraska.

   - If Salemart locates in city A and Bestdeal in city B, then Salemart can expect an annual profit of $50,000 more than Bestdeal's annual profit.
   - If both locate in city A, they expect equal profits.
   - If Salemart locates in city B and Bestdeal in city A, then Bestdeal's profits will exceed Salemart's by $25,000.
   - If both companies locate in city B, then Salemart's profits will exceed Bestdeal's by $10,000.

   What are the best strategies in this situation and what is the saddle point of the game?

7. Jon and Gretchen each have three dimes. They both hold one, two, or three coins in a clenched fist and open their fists together.

   - If they both are holding the same number of coins, Jon will take the coins that Gretchen is holding.
   - If they are holding different numbers of coins, then Gretchen will take the coins that Jon is holding.

a. Write the payoff matrix from Jon's point of view.

b. Does this game have a saddle point? If so, what are the best strategies for Jon and Gretchen?

8. Mike is going home to see his wife Nancy when he suddenly remembers that today may be a special anniversary. He always brings her a single red rose on this occasion. But he's not sure. Maybe the anniversary is next week. What should he do?

   - If it is their anniversary and he doesn't bring a rose, then he'll be in big trouble. On a scale from 0 to 10, he'd score a –10.
   - If he doesn't bring a rose and it isn't their anniversary, Nancy won't know anything about his frustration and he'll score a 0.
   - If he brings a rose and it is not their anniversary, then Nancy will be suspicious that something funny is going on but he'll score about a 2.
   - If it is their special anniversary and he brings a rose, then Nancy will be expecting it and he'll score a 5.

   Write a payoff matrix for this situation. What is Mike's best strategy?

9. School board and Teacher Education Association representatives are meeting to negotiate a contract. Each side can either threaten (reduction in staff or strike), refuse to negotiate, or negotiate willingly. Each side decides its strategy before coming to the negotiation table. The following payoff matrix gives the percentage pay increases for the teachers that would result from each combination of strategies. Find the best strategies for each side.

		School Board		
		Threaten	Refuse	Negotiate
Teachers	Threaten	5	4	3
	Refuse	3	0	2
	Negotiate	4	3	2

Projects

10. Research and report to the class on the life and work of John von Neumann.

11. Research and report to the class on the development of game theory during World War II.

12. Find and report to the class on applications of game theory in foreign policy, political science, economics, or business.

Lesson 7.5

Game Theory, Part 2

The games considered in the previous lesson were strictly determined. In this lesson, games in which there is no single best strategy for each player are introduced.

Explore This

Look again at the game of the previous lesson. Suppose that Sol, knowing that he will lose if Tina plays rationally, proposes changing the game. The new rules he suggests are that he will win four pennies if both coins are heads and one penny if both coins are tails. He will lose two pennies if he shows heads and Tina shows tails, and three pennies if he shows tails and Tina shows heads.

The new payoff matrix is:

$$\begin{array}{ccc} & & \text{Tina} \\ & & \begin{array}{cc}\text{Heads} & \text{Tails}\end{array} \\ \text{Sol} & \begin{array}{c}\text{Heads} \\ \text{Tails}\end{array} & \left[\begin{array}{cc} 4 & -2 \\ -3 & 1 \end{array}\right]. \end{array}$$

Here is the same matrix with the row minimums and column maximums.

		Tina Heads	Tina Tails	Row minimums
Sol	Heads	4	−2	(−2)
	Tails	−3	1	−3
Column maximums		4	(1)	

The maximin is −2 and the minimax is 1. Since the maximin does not agree with the minimax, the game is not strictly determined.

Play the game with a partner. What is the best strategy for Sol?

Mixed Strategies

The best strategy for either player is to display a mixture of heads and tails and keep the other player guessing. One way to do this would be to flip the coin and allow it to appear heads or tails at random. But such a strategy would cause heads and tails to appear in roughly equal portions, and it is not clear that this would be best for either player.

Using another strategy, Sol could roll a die and show heads if one, two, three, or four spots appeared, and tails otherwise. He might reason that this would benefit him because he would show heads two-thirds of the time, and he wins the most if two heads appear.

Consider what will happen if Sol and Tina each decide to flip their coins. The probability of heads is .5, as is the probability of tails. Because Sol's flip and Tina's flip are made independently, the probability of both showing heads or both showing tails is $.5 \times .5 = .25$. The same is true for the cases in which one shows a head and the other shows a tail (see Figure 7.4).

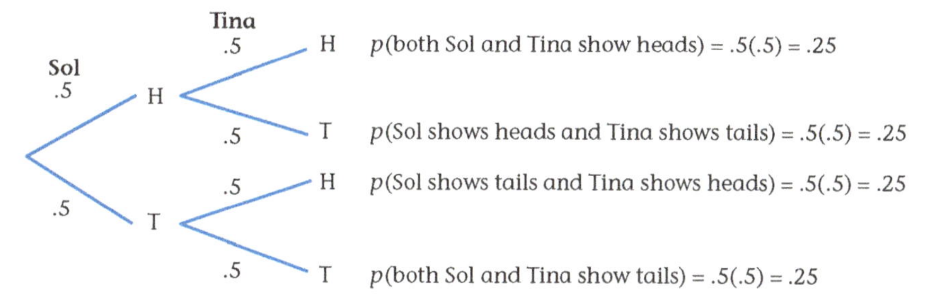

Figure 7.4. Probabilities when Sol and Tina flip their coins.

The probability distribution for Sol's winnings for this case is shown in the following table.

Outcome	HH	HT	TH	TT
Probability	.25	.25	.25	.25
Amount won	4	−2	−3	1

The expected payoff of the game for Sol is .25(4) + .25(−2) + .25(−3) + .25(1) = 1.00 − .50 − .75 + .25 = 0.

Since Tina's payoffs are the opposite of Sol's, her expected payoff is .25(−4) + .25(2) + .25(3) + .25(−1) = −1.00 + .50 + .75 − .25 = 0. If both players display heads and tails in equal proportions in this way, the game is **fair** because their expected payoffs are equal.

But suppose that Tina decides to play heads 40% of the time, while Sol continues flipping his coin. The probability of both heads is now $.5 \times .4 = .2$, while the probability of both tails is $.5 \times .6 = .3$. The probability that Sol shows heads and Tina shows tails is $.5 \times .6 = .3$ and that Sol shows tails and Tina shows heads is $.5 \times .4 = .2$ (see Figure 7.5).

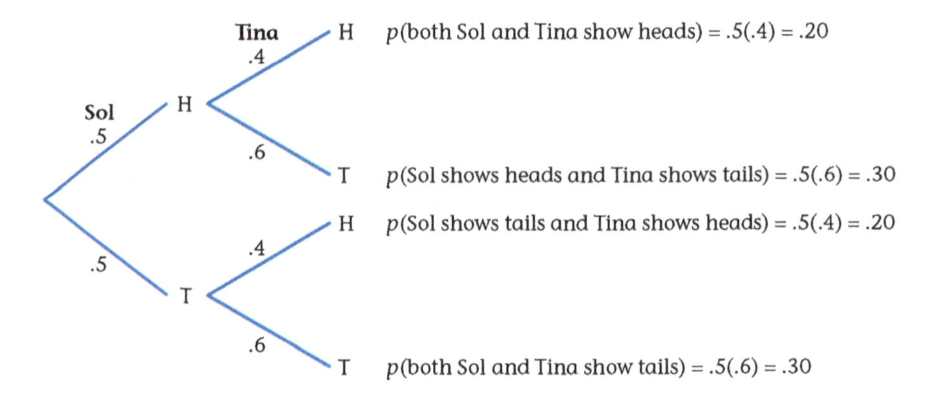

Figure 7.5. Probabilities when Sol flips his coin and Tina shows heads 40% of the time.

The probability distribution for Sol's winnings now looks like this.

Outcome	HH	HT	TH	TT
Probability	.2	.3	.2	.3
Amount won	4	−2	−3	1

The expected payoff for Sol is now .2(4) + .3(–2) + .2(–3) + .3(1) = .8 – .6 – .6 + .3 = –0.1. This means he will lose 0.1 pennies per play, or 1 penny every 10 plays. Tina has an advantage and the game is no longer fair!

Using this new strategy, Tina can gain an advantage over Sol if she knows he will display heads and tails in equal proportions. But she does not know that Sol is going to do this. So how can she decide her best mixture of strategies? How can Sol decide what is best for him?

Now reconsider the game from Sol's point of view, and suppose that Tina plays heads every time while Sol continues to flip his coin. The outcomes for this combination are shown in Figure 7.6. Sol's expected payoff is now .5(4) + 0(–2) + .5(–3) + 0(1) = .20 – 1.5 = –1.3.

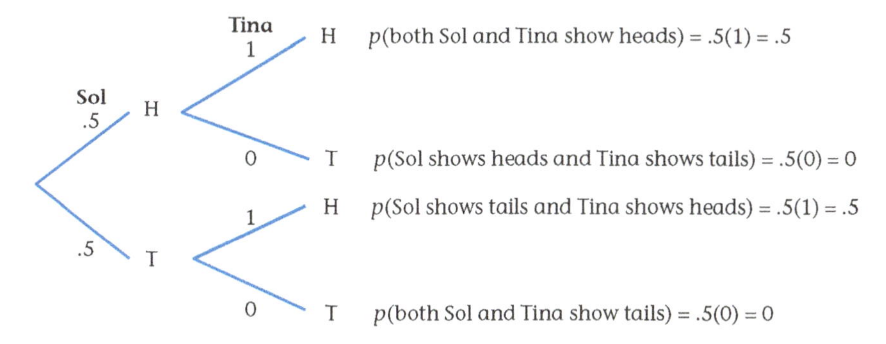

Figure 7.6. Probabilities when Sol flips his coin and Tina always plays heads.

If Tina decides to play tails each time while Sol continues to flip his coin (see Figure 7.7), Sol's expected payoff is 0(4) + .5(–2) + 0(–3) + .5(1) = –1.0 + .5 = –0.5.

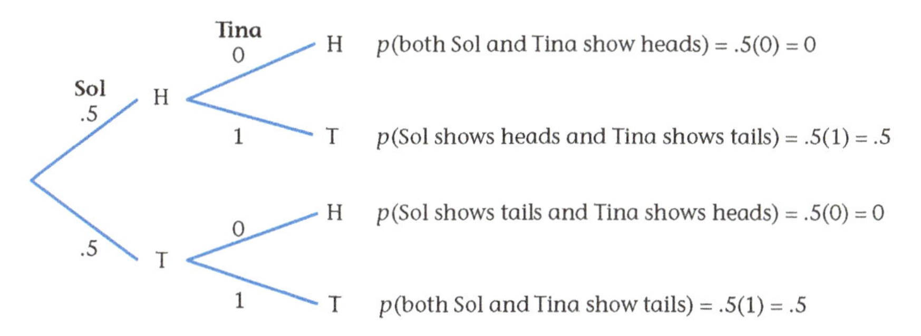

Figure 7.7. Probabilities when Sol flips his coin and Tina always plays tails.

Another way to show these calculations is to write the probabilities of Sol's displaying heads and tails in a row matrix and find the matrix product.

$$[.5 \quad .5]\begin{bmatrix} 4 & -2 \\ -3 & 1 \end{bmatrix} = [.5(4) + .5(-3) \quad .5(-2) + .5(1)]$$
$$= [.2 - 1.5 \quad -1 + .5] = [-1.3 \quad -0.5]$$

Suppose Sol switches to displaying heads 60% of the time. Then this matrix product is

$$[.6 \quad .4]\begin{bmatrix} 4 & -2 \\ -3 & 1 \end{bmatrix} = [.6(4) + .4(-3) \quad .6(-2) + .4(1)]$$
$$= [2.4 - 1.2 \quad -1.2 + .4] = [1.2 \quad -0.8].$$

This means that if Sol displays heads 60% of the time, he will gain 1.2 pennies per play if Tina always displays heads and lose 0.8 pennies per play if Tina always displays tails.

In general, if the probability Sol will display heads is p, his expected winnings per play, if Tina displays all heads or all tails, are

$$[p \quad 1 - p]\begin{bmatrix} 4 & -2 \\ -3 & 1 \end{bmatrix} = [4p - 3(1 - p) \quad -2p + 1(1 - p)].$$

Because it is not very likely that Tina will display all heads or all tails, Sol's best strategy is to act in such a way that the two expected payoffs are balanced or equalized. To find the value of p that does this, set the two expected payoffs equal to each other and solve the resulting equation.

$$4p - 3(1 - p) = -2p + 1(1 - p)$$
$$4p - 3 + 3p = -2p + 1 - p$$
$$7p - 3 = -3p + 1$$
$$10p = 4$$
$$p = .4$$
$$1 - p = .6$$

Sol's best strategy is to display heads four-tenths of the time and tails six-tenths of the time.

Technology Note

One way Sol could accomplish displaying heads four-tenths of the time and tails six-tenths of the time is to generate a random number on a calculator. If the random number is less than or equal to 0.4, he should display heads.

Tina's best strategy can be determined in a similar way. Call the probability that she displays heads q. Because she is the column player, multiply the payoff matrix times a column matrix to obtain her expected payoffs if Sol plays either all heads or all tails.

$$\begin{bmatrix} 4 & -2 \\ -3 & 1 \end{bmatrix}\begin{bmatrix} q \\ 1-q \end{bmatrix} = \begin{bmatrix} 4q - 2 + 2q \\ -3q + 1 - q \end{bmatrix}$$

Equate the two entries in the resulting matrix and solve to find Tina's best strategy.

$$\begin{aligned} 4q - 2 + 2q &= -3q + 1 - q \\ 6q - 2 &= -4q + 1 \\ 10q &= 3 \\ q &= .3 \\ 1 - q &= .7 \end{aligned}$$

In this case, Tina's best strategy is to display heads three-tenths of the time and tails seven-tenths of the time.

If both players pursue these strategies, the probability that a pair of heads will appear is .4(.3) = .12, or 12% of the time, and that a pair of tails will appear is .6(.7) = .42, or 42% of the time. The probability that Sol shows heads and Tina shows tails is .4(.7) = .28 or 28% of the time, and that Sol shows tails and Tina shows heads is .6(.3) = .18, or 18% of the time (see Figure 7.8).

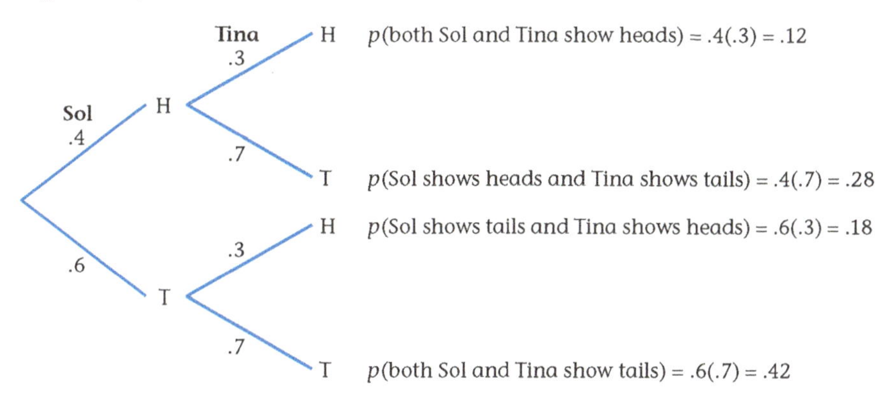

Figure 7.8. Probabilities when both Sol and Tina play their best strategies.

The resulting probability distribution from Sol's point of view is below.

Outcome	HH	HT	TH	TT
Probability	.12	.28	.18	.42
Amount won	4	−2	−3	1

Sol's expected payoff for the game is $.12(4) + .28(-2) + .18(-3) + .42(1) = .48 - .56 - .54 + .42 = -0.2$. If both players pursue their best strategy, the game favors Tina. She can expect to win 0.2 of a penny per play, or 2 pennies every 10 plays, from Sol.

Using Matrices to Model the Expected Payoff

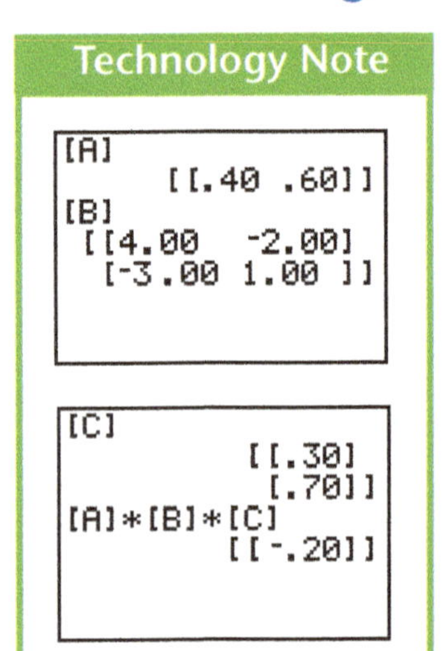

You can model the expected payoff for a game easily by using matrices and a calculator. To do this, form a row matrix A using the probabilities for the row player and a column matrix C using the probabilities of the column player. The expected payoff for the game for the row player equals the matrix product ABC, where B is the payoff matrix for the game.

In the preceding example, in which both Sol and Tina play their best strategies,

$$A = [.4 \quad .6],\ C = \begin{bmatrix} .3 \\ .7 \end{bmatrix},\ \text{and}$$

$$ABC = [.4 \quad .6]\begin{bmatrix} 4 & -2 \\ -3 & 1 \end{bmatrix}\begin{bmatrix} .3 \\ .7 \end{bmatrix} = -0.2.$$

A Graphical Solution for Sol's Best Strategy

Sol's search for a best strategy can be visualized graphically:

1. Draw a horizontal line to represent the probability of Sol's displaying heads. Scale this axis in tenths from 0 to 1 (see Figure 7.9).

2. Draw vertical axes at each end of the horizontal axis and scale them from the minimum amount Sol can win (−3 in this case) to the maximum amount (4 in this case).

3. Draw a diagonal line to represent what happens if Tina always displays heads. To do this, notice that if Tina displays heads and Sol displays tails, Sol will lose 3 cents. Place a dot at –3 on the vertical axis on the left, where the probability of Sol's displaying heads is 0. Similarly, if Sol displays heads and Tina displays heads, he will win 4 cents. Place a dot at the 4 on the vertical axis on the right where the probability of Sol's displaying heads is 1. Connect the two dots with a diagonal line. Sol's expected winnings for his various strategies for displaying heads when Tina always displays heads can be read from this line (see line 1 in Figure 7.9).

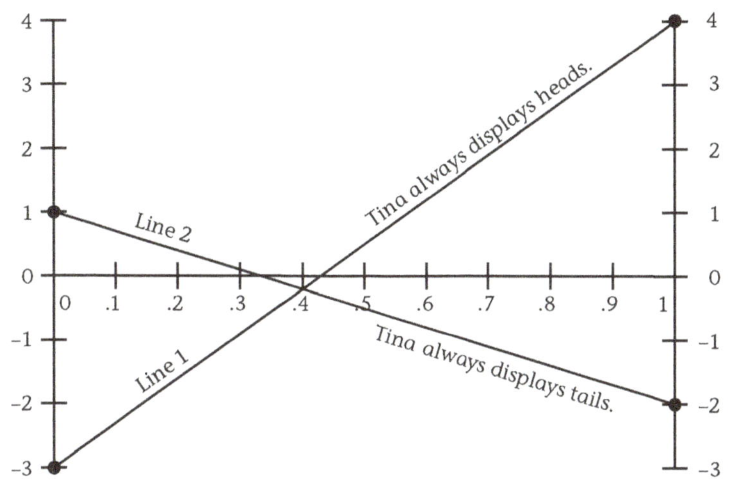

Figure 7.9. Sol's best strategy for displaying heads.

4. Repeat the procedure in step 3 to draw a diagonal line that shows what happens if Tina always displays tails. Place a dot at 1 on the left vertical axis since Sol wins 1 cent when both he and Tina show tails. Similarly, place a dot at –2 on the vertical axis on the right since Sol loses 2 cents if he displays heads and Tina displays tails. Connect the two dots. Sol's expected winnings for his various strategies when Tina always displays tails can be read from this line (see line 2 in Figure 7.9).

Since Sol's best strategy is to act in such a way that the two expected payoffs are balanced or equalized, the intersection of the two lines lies directly below Sol's best strategy for displaying heads. His expected payoff for playing this strategy can be read from the vertical axes.

To calculate the exact values for Sol's best strategy and his expected payoff from the graph, let x represent the probability of Sol's displaying heads and P represent the payoff. Then the equations of line 1 and line 2 in slope-intercept form are $P = 7x - 3$ and $P = -3x + 1$, respectively. Setting these two equations equal and solving for x, we get the following.

$$\begin{aligned} 7x - 3 &= -3x + 1 \\ 10x &= 4 \\ x &= .4 \end{aligned}$$

Substitute .4 for x in the first equation.

$$P = 7x - 3 = 7(.4) - 3 = 2.8 - 3 = -0.2$$

Exercises

1. Suppose that in the example of this lesson, Sol decides to return to flipping his coin while Tina continues to pursue her best strategy of playing heads three-tenths of the time.
   a. Set up a tree diagram to compute the probabilities of each of the four outcomes for this game.
   b. What is the probability that both Sol and Tina will show heads?
   c. What is the probability that Tina will show tails and Sol will show heads?
   d. What is the probability that Tina will show heads and Sol will show tails?
   e. What is the probability that both Sol and Tina will show tails?
   f. Write a probability distribution chart for Sol's winnings.
   g. Calculate Sol's expected payoff for this game. Explain what this means in terms of pennies won or lost.
   h. How does this payoff compare with Sol's expected payoff if he plays his best strategy as computed in this lesson?
2. Use matrices and a calculator as shown on page 424 to find the expected payoffs for Sol in parts a–c.
   a. Suppose that in the example of this lesson, Sol decides to play heads three-fourths of the time while Tina continues to pursue her best strategy of playing heads three-tenths of the time. Find Sol's expected payoff for this situation.

b. Choose two or three other strategies for Sol to play while Tina continues to pursue her best strategy of playing heads three-tenths of the time. Compute Sol's expected payoff for these strategies.

c. Suppose now that Sol returns to using his best strategy of playing heads four-tenths of the time while Tina plays a variety of strategies. Choose three or four different strategies for Tina to play while Sol plays his best strategy and find Sol's expected payoff in each case.

d. Compare your results with your classmates' results for Sol's expected payoff in parts a–c. Make a conjecture based on your observations.

3. Suppose that Sol and Tina change their game so that the payoffs to Sol are

$$\begin{array}{ccc} & & \text{Tina} \\ & & \begin{array}{cc} \text{Heads} & \text{Tails} \end{array} \\ \text{Sol} & \begin{array}{c} \text{Heads} \\ \text{Tails} \end{array} & \begin{bmatrix} 3 & -2 \\ -2 & 1 \end{bmatrix} \end{array}$$

a. Use the row matrix $[p \quad 1 - p]$ to find Sol's best strategy for this game.

b. Use the column matrix $\begin{bmatrix} q \\ 1 - q \end{bmatrix}$ to find Tina's best strategy for this game.

c. Set up a tree diagram to compute the probabilities of each of the four outcomes for this game.

d. Prepare a probability distribution chart for Sol's winnings.

e. Find Sol's expected payoff for this game.

f. Interpret your answer in part e in terms of how many pennies Sol can expect to win or lose over a number of games.

g. Construct a graph showing Sol's best strategy for playing heads in this game.

h. Find the equations of the lines in part g. Set these equations equal and find Sol's best strategy and his expected payoff for this game.

4. The procedure outlined in this lesson is designed to determine the best mixture of strategies when a game is not strictly determined. Therefore, you should always inspect a game to see whether it is strictly determined and apply the saddle point technique of the previous lesson if it is. It is, however, easy to forget to do this. To see what will happen if you attempt to determine a mixture of strategies for a game that is strictly determined, apply the techniques of this lesson to the strictly determined game that Sol and Tina were playing in the last lesson and try to find the best mixture of strategies for each of them. Use the following payoff matrix.

		Tina	
		Heads	Tails
Sol	Heads	3	–1
	Tails	–1	–2

5. a. For the game defined by the following matrix, determine the best strategies for both players.

		Player B	
		C1	C2
Player A	R1	1	3
	R2	4	2

   b. Add 5 to each element in the matrix given in part a. How does this affect the best strategies?

   c. Multiply each element in the matrix in part a by 3. How does this affect the best strategies?

   d. Make a conjecture based on the results of parts b and c.

   e. Challenge: Use algebra to prove your conjecture.

6. In a game known as Two-Finger Morra, two players simultaneously hold up either one or two fingers. If they hold up the same number of fingers, player 1 wins from player 2 the sum (in pennies) of the digits held up on both players' hands. If they hold up different numbers, then player 2 wins the sum from player 1.

a. Write the payoff matrix for this game.

b. Find the best strategy for each player.

c. Find the expected payoff for the row player.

d. Is this a fair game? Explain your answer.

7. In another version of the game in Exercise 6, if the sum of the fingers held out by the two players is even, player 1 will win 5 cents. If the sum is odd, player 2 will win 5 cents.

   a. Write the payoff matrix for this version.

   b. Find the best strategy for each player.

   c. Find the expected payoff for the row player.

   d. Is this a fair game? Explain your answer.

8. A group of parents are in an uproar about a new social studies program that the school district has adopted. They are seeking to have the program removed from the curriculum. A second group of parents believe the new program is a solid choice and are organizing in favor of keeping it. In order to bring the issue before the voters in the town, the opposing group must collect 400 supporting signatures from registered voters. Both sides are campaigning vigorously by making telephone calls, sending out mailings, and going door to door to contact voters. The local newspaper has estimated the number of signatures that the opposing group is expected to collect with each combination of strategies.

   a. What are the best strategies for both groups of parents? (Hint: Use the concept of dominance to eliminate a row and column.)

		Group in favor		
		Phone	Mail	Door
Group against	Phone	150	75	100
	Mail	350	300	200
	Door	500	100	400

   b. If both follow their best strategies, can the opposing group expect to gather enough signatures to get the issue on the ballot?

9. Two rival TV networks compete for prime time audiences by showing comedy, drama, and sports. The following matrix gives the payoffs for network A in terms of percentages of regular viewers who watch its channel for various combinations of programs. Find the best strategy for each network and the expected payoff for network A.

		Network B		
		Comedy	Drama	Sports
Network A	Comedy	10	50	20
	Drama	40	30	50
	Sports	30	20	60

10. In a campaign for student council president at Northeast High the top two candidates, Betty and Bob, are each making two promises about what they will do if they are elected. The payoff matrix in terms of the number of votes Betty will gain follows. What is the best strategy for each candidate and what is Betty's expected payoff?

		Bob	
		A	B
Betty	1	200	100
	2	50	180

Project

11. The games you studied in this and the previous lesson are known as *zero-sum games*, because one person's loss is the other's gain. If, for example, Sol wins $2, then Tina loses the same amount. In some games, a particular outcome may be worth 2 to one player, but not –2 to the other. Examples of such games include the prisoner's dilemma, chicken, and arms races between countries. Research and report on games that are not zero sum.

Computer/Calculator Exploration

12. Write a graphing calculator (or computer) program using matrices to find the expected payoff for the row player in a two-person game that is not strictly determined.

Chapter Extension

A Look at a Dominance Matrix

In this example a dominance matrix is used to examine "pecking order" behavior among five neighborhood cats: Bruiser, Quince, Wruin, Pebbles, and Yac. The purpose of using this simple struggle for top cat is to give you an understanding about how this model can be used to measure and compare the dominance of one person over another in political or business situations.

Close observation of the behavior of the cats reveals that there is a definite sense of who is allowed to hunt near the choicest mouse holes when more than one cat is in the field.

- Bruiser lives up to his name and chases away Wruin, Yac, and Pebbles.
- Feisty little Quince stands up to Bruiser and lets Yac know, in no uncertain terms, that she is the boss cat when their paths cross.
- Wruin dominates her little sister, Quince, and will not tolerate Pebbles.
- Pebbles always picks on Quince and Yac.
- Finally, Yac, even though he is the smallest, somehow manages to intimidate Wruin.

The directed graph below illustrates this furry dominance. The direction indicates who is the dominant cat for each possible pair.

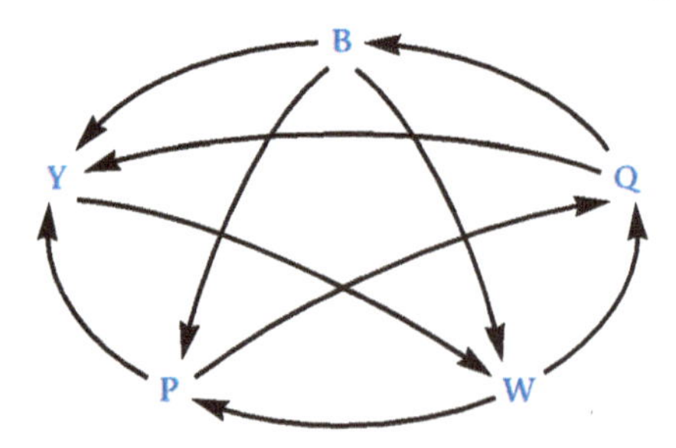

A 5 by 5 **dominance matrix** D can also be used to model this situation where a 1 represents the dominance of one cat over another and 0s otherwise.

$$D = \begin{array}{c} \\ B \\ Q \\ W \\ P \\ Y \end{array} \begin{array}{c} \begin{array}{ccccc} B & Q & W & P & Y \end{array} \\ \begin{bmatrix} 0 & 0 & 1 & 1 & 1 \\ 1 & 0 & 0 & 0 & 1 \\ 0 & 1 & 0 & 1 & 0 \\ 0 & 1 & 0 & 0 & 1 \\ 0 & 0 & 1 & 0 & 0 \end{bmatrix} \end{array}$$

Notice that the entries along the main diagonal in the dominance matrix are all zero. This represents the fact that a cat cannot dominate itself. Notice, also, that if entry D_{ij} is a 1, then entry D_{ji} is a 0. This is true since only one cat can dominate the other. They cannot be mutually dominant.

It can also be seen by looking at the matrix that the relationship among the cats is not transitive. For example, Bruiser dominates Wruin, and Wruin dominates Quince, but Bruiser does not dominate Quince.

Since the relationship among the cats is not transitive, it is not immediately apparent who is the most powerful. One way to decide this issue is to look at the dominance matrix and sum the numbers in each row. If we do this, the result is called an **authority column matrix**.

$$A = \begin{matrix} \text{Bruiser} \\ \text{Quince} \\ \text{Wruin} \\ \text{Pebbles} \\ \text{Yac} \end{matrix} \begin{bmatrix} 3 \\ 2 \\ 2 \\ 2 \\ 1 \end{bmatrix}$$

This places Bruiser at the top of the pecking order in dominating three of the other cats. Quince, Wruin, and Pebbles tie. And poor Yac lands at the bottom with the least amount of power.

This *direct* measure may be all right for determining the top cat in the neighborhood. However, if we were using this system to examine the relative power or influence over others among a group of people, we would probably be interested in indirect measures of influence as well. For example, if B has direct influence over P and P has direct influence over Q, then there is a sense that B may have some *indirect* influence over Q. This indirect, or second-order, influence should be taken into consideration in determining the degree of B's power.

Return now to the dominance matrix D for this group of cats. Suppose you square this matrix.

$$D^2 = \begin{matrix} \\ B \\ Q \\ W \\ P \\ Y \end{matrix} \begin{matrix} \begin{matrix} B & Q & W & P & Y \end{matrix} \\ \begin{bmatrix} 0 & 2 & 1 & 1 & 1 \\ 0 & 0 & 2 & 1 & 1 \\ 1 & 1 & 0 & 0 & 2 \\ 1 & 0 & 1 & 0 & 1 \\ 0 & 1 & 0 & 1 & 0 \end{bmatrix} \end{matrix}$$

What do the entries in this matrix represent? Look at the entry in row 1 column 2. This value is found by multiplying the components of row 1 of matrix D by the corresponding components in the second column and summing the five products.

$$(0 \times 0) + (0 \times 0) + (1 \times 1) + (1 \times 1) + (1 \times 0) = 1 + 1 = 2$$

Notice that the entries in D^2 can only be nonzero if both factors in at least one product are equal to 1. In the first nonzero product, the first 1 comes from the fact that Bruiser dominates Wruin, and the second 1 comes from the fact that Wruin dominates Quince. And since this product is nonzero, we can deduce that Bruiser has an indirect second-order influence over Quince through Wruin.

Similarly, the second nonzero product indicates that Bruiser dominates Pebbles, and Pebbles dominates Quince. This shows that Bruiser also has an indirect second-order influence over Quince through Pebbles. Thus the 2 in row 1, column 2 of D^2 indicates that Bruiser has two second-order influences over Quince. We can verify this conclusion by looking at the digraph. One edge points from B to W and another points from W to Q. A second indirect path points from B to P and from P to Q.

You can use a similar argument to show that D^3 will represent the number of third-order influences that exist for each cat.

$$D^3 = \begin{array}{c} \\ B \\ Q \\ W \\ P \\ Y \end{array} \begin{array}{c} \begin{array}{ccccc} B & Q & W & P & Y \end{array} \\ \begin{bmatrix} 2 & 2 & 1 & 1 & 3 \\ 0 & 3 & 1 & 2 & 1 \\ 1 & 0 & 3 & 1 & 2 \\ 0 & 1 & 2 & 2 & 1 \\ 1 & 1 & 0 & 0 & 2 \end{bmatrix} \end{array}$$

Now if you add all the entries in corresponding rows from the three matrices D, D^2, and D^3 (omitting the diagonal of matrix D^3), the result will be an influence matrix showing the total number of direct first-order and indirect second- and third-order influences exercised by each cat over the others.

$$I = \begin{array}{l} \text{Bruiser} \\ \text{Quince} \\ \text{Wruin} \\ \text{Pebbles} \\ \text{Yac} \end{array} \begin{bmatrix} 17 \\ 13 \\ 13 \\ 11 \\ 7 \end{bmatrix}$$

The resulting matrix indicates that Bruiser is still top cat and Yac remains at the bottom, but at least one of the ties has been eliminated. The question now is where do you stop when you are adding up dominance matrices in this way? Also, is it perhaps more reasonable to give less weight to the second-, third-, and higher-order influence matrices than we give to the direct-order matrix? One way to accomplish this is by dividing D^2 by 2 and D^3 by 3, and so on, if higher-order influence matrices are used in determining the power of each individual member of the group.

The case of the dominant cat may not be of great interest to persons other than the cat owners. However, the model developed in the cat problem can be applied to groups of people such as legislators or corporate leaders to determine who is the most influential or who has the most power.

Chapter 7 Review

1. Write a summary of what you think are the important points of this chapter.
2. Suppose that a three-sector economy has this consumption matrix.

$$C = \begin{array}{c} \\ A \\ B \\ C \end{array} \begin{array}{c} \begin{array}{ccc} A & B & C \end{array} \\ \begin{bmatrix} 0.1 & 0.2 & 0.3 \\ 0.1 & 0.3 & 0.2 \\ 0.2 & 0.1 & 0.2 \end{bmatrix} \end{array}$$

a. Draw a weighted digraph for this economy.

b. A production matrix, P, follows. Find the internal consumption matrix product CP. Find the external demand matrix D, where $D = P - CP$.

$$P = \begin{bmatrix} 8 \\ 12 \\ 15 \end{bmatrix}$$

c. An external demand matrix, D, follows. Find the production matrix P for this economy. Recall that $P = (I - C)^{-1}D$.

$$D = \begin{bmatrix} 6 \\ 8 \\ 12 \end{bmatrix}$$

3. Mike and Brit are playing poker for pennies. Mike is holding a very poor hand and is considering bluffing or not bluffing. Brit can either call or not call the bluff. The payoff matrix for this situation is shown below. Over the course of several games in which Mike comes up with a poor hand, what should his strategy be?

		Brit	
		Call	Not call
Mike	Bluff	–10	10
	Not bluff	–2	0

4. Mike and Brit soon get bored playing the game described in Exercise 3. They each draw two cards from the deck. Mike draws a 4 of spades and an ace of hearts. Brit draws a 3 of clubs and a 2 of diamonds. They make up a new game to play with the following rules. Each player will show one card. If both cards shown are the same color, Brit will pay Mike the sum of the face value of the cards in pennies. If the cards shown are of different colors, Mike will pay Brit the sum of the face values shown. (Note: The value of an ace is 1.)

   a. Write a payoff matrix for the game.

   b. Find the best strategies for both Mike and Brit.

   c. Use a probability tree to calculate the probability of each of the four possible outcomes when Mike and Brit play their best strategies.

   d. Set up a probability distribution table for this game.

   e. Find the expected payoff of the game for Mike. Explain what this means in terms of pennies won or lost.

5. The discrete mathematics teacher has three class starter activities, one of which she uses to begin class every day: a pop quiz, a quickie review, and a small-group problem-solving activity. She never uses the same activity two days in a row.

   - If she gave a pop quiz yesterday, she will toss a coin, and do a quickie review if it comes up heads.
   - If she used a review, she will toss two coins and switch to problem solving if two heads come up.
   - If she did a problem-solving activity, then she will toss three coins, and if three heads come up, she gives a pop quiz again.

The transition matrix for this scheme is

$$\begin{array}{c} \\ Q \\ R \\ P \end{array} \begin{array}{c} \begin{array}{ccc} Q & R & P \end{array} \\ \begin{bmatrix} 0 & \frac{1}{2} & \frac{1}{2} \\ \frac{3}{4} & 0 & \frac{1}{4} \\ \frac{1}{8} & \frac{7}{8} & 0 \end{bmatrix} \end{array}.$$

a. If the teacher gives a quiz on Monday, what is the probability that she will give another quiz on Friday (4 days later)?

b. In the long run, how often should the students expect that the teacher will start class with a quiz?

c. In the long run, what activity will the teacher use most often to begin class, and how frequently will she use it?

6. The Super X sells three kinds of sandwiches that many of the students at Southeast High especially like for lunch—Super X Original, Italian Special, and Barbecue Beef. The Super X clerk observed that the same students were coming in for sandwiches for lunch every school day and that the kind of sandwich that each student purchased depended on what he or she had ordered on the previous visit. He conducted a survey and found the following:

- Of the students who ordered the Original on their last visit, 20% ordered it again the next time, whereas 25% switched to Italian and 55% switched to Barbecue Beef.
- Of the students who ordered the Italian sandwich the last time, 35% did so again the next time, but 45% switched to the Original and 20% switched to the Barbecue Beef.
- Of the students who got the Barbecue Beef the last time, 55% ordered it the next time, 20% switched to the Original, and 25% switched to Italian.

a. Set up the transition matrix for this Markov chain.

b. If the same students tend to buy Super X sandwiches for lunch every day, what is the probability that a student who buys the Italian sandwich on Monday will have it again on Wednesday?

c. In the long run, what percentage of the orders will be for the Original? For the Italian? For the Barbecue Beef?

7. A certain economy consists of three industries: transportation, petroleum, and agriculture.

- The production of $1 million worth of transportation requires an internal consumption of $0.2 million worth of transportation, $0.4 million worth of petroleum, and no agriculture.
- The production of $1 million worth of petroleum requires an internal consumption of $0.3 million worth of transportation, $0.2 million worth of petroleum, and $0.3 million worth of agriculture.
- The production of $1 million worth of agriculture requires an internal consumption of $0.3 million worth of transportation, $0.2 million worth of petroleum, and $0.25 million worth of agriculture.

a. Draw a weighted digraph for this economy.

b. Write a consumption matrix, C, representing this information.

c. On what sector of the economy is transportation the most dependent? The least dependent?

d. If the agriculture sector has an output of $5.4 million, what is the input in dollars from petroleum? From agriculture?

e. A production matrix, P, in millions of dollars follows. Find the internal consumption matrix product CP and the external demand matrix D.

$$P = \begin{matrix} \text{Transportation} \\ \text{Petroleum} \\ \text{Agriculture} \end{matrix} \begin{bmatrix} 20 \\ 25 \\ 15 \end{bmatrix}$$

f. An external demand matrix, D, in millions of dollars, follows. How much must each sector produce to meet this demand?

$$D = \begin{matrix} \text{Transportation} \\ \text{Petroleum} \\ \text{Agriculture} \end{matrix} \begin{bmatrix} 4.6 \\ 5.2 \\ 3.0 \end{bmatrix}$$

8. Two computer companies (1 and 2) are competing for sales in two large school districts (A and B). The following payoff matrix shows the differences in sales for companies 1 and 2 in hundreds of thousands of dollars if they focus their full sales force on either school district. Find the best strategy for each company.

		Computer company 2	
		A	B
Computer company 1	A	3	7
	B	–7	–3

9. Suppose that in the final days of a political campaign for mayor, the Democrats and Republicans are planning their strategies for winning undecided voters to their political camps. The Democrats have decided on two strategies, plan A and plan B. The Republicans plan to counter with plans C and D.

   The following matrix gives the payoff for the Democrats of the various combinations of strategies. The numbers represent the percentage of the undecided voters joining the Democrats in each case.

		Republicans	
		Plan C	Plan D
Democrats	Plan A	30	60
	Plan B	50	40

   a. Find the best strategies for both parties.

   b. Find the expected payoff for the Democrats.

10. A manufacturing company has divisions in Massachusetts, Nebraska, and California. The company divisions use goods and services from each other as shown in the following consumption matrix C.

$$C = \begin{array}{c} \\ MA \\ NE \\ CA \end{array} \begin{array}{c} \begin{array}{ccc} MA & NE & CA \end{array} \\ \begin{bmatrix} 0.04 & 0.02 & 0.03 \\ 0.03 & 0.01 & 0.05 \\ 0.01 & 0.02 & 0.04 \end{bmatrix} \end{array}$$

a. Draw a weighted digraph to model this situation.

b. Find the total production needed to meet a final consumer demand of $50,000 from Massachusetts, $30,000 from Nebraska, and $40,000 from California.

c. What will the internal consumption be for each division to meet the demands in part b?

11. Two competing dairy stores choose daily strategies of raising, not changing, or lowering their milk prices. The following payoff matrix shows the percentage of customers who go from store A to store B for each combination of strategies. What should each store do?

		Store B		
		Raise	No change	Lower
Store A	Raise	4	–1	–4
	No change	2	1	–2
	Lower	5	2	3

Bibliography

Bittinger, M. L., and J. C. Crown. 1989. *Finite Mathematics*. Reading, MA: Addison Wesley.

Bogart, Kenneth P. 1988. *Discrete Mathematics*. Lexington, MA: D. C. Heath and Company.

Brandenburger, A. M., and B. J. Nalebuff. 1996. *Co-opetition*. Garden City, NY: Doubleday.

Brams, S. J. 1985. *Rational Politics: Decisions, Games, and Strategy*. New York: Congressional Quarterly.

Case, J. H. 1979. *Economics and the Competitive Process*. New York: New York University Press.

COMAP. 2013. *For All Practical Purposes: Mathematical Literacy in Today's World*. 9th ed. New York: W. H. Freeman.

Cozzens, M. B., and R. D. Porter. 1987. *Mathematics and Its Applications*. Lexington, MA: D. C. Heath and Company.

Dixit, A. K., and B. J. Nalebuff. 1991. *Thinking Strategically. The Competitive Edge in Business, Politics, and Everyday Life*. New York: W. W. Norton & Company.

Keller, M. K. 1983a. *Food Service Management and Applications of Matrix Methods*. Lexington, MA: COMAP, Inc.

Keller, M. K. 1983b. *Markov Chains and Applications of Matrix Methods: Fixed Point and Absorbing Markov Chains*. Lexington, MA: COMAP, Inc.

Kemeny, J. G., J. L. Snell, and G. L. Thompson. 1957. *Introduction to Finite Mathematics*. Englewood Cliffs, NJ: Prentice Hall.

Leontief, W. 1986. *Input-Output Economics*. New York: Oxford University Press.

Mauer, S. B., and A. Ralston. 2004. 3rd ed. *Discrete Algorithmic Mathematics*. Wellesley, MA: A. K. Peters Ltd.

Nasar, Sylvia. 1998. *A Beautiful Mind*. New York, NY: Simon & Schuster.

National Council of Teachers of Mathematics. 1988. *Discrete Mathematics Across the Curriculum K-12*. Reston, VA: National Council of Teachers of Mathematics.

North Carolina School of Science and Mathematics. 1988. *New Topics for Secondary School Mathematics: Matrices*. Reston, VA: National Council of Teachers of Mathematics.

Poundstone, William. 1993. *Prisoner's Dilemma*. Garden City, NY: Doubleday.

Rapoport, A. 1973. *Two-Person Game Theory: The Essential Ideas*. Ann Arbor: University of Michigan Press.

Rapoport, A., and A. M. Chammah. 1965. *Prisoner's Dilemma: A Study in Conflict and Cooperation*. Ann Arbor: University of Michigan Press.

Ross, K. A., and C. R. B. Wright. 1985. *Discrete Mathematics*. 5th ed. Englewood Cliffs, NJ: Prentice Hall.

Straffin, P. 1993. *Game Theory and Strategy*. Washington, D.C.: Mathematical Association of America.

Tuchinsky, P. M. 1989. *Management of a Buffalo Herd*. Lexington, MA: COMAP, Inc.

Wheeler, R. E., and W. D. Peebles. 1987. *Finite Mathematics with Applications to Business and the Social Sciences*. Monterey, CA: Brooks/Cole.

Williams, J. D. 1982. *The Compleat Strategyst*. Mineola, NY: Dover Publications.

Zagare, F. C. 1985. *The Mathematics of Conflict*. Lexington, MA: COMAP, Inc.

CHAPTER

Recursion

8

Recursion is a process that creates new objects from existing ones that were created by the same process. The recurrence relations you wrote in previous chapters are an example: They enable you to calculate new values from existing ones that were calculated with the same formula.

Recursive processes can be geometric; although the computer programs used to implement geometric recursion do so by performing numerical calculations. Fractal images are among the best-known examples of geometric recursion. For example, fractal techniques are used to create artificial landscapes for movies.

However, the results produced by recursive processes can change from regular and predictable to chaotic when only slight modifications are made. This fact has given rise to a branch of mathematics called chaos theory, which has been used to provide a theoretical explanation for the unexpected collapse of stable systems such as large power grids.

- How can recursion be used to create appealing images?
- How can recursion help people plan their financial futures?
- How can slight changes in a recursive process change behavior from predictable to chaotic?

In this chapter, you will use recursive modeling to answer these and many other important questions.

Lesson 8.1

Introduction to Recursive Modeling

The simplest recursive processes are numerical: one number in a list is determined by applying simple mathematical calculations to one or more of the preceding numbers. This is the type of recursion you have considered in previous chapters and is also the type with which this chapter begins.

Reconsider a problem that you first saw in Lesson 2.6. Luis and Britt examine the number of handshakes that occur when every person in a group shakes hands with every other person. The following is a table similar to the one you made in Lesson 2.6.

Number of People in the Group	Number of Handshakes
1	0
2	1
3	3
4	6
5	10

When a new person enters a group in which everyone has shaken hands, the new person has to shake hands with each of the people who are already in the group. Thus, the number of handshakes in a group of n people is $n - 1$ more than the number of handshakes in a group of $n - 1$ people. If H_n represents the number of handshakes in a group of n people, this recurrence relation can be expressed symbolically as $H_n = H_{n-1} + (n - 1)$.

Your work with this recurrence relation included writing a formula called a **solution to the recurrence relation** and using mathematical induction to prove the formula correct. In this case, the solution, which is also called a **closed-form solution**, is $H_n = \frac{n(n-1)}{2}$.

Closed-form solutions are useful because, unlike recurrence relations, they calculate a value directly. For example, you can find the number of handshakes in a group of ten people without knowing the number of handshakes in a group of nine people. However, closed-form solutions can be difficult to find. If such solutions are found by trial and error, mathematical induction can be used to prove their validity.

There are techniques other than trial and error that can be used to find closed-form solutions. For example, the counting methods in Chapter 6 are useful for certain kinds of recurrence relations. The handshake problem requires that every pair of people shake hands. In a group of n people, there are $C(n, 2)$ ways of selecting a pair, and so there are $C(n, 2) = \frac{n!}{(n-2)!2!}$ handshakes. But $n! = n(n-1)(n-2)!$, so the counting solution is equivalent to the solution you hypothesized and proved in Chapter 2.

The closed-form calculation of the number of handshakes in a group of, say, 100 people is a simple one: $100 \times \frac{99}{2} = 4{,}950$. Obtaining this solution with a recurrence relation requires extending the table on page 444 to 100 rows.

Handshakes Matter for First Impressions, Brain Study Confirms

HealthDay News
November 2, 2012

A new study provides scientific evidence that a handshake helps you make a good first impression.

University of Illinois researchers used functional MRI scans to monitor brain activity in volunteers who watched and rated videos of people interacting in a business setting. The scans showed a positive "social evaluation" brain response when the volunteers saw the people in the videos shake hands.

It's not just a handshake that promotes positive feelings, but a firm, confident, yet friendly handshake, noted study co-leader Florin Dolcos. "In a business setting this is what people expect," he said. "Not a very long time ago you could get a loan based on a handshake. So it conveys something very important, very basic. Yet the science underlying this is so far behind. We knew these things intuitively but now we also have the scientific support."

Extending a table to 100 or more rows is a tedious task when done by hand. Fortunately, there are several ways to apply technology to the problem. Indeed, the speed of computers has made recursive methods much more useful today than they were a few decades ago.

One type of technology that is very useful for recursive modeling is a computer *spreadsheet*. A spreadsheet is a matrix consisting of columns labeled with the letters A, B, C, . . . and rows labeled with the numerals 1, 2, 3, ... A particular location in the spreadsheet is called a *cell* and is denoted by its column letter and row number, such as A1 or C5. Cells may contain verbal information, numeric values, or formulas that refer to other cells. Spreadsheets have copy features that allow formulas to be copied so that tables can be generated rapidly.

Another way technology can be used in recursive modeling is by developing a calculator or computer program that generates values from the relation. Programming requires that an appropriate algorithm be adapted to the language used by the calculator or computer. The following is an algorithm for the handshake problem that can be adapted to a calculator or computer. The variable N represents the number of people in the group, and H represents the number of handshakes.

1. Store the number 1 for variable N and the number 0 for variable H.
2. Display N and H.
3. Add 1 to N and store the result as the new value of N.
4. Add $N - 1$ to H and store the result as the new value of H.
5. Repeat steps 2 through 4.

Step 4 of this algorithm uses the recurrence relation to calculate the number of handshakes. The closed form can also be used in this step. To do so, replace step 4 with "store $\frac{N(N-1)}{2}$ as the new value of H."

Some calculators have special functions designed to generate values from recurrence relations.

Using Computer Spreadsheets with Recurrence Relations

To model the handshake problem with a spreadsheet, type suitable labels in the first row. In cell A2, type initial value of 1 for the number of people. In cell B2, type the initial value of 0 for the number of handshakes. In cell A3, type the formula A2 + 1. In cell B3, type the formula B2 + A2, which is equivalent to the recurrence relation $H_n = H_{n-1} + (n - 1)$. You can fill the remaining rows by copying row 3. Note that most spreadsheets require the initial character of a formula to be either = or +. The completed spreadsheet is shown here in two ways: with formulas and with numeric results.

	A	B	C
1	Number of people	Number of handshakes	Closed form
2	1	0	= A2*(A2 – 1)/2
3	= A2 + 1	= A2 + B2	= A3*(A3 – 1)/2
4	= A3 + 1	= A3 + B3	= A4*(A4 – 1)/2
5	= A4 + 1	= A4 + B4	= A5*(A5 – 1)/2
6	= A5 + 1	= A5 + B5	= A6*(A6 – 1)/2
7	= A6 + 1	= A6 + B6	= A7*(A7 – 1)/2
8	= A7 + 1	= A7 + B7	= A8*(A8 – 1)/2
9	= A8 + 1	= A8 + B8	= A9*(A9 – 1)/2
10	= A9 + 1	= A9 + B9	= A10*(A10 – 1)/2

	A	B	C
1	Number of people	Number of handshakes	Closed form
2	1	0	0
3	2	1	1
4	3	3	3
5	4	6	6
6	5	10	10
7	6	15	15
8	7	21	21
9	8	28	28
10	9	36	36

Writing Programs for Recurrence Relations

On the left is a computer algorithm written in BASIC that generates a table for the handshake problem. On the right is a similar calculator algorithm for Texas Instruments graphing calculators.

```
10 N = 1:H = 0          1 →N:0 →H
20 PRINT N, H           Lbl A
30 N = N + 1            Disp N, H
40 H = H + N − 1        N + 1 →N:H + N − 1 →H
50 GO TO 20             Goto A
```

Because these algorithms do not end, a statement should be added to terminate the table at some value. For example, to stop the program after calculation of the number of handshakes in a group of ten people, add the line 45 IF N>10 THEN STOP to the BASIC algorithm or the two lines If N>10 and Stop before the last line of the calculator algorithm. Note that the calculator algorithm does not display paired values of *N* and *H* on a single line. One way to remedy this inconvenience is to store the values in a 1 × 2 matrix and display the matrix.

Calculators with Recursion Features

The following screens demonstrate the recursion features of one type of graphing calculator. The left screen shows the entry of the handshake recurrence relation, which is done after the calculator is placed in its sequence mode. The right screen shows the resulting table, which appears after the table's initial value and increment have been set.

```
Plot1 Plot2 Plot3
 nMin=1
·.u(n)=u(n−1)+(n−
1)
 u(nMin)={0}
·.v(n)=
 v(nMin)=
·.w(n)=
```

n	u(n)
1	0
2	1
3	3
4	6
5	10
6	15
7	21

u(n)=u(n−1)+(n−...

Exercises

1. Consider a variation of this lesson's handshake problem. There are an equal number of men and women in Luis and Britt's group and each person shakes hands with all members of the opposite sex.

   a. Draw a graph for a group of four men and four women in which the vertices represent the men and women in the group and the edges represent the handshakes. Recall your work in graph theory. What kind of a graph is this?

   b. If there is only one man and one woman in the group, how many handshakes are made? With two couples? With three couples?

   c. Complete the following table to investigate the number of handshakes that are made.

Number of Couples	Number of Handshakes	Recurrence Relation
1		
2		
3		
4		
5		

   d. Assume that there are H_{n-1} handshakes for $n - 1$ couples and that another couple joins the party. How many additional handshakes are now possible?

   e. Write a recurrence relation that describes the relationship between the number of handshakes (H_n) for n couples and the number of handshakes (H_{n-1}) for $n - 1$ couples.

2. Consider another variation of the handshake problem in which each man shakes hands with each of the women *except* his date.

   a. Make a table showing the number of handshakes that occur when there is one couple. Two couples. Three couples. Four couples.

   b. Assume that you know the number of handshakes with $n - 1$ couples. How many additional handshakes are made when the nth couple arrives?

c. Write a recurrence relation for the total number of handshakes (H_n) when there are n couples.

3. a. Write a recurrence relation to describe each of the following patterns. (Note: Do not give closed-form formulas.)

   i. 1, 4, 7, 10, 13, . . .
   ii. 1, 2, 4, 8, 16, 32, . . .
   iii. 1, 3, 6, 10, 15, 21, 28, . . .
   iv. 1, 2, 6, 24, 120, 720, . . .

   b. What is the next term in each of the patterns in part a?

Mathematician of Note

Leonardo of Pisa, an Italian mathematician who lived from about 1170 to about 1250 and is also known as Fibonacci, popularized the sequence in Exercise 4 in his writings. Today, the sequence is famous for its many applications.

4. The ability to recognize patterns is considered a mark of intelligence. Thus, most IQ tests include questions about numerical patterns. For example, a question on an IQ test gives the sequence 1, 2, 3, 5, 8, 11 and asks which of the numbers does not belong.

   a. Explain why the correct answer is that the last number, 11, does not fit the pattern.

   b. Write a recurrence relation that describes the pattern.

5. You cannot use a recurrence relation to generate terms unless you have an initial value. For example, $t_n = 2t_{n-1} - 3$ has terms 5, 7, 11, 19, . . . if the initial value t_1 is 5. But if the initial value is 6, then the terms are 6, 9, 15, 27, ... Notice, however, if the initial value is 3, then all the terms are 3. An initial value for which all the terms of the recurrence relation are the same is called a **fixed point**.

   Find the fixed point for each of the following recurrence relations if one exists.

   a. $t_n = 2t_{n-1} - 4$

   b. $t_n = 3t_{n-1} + 2$

   c. $t_n = 2t_{n-1}$

   d. $t_n = t_{n-1} + 3$

6. If two rays have a common endpoint, one angle is formed. If a third ray is added, three angles are formed. See the following figure.

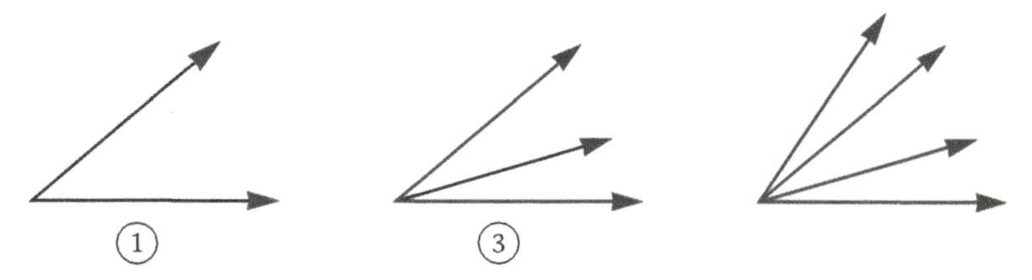

a. How many angles are formed if a fourth ray is added? A fifth ray?

b. Write a recurrence relation for the number of angles formed with n rays.

c. Write a closed-form solution.

d. Use your closed-form solution to find the number of angles formed by ten rays.

7. For the original handshake problem in which everyone shakes hands with everyone else, construct a table for one through eight people in the following manner.

First column: term number

Second column: number of handshakes

Third column: differences of successive numbers from column 2

Fourth column: differences of successive numbers from column 3

a. What do you notice about the fourth column?

b. What degree is the polynomial that was obtained for the closed-form solution of the handshake problem? Compare your answer with the number of difference columns in your table.

8. Consider the closed-form polynomial $S_n = 4n^3 - 3n + 2$.

a. Make a table, as in Exercise 7, for $n = 1, \ldots, 8$. Include difference columns until the numbers in the last difference column are equal.

b. How many difference columns did you need?

c. How does the number of difference columns compare with the degree of the closed-form polynomial?

9. Let V_n be the number of vertices in a complete binary tree. (A binary tree is complete if each vertex of the tree has either two or no children.) Level 0 is the root of the tree. The first three trees follow.

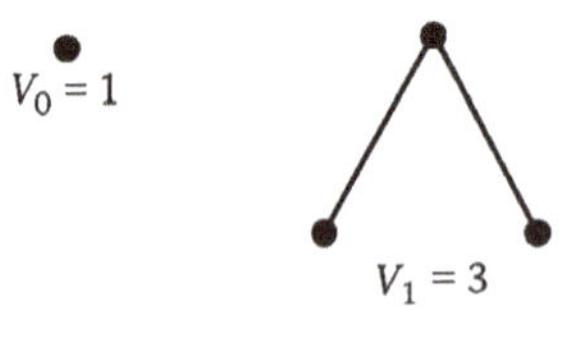

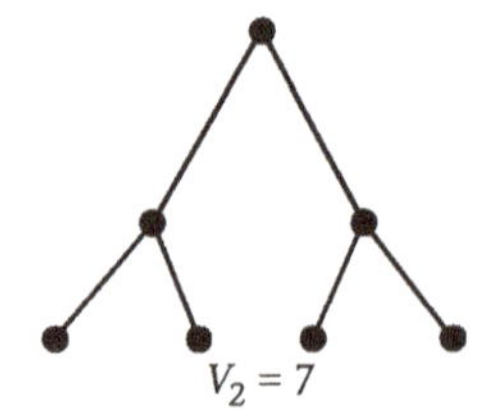

a. Make a table for $V_0, \ldots, V_6$.

b. Write a recurrence relation to describe V_n.

In Exercise 9, it is convenient to begin the table with V_0 because the root of the tree is considered level 0. There are other cases in which the initial value of the recurrence relation is labeled with the subscript 0. It is, for example, a useful practice when working with time intervals as in Exercises 10 and 11.

Africanized Bees Found in East Tennessee

Associated Press
April 10,2012

Tennessee agricultural officials say a colony of partially Africanized bees, which are more aggressive than domestic honeybees, have been found in East Tennessee.

The Tennessee Agriculture Department said that the bees were found last week in a colony belonging to a beekeeper in Monroe County. State Apiarist Mike Studer said these bees have been found in other states, but this is the first time they have been found in Tennessee.

The biggest difference between Africanized honeybees and European honey bees is their behavior, which includes fiercely defending their nests.

10. Since the 1980s, residents of the southwestern United States have been concerned about an influx of African "killer" bees. In 1987, the number of African bees was estimated at 5,000. It was also estimated that the population would increase at a rate of 12% annually. Let B_n be the number of African bees in Texas each year, where $n = 0$ corresponds to the beginning of the year 1987. Then B_1 indicates "the end of year 1."

a. Make a table with entries for $B_0, \ldots, B_4$.

b. Write a recurrence relation for B_n.

c. Use a spreadsheet or calculator to determine when the population of African bees is predicted to exceed 100,000. In what year is this predicted to occur?

11. Susie puts $800 in a bank account that pays 5% interest per year, compounded yearly. Let n be the number of years she leaves the money in the bank, let A_0 be the initial amount of money ($800), and let A_n be the amount of money in the bank at the end of n years.

    a. By creating a table, find out how many years it takes for Susie's money to double.

    b. Write a recurrence relation for this situation.

12. The term *fractal* was invented in the 1970s by Benoit Mandelbrot to describe a class of geometric objects with unusual properties. One of those properties is that a fractal contains parts that resemble the whole. The following sequence of figures demonstrates the construction of such a "self-similar" object.

A short algorithm describes the process used to create the figures:

1. Draw a line segment.
2. For each line segment in the current figure, erase the middle third and draw a "peak" by constructing an equilateral triangle having the erased portion as a base.
3. Repeat step 2.

a. How does the length of a segment in one figure of the sequence compare with the length of a segment in the preceding figure?

b. How does the number of segments in one figure of the sequence compare with the number of segments in the preceding figure?

c. Write a recurrence relation that describes the relationship between the total length of the segments in figures n and $n - 1$.

Mathematician of Note

Benoit Mandelbrot (1924–2010)

Mandelbrot was Sterling Professor of Mathematical Sciences at Yale University.

Computer/Calculator Explorations

13. Some computer drawing utilities such as Geometer's Sketchpad support recursion. Use a utility with recursion features to construct figures like those in Exercise 12.

14. The Logo computer language supports recursion. Use Logo to construct figures like those in Exercise 12.

15. Perform this spreadsheet experiment: Type the number 1 in cell A1. In cell A2 type the formula A1, and in cell B2 type the formula A1 + B1. Copy the formula in cell A2 into cell A3; copy the formula in cell B2 into cells B3, C3, D3, ... (Stop at a convenient cell in row 3.) Copy row 3 into several of the rows that follow it. Identify the results and explain why they are produced by this procedure.

Projects

16. Research and report on several current examples of recursive modeling not discussed in this chapter. (You may want to begin your research now and continue it as this chapter progresses.)

17. Research the Fibonaccii sequence (see Exercise 4). Prepare a report on its history and several of its current modeling applications.

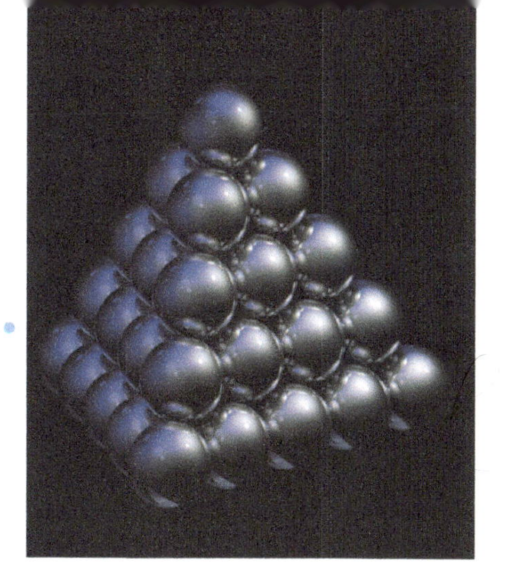

Lesson 8.2

Finite Differences

Previous lessons discussed two approaches to the problem of finding a closed-form solution to the handshake problem:

1. Trial and error followed by an induction proof of the hypothesized formula
2. Counting techniques

This lesson considers an approach known as the method of finite differences, which can be used to find a closed-form solution to the handshake problem and a variety of other problems.

Recall that the handshake problem is described recursively by $H_1 = 0$, $H_n = H_{n-1} + (n - 1)$. The following is a table generated by this recurrence relation. The third column contains the differences between successive values in the second column. The fourth column contains the differences between successive values in the third column.

Number of People	Number of Handshakes	Differences	
		First	Second
1	0	—	—
2	1	1	—
3	3	2	1
4	6	3	1
5	10	4	1
6	15	5	1
7	21	6	1
8	28	7	1

The constant second differences indicate that the closed-form solution for this recurrence relation is a second-degree polynomial, which has the general form $an^2 + bn + c$.

Consider what happens when the general second-degree polynomial is evaluated for consecutive integral values of n, and first and second differences are found. The following table shows the results.

Value of n	Value of Polynomial	Differences: First	Differences: Second
1	$a + b + c$	—	—
2	$4a + 2b + c$	$3a + b$	—
3	$9a + 3b + c$	$5a + b$	$2a$
4	$16a + 4b + c$	$7a + b$	$2a$
5	$25a + 5b + c$	$9a + b$	$2a$

Notice that the second differences are not only constant, but also the value of the difference is twice the value of the coefficient of n^2. In the case of the handshake problem, this result means that the constant difference of 1 indicates that one term of the closed-form solution is $\frac{1}{2}n^2$.

The remaining terms of the closed-form solution can be found by substituting values from the table into the polynomial $H_n = \frac{1}{2}n^2 + bn + c$.

Although the method just demonstrated works well when the closed-form solution is second degree, it is much more tedious for degrees higher than 2. The following alternative method uses technology and is therefore easier to extend to higher degrees.

Reconsider the handshake problem, a situation in which you know the solution is second degree: $H_n = an^2 + bn + c$. Since there are three values that you need to know (a, b, and c), select any three pairs of values from your table. The first three are convenient because of their relatively small values. Form three equations by substituting these three pairs into the general second-degree polynomial $H_n = an^2 + bn + c$.

When $n = 1$, $0 = a + b + c$.

When $n = 2$, $1 = 4a + 2b + c$.

When $n = 3$, $3 = 9a + 3b + c$.

Solve this system using the matrix techniques in Chapter 7.

$$\begin{bmatrix} 1 & 1 & 1 \\ 4 & 2 & 1 \\ 9 & 3 & 1 \end{bmatrix}^{-1} \times \begin{bmatrix} 0 \\ 1 \\ 3 \end{bmatrix} = \begin{bmatrix} 0.5 \\ -0.5 \\ 0 \end{bmatrix}$$

The finite differences method can be used whenever the differences in consecutive values of the recurrence relation become constant in a finite number of columns. The degree of the closed-form solution is the same as the number of columns needed to achieve the constant differences. The number of equations in the system needed to find the closed-form solution is 1 more than its degree.

A Finite Differences Example

Consider a stack of cannonballs at Fort Recurrence (see Figure 8.1).

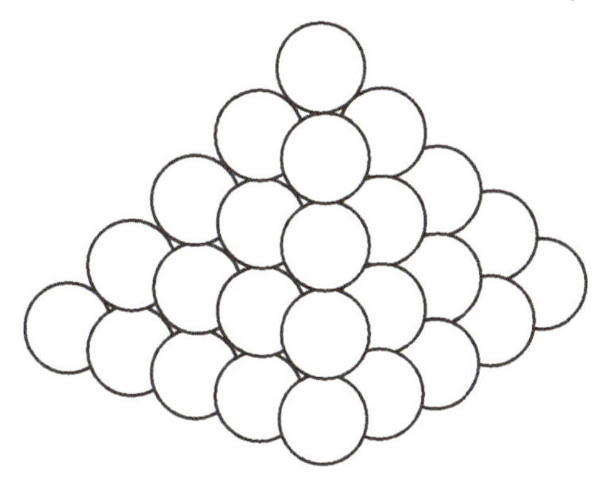

Figure 8.1 Cannonballs at Fort Recurrence.

The following table displays the number of cannonballs in a pyramid of n layers.

Number of Layers	Number of Cannonballs	Differences		
		First	Second	Third
1	1	—	—	—
2	5	4	—	—
3	14	9	5	—
4	30	16	7	2
5	55	25	9	2
6	91	36	11	2

The recurrence relation that describes the number of cannonballs in a stack of n layers is $C_n = C_{n-1} + n^2$. The constant differences in the third column indicate that the closed-form solution is third degree: $C_n = an^3 + bn^2 + cn + d$. The system created by this general third-degree polynomial and the first four values in the table is:

When $n = 1$, $1 = a + b + c + d$.

When $n = 2$, $5 = 8a + 4b + 2c + d$.

When $n = 3$, $14 = 27a + 9b + 3c + d$.

When $n = 4$, $30 = 64a + 16b + 4c + d$.

The matrix solution is

$$\begin{bmatrix} 1 & 1 & 1 & 1 \\ 8 & 4 & 2 & 1 \\ 27 & 9 & 3 & 1 \\ 64 & 16 & 4 & 1 \end{bmatrix}^{-1} \times \begin{bmatrix} 1 \\ 5 \\ 14 \\ 30 \end{bmatrix} = \begin{bmatrix} 0.3333 \\ 0.5 \\ 0.1667 \\ 0 \end{bmatrix}.$$

The closed-form solution, therefore, is $C_n = \frac{1}{3}n^3 + \frac{1}{2}n^2 + \frac{1}{6}n$. Note that unlike the case in which the solution is second degree, the coefficient of the first term is not one-half the constant difference.

Unfortunately, the finite difference method does not apply to recurrence relations that never achieve constant differences. In such cases, other methods that are described in later lessons of this chapter are often successful.

This lesson's exercises investigate several situations that can be modeled with recurrence relations in which the differences eventually become constant.

Including Differences in a Spreadsheet

If your spreadsheet already contains the number of people and the number of handshakes in columns A and B, then adding columns for differences requires very little effort. You can add a difference column by typing one additional formula and then copying it into as many cells as necessary. If, for example, the spreadsheet has the number of handshakes for a group of 1 in cell B2, for a group of 2 in cell B3, and so forth, place the first difference in cell C3 by typing the formula B3 – B2. Copy this formula into other cells of column C. Since the values in column C are not constant, copy the same formula into the cells of column D starting in cell D4. Because the values in column D are constant, you can stop.

The first spreadsheet below shows the formulas; the second shows the values that result.

	A	B	C	D
1	Number of people	Number of handshakes	First differences	Second differences
2	1	0		
3	= A2 + 1	= B2 + A2	= B3 – B2	
4	= A3 + 1	= B3 + A3	= B4 – B3	= C4 – C3
5	= A4 + 1	= B4 + A4	= B5 – B4	= C5 – C4
6	= A5 + 1	= B5 + A5	= B6 – B5	= C6 – C5
7	= A6 + 1	= B6 + A6	= B7 – B6	= C7 – C6
8	= A7 + 1	= B7 + A7	= B8 – B7	= C8 – C7
9	= A8 + 1	= B8 + A8	= B9 – B8	= C9 – C8
10	= A9 + 1	= B9 + A9	= B10 – B9	= C10 – C9

	A	B	C	D
1	Number of people	Number of handshakes	First differences	Second differences
2	1	0		
3	2	1	1	
4	3	3	2	1
5	4	6	3	1
6	5	10	4	1
7	6	15	5	1
8	7	21	6	1
9	8	28	7	1
10	9	36	8	1

Difference Columns on a Graphing Calculator

Some graphing calculators have a function that calculates the differences between successive pairs of values in a list. Note that the calculator used to create the following screens places a given difference opposite the first member of the pair rather than the second.

L1	L2	L3 3
1	0	------
2	1	
3	3	
4	6	
5	10	
6	15	
7	21	

L3 =ΔList(L2)

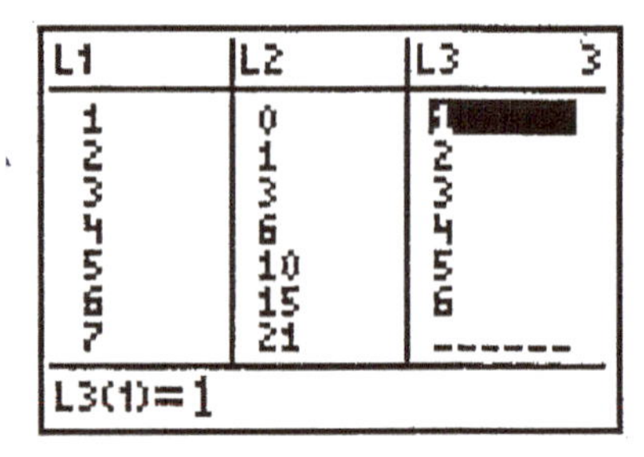

Exercises

1. Use finite differences to determine the degree of the closed-form formula that generates the given sequence.

   a. −3, −2, 3, 12, 25, 42, 63, 88, 117, 150, 187, 228, 273, 322, . . .

   b. 0.29, 0.52, 0.75, 0.98, 1.21, 1.44, 1.67, 1.90, 2.13, 2.36, 2.59, . . .

   c. 0, −2, −2, 0, 4, 10, 18, 28, 40, 54, 70, 88, 108, 130, 154, . . .

   d. 1, 3, 9, 27, 81, 243, 729, 2187, 6561, 19683, 59049, 177147, . . .

2. For each part of Exercise 1, determine the closed-form formula that generates the sequence.

3. a. Write a recurrence relation for the number of edges T_n in a complete graph with n vertices, K_n.

   b. For your recurrence relation in part a, what is the initial condition? (That is, how many edges are in a graph with one vertex?)

   c. Use finite difference techniques to determine a closed-form formula for the number of edges in a K_n graph.

4. $a_1 = 1$ and $a_n = 3a_{n-1} - 5$

   a. Find the first few (six to eight) terms.

   b. Find the fixed point for this recurrence relation. (Hint: When a recurrence relation has a fixed point, all the terms are the same. Replace a_n and a_{n-1} with a single variable such as x, then solve. Check your solution by using it as an initial value in the recurrence relation.)

5. A triangle has no diagonal, a quadrilateral has two diagonals, and a pentagon has five diagonals.

   a. Write a recurrence relation for the number of diagonals in an n-sided polygon.

   b. Use finite difference techniques to find a closed-form formula for the number of diagonals in an n-sided polygon.

6. An auditorium has 24 seats in the front row. Each successive row, moving toward the back of the auditorium, has 2 additional seats. The last row has 96 seats.

   a. Create a table with a column for the number of the row and a column for the number of seats in that row. Complete at least the first six entries in the table.

   b. Write a recurrence relation for the number of seats in the nth row.

   c. Find a closed-form solution for the number of seats in the nth row. (One way to do this is to use finite differences techniques.)

   d. How many rows are in the auditorium? Explain.

   e. Add a third column, "Total seats," to your table from part a. Complete at least the first six sums in this column.

   f. Write a recurrence relation for the total number of seats in the first n rows of the auditorium.

   g. Write a closed-form formula for the total number of seats in the first n rows of the auditorium.

7. A house purchased in 2000 increased in value at the rate of 5% per year.

   a. If the original cost of the house was $88,000, calculate the value of the house each year from 2000 to 2015. (A spreadsheet might be nice to use here.)

   b. Write a recurrence relation for the value of the house at the end of the *n*th year since 2000.

   c. Calculate the finite differences for your numbers in part a. Do you eventually obtain constant differences?

8. Since 2010, a herd of 50 deer has been increasing at the rate of approximately 4% per year.

   a. Make a table that gives the number of deer at the end of each year ($T_0 = 50$).

   b. If the herd's habitat can provide food for a maximum of 325 deer, in what year will there not be enough food?

   c. Write a recurrence relation for the number of deer at the end of the *n*th year.

   d. Calculate the finite differences for your table in part a. Do you eventually obtain constant differences?

Idaho Officials Fret Over Deer Collision Rise

Claims Journal
November 28, 2012

Idaho's wildlife managers are starting to worry about the growing number of collisions involving motorists and deer, elk and moose.

Studies show more than 5,000 of the animals were killed by vehicles on Idaho roadways last year, and that total could be higher since many collisions aren't reported to law enforcement or insurance companies.

Now, Idaho Fish and Game Department managers say it's time to ramp up monitoring and wildlife crossing programs.

Wildlife crossings, which are showing positive results in other states, might become more commonplace along known animal migration corridors in Idaho.

9. Investigate the number of squares in a stair-step design. If possible, find a recurrence relation and a closed-form formula for the number of squares (S_n) at stage n.

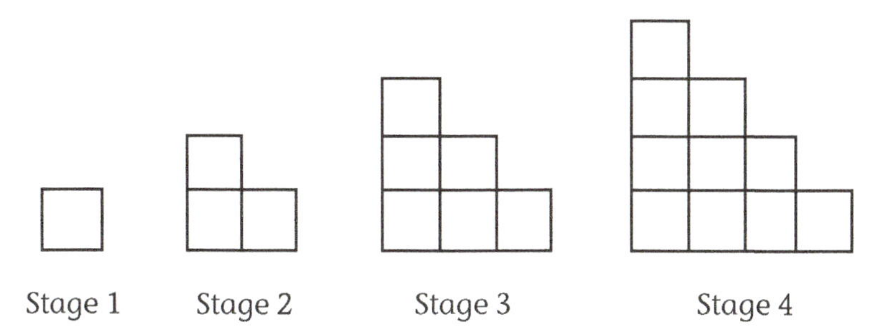

10. This lesson includes an analysis of second-degree polynomials that uncovered a connection between the leading coefficient of a second-degree closed-form solution and the constant difference. Perform a similar analysis for the third-degree polynomial. How is the leading coefficient related to the constant difference?

Computer/Calculator Explorations

11. Graphing calculators have statistical functions that fit various kinds of mathematical functions to a set of data. Many of these calculators include several kinds of polynomials in this collection of functions. Prepare a report on the polynomial-fitting capabilities in your calculator and show how they can be used to find closed-form polynomial solutions to recurrence relations for which differences become constant.

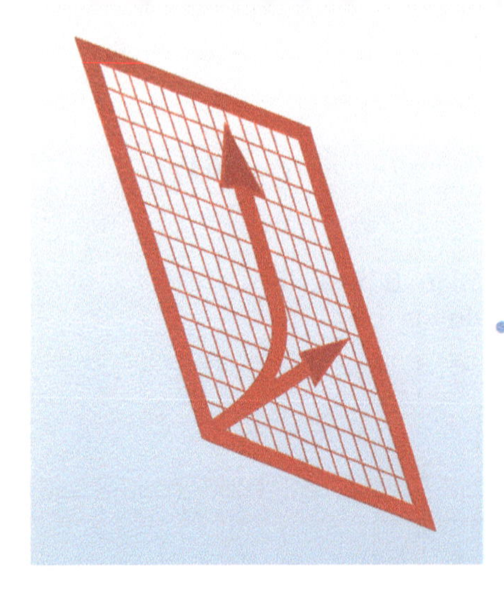

Lesson 8.3

Arithmetic and Geometric Recursion

Counting techniques and finite differences are two methods that can be used to find closed-form solutions for recurrence relations. However, there are many kinds of recurrence relations, and no method is capable of finding a closed-form solution for all of them.

Two of the most common types of recurrence relations are those in which a term is generated by either adding a constant to the previous term or multiplying the previous term by a constant. The first type is called **arithmetic**, and the constant is called the **common difference**. The second type is called **geometric**, and the constant is called the **common ratio**. This lesson considers arithmetic and geometric recurrence relations and a few of their applications.

A surprising fact about these two types of recurrence relations is that a geometric recurrence relation with a common ratio larger than 1 and a positive first term will eventually grow to a larger value than an arithmetic recurrence relation, even if the latter's first term and common difference are relatively large. For example, consider a job that employs you for 30 days and in which you have a choice of two methods of payment. Method 1 pays \$5,000 the first day and raises the daily pay by \$10,000 each day after that. Method 2 pays only \$0.01 the first day but doubles the amount you are paid each successive day. The questions are, of course: Which salary is better? How much better?

The first method of payment is arithmetic, and the common difference is $10,000. The second method is geometric, and the common ratio is 2. If P_n is the payment on the nth day, the arithmetic recurrence relation is $P_n = P_{n-1} + 10{,}000$; the geometric recurrence relation is $P_n = 2P_{n-1}$.

To determine the amount by which one salary is better than the other, you must find the total amount each pays over the 30-day period. If T_n represents the total, then a recurrence relation that describes the total is $T_n = T_{n-1} + P_n$.

Of course, all these questions can be answered by using a computer or calculator to generate a comparative table for the entire 30-day period. However, there are formulas that can find all the relevant information directly. Since these formulas have many modeling applications, this lesson next discusses how the formulas can be developed.

Formulas for Arithmetic and Geometric Terms

In method 1, the pay on the first day is $5,000, the pay on the second day is $5,000 + $10,000, and the pay on the third day is $5,000 + $10,000 + $10,000 = $5,000 + 2($10,000). Therefore, the pay on the nth day is $5,000 plus $n - 1$ raises of $10,000 each. Thus, the formula for the pay on the nth day is $5,000 + ($n - 1$)$10,000.

In general, the nth term of an arithmetic recurrence relation is found by adding $n - 1$ common differences to the first term: $t_n = t_1 + (n - 1)d$.

In method 2, the pay on the first day is $0.01, the pay on the second day is $0.01(2), and the pay on the third day is $0.01(2)(2) = $0.01($2^2$). Therefore, the pay on the nth day is $0.01 doubled $n - 1$ times. Thus, the formula for the pay on the nth day is $0.01($2^{n-1}$).

In general, the nth term of a geometric recurrence relation is found by multiplying the first term by the common ratio $n - 1$ times: $t_n = t_1(r^{n-1})$.

With these formulas, the comparison of wages on the 30th day is a simple matter:

Method 1: t_{30} = $5,000 + (30 − 1)$10,000 = $295,000.

Method 2: t_{30} = $0.01($2^{30-1}$) = $5,368,709.12.

The trend lines in Figure 8.2 show the wages for each method over the 30-day period.

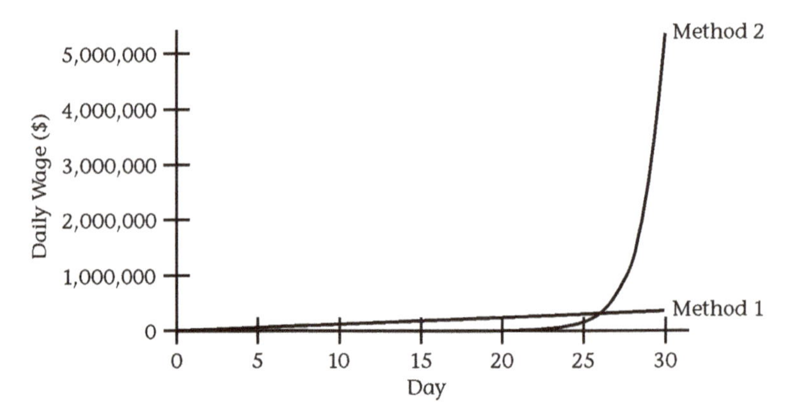

Figure 8.2. Daily wages by two methods.

Sums of Arithmetic and Geometric Terms

A complete comparison of the two methods requires the total of the 30 daily wages for each method. To examine the total pay for method 1, consider the general arithmetic recurrence relation with first term t_1 and common difference d.

Term Number	Term	Sum of First n Terms	Differences: First	Differences: Second
1	t_1	t_1	—	—
2	$t_1 + d$	$t_1 + (t_1 + d) = 2t_1 + d$	$t_1 + d$	—
3	$t_1 + 2d$	$(t_1 + 2d) + (2t_1 + d) = 3t_1 + 3d$	$t_1 + 2d$	d
4	$t_1 + 3d$	$(t_1 + 3d) + (3t_1 + 3d) = (4t_1 + 6d)$	$t_1 + 3d$	d

The constant second differences indicate that the closed-form solution is a second-degree polynomial; $t_n = an^2 + bn + c$. The related system of equations created from the first three pairs in the table is:

When $n = 1$, $t_1 = a + b + c$.

When $n = 2$, $2t_1 + d = 4a + 2b + c$.

When $n = 3$, $3t_1 + 3d = 9a + 3b + c$.

Matrices can be used to solve the system:

$$\begin{bmatrix} 1 & 1 & 1 \\ 4 & 2 & 1 \\ 9 & 3 & 1 \end{bmatrix}^{-1} \begin{bmatrix} & t_1 & \\ 2t_2 & + & d \\ 3t_1 & + & 3d \end{bmatrix} =$$

$$\begin{bmatrix} 1 & 1 & 1 \\ 4 & 2 & 1 \\ 9 & 3 & 1 \end{bmatrix}^{-1} \begin{bmatrix} 1 & 0 \\ 2 & 1 \\ 3 & 3 \end{bmatrix} \begin{bmatrix} t_1 \\ d \end{bmatrix} =$$

$$\begin{bmatrix} 0 & 0.5 \\ 1 & -0.5 \\ 0 & 0 \end{bmatrix} \begin{bmatrix} t_1 \\ d \end{bmatrix} = \begin{bmatrix} & 0.5d \\ t_1 & -0.5d \\ & 0 \end{bmatrix}$$

In general, the sum of the first n terms of an arithmetic recurrence relation is $0.5dn^2 + (t_1 - 0.5d)n$.

The sums in method 2 do not generate constant differences in a finite number of steps, and, therefore, a closed-form solution cannot be found by the finite difference method. It can, however, be found by other means.

Consider the general geometric recurrence relation with first term t_1 and common ratio r. The sum of the first n terms, S_n, is

$$S_n = t_1 + t_1r + t_1r^2 + \ldots + t_1r^{n-1}.$$

Multiply this equation by r:

$$rS_n = r(t_1 + t_1r + t_1r^2 + \ldots + t_1r^{n-1}).$$

Distribute r on the right side of the equation, and subtract this equation from the original equation:

$$\begin{aligned} S_n &= t_1 + t_1r + t_1r^2 + \ldots + t_1r^{n-1} \\ rS_n &= \quad\quad t_1r + t_1r^2 + t_1r^3 + \ldots + t_1r^n \\ S_n - rS_n &= t_1 - t_1r^n. \end{aligned}$$

Now factor both sides and divide by $(1 - r)$:

$$S_n(1 - r) = t_1(1 - r^n),$$

or

$$S_n = \frac{t_1(1-r^n)}{1-r}.$$

In general, the sum S_n of the first n terms of a geometric recurrence relation is $\frac{t_1(1-r^n)}{1-r}$.

With the arithmetic and geometric sum formulas, comparison of the total wages is a simple matter:

Method 1: $S_{30} = 0.5 \times 10{,}000 \times 30^2 + (5{,}000 - 0.5 \times 10{,}000) \times 30 = \$4{,}500{,}000$.

Method 2: $S_{30} = 0.01(1 - 2^{30})/(1 - 2) = \$10{,}737{,}418.23$.

Lighthammer Continues Geometric Growth

Clarinet News
April 9, 2003

Lighthammer, the leading supplier of manufacturing intelligence solutions, announced first quarter 2003 results ending March 31, 2003 showing an increase of 385% over first quarter 2002. Russ Fadel, Lighthammer CEO commented, "We are pleased with the tremendous year-to-year growth of our business particularly with the weak economy and the political uncertainty created by the war in Iraq. Our geometric growth is directly attributed to the operational performance improvement enabled by our manufacturing intelligence solution."

Including Sums in Spreadsheets

If your spreadsheet has the term numbers and the terms of a recurrence relation in columns A and B, respectively, adding a column for the sum requires entering a single formula, then copying it into as many cells as you need. For example, if the first term of the recurrence relation is in cell B1, the second term is in B2, and so forth, type B1 in cell C1 and the formula C1 + B2 in cell C2. Copy the formula into cells C3, C4, and so on.

Including Sums in Programs

Lesson 8.1 gives an algorithm for generating terms of the recurrence relation $H_n = H_{n-1} + (n - 1)$ and implementations for the BASIC language and a calculator language. To include sums in the algorithm, introduce a variable (S) for the sum, give it an initial value equal to the first term, and have it accumulate values of the terms as the terms are generated. Below are Lesson 8.1's implementations with instructions to accumulate sums.

```
10 N = 1:H = 0:S = H
20 PRINT N, H, S
30 N = N + 1
40 H = H + N - 1
50 S = S + H
60 GO TO 20
```

```
1 → N:0 → H:H → S
Lbl A
Disp N, H, S
N + 1 → N:H + N - 1 → H:S + H → S
Goto A
```

Keep in mind that neither algorithm terminates unless additional instructions are added.

Sums on a Graphing Calculator

Some graphing calculators have functions that automatically calculate sums. The screen on the left shows the closed-form solution $t_n = 0.01 \times 2^{n-1}$ as it is used to calculate the 30 daily wages by method 2. The results are stored in one of the calculator's lists. The second screen shows the sum function as it is used to find the sum of the 30 daily wages.

```
seq(.01*2^(X-1),
X,1,30)→L1
{.01 .02 .04 .0...
```

```
seq(.01*2^(X-1),
X,1,30)→L1
{.01 .02 .04 .0...
sum(L1)
        10737418.23
```

In this lesson's exercises you have the opportunity to use your knowledge of arithmetic and geometric terms and sums to model a variety of situations.

Exercises

1. Consider the following sequences:

   i. 2, 5, 8, 11, 14, . . .
   ii. 64, 32, 16, 8, 4, 2, 1, . . .
   iii. 10, 12, 14.4, 17.28, 20.736, . . .
   iv. 2, 3, 5, 8, 13, 21, . . .
   v. $\frac{3}{10}, \frac{3}{100}, \frac{3}{1000}, \frac{3}{10{,}000}, \ldots$
   vi. 3, 4, 6, 9, 13, 18, . . .

   a. Which of the sequences are geometric, which are arithmetic, and which are neither?

   b. Write a recurrence relation for each sequence.

   c. For those sequences that are arithmetic or geometric, write a closed-form formula for the nth term.

2. Consider the general recurrence relation for a geometric sequence, $t_n = rt_{n-1}$. Find the fixed point for this recurrence relation if one exists. (Recall the hint in Exercise 4b, page 461, of Lesson 8.2.)

3. Consider the general recurrence relation for an arithmetic sequence, $t_n = t_{n-1} + d$. Find the fixed point for this recurrence relation if one exists.

4. 900 telephone numbers charge the caller for the call. Consider a 900 teen chat line that charges \$2 for the first minute and \$0.95 for each additional minute.

   a. What recurrence relation does the statement of the problem suggest? (Make a table if necessary.)

   b. Find a closed-form formula to describe the cost of a call.

   c. Assume that 5,000 teens use the line each week and talk for an average of 15 minutes. How much income is produced? Suppose also that the bulk of the cost to the company operating the line is the cost of the long distance line, which averages \$3 for a 15-minute call. How much profit does the company make each week?

5. George deposits \$5,000 in the bank at an annual interest rate of 4.8% compounded yearly.

   a. Write a recurrence relation for the amount of money in the account at the end of n years.

   b. Write a closed-form formula for the amount of money in the account at the end of n years.

   c. How much money is in George's account at the end of 3 years?

   d. The bank decides to become competitive with other banking institutions in the city by offering an interest rate of 4.8% that compounds monthly. What monthly interest rate does George receive now?

   e. Write a recurrence relation for the amount of money in the account at the end of n months.

f. Write a closed-form formula for the amount of money in the account at the end of n months.

g. Find the amount in George's account at the end of 3 years (36 months).

h. Compare the amount of money in George's account at the end of 3 years when the interest is compounded yearly with the amount at the end of 3 years when the interest is compounded monthly.

6. In Exercise 5, suppose George is given the choice of investing his \$5,000 at 4.8% annual interest compounded monthly or at 5.0% annual interest compounded yearly. Compare the two methods of investment at the end of 1 year. At the end of 2 years. At the end of 3 years.

7. Find the fifteenth term and the sum of the first 15 terms for the sequences in parts a to c.

   a. 4, 9, 14, 19, . . .

   b. 45.75, 47, 48.25, 49.5, . . .

   c. 3650, 3623, 3596, 3569, 3542, . . .

   d. From your work in other mathematics courses, you may be familiar with the use of the Greek letter Σ (sigma) to indicate sums. For example, if you are summing the first five terms of the sequence whose terms are generated by $2n - 3$, the sum can be indicated in this way: $\sum_{1}^{5}(2n - 3)$. The term numbers of the first and last terms that are included in the sum are written at the bottom and at the top of sigma, respectively, and the formula for the nth term is written in parentheses to the right. Use this type of summation notation to represent each of the sums you found in parts a to c.

8. The number of deer on Fawn Island is currently estimated at 500 and is increasing at a rate of 8% per year. At the present time, the island can support 4,000 deer, but acid rain is destroying the vegetation on the island and the number of deer that can be fed is decreasing by 100 per year. In how many years will there not be enough food for the deer on Fawn Island? Explain how you got your answer.

9. The cost of n shirts selling for $14.95 each is given by $C_n = 14.95n$.

   a. What is the equivalent recurrence relation?

   b. Is the recurrence relation arithmetic, geometric, or neither? If it is arithmetic or geometric, what are the first term and the common difference or ratio?

10. At 5.5% compounded yearly, what amount must Bill's parents invest for Bill at age 10 if they want Bill to be a millionaire when he reaches age 50?

11. To double your investment at the end of 11 years, what annual interest rate do you need to receive?

12. The following table gives data relating temperature in degrees Fahrenheit to the number of times a cricket chirps in one minute. Although the data are not quite a perfect arithmetic sequence, they are close.

Temperature (°F)	50	52	55	58	60	64	68
Chirps per minute	40	48	60	73	80	98	114

   a. Assuming that the data form an arithmetic sequence, with the temperature as the term number and the chirps per minute as the term value, what is the common difference per degree rise in temperature?

   b. What is the temperature when a cricket is chirping 110 times per minute?

   c. At what temperature does a cricket stop chirping?

By permission of John L. Hart FLP, and Creators Syndicate, Inc.

13. A ball is dropped from a height of 8 feet and rebounds on each bounce to 75% of its height on the previous bounce.

    a. How high does the ball reach after the sixth bounce?

    b. What is the total distance that the ball has traveled just before the seventh bounce?

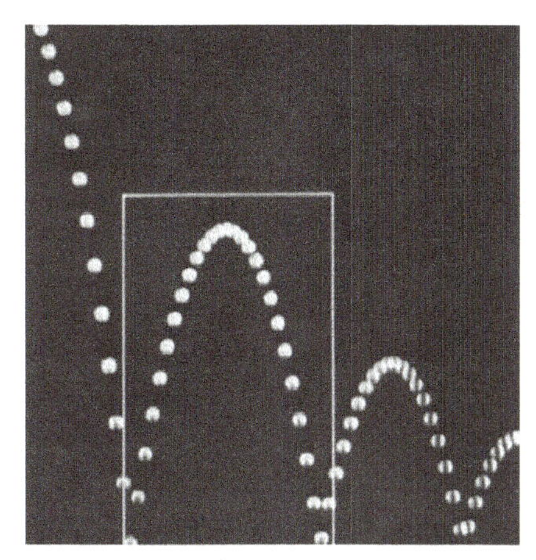

A rubber ball rebounds to a fraction of the height of the previous bounce, as shown in this time-lapse photograph.

14. The tenth term of an arithmetic sequence is 4, and the twenty-fifth term is 20. Find the first term and the common difference for this arithmetic sequence.

15. The average cost of a year of college education in a public university is about $16,500. The cost is increasing an average of 5% per year.

    a. If the current rate continues, what will be the cost of a year of college education in 30 years?

    b. Ten years from now you start saving for the college education of a child. The best interest rate you can get is 6% compounded annually. What amount would you have to put in the bank in order to pay for a year of college 20 years later?

The Rising Cost of College

Evansville Courier & Press
November 26, 2012

Many parents dream of the day their child strides to the stage to receive a college diploma. On the other hand, many parents dread the day that college tuition bills arrive. Despite steep tuition costs, a college education may be a sound investment that's within the reach of many families. All it takes is a degree of planning.

Tuition costs have been increasing at an average of 5 percent annually in recent years. As a result, four years at a public college or university for today's newborns could tip the scales at more than $145,000. Given these statistics, it is easy to believe that in 2009, contemporary college students graduated with an average $24,000 in debt 1. Unheard of in previous generations, this debt burden is now commonplace.

16. Over time, the number of bacteria in a culture grows geometrically. There are 600 bacteria at $t = 0$ and 900 bacteria at $t = 3$ hours.

    a. What is the approximate common ratio for this sequence?

    b. If the growth rate continues, approximately how many bacteria are there after 10 hours?

    c. At what time will there first be 50,000 bacteria in the culture?

17. A 50-meter-long pool is constructed with the shallow end 0.85 meter deep. For each meter of length, starting at the shallow end, the pool deepens by 0.06 meter.

    a. How deep is the deepest part of the pool? Explain.

    b. A rope is to be placed across the pool where the pool depth is 1.6 meters in order to mark the end of the shallow section. How far from the shallow end of the pool should the rope be placed?

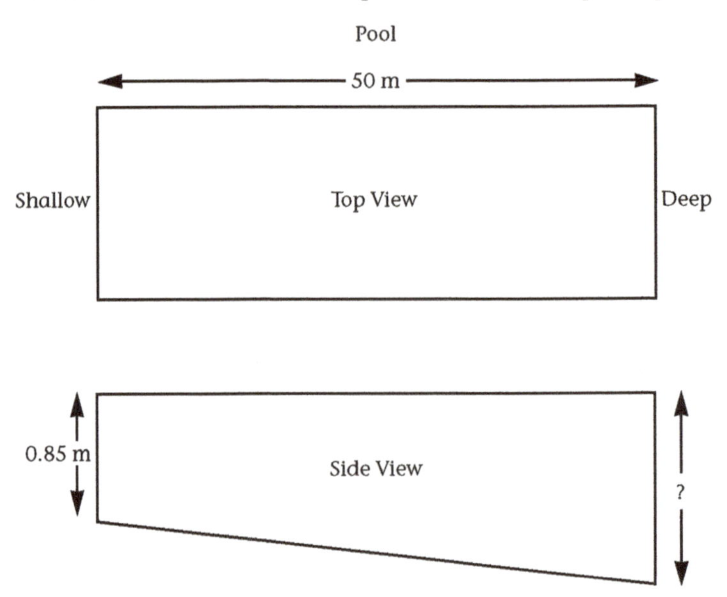

18. A company has sales of \$6 million one year and \$8 million the next year. The company expects to experience "geometric growth" at a similar rate in succeeding years.

    a. Write a recurrence relation that describes the company's growth and find a solution to the recurrence relation.

b. Predict the company's sales 5 years from now and 10 years from now.

c. Do you think the predictions are realistic? Explain.

19. Read the news article on page 468.

a. If a company has a 385% revenue increase in one year, what is the geometric growth rate?

b. If a company has \$200 million annual revenue and then has two years of growth at the rate you found in part a, what is the current revenue?

Computer/Calculator Explorations

20. The fractal curve construction described in Exercise 12 of Lesson 8.1 on page 453 can be applied to the sides of an equilateral triangle to produce a fractal that is often called a snowflake curve. At right is one stage of such a construction. Adapt the procedure you developed in either Exercise 13 or Exercise 14 in Lesson 8.1 to produce several snowflake curves.

Projects

21. Some geometric recurrence relations can be used to generate infinite sequences with finite properties. For example, the recurrence relation $t_n = \frac{1}{2} t_{n-1}$ with $t_1 = 1$ generates the sequence $1, \frac{1}{2}, \frac{1}{4}, \frac{1}{8}, \ldots$. The terms of this sequence approach 0, and the sequence of sums: $1, 1 + \frac{1}{2}, 1 + \frac{1}{2} + \frac{1}{4}, \ldots$ approaches 2. Research and report on conditions for which the sequence of terms and the sequence of sums of a geometric recurrence relation approach a finite value. Apply the results of your investigation to the perimeter and area of a snowflake curve based on an equilateral triangle with sides of length 1 (see Exercise 20).

22. In this lesson, the closed-form formulas for terms and sums of arithmetic and geometric recurrence relations are proved by algebraic symbol manipulation. Prepare a report showing how these formulas can be proved by mathematical induction.

Lesson 8.4

Mixed Recursion, Part 1

The previous lesson considered arithmetic and geometric recurrence relations. This lesson examines recurrence relations that have both a geometric and arithmetic component.

Consider a puzzle called Towers of Hanoi, which involves three pegs and disks of varying sizes stacked from largest to smallest on one of the pegs (see Figure 8.3).

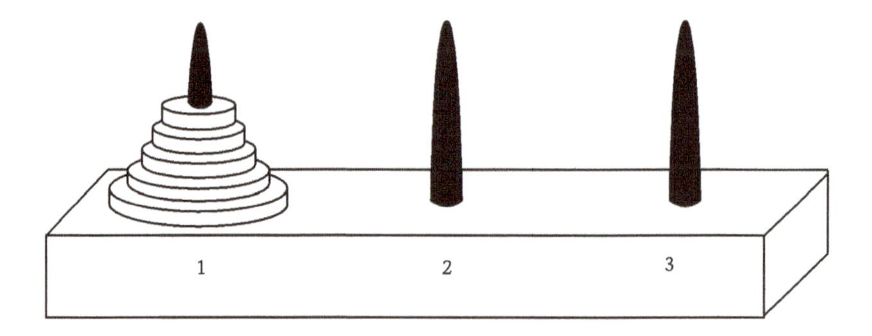

Figure 8.3. Towers of Hanoi.

The puzzle's goal is to move the disks from the first peg to the third peg in as few moves as possible. Disks may be placed temporarily on the middle peg or moved back to the first peg, but only one disk may be moved at a time and a disk may never be placed on top of one that is smaller than it.

Explore This

Artwork appearing on the box for the original puzzle, first marketed in 1883 by the French mathematician Edouard Lucas. The text translates: "The Tower of Hanoi, Authentic Brain Teaser of the Annamites, A Game Brought Back from Tonkin by Professor N. Claus (of Siam), Mandarin of the College of Li-Sou-Stian!"

Cut several round or square pieces of different sizes from paper or poster board. Use the pieces as disks. Number three spots on a large piece of paper and stack several of the pieces from largest to smallest on spot number one. You now have a rudimentary version of the Towers of Hanoi puzzle.

At the direction of your instructor, team up with one or more members of your class and try to solve the puzzle. Keep a record of the fewest moves in which puzzles with different numbers of disks can be completed in a table like the following.

Number of Disks	Fewest Moves to Complete Puzzle
1	1
2	
3	
4	
5	
...	

As you progress, look for patterns in the data as well as in the way you are completing the puzzle. If M_n represents the fewest moves in which a puzzle of n disks can be completed, look for a recurrence relation that describes the relationship between M_n and M_{n-1}. If time permits, conjecture a closed-form formula for M_n.

Your investigation of the Towers of Hanoi puzzle should have concluded that neither an arithmetic nor a geometric recurrence relation describes the number of moves: You cannot generate a term of the sequence by adding a constant or by multiplying by a constant; you need to do both.

A **mixed recurrence relation** is one in which both multiplying by a constant and adding another constant are used to generate a term from the previous. The general form of a mixed recurrence relation is $t_n = at_{n-1} + b$. Many common situations can be modeled with mixed recurrence relations. Examples include financial applications such as annuities and loan repayment, the way an object cools or heats, and the spread of diseases.

To use mixed recurrence relations, you need to be able to recognize that a given data set can be modeled with a mixed recurrence relation. Consider a table of data generated by the mixed relation $t_n = 2t_{n-1} + 3$ with $t_1 = 4$.

n	t_n	Differences
1	4	—
2	$2(4) + 3 = 11$	$11 - 4 = 7$
3	$2(11) + 3 = 25$	$25 - 11 = 14$
4	$2(25) + 3 = 53$	$53 - 25 = 28$
5	$2(53) + 3 = 109$	$109 - 53 = 56$

Note that the values in the difference column grow by a factor of $a = 2$. This characteristic of data generated by a mixed recurrence relation makes it possible to identify situations in which a mixed recurrence relation is an appropriate model and to find the values of a and b.

When data are generated by a mixed recurrence relation, the ratio of any difference to the one preceding it is the same as the value of a in $t_n = at_{n-1} + b$.

For example, consider the following table showing the increase in the value of a house over several years after its purchase.

Year	Value	Differences
0	120,000	—
1	122,000	2,000
2	124,200	2,200
3	126,620	2,420

The ratio of the second difference to the first is $\frac{2{,}200}{2{,}000} = 1.1$. The ratio of the third difference to the second is $\frac{2{,}420}{2{,}200} = 1.1$. Therefore, a mixed recurrence relation is appropriate, and the value of a is 1.1.

The recurrence relation that models the value of the house after n years is $V_n = 1.1V_{n-1} + b$. The value of b can be found by substituting two successive values from the second column of the table into this equation. For example, if the first two values are used, the result is $122{,}000 = 1.1(120{,}000) + b$. Solving for b gives $b = 122{,}000 - 1.1(120{,}000)$, or $b = -10{,}000$. The completed model for the value of the house after n years is $V_n = 1.1V_{n-1} - 10{,}000$.

A Mixed Recursion Example

Many people save money by making regular investments—from each paycheck, for example. Often such savings are part of a retirement plan that shelters current income from taxes. (Such plans are called Individual Retirement Accounts, or IRAs.) A plan to which you make regular contributions in order to have income at retirement is called an **annuity**. Annuities can be modeled with mixed recurrence relations.

Consider an annuity that draws 6% annual interest compounded monthly, to which monthly deposits of $200 are made. The monthly interest rate of $\frac{0.06}{12} = 0.005$. The following table shows the growth during the first few months.

Month	Balance ($)
0	200
1	1.005(200) + 200 = 401
2	1.005(401)+ 200 = 603.01
3	1.005 (603.01) + 200 = 806.03

If B_n represents the balance at the end of the nth month, then $B_n = 1.005B_{n-1} + 200$. This mixed recurrence relation can be used to create a table that tracks the growth of the plan over many months or years.

Annuities on Spreadsheets

Spreadsheet techniques discussed in previous lessons can be used to create annuity models. However, it is useful to create the spreadsheet in a way that allows monthly payments and interest rates to be changed easily so you can explore various options.

Create a spreadsheet in which a monthly payment of $200 and an annual interest rate of 0.06 are stored in cells A1 and B1. Type column headings of "Month" and "Balance" in the next row. In cell A3 type 0 and in cell B3 type the simple formula A1. In cell A4 type the formula A3 + 1. In cell B4

How to Ensure Regular Flow of Money in Retirement Years

Business Day
November 6, 2012

You need regular flow of money during your retirement years. One of the ways to get it is to invest your retirement funds with annuity providers who make monthly payments for a pre-defined number of years. The insurance company invests the money and investors, in turn, are paid at regular intervals.

There are two types of annuities - immediate and deferred. In immediate annuity, the person invests a lump sum and the insurance company starts paying back immediately.

Immediate annuities are suitable for investors who have retired or are nearing retirement, those needing steady income from the money saved and those who are worried about outliving their savings. In deferred annuity, a person invests systematically for a number of years and allows the investments to grow with time, without attracting any tax.

type the formula (1 + $B$1/12)*B3 + $A$1. This formula is the key to the spreadsheet model. The dollar sign causes the referenced columns and rows to be fixed so that the spreadsheet does not change them when the formula is copied into other cells.

	A	B
1	200	0.06
2	Month	Balance
3	0	= A1
4	= A3 + 1	= (1 + $B$1/12)*B3 + $A$1
5		
6		

Complete the spreadsheet by copying the formulas in row 4 into as many rows as you like.

Annuities on a Graphing Calculator

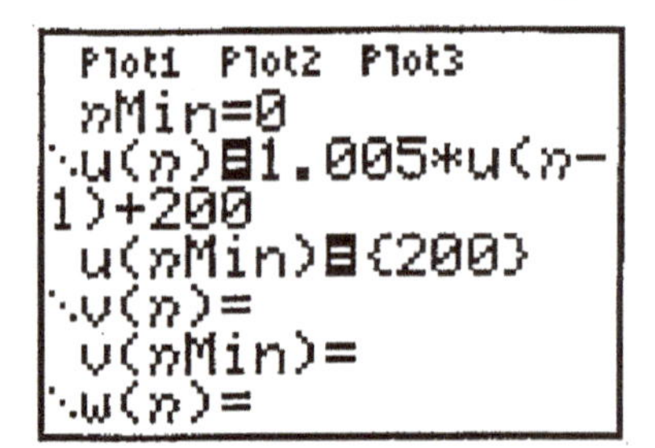

n	u(n)
0	200
1	401
2	603.01
3	806.02
4	1010.1
5	1215.1
6	1421.2

n=0

The screens above show how the annuity example in this lesson is modeled on one type of graphing calculator and the table that results. The table can be scrolled to any desired value. However, annuities are often active for many years, and scrolling to, say, the tenth year (120th month) can be tedious. This calculator allows calculation of any term on the home screen (below).

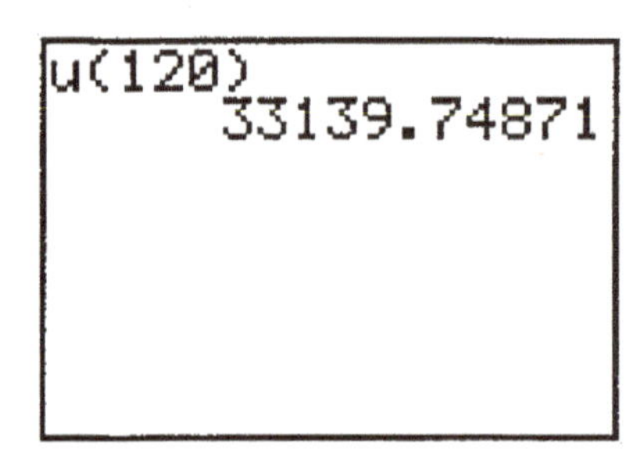

Exercises

1. Use the annuity example of this lesson with a monthly addition of $200 and an annual rate of 6% compounded monthly.

   a. Extend the table to 20 years (240 months). Record the balance at the end of 20 years.

   b. Determine the portion of the balance after 20 years that was paid into the account by its owner.

   c. Determine the portion of the balance after 20 years that was paid in interest.

   d. Compare the amount in the account after 30 years with the amount after 20 years.

   e. Change the interest rate to 8% compounded monthly and determine the amount in the account after 20 years. How does it compare with the balance after 20 years when the interest is 6%?

   f. Change the monthly addition to $300 and determine the amount in the account paying 6% after 20 years. How does the balance compare with the balance after 20 years in the 6% account with $200 monthly additions?

2. Below is a table similar to the one you made for the Towers of Hanoi puzzle.

Number of Disks	Number of Moves
1	1
2	3
3	7
4	15
5	31

The recurrence relation that describes these data is $M_n = 2M_{n-1} + 1$. You may have noticed that the number of moves always seems to be 1 smaller than a power of 2 and conjectured the closed-form formula $M_n = 2^n - 1$. In this exercise, assume that this is the correct formula.

   a. The original version of the puzzle was accompanied by a legend that at the beginning of time God created a temple in which a group of priests are working tirelessly to complete a 64-disk version of the puzzle. Find the number of moves necessary to complete the 64-disk version.

b. The legend says that the priests move one disk a second and work nonstop in shifts; when the puzzle is completed, God will end the world. Find the number of years from the beginning of creation until the end of the world according to this legend.

3. The closed-form formula for the Towers of Hanoi problem, $M_n = 2^n - 1$, can be proved by mathematical induction.

   a. What initial step is necessary?

   b. The proof assumes that the formula works for a puzzle with k disks and then attempts to prove the formula must also work for a puzzle with $k + 1$ disks. Rewrite the assumption and goal of the induction proof in terms of the formula.

   c. Complete the proof by applying the recurrence relation.

4. Marty had $2,000 in her savings account when she graduated from college and began work. She converted the savings account to an annuity into which she deposits $100 each month. She expects the annuity to earn 0.5% monthly.

   a. Complete this table.

n (in Months)	t_n
0	2,000
1	1.005(2,000) + 100 = 2,110
2	
3	

   b. Find the recurrence relation for the amount of money at the end of n months.

   c. Use your recurrence relation to find the amount in Marty's account at the end of the fourth month; at the end of the fifth month.

5. The Hamadas borrow $12,000 to buy a boat. The yearly interest on the loan is 8.4% and the monthly payment is $280.

   a. Complete the table.

t (in Months)	t_n
0	12,000
1	1.007(12,000) – 280 = 11,804
2	
3	

   b. Write a recurrence relation for the loan balance at the end of n months.

   c. Use a spreadsheet, calculator, or computer program to extend the table in part a until the loan is paid off. The values that you are calculating form what is known as an *amortization schedule*.

   d. Explore what happens to the Hamadas' loan if the interest rate and monthly payment remain the same but they borrow $40,000. What is the amount in the loan at the end of 1 month? At the end of 2 months? At the end of n months? What is the $40,000 called?

6. Newton's law of cooling says that over a fixed time period the change in temperature of an object is proportional to the difference between the temperature of the object and its surrounding environment. If t is the temperature of the object, s is the temperature of the surroundings, and a is the constant of proportion, then $t_n - t_{n-1} = a(t_{n-1} - s)$.

   A cup of cocoa is brewed to a temperature of 170°F. When set in a room with temperature 70°F, the temperature of the cocoa drops to 162°F in 1 minute.

   a. Write a recurrence relation for the temperature of the cocoa after n minutes.

   b. Simplify the recurrence relation you wrote in part a to the form $t_n = at_{n-1} + b$.

   c. Use the recurrence relation to find the temperature of the hot cocoa after 2 minutes.

   d. Find the fixed point for this recurrence relation. What is the significance of the fixed point in this situation?

7. To help him finish his final year of college, Sam took out a loan of \$12,000. At the end of the first year after he graduated, there was an \$11,200 balance, and at the end of the second year, \$10,320 remained. The amount of money left at the end of n years can be modeled by the mixed recurrence relation $t_n = at_{n-1} + b$.

   a. The information stated above is summarized in the following table.

n	t_n
0	12,000
1	11,200
2	10,320

   Use the general form of a mixed recurrence relation and the data in the table to write a system of equations. Solve for a and b. What is the recurrence relation for the amount of money in Sam's account after n years?

   b. What is the balance owed on the loan at the end of the third year? At the end of the fourth year?

   c. What is the rate of interest on this loan?

   d. Find the fixed point for this recurrence relation. What is the significance of this amount of money?

8. Suppose a college's tuition over the past 3 years has risen from \$12,000 to \$12,900, to the present cost of \$13,845. Use a mixed recurrence relation of the form $t_n = at_{n-1} + b$ to predict next year's tuition.

9. A virus is spreading through Central High. In the following table, n represents a given period of time, and t_n represents the number of people exposed to the virus at the end of the time period.

n	t_n
5	500
6	750
7	900
8	990
9	1,044

a. Write a recurrence relation that describes these data.

b. Use the recurrence relation to find the number of people exposed to the virus at the end of time period 10. At the end of time period 4.

10. Models for exposure to disease often assume that the number of people exposed during a given time interval is directly proportional to the number of people not yet exposed at the beginning of the time interval. In other words, if t_n represents the number of people exposed during time period n, P represents the total population, and k is the constant of proportion, then $t_n - t_{n-1} = k(P - t_{n-1})$.

    a. Find values of k and P that show this recurrence relation is equivalent to the one you found in Exercise 9.

    b. What is the fixed point for your recurrence relation in Exercise 9? What is the significance of the fixed point in this situation?

Flu is Widespread in 11 States

WebMD
January 14, 2011

While only four states have high levels of flu activity, the seasonal illness is widespread across 11 states. And there's evidence of outbreaks in several U.S. cities.

Four children died of flu in the first week of January, the latest week for which CDC data is available. So far, the 2010-2011 flu season has claimed the lives of eight children. There were 282 U.S. pediatric deaths during last year's flu pandemic, 133 in the 2008-2009 season, and 88 in the 2007-2008 season.

Hospitals in 122 U.S. cities report that deaths from "pneumonia and influenza" -- a statistic that reflects flu activity -- are at the epidemic level. They've been on the upswing since just before Christmas.

While the CDC data show that flu definitely has arrived, they do not predict where, when, or even whether the disease will flare up or die down. In states with high flu activity, flu may be limited to one or two large cities. And in states with widespread flu activity, there may be few if any hot spots.

11. The terms of a recurrence relation can be graphed on a rectangular coordinate system in which term numbers are placed on the x-axis and the values of the terms on the y-axis.

    a. Prepare such a graph for the mixed recurrence relation $t_n = 0.5t_{n-1} + 1$ with $t_1 = 4$. Show at least the first ten terms.

    b. What does the graph tell you about the long-term behavior of the recurrence relation?

Computer/Calculator Explorations

12. People often use annuities to save for a goal. Consider someone who wants to have $10,000 at the end of 5 years and expects to earn 6% annual interest compounded monthly. What monthly deposit should this person make? Use a spreadsheet to experiment with different monthly deposits until you find one that does the job and report to your class on the results.

13. Automobile manufacturers occasionally have promotions in which the customer is offered a choice between low-interest financing and a cash rebate. Select a promotion of that type that has recently been offered in your area (you may have to call a few dealerships for the information) and a car to which the promotion applies.

    a. Construct a spreadsheet that compares the two options. In other words, track both the low-interest loan on the car's value after the required down payment and a regular loan on the car's value after the rebate and the required down payment. Which is the better choice?

    b. Construct a spreadsheet that compares taking the low-interest loan with paying cash less the rebate for the car. Is it better to take the loan and put the cash in an account, such as a certificate of deposit that pays a relatively high interest rate, or to pay cash for the car to get the rebate? (Check with a bank to get current interest rates.)

14. Conduct an investigation into the advantages and disadvantages of short-term versus long-term loans for a car or a house. For example, compare the payments and total cost of a 30-year home loan versus a 15-year home loan.

15. Mortgage companies sometimes offer a home-loan repayment plan with payments every four weeks instead of monthly, resulting in a shorter loan term. Contact mortgage companies in your area for details of such a plan. Compare the total cost of both plans for a typical home loan.

Lesson 8.5

Mixed Recursion, Part 2

The previous lesson considers several real-world situations that can be modeled with mixed recurrence relations. The long duration of some applications such as annuities often means that tables of several hundred rows must be created. Although calculators and computers make construction of large tables feasible, a general closed-form solution for mixed recurrence relations could further decrease the time required to answer important questions. This lesson develops the closed-form solution for mixed recurrence relations.

As an example, consider the mixed recurrence relation $t_n = 2t_{n-1} - 3$ with $t_1 = 4$. Recall that the fixed point of a recurrence relation can be found by equating t_n and t_{n-1}. In this case, solving the equation $x = 2x - 3$ gives a fixed point of 3. Now consider a table of the first four terms of this recurrence relation. Notice, as shown in the third column, that subtracting the fixed point from each term produces a sequence of powers of 2.

n	t_n	$t_n - 3$
1	4	1
2	5	2
3	7	4
4	11	8

Therefore, the closed-form solution is $t_n = 2^{n-1} + 3$.

Drawing a conclusion from a single example is unwise, so consider the same recurrence relation, but with $t_1 = 8$. The following table shows the first four terms.

n	t_n	$t_n - 3$
1	8	5
2	13	10
3	23	20
4	43	40

Compare the third column of this table with the third column of the previous table. The powers of 2 are there, but each has a multiplier of 5, which is the difference between the first term and the fixed point. Therefore, the closed-form solution is $t_n = 5(2^{n-1}) + 3$. Keep in mind that 5 is the difference between the first term and the fixed point, 2 is the multiplier in the original recurrence relation, and 3 is the fixed point.

If there is uncertainty about the validity of a closed-form formula, mathematical induction can be used to prove that the formula is correct. Consider the closed-form formula $t_n = 5(2^{n-1}) + 3$.

First, establish a base case. That is, be sure that the formula produces the correct first term: $5(2^{1-1}) + 3 = 5(1) + 3 = 8$.

Now you must show that if $t_n = 5(2^{n-1}) + 3$ generates the nth term, then $t_{n+1} = 5(2^{(n+1)-1}) + 3 = 5(2^n) + 3$ generates the $n + 1$st term.

The term t_{n+1} is generated by multiplying the previous term by 2 and subtracting 3: $t_{n+1} = 2t_n - 3$, but $t_n = 5(2^{n-1}) + 3$, so by substitution $t_{n+1} = 2[5(2^{n-1}) + 3] - 3$. Simplify this expression:

$$2[5(2^{n-1}) + 3] - 3$$
$$= 5(2)(2^{n-1}) + (3)2 - 3$$
$$= 5(2^n) + 6 - 3$$
$$= 5(2^n) + 3.$$

Thus, mathematical induction guarantees that $t_n = 5(2^{n-1}) + 3$ generates all the terms of $t_n = 2t_{n-1} - 3$ with $t_1 = 8$.

It is tempting to base a general conclusion on the previous example: a closed-form solution for a recurrence relation of the type $t_n = at_{n-1} + b$ is $t_n = (t_1 - p)(a^{n-1}) + p$, where p, the fixed point, is $\frac{b}{1-a}$, which is obtained by solving the equation $x = ax + b$. However, the previous induction proof applies only to one specific recurrence relation and, therefore, does not establish a general formula. You will do that in this lesson's exercises.

An Annuity Example

Consider the annuity example of Lesson 8.4 (see "A Mixed Recursion Example" on page 479). The account pays 6% annual interest compounded monthly and has monthly deposits of \$200. The recurrence relation for the balance at the end of the nth month is

$$B_n = \left(1 + \frac{0.06}{12}\right)B_{n-1} + 200 \text{ or}$$

$$B_n = 1.005\, B_{n-1} + 200.$$

The following table tracks the account for the first few months.

Month	Balance
0	200
1	1.005(200) + 200 = 401
2	1.005(401) + 200 = 603.01
3	1.005(603.01) + 200 = 806.03

Note that $t_0 = 200$, $t_1 = 401$, $a = 1.005$, and $b = 200$. Therefore, the fixed point is $\frac{200}{1-1.005} = -40{,}000$.

Substituting for t_1, a, and b in the general closed-form solution for mixed recurrence relations, $t_n = (t_1 - p)(a^{n-1}) + p$, gives

$$t_n = (401 + 40{,}000)(1.005^{n-1}) - 40{,}000, \text{ or}$$

$$t_n = 40{,}401(1.005^{n-1}) - 40{,}000.$$

Determining the amount in the account after, say, 20 years requires evaluating the closed-form solution for $n = 240$.

$$40{,}401(1.005^{240-1}) - 40{,}000 = \$93{,}070.22$$

Retirement Planning a Must for Young Workers

Houston Chronicle
May 25, 2003

At the height of the technology boom in 1999, Jessica Lontz was earning \$150,000 as a recruiter for the Philadelphia area's tech sector.

"I was great at throwing money down the drain," said Lontz, a Philadelphia resident who is now 27 and making a third of what she earned in 1999. "In retrospect, I should have saved a ton more than I did. It's almost comical now that I look back at it."

In terms of saving for retirement, Lontz is typical of workers in their 20s. According to a Fidelity Investments report published in 2000, less than half of workers in their 20s participated in their company's retirement savings plan.

"The longer it takes for you to get started, the less you'll wind up with in retirement income," said Jack VanDerhei, an expert on 401(k) plans and a professor at Temple University's Fox School of Business and Management.

This is because earnings on money invested annually in a tax-deferred account are compounded. For example, if a 25-year-old starts saving \$1,000 a year for 40 years in an account with an average annual return of 7 percent, he would have about \$214,000 by age 65, according to Vanguard Group, the mutual-fund giant in Malvern, PA.

Waiting until age 35 to start saving \$1,000 annually, the same person would have amassed about \$100,000 by 65—less than half what he could have had if he had started saving just 10 years earlier, Vanguard said.

A commonly asked question about an annuity involves the time required for it to reach a certain amount. For example, to determine when this annuity reaches \$200,000, solve the equation $200{,}000 = 40{,}401(1.005^{n-1}) - 40{,}000$. Doing so requires a little algebra.

$$40{,}401(1.005^{n-1}) - 40{,}000 = 200{,}000$$

$$40{,}401(1.005^{n-1}) = 200{,}000 + 40{,}000$$

$$(1.005^{n-1}) = \frac{240{,}000}{40{,}401}$$

$$(n-1)\log(1.005) = \log\left(\frac{240{,}000}{40{,}401}\right)$$

$$n - 1 = \frac{\log\left(\frac{240{,}000}{40{,}401}\right)}{\log(1.005)}$$

$$n = \frac{\log\left(\frac{240{,}000}{40{,}401}\right)}{\log(1.005)} + 1$$

The solution can be evaluated on a calculator to obtain about 358 months, or just under 30 years.

Technology Note

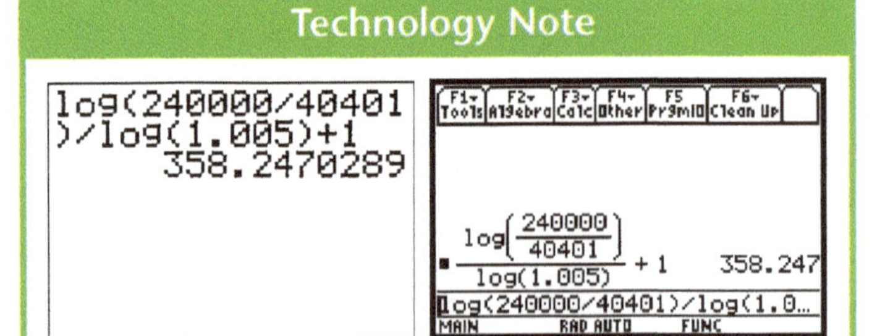

Evaluation of the example's solution on two graphing calculators. The expression is typed the same way on both calculators, but the second displays it the way it is usually written.

In the following exercises, the general closed form for mixed recurrence relations is used to model situations similar to those in Lesson 8.4. Use the closed form to solve the problems, and use the recurrence relation together with either a spreadsheet or calculator to check your answers.

Exercises

1. Find the fixed point and the closed form for each of these recurrence relations. Use the closed form to find the 100th term.

   a. $t_1 = 1, t_n = 2t_{n-1} + 3$

   b. $t_1 = 5, t_n = 3t_{n-1} - 7$

   c. $t_1 = 2, t_n = 4t_{n-1} - 5$

   d. $t_1 = -4, t_n = t_{n-1} + 2$

2. An annuity plan pays an annual rate of 8% compounded monthly and includes monthly additions of $150.

   a. Write the recurrence relation for B_n, the balance at the end of the nth month.

   b. Use your recurrence relation to build a table showing B_n for $n = 0, 1, 2, 3, 4$.

   c. Find the fixed point for the recurrence relation.

   d. Use the fixed point and the value of B_1 to write the closed form.

   e. Use the closed form to find the account balance at the end of 30 years.

   f. Use the closed form to determine the amount of time that it takes the account to grow to $500,000.

   g. Suppose the owner of the account wants it to reach $500,000 in 30 years. What monthly additions are required? (Hint: You must find the value of b in $B_n = \left(1 + \frac{0.08}{12}\right)B_{n-1} + b$. Write an expression for the fixed point, leaving b as an unknown. Write the appropriate closed form, set it equal to 500,000 when $n = 360$, and solve for b.)

3. Jilian borrows $12,000 to buy a car. The annual interest rate is 12% compounded monthly and the monthly payments are $260.

   a. Write a recurrence relation for the unpaid balance at the end of the nth month.

   b. Use the recurrence relation to tabulate the unpaid balance at the end of months 0, 1, 2, 3, and 4.

c. Find the fixed point for the recurrence relation.

d. Find the closed form.

e. Use the closed form to determine the unpaid balance at the end of 2 years.

f. Use the closed form to determine the amount of time needed to repay the loan. That is, set the closed form equal to 0 and solve for n. Round your answer to the nearest whole number of months.

g. Multiply your previous answer by the monthly payment and determine the amount Jilian really pays for her car. What is the total amount of interest that she pays?

h. Suppose Jilian wants to pay for the car in 3 years. What is her monthly payment? (See the hint in part g of Exercise 2.)

i. Interpret the fixed point you found in part c. That is, what is the meaning of the fixed point's dollar value in Jilian's situation?

4. In this lesson the closed-form solution for $t_n = at_{n-1} + b$ is given as $t_n = (t_1 - p)(a^{n-1}) + p$, where p is the fixed point $\frac{b}{1-a}$. Mathematical induction was used to prove this formula, but only for a specific case. This exercise uses mathematical induction to prove that this closed form is correct for all mixed recurrence relations.

a. To begin, verify that the closed form works for t_1: Replace n with 1 in $(t_1 - p)(a^{n-1}) + p$ and show that this really is t_1.

b. The next step is to show that whenever the closed form generates the correct value of t_n, it also generates the correct value of t_{n+1}. To begin this process, write the closed form for t_n and for t_{n+1}.

c. In a mixed recurrence relation, a term is generated by multiplying the previous term by a and adding b. Generate the $(n + 1)$th term by multiplying the closed form for the nth term by a and adding b.

d. Algebraically simplify the previous expression until it matches the closed form for the $(n + 1)$th term. (Hint: You might find it helpful to replace the second occurrence of the fixed point with $\frac{b}{1-a}$, but not the first.)

5. In Exercise 6 from Lesson 8.4 (page 483) you used Newton's law of cooling to model a cup of cocoa that drops from 170°F to 162°F in 1 minute in a room with temperature 70°F.

   a. Rewrite the recurrence relation and recalculate the fixed point for the recurrence relation.

   b. Write the closed form for the temperature of the cocoa after n minutes.

   c. Use the closed form to find the temperature of the cocoa after 5 minutes.

   d. Use the recurrence relation to create a table showing the temperature each minute through the first 5 minutes. Compare the last entry of the table with your answer to part c.

   e. Would it make sense to try to use the closed form to determine the time needed for the cocoa to freeze? Try to do so. What happens?

6. The following data represent the number of people at Central High who have heard a rumor.

Number of Hours after Rumor Began	Number of People Who Have Heard It
1	80
2	240
3	320
4	360

   a. Write a recurrence relation for the number of people who have heard the rumor after n hours.

   b. Find the fixed point for the recurrence relation.

   c. Use the fixed point to write the closed form.

   d. Use the closed form to determine the number of people who have heard the rumor after 10 hours.

   e. The way in which rumors and other information spread through a population is similar to the way in which disease spreads through a population. Compare this exercise with Exercises 9 and 10 of Lesson 8.4 (pages 484–485) and explain the significance of the fixed point you found in part c.

7. People often express surprise at figures like those in the news article on page 489.

   a. Check the claim in the article for a 25-year-old person who saves \$1,000 a year. Find the account balance when the person is 65 and explain how you used your knowledge of recurrence relations.

   b. Check the claim for the person who starts saving at age 35. Be sure to explain how you used your knowledge of recurrence relations.

   c. If a person has saved \$214,000 for retirement, what can the person withdraw each month without decreasing the account balance if the account is earning 8% annual interest compounded monthly when the person retires?

8. Sasha is 21 and just graduated from college. She is considering two plans for saving for retirement after age 65.

   - Plan A: Save \$2,000 a year for 5 years starting now. Make no more payments after 5 years, but leave the money in the account.
   - Plan B: Wait until age 45, then save \$2,000 a year until retirement.

   Sasha thinks that by managing the investment carefully, she can average 9% a year interest. She is leaning toward Plan B because she thinks it yields more money at retirement and because it allows her to have more spending money now. What do you think? Explain.

9. The fixed point of a mixed recurrence relation cannot be calculated if $a = 1$ because the denominator of $\frac{b}{1-a}$ is 0. Discuss what to do when this happens.

Modeling Projects

10. Design your own savings plan. State the age at which you plan to start saving, the age at which you would like to retire, and the amount of money you would like to have in your account on retirement. Contact investment specialists in your area to determine an approximate rate of return on current annuities. Determine the amount you need to pay into the account on a regular basis to achieve your goal. Discuss the amounts you can withdraw from the account to meet regular expenses without depleting the account and if the account is depleted gradually over a reasonable life expectancy. (Keep a copy of the results. One day you will be glad you took this course.)

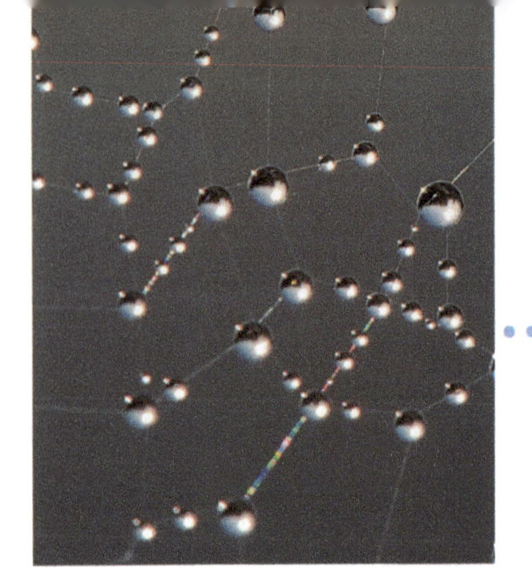

Lesson 8.6

Cobweb Diagrams

In the last two lessons of this chapter you analyzed mixed recurrence relations numerically and algebraically. Your understanding of these recurrence relations will benefit from a visual model for their behavior, which is the topic of this lesson.

As an example, consider the recurrence relation $t_n = 2t_{n-1} - 1$ with $t_1 = 3$. The first four terms are shown in the following table.

n	t_n
1	3
2	5
3	9
4	17

Although it may seem obvious, it is important to realize that a given term can be thought of as t_n or t_{n-1}. The first term 3, for example, is t_n if you are thinking of n as 1, but it is t_{n-1} if you are thinking of n as 2. As you build a table of values for a recurrence relation, each term first takes on the role of t_n, then the role of t_{n-1}, and then fades into history, something like an officer of an organization who serves a term as president, then a term as past president, and finally disappears from office. This succession of terms can be visualized by graphing the recurrence relation $t_n = 2t_{n-1}$ as separate functions $y = t_n$ and $y = 2t_{n-1} - 1$, or $y = x$ and $y = 2x - 1$. Associate the x-axis with t_n and the y-axis with t_{n-1}.

The following algorithm describes the process of graphing the generation of successive terms.

1. Graph the lines $y = x$ and $y = 2x - 1$.
2. Locate the first term of the recurrence relation on the x-axis.
3. Draw a vertical line from the first term to the line $y = 2x - 1$. You can think of this step as representing the substitution of the first term into $2t_{n-1} - 1$ in order to generate the second term.
4. Draw a horizontal line from $y = 2x - 1$ to $y = x$. You can think of this line as representing the term's transition from the role of t_n to the role of t_{n-1} in preparation for generation of the next term.
5. Draw a vertical line upward from $y = x$ to $y = 2x - 1$.
6. Repeat steps 4 and 5 for as many terms as desired.

The graph that results is sometimes called a cobweb diagram (see Figure 8.4).

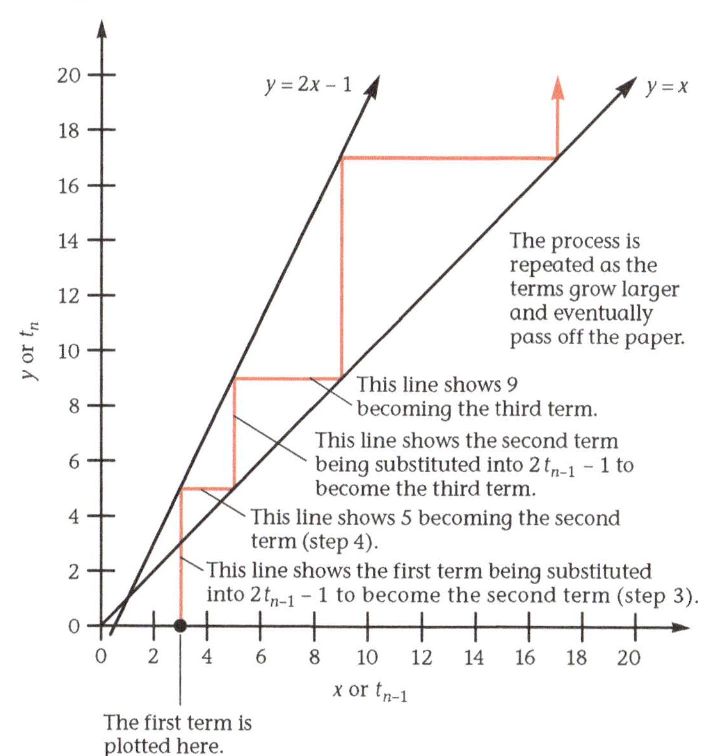

Figure 8.4. Cobweb diagram for the recurrence relation $t_n = 2t_{n-1} - 1$ with $t_1 = 3$.

Note that the intersection of the lines $y = x$ and $y = 2x - 1$ represents the fixed point. If the first term is 1, the first vertical line segment hits the point of intersection and the process can go nowhere.

Cobwebs on Graphing Calculators Without Recursion Features

The following is a generic cobweb diagram algorithm that can be adapted to any graphing calculator.

1. Set a suitable graphing range.
2. Graph the line $y = x$ as the calculator's first function ($Y1$) and the line whose equation is found by replacing t_n with y and t_{n-1} with x in the recurrence relation as the calculator's second function ($Y2$). (Be sure all other functions and plots are turned off.)
3. Input the value of the initial point for variable A.
4. Replace X with the value of A.
5. Draw a line from the point $(X, 0)$ to $(X, Y2)$.
6. Pause and wait for the user to press ENTER.
7. Draw a line from $(X, Y2)$ to $(Y2, Y2)$.
8. Pause and wait for the user to press ENTER.
9. Replace X with the current value of $Y2$.
10. Draw a line from (X, X) to $(X, Y2)$.
11. Pause and wait for the user to press ENTER.
12. Go to Step 7.

The first three steps can be included in the program or performed before the program is run. The specific commands to implement the algorithm vary with the calculator model. Consult your calculator manual or talk with someone knowledgeable about your calculator's programming features if you are unsure of what to do.

Cobwebs on Graphing Calculators with Recursion Features

Some graphing calculators have a special graphing mode for cobweb diagrams. With the calculator in the sequence mode, the recurrence relation is entered (left screen). After the calculator is placed in a special

```
Plot1 Plot2 Plot3
nMin=1
\u(n)=2u(n-1)-1
 u(nMin)={3}
\v(n)=
 v(nMin)=
\w(n)=
 w(nMin)=
```

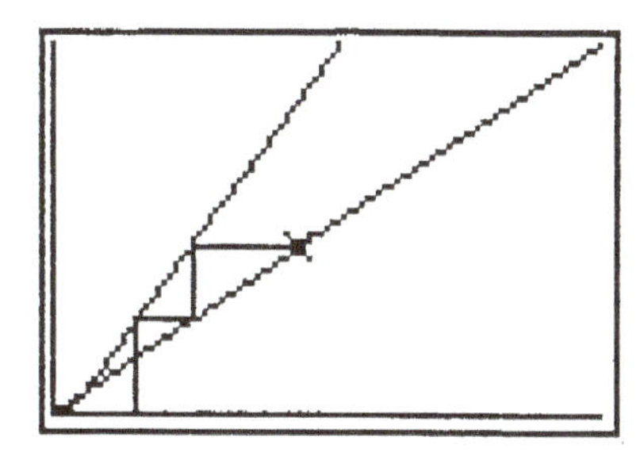

web graphing mode, it draws the line $y = x$ and a line representing the recurrence relation (right screen). A segment of the cobweb is drawn each time the user presses a designated key.

The behavior exhibited in this lesson's cobweb diagram example is not the only kind of behavior that can occur. The following exercises explore other kinds of behavior.

Exercises

1. Construct a cobweb diagram for the indicated number of terms of each recurrence relation. Find all the terms before beginning the graph so that you can choose a suitable scale for the axes.

   a. $t_n = 3t_{n-1} - 8$ with $t_1 = 5$, four terms.

   b. $t_n = 5 - t_{n-1}$ with $t_1 = 3$, four terms.

   c. $t_n = 9 - 0.5t_{n-1}$ with $t_1 = 2$, four terms.

   d. $t_n = 0.5t_{n-1} + 6$ with $t_1 = 2$, four terms.

2. Find the fixed point for each of the recurrence relations in Exercise 1 and mark the fixed point on your cobweb diagram. Mathematicians categorize some fixed points as *repelling*, and others as *attracting*. Which fixed points in Exercise 1 do you think are attracting. Which are repelling? Which appear to be neither?

3. Experiment with mixed recurrence relations $t_n = at_{n-1} + b$. Try various values of a and b. When is a fixed point attracting and when is it repelling? When is it neither?

4. The following figure is a cobweb diagram for a recurrence relation.

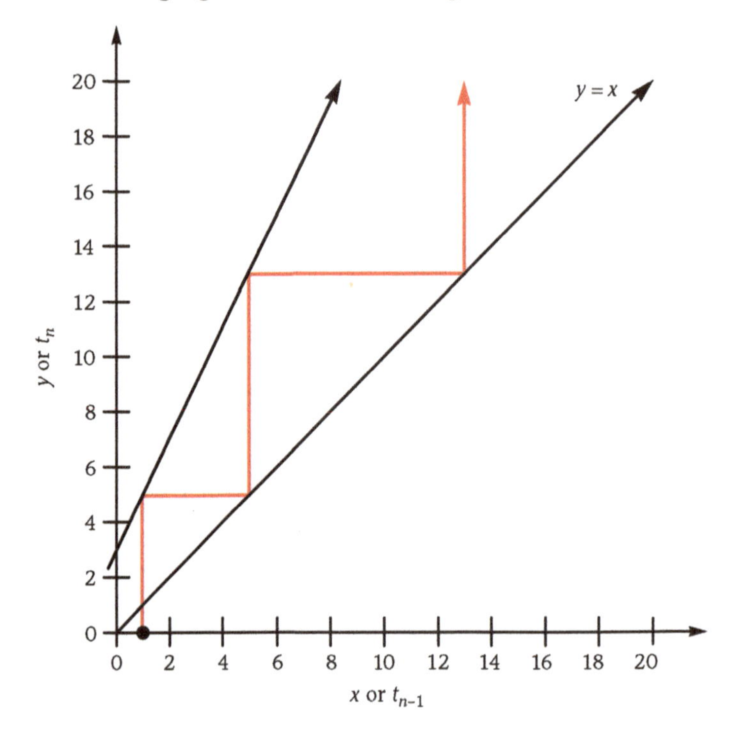

a. What is the first term of the recurrence relation?

b. How many terms can be determined from the diagram?

c. Make a table showing all the terms that can be read from the diagram.

d. Write the recurrence relation.

e. Find the fixed point of the recurrence relation from the graph. Show how algebra can be used to find the fixed point.

5. The deer population, estimated at 12,000, in a region has grown so large that the deer are becoming pests. Wildlife biologists estimate that the population is growing at a rate of 4% a year. Officials want to issue a sufficient number of hunting permits so that the population is decreased to 10,000 over the next 10 years.

a. Recommend a number of permits that accomplishes this goal.

b. If officials want to hold the population constant when it reaches 10,000, how many permits should they issue?

c. Do you think this is a reasonable plan? Explain.

6. Consider the recurrence relation $t_n = 4 - 0.5(t_{n-1})^2$ with $t_1 = 0$. This recurrence relation is called *second degree* because squaring is used in the calculation of a term from the previous term.

   a. Make a table showing the first 6 terms.

   b. Draw a cobweb diagram displaying the first 6 terms.

   c. Change the first term to 1. Use a spreadsheet or calculator to explore the behavior of the first 20 terms. Also explore the related cobweb diagram.

   d. Change the first term to 2. What happens?

   e. Change the first term to 5. What happens?

   f. Use algebra to find the fixed points. (Because this recurrence relation is second degree, it has two.)

   g. You have tried this recurrence relation with first terms of 0, 1, 2, and 5. In which cases were the terms attracted to a fixed point and in which cases were they repelled? Were there any cases in which neither seemed to happen or in which it wasn't possible to tell?

7. A model that is sometimes used to represent the growth of a population in an environment that is capable of supporting only a limited number (the environment's carrying capacity) of the species says that if the uninhibited growth rate of the population is r (in decimal form) and the maximum number the environment can support is m, then the recurrence relation that describes the total number of the species t_n in a given time period is

$$t_n = \left[1 + r\left(1 - \frac{t_{n-1}}{m}\right)\right]t_{n-1}.$$

An important idea reflected in this model is that the uninhibited growth rate is reduced as the population approaches the maximum number the environment can support.

Rangers Round Up Badlands Buffalo

KELO
September 21,2012

The herd is ready to run at Custer State Park ahead of the annual Buffalo Roundup, which starts Monday. Preparations at the park are well underway for the animals and the visitors.

Around 1,200 bison call Custer State Park home. More than a thousand of them will wow spectators at the 47th Annual Buffalo Roundup.

Last year's roundup drew in more than 14,000 spectators. And with numbers like that it's easy to forget that the roundup's real purpose is herd management.

Park staff will cull hundreds of bison from the herd in the weeks after the roundup for sale at the buffalo auction this November.

Over the past 47 years, the auction has played a huge role in revitalizing the overall buffalo population.

The park was planning on selling fewer buffalo in an effort to build up the herd's numbers, which is typically closer to 1,500 head. But that's changed due to the summer's drought conditions.

The population of a particular species of animal has an annual growth rate of 10%, and the environment is capable of supporting 10 (in thousands) of the animals.

a. Write the recurrence relation for the number of animals after n years.

b. Use a table or a cobweb diagram to explore the growth of the population if it currently is 5,000. Describe the results.

c. Use your table or cobweb diagram to experiment with different initial populations. Be sure to include one that is over 10. Describe the results.

d. What are the fixed points of the recurrence relation? Explain their significance in this situation.

Computer/Calculator Explorations

8. A fractal related to the snowflake curve (see Exercise 20 in Lesson 8.3 on page 475) is called a Sierpinski gasket. Like the snowflake curve, it is based on an equilateral triangle. It is constructed by connecting the midpoints of the sides of the triangle and removing the triangle that results. The same process is applied to the remaining triangles, and so forth. The first three stages follow.

Use a computer drawing utility or Logo to develop a recursive procedure that draws several of these figures. Investigate their properties. What, for example, can you conclude about the areas and perimeters of these figures?

Projects

9. The *order* of a recurrence relation is the difference between the highest and lowest subscripts. Nearly all the recurrence relations in this chapter are first order. The recurrence relation $t_n = t_{n-1} + t_{n-2}$ is second order because the difference between n and $n - 2$ is 2. Investigate some recurrence relations of order higher than 1. Include a few applications of these recurrence relations in your report.

10. A recent mathematical topic related to mixed recurrence relations is called *chaos*. Prepare a report on chaos theory. Include the role of mixed recurrence relations and a few applications in your report.

Chapter Extension

Fractal Dimensions

Fractal art is familiar to almost everyone. Its appealing images adorn t-shirts, posters, company logos, and advertisements. But fractals often do their work discreetly: They are used to create realistic landscapes for movies, to compress data for storage in computers, to estimate the lengths of coastlines, and to create models of biological structures.

A fractal is a figure whose dimension is not a whole number. This fact seems strange to anyone who has never thought about the possibility of a fractional dimension. To understand the concept, it's necessary to think about some familiar objects in a slightly different way.

Consider a line segment, a square, and a cube. Their dimensions are 1, 2, and 3, respectively. Subdivide their sides into, say, 3 equal parts (see Figure 8.5).

Each division forms several objects similar in shape to the original: the segment is divided into 3 segments, the square is divided into 9 squares, and the cube is divided into 27 cubes. The number of similar pieces can be counted by raising 3 to a power: $3^1 = 3$ segments, $3^2 = 9$ squares, and $3^3 = 27$ cubes. The exponent equals the figure's dimension.

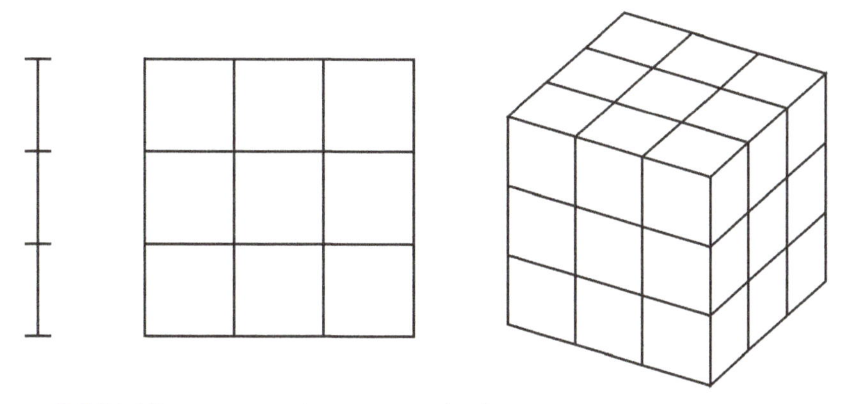

Figure 8.5 Dividing a segment, square, and cube.

Consider a similar analysis of the snowflake curve. The first three steps of its construction are shown in Figure 8.6.

Figure 8.6 The first three steps of the snowflake curve construction.

The second stage is constructed by dividing the first stage into three equal pieces, eliminating one of them, and adding two new ones. The second stage has four parts similar to the first stage; the same can be said of the third compared to the second. The fractal dimension is the power d of 3 that equals 4: $3^d = 4$. Solving for d gives the fractal dimension of the snowflake curve.

$$3^d = 4$$

$$d \log 3 = \log 4$$

$$d = \frac{\log 4}{\log 3}$$

The snowflake curve's dimension is approximately 1.26.

This procedure works for the snowflake curve because it contains copies of itself, in other words, because it is self-similar. To find the fractal dimension of an irregular object such as a coastline, a different procedure is used.

Place a large square over a map of the coastline. Bisect the sides of the square and draw lines dividing the square into 4 smaller squares. Count the number of smaller squares that the coastline intersects. Find the quotient of the log of the number of small squares the coastline intersects and the log of the number of segments into which the sides of the original square are divided (2 in this case).

The process is repeated with a larger number of divisions of the sides of the original square. The ratio

$$\frac{\log(\text{number of squares containing coastline})}{\log(\text{number of segments})}$$

approaches the fractal dimension of the coastline as the number of divisions increases. (Coastline dimensions are usually between 1.15 and 1.25.) Figure 8.7 shows one step in the application of the method.

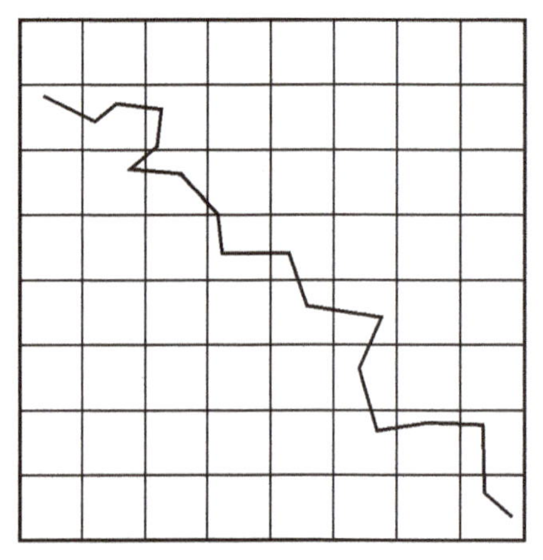

The large square is divided into $8 \times 8 = 64$ smaller squares. The shape passes through 15 of the smaller squares. Therefore, the quotient is $\frac{\log 15}{\log 8} \oplus 1.30$.

Figure 8.7. A step in the calculation of a fractal dimension.

Fractal Dimension Analysis Aids Breast Cancer Prognosis

Science Daily
April 25, 2011

Cancer researchers at the University of Calgary are investigating a new tool to use for the prognosis of breast cancer in patients. This new digital tool will help give patients a more accurate assessment of how abnormal and aggressive their cancer is and help doctors recommend the best treatment options.

Currently, a useful factor for deciding the best treatment strategy for early-stage breast cancer is tumor grade, a score assigned by a pathologist based on how abnormal cancer cells from a patient tissue sample look under the microscope. However, tumor grade is somewhat subjective and can vary between pathologists. Hence, there is a need for more objective methods to assess cancer tissue, which could improve risk assessment and therapeutic decisions.

Using a mathematical computer program developed at the U of C, Mauro Tamabsco and his team used fractal dimension analysis to quantitatively assess the degree of abnormality and aggressiveness of breast cancer tumors obtained through biopsy. Fractal analysis of images of breast tissue specimens provides a numeric description of tumor growth patterns as a continuous number between 1 and 2. This number, the fractal dimension, is an objective and reproducible measure of the complexity of the tissue architecture of the biopsy specimen. The higher the number, the more abnormal the tissue is.

According to the team's published study, this novel method of analysis is more accurate and objective than pathological grade.

Chapter 8 Review

1. Write a summary of what you think are the important points of this chapter.
2. For each of the following, write a recurrence relation to describe the pattern, find a closed-form solution, and find the 100th term.
   a. 2, 6, 10, 14, . . .
   b. 3, 8, 23, 68, . . .
   c. 3, 6, 12, 24, . . .
   d. –1, 1, 5, 13, . . .
3. Which of the sequences in Exercise 2 are arithmetic? Which are geometric?
4. Find the fixed point for each of the following recurrence relations if one exists. Check your fixed point by using it and the recurrence relation to write the first four terms.
   a. $t_n = 5t_{n-1}$
   b. $t_n = 5t_{n-1} + 3$
   c. $t_n = t_{n-1} - 3$
   d. $t_n = 3 - 2t_{n-1}$
5. In Exercise 4, find a closed-form formula for each recurrence relation if the first term is 2. Use the closed form to find the 100th term.

6. Which of the recurrence relations in Exercise 4 are arithmetic? Which are geometric?

7. According to the United States Department of Agriculture, sales of conventional milk in the U. S. declined 1.8% in 2012 over 2011 sales, while sales of organic milk increased 3.0%. Assume that these trends continue indefinitely.

   a. Write a recurrence relation that describes the annual growth of the national organic dairy market.

   b. Write a recurrence relation that describes the annual growth of the conventional dairy market.

What Does "Organic Milk" Mean?

Consumer Reports News
September 22, 2010

Earlier this year, the Department of Agriculture clarified the amount of time cows that produce organic milk must spend grazing on grass. By July 2011, they must have year-round access to the outdoors, access to pasture during the grazing season, and a specified minimum intake from pasture grown without synthetic herbicides and pesticides. They also must not have been treated with hormones or antibiotics.

Can people taste the difference between organic and non-organic versions? To find out, the testers at Consumer Reports bought eight brands of 2 percent milk, four conventional and four organic; poured samples into cups; and had 112 employees taste all the milks. The staffers were no better than random guessers at telling which were organic. That's not surprising, because feed and pasteurization type are apt to play a big role in taste.

You can find out where the milk you buy comes from by going to www.whereismymilkfrom.com and typing in a code from the carton.

8. Consider the following sequence of squares.

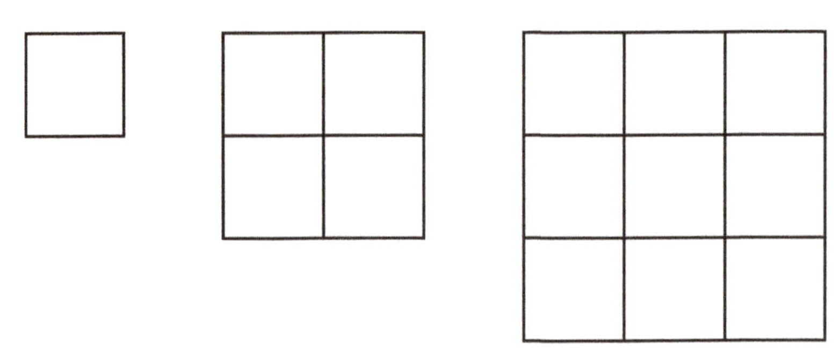

a. Let S_n represent the total number of squares of all sizes in the figure with sides n units long. For example, $S_2 = 5$ because there are 4 small squares and 1 large one in the second figure. Complete the following table by counting squares and drawing additional figures in the sequence.

n	S_n
1	1
2	5
3	
4	
5	

b. Add difference columns to your table.

c. Find a closed-form formula for S_n. Use your formula to find the total number of squares of all sizes on a checkerboard.

9. Use finite differences to determine the degree of the closed-form polynomial that was used to generate this sequence and show how to find the closed form by solving an appropriate system.

–5, –2, 3, 10, 19, 30, 43, . . .

10. The following table shows the number of gifts given on the nth day of Christmas and the total number of gifts given through the first n days as described in the song, "The Twelve Days of Christmas."

Day	Gifts on That Day	Total Number of Gifts
1	1	1
2	$1 + 2 = 3$	$1 + 3 = 4$
3	$1 + 2 + 3 = 6$	$4 + 6 = 10$
4		
5		
6		

a. Complete the table through the sixth day.

b. Write a recurrence relation for the number of gifts given on the nth day, G_n, and a recurrence relation for the total number of gifts given through the nth day, T_n.

c. Find a closed-form formula for G_n and for T_n.

11. In 2014 the cost of a first-class letter was \$0.49 for the first ounce and \$0.21 for each additional ounce or fraction thereof.

a. Write a recurrence relation to describe the amount of postage P_n on a letter that weighs between $n - 1$ and n ounces.

b. Find a closed-form formula for P_n.

12. Roberto deposits \$1,000 in an account paying 4.8% annual interest compounded monthly.

a. Complete the following table showing the balance in Roberto's account at the end of the first few months.

Month	Balance
0	\$1,000
1	
2	
3	

b. Write a recurrence relation for the balance in Roberto's account at the end of the nth month.

c. Find a closed-form formula for the balance at the end of the nth month.

d. Determine the number of years that it takes for the amount in Roberto's account to double.

13. Joan deposits \$5,000 in an annuity account to which she makes \$100 monthly additions. The account pays 6.4% annual interest compounded monthly.

    a. Complete the following table showing the balance in Joan's account at the end of the nth month.

Month	Balance
0	\$5,000
1	
2	
3	

    b. Write a recurrence relation for the amount in Joan's account at the end of the nth month.

    c. Find a closed-form formula for the amount in Joan's account at the end of the nth month.

    d. Find the balance in Joan's account at the end of 5 years.

    e. How long does it take Joan's account to grow to \$50,000?

14. A club at Central High wants to raise money by selling t-shirts with the school logo printed on them. The supplier of t-shirts charges a \$50 setup fee and \$3.50 per shirt.

    a. Write a recurrence relation that describes the cost (in dollars) of n shirts.

    b. Write a closed-form formula for the cost (in dollars) of n shirts.

    c. If the club orders \$100 worth of shirts, what is the cost per shirt?

15. Martha borrows \$11,000 to buy a car. Her loan carries an annual interest rate of 9.6% compounded monthly and her monthly payments are \$230.

    a. Write a recurrence relation for the unpaid loan balance at the end of the nth month.

    b. Write a closed-form formula for the unpaid balance at the end of the nth month.

    c. How long does it take Martha to pay for her car?

    d. What is the total amount of interest that Martha pays?

    e. If Martha wants to pay off the loan in 3 years, what monthly payments does she make?

16. A patient is injected with a radioactive dye for an x-ray procedure. The radioactivity measures 1,000 cpm (counts per minute) at the initial injection and declines by 90% each minute afterward.

    a. Write a recurrence relation for the amount of radioactivity after n minutes.

    b. Write a closed-form formula for the amount of radioactivity after n minutes.

    c. If the x-ray cannot be taken when the count is below 500 cpm, during how many minutes can the x-ray be taken?

17. The following data show the change in value of a painting over a period of years.

Year	Value
1	\$8,000
2	\$16,000
3	\$28,000
4	\$46,000

    a. Write a recurrence relation to describe the value of the painting after n years.

    b. Find a closed-form formula for the value of the painting after n years.

18. Create a cobweb diagram that displays the first four terms of $t_n = 0.5t_{n-1} + 12$ with $t_1 = 12$.

19. The absorption and elimination of medicine by the human body can be modeled with a mixed recurrence relation. Suppose a medication has a recommended dosage of 500 milligrams every 4 hours and that a person's body eliminates about 60% of the medication in a 4-hour period. (Actual rates vary with the medication and the individual's weight and metabolism.)

    a. Write a recurrence relation to model the amount of the medication in the person's body after n 4-hour periods.

    b. Explore the behavior of the medication over several days. Describe your findings.

c. On the basis of your answer to part b, describe what you expect to find in a cobweb diagram. Construct one to confirm your answer.

d. Doctors sometimes advise a patient to take a double dose the first time they take a medication. On the basis of your analysis in parts a to c, why do you think that is?

20. Swimming pools require chlorine to inhibit the growth of algae and bacteria. But too much chlorine is not healthy for swimmers. The recommended concentration is between 1 and 3 ppm (parts per million). Reaction to sunlight and other factors cause chlorine to dissipate.

a. A pool currently has a chlorine concentration of 2 ppm. Intense sunlight causes 20% of the chlorine to be lost each day. To compensate, the pool owner adds 1 ppm per day. Write a recurrence relation that describes the concentration after n days.

b. What happens to the chlorine concentration in part a if the owner continues to add 1 ppm daily over many days?

c. Recommend a daily addition for the pool in part a. Explain your thinking.

Bibliography

Albright, Brian. 2009. *Mathematical Modeling with Excel.* Jones & Bartlett Publishers.

Arganbright, Deane. 1985. *Mathematical Applications of Electronic Spreadsheets.* New York: McGraw-Hill.

Arganbright, Deane, and Erich Neuwirth. 2003. *The Active Modeler: Mathematical Modeling with Microsoft Excel.* Pacific Grove, CA: Duxbury Press.

Cannon, Lawrence O., and Joe Elich. 1993. "Some Pleasures and Perils of Iteration." *Mathematics Teacher*, March, pp. 233–239.

Davis, Morton D. 2001. *The Math of Money: Making Mathematical Sense of Your Personal Finances*. New York: Copernicus Books.

Devaney, Robert L. 1990. *Chaos, Fractals, and Dynamics: Computer Experiments in Mathematics.* Menlo Park, CA: Addison Wesley.

Feldman, David P. 2012. *Chaos and Fractals: An Elementary Introduction.* New York: Oxford University Press.

Gardner, Martin. 1995. *New Mathematical Diversions: More Puzzles, Problems, Games, and Other Mathematical Diversions.* Rev. ed. Washington, DC: Mathematical Association of America.

Gleick, James. 2008. *Chaos: Making a New Science.* New York: Viking Penguin.

Hofstadter, Douglas R. 1999. *Gödel, Escher, Bach: An Eternal Golden Braid.* 20th Anniversary ed. New York: Basic Books.

Mandelbrot, Benoit B. 1983. *The Fractal Geometry of Nature.* New York: W. H. Freeman.

Meyer, Rochelle Wilson, and Walter Meyer. 1990. *Play It Again Sam: Recurrence Equations in Mathematics and Computer Science.* Arlington, MA: COMAP, Inc.

Peitgen, Heinz-Otto, Hartmut Jurgens, and Dietmar Saupe. 1992. *Fractals for the Classroom, Part One: Introduction to Fractals and Chaos.* New York: Springer Verlag.

Peterson, Ivars. 2001. *Fragments of Infinity: A Kaleidoscope of Math and Art,* New York: John Wiley & Sons.

Pickover, Clifford A, ed. 1996. *Fractal Horizons: The Future Use of Fractals.* New York: St. Martin's Press.

Pickover, Clifford. 2001. *Computers, Pattern, Chaos, and Beauty.* New York: Dover.

Sandefur, James T. 1990. *Discrete Dynamical Systems: Theory and Applications.* Oxford: Oxford University Press.

Sandefur, James T. 2003. *Elementary Mathematical Modeling: A Dynamic Approach.* Belmont, CA: Brooks/Cole.

Seymour, Dale, and Margaret Shedd. 1973. *Finite Differences: A Pattern-Discovery Approach to Problem Solving.* Palo Alto, CA: Dale Seymour.

Stewart, Ian. 2002. *Does God Play Dice?: The Mathematics of Chaos.* 2nd ed. New Yark: Wiley-Blackwell.

Codes

CHAPTER 9

Codes seem to be everywhere. There are universal product codes (UPC) on everything you buy. There are zip codes on everything you get in the mail. You use codes to do bank transactions such as ATM deposits and withdrawals. You probably have several access codes for Internet accounts.

A code is nothing more than a symbolic way to represent information. Its purpose might be to improve the flow of information, as is the case with UPC codes and ZIP codes. But a code's purpose can also be to protect private information, as is the case with bank codes and Internet access codes.

- How can a code keep information private, but also allow the information to be readable by intended recipients of the information?
- How can a code that improves the flow of information detect errors that can occur when the information is transmitted electronically?
- Is it possible for codes to detect errors and also correct them?

The science of coding is primarily a type of mathematical modeling that has provided answers to these and many other important questions that have arisen in the Information Age. You will see how these and other modeling questions have been resolved in this chapter.

Lesson 9.1

A Coding Activity

One of the first Information Age codes to appear in the United States was the ZIP (zone improvement plan) code, which was introduced in 1963 to improve the flow of mail. The original zip codes were 5-digit numbers. (They were extended to 9 digits in 1983.)

- The first digit identified one of 10 geographic regions.
- The second and third digits identified a mail distribution facility called a sectional center.
- The fourth and fifth digits identified a local post office. (Figure 9.1)

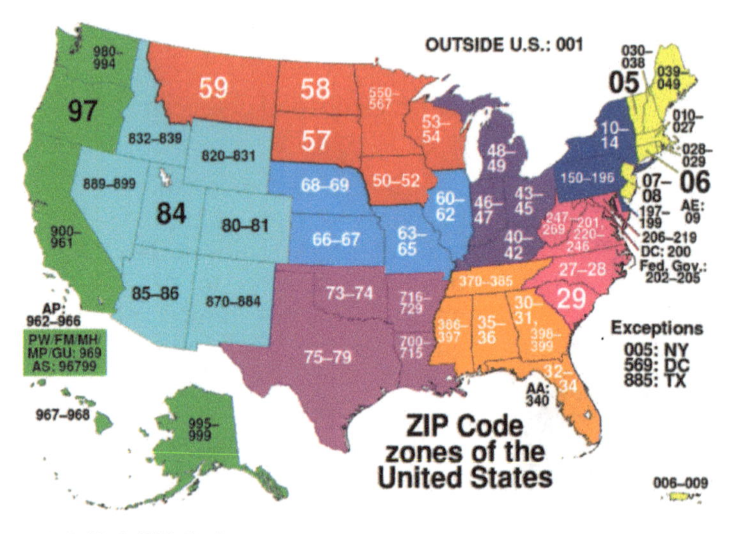

Figure 9.1. A U. S. ZIP Code map.

The use of a numeric code to improve information flow was so unusual in 1963 that the United States Postal Service conducted a publicity campaign that featured a cartoon character called Mr. ZIP (Figure 9.2).

Figure 9.2. Mr. ZIP.

Numeric ZIP codes are written and read by people. As postal automation advanced, ZIP codes were themselves coded into a series of long and short bars called a Postnet code in order to make them readable by machines (Figure 9.3).

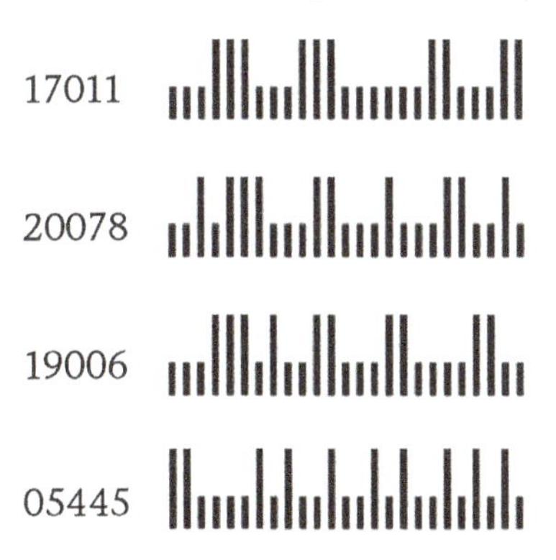

Figure 9.3. Three ZIP codes and their Postnet representations.

Analyze the Postnet codes in Figure 9.3. Your goal in this activity is to determine the Postnet representation for each of the digits 0–9. When you have finished your analysis, use the results to answer the questions in the exer cises.

A Golden Anniversary for the ZIP Code

Lod Angeles Daily News
June 30, 2013

Postman John Garcia was shouldering his mailbag through Arcadia half a century ago when he noticed five new numbers inscribed at the bottom of a letter: 91006.

"We just thought it was another scheme to give somebody work," recalled Garcia, 72, a United States Postal Service carrier in Arcadia since 1959. "But it turned out it made the nation's postal system work.

Launched on July 1, 1963, the mundane ZIP code on the tail of every address has guided letters and packages to every nook and lane in the nation.

The Zoning Improvement Plan not only streamlined a U.S. Post Office Department groaning beneath a mountain of mid-century mail. It also transformed the modern postal service, leading to the automatic sorting of hundreds of billions of letters and parcels now optically scanned and bar-coded to arrive at every home and business – a task once exclusively performed by hand.

While a postal clerk could sort 600 letters an hour by hand, a bar-code sorter can race through 30,000, thanks to the ZIP code.

A recent IBM analysis placed a value of today's ZIP code benefits at $9.5 billion a year.

Exercises

1. What digit is represented by this Postnet code? Explain.

2. A binary number is based on powers of 2 and composed of ones and zeros. For example, the binary number 101 has a numeric value of

$$1 \times 2^2 + 0 \times 2^1 + 1 \times 2^0 = 4 + 0 + 1 = 5.$$

Are Postnet representations the same as binary numbers? Explain.

3. Postnet codes use groups of five bars with three short and two long for a reason. Use your knowledge of counting from Chapter 6 to discuss why you think this choice was made.

4. Scanners that read Postnet codes are programmed to decode them and direct mail to its proper destination. But errors can occur. For example, a smudge on an envelope might cause the scanner to read a group of five bars as four longs and one short. What do think would happen in such a case?

5. A Postnet representation of a ZIP code contains information than just the ZIP code. A machine that prints a Postnet code adds:

   - An extra long bar at the beginning and another extra long bar at the end. These bars serve as frames for the ZIP code and are ignored by the scanner.
   - An extra digit at the right end of the ZIP code that is determined by the code.

   For example, here is the Postnet representation for the ZIP code 58501:

   What is the extra digit in the Postnet code for 58501?

6. Here are three more zip codes and their Postnet representations. Find the extra digit for each.

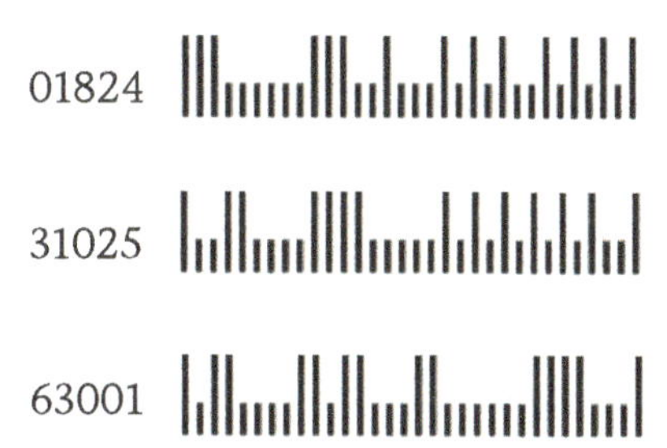

7. Examine the extra digits you found in Exercises 5 and 6. Do you notice a pattern?

8. The ZIP+4 system, which uses 9-digit ZIP codes, has the same Postnet representation as 5-digit ZIP codes.

   a. Find the ZIP code and the extra digit in this Postnet representation of a 9-digit ZIP code.

   b. Does the extra digit you found confirm your conclusion in Exercise 7? Explain.

9. It is fairly common for a barcode scanner to misread the bars, which often results in a human being typing the numeric code manually. Scanners are programmed to know when an error occurs, and barcodes are designed with this in mind. In this lesson, you have seen two ways in which the design of Postnet barcodes helps detect errors. Explain.

Projects

10. The United States Postal Service is replacing The Postnet barcode system with a new system called Intelligent Mail barcode. Research and report on this new system. Explain how it differs from Postnet and the reasons for the change.

Lesson 9.2

Error Detection Models

Errors are a fact of coding life—regardless of whether the numeric form of a code is written and read by humans or the bar form of a code is written and read by machines.

Humans are prone to certain kinds of errors when reading and writing code numbers. Two of the most common are:

- single-digit errors, and
- transposition of two adjacent digits.

Machine errors are also common. Perhaps you've experienced such an error when the scanner at a retail checkout cannot read a bar code, resulting in manual entry by a clerk. Machine errors have many causes, such as a smudge on a product's bar code. The general term for machine errors is noise.

Almost all modern codes use some type of error-detection model. The models vary, as do the kinds of errors they can detect. In this lesson you will learn about a few of these systems and examine how they work.

Check Digits

A **check digit** is an extra digit that is appended a numeric code in order to help detect errors.

The error-detection model that is used for ZIP codes is the same for 5- and 9-digit zip codes:

- Find the sum of the digits in the ZIP code.
- Choose a check digit to make the sum divisible by 10.

For example, the sum of the digits of the ZIP code 68601 is $6 + 8 + 6 + 0 + 1 = 21$. Choose 9 for the check digit because $21 + 9 = 30$, which is divisible by 10. Thus, the ZIP code with the check digit included is 686019 (Figure 9.4).

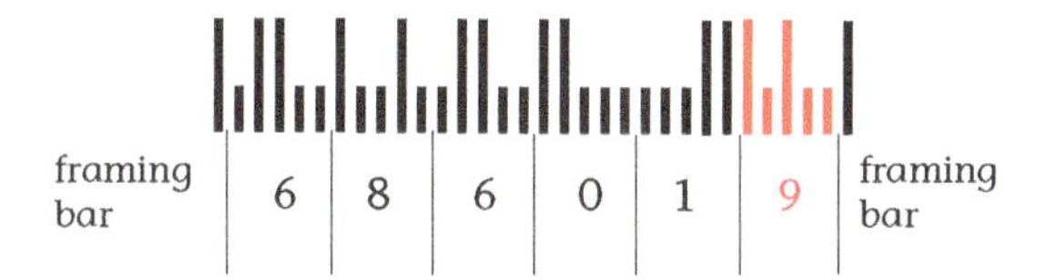

Figure 9.4. ZIP code 68601 and its check digit.

The error-detection model used for ZIP codes detects single-digit errors. For example, if 686019 is misread as, say, 786019, the sum is no longer divisible by 10, and the error is detected. However, this model does not detect transposition errors. For example, if 686019 is misread as, say, 866019, the sum is still divisible by 10.

Divisibility Tests

Error-detection models for codes are often built around divisibility tests. That is, a check digit is chosen to make a number derived from the code (such as the sum of its digits) divisible by some integer.

In the ZIP-code model, the divisibility test is easy because a number is divisible by 10 only if its last digit is 0. But other models use divisibility by numbers for which divisibility is not so obvious. For example, the model for ISBN (International Standard Book Numbers) uses divisibility by 11.

The following table summarizes some common divisibility tests.

Divisible by	Test
2	The last digit is even
3	The sum of the digits is divisible by 3
5	The last digits is 0 or 5
9	The sum of the digits is divisible by 9
10	The last digit is 0
11	Truncate the number by dropping the last digit. The difference between the truncated number and the last digit is divisible by 11.

The test for divisibility by 11 is more complicated than the others in the table. As an example, consider the number 97515. Truncate by dropping 5 to get 9751. Find the difference: 9751 − 5 = 9746. It's not clear

South Jersey Township Uses Barcode Technology to Alert Residents

Press of Atlantic city
June 24, 2013

The police department will be able to immediately contact homeowners and business owners in case of an emergency.

A personalized barcode on the door frame of a house or business helps police access a database that stores the name and cell phone number, as well as any emergency contacts or other pertinent information from the homeowner.

A one-time fee of $6 pays for the barcode and storage of information. Though it is accessible via an app, it doesn't mean that anyone who downloads the app has access. The system provides personal logins and passwords for each officer, which also allows the department to keep track of which officer has scanned which barcode.

without the aid of a calculator if 9746 is divisible by 11, so apply the test again to 9746: 974 – 6 = 968. Apply the test a third time to 968: 96 – 8 = 88. Since 88 is divisible by 11, so is 97515.

Of course, when a calculator is available, any divisibility check can be done quickly without the aid of special tests.

Confirming Divisibility Tests

Visibility tests can be confirmed algebraically; doing so helps to understand why they work.

As an example, consider a 4-digit number and the test for divisibility by 9. Use a, b, c, and d for the number's digits (from left to right) and assume that the sum of the digits is divisible by 9. If the sum of the digits is divisible by 9, $a + b + c + d = 9n$, where n is some integer.

You need to show that the original number is divisible by 9. But you cannot represent the number algebraically by writing abcd since this is the product of the four digits. Since a is the leftmost digit of a 4-digit number, its proper algebraic representation is 1000a. Thus, b's representation is 100b, and c's is 10c. The proper algebraic representation for the 4-digit number is therefore:

$$1000a + 100b + 10c + d.$$

Now that the algebraic representations are established, rephrase the problem: Given that $a + b + c + d = 9n$, show that $1000a + 100b + 10c + d$ is also a multiple of 9.

Since $a + b + c + d = 9n$, $d = 9n - a - b - c$. Substitute for d in $1000a + 100b + 10c + d$:

$$1000a + 100b + 10c + 9n - a - b - c = 999a + 99b + 9c + 9n$$
$$= 9(111a + 11b + c + n).$$

$111a + 11b + c + n$ must be an integer since a, b, c, and n are integers. So $1000a + 100b + 10c + d$ is divisible by 9 because it is the product of 9 and another integer.

Exercises

1. What is the check digit for the 9-digit ZIP code 01730–1459.
2. What ZIP code is represented by this Postnet code, and what is the check digit?

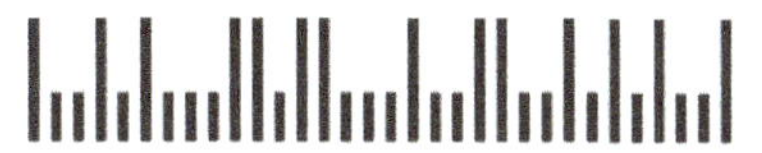

3. Consider this Postnet code.

   a. Explain why this Postnet code does not represent a legitimate ZIP code.

   b. In this case, the machine that reads this Postnet code can use the check digit to correct the error. Explain.

4. The Universal Product Code (UPC) was first used in 1974 for groceries. It has since been expanded to most retail items. The bar form of this code is composed of light and dark bars of varying thicknesses that identify a code's digits to a scanner (Figure 9.5). The numeric form of the code has twelve digits divided into four groups:
   - The first group is a single digit that represents the type of product. For example, 0 represents general merchandise.
   - The second group has five digits that identify the manufacturer.
   - The third group also has five digits, which are assigned by manufacturers to identify each of their products.
   - The fourth group is a single digit: the check digit.

Figure 9.5. A Universal Product Code and its bar representation

As with ZIP codes, the UPC error-detection model is based on divisibility by 10. However, the calculation is slightly more complex: The first digit of the numeric UPC is multiplied by 3, the second by 1, the third by 3, the fourth by 1, and so forth. For example, the check digit for the UPC 0 70734 00008 is the digit n that makes this sum divisible by 10:

$$3(0) + 1(7) + 3(0) + 1(7) + 3(3) + 1(4) + 3(0) + 1(0) + 3(0) + 1(0) + 3(8) + n$$

a. Explain why this check digit is 9.

b. What is the check digit for the UPC 0 71436 80753?

c. Suppose the UPC 0 70734 00008 9 is erroneously written with the second and third digits transposed: 0 07734 00008 9. Would this error be detected? Explain.

d. Is 0 17250 30102 3 a valid UPC? Explain.

e. If the third and fourth digits of the UPC in part (d) are transposed, is this error detected? Explain.

5. The use of International Standard Book Numbers (ISBN) began in the 1970s. The original ISBNs had 10 digits. In 2007, the system was revised to 13 digits. This exercise considers the original 10-digit numbers. A 10-digit ISBN has four groups:

- The first group identifies the language of the country in which the book is published (0 or 1 indicates an English-speaking country).
- The second group identifies the publisher.
- The third group is assigned by publishers to identify each of their books.
- The last group is a single digit: the check digit.

The ISBN error-detection model is based on divisibility by 11. If the first nine digits of an ISBN are a, b, c, d, e, f, g, h, and i, a check digit n is chosen so that this sum is divisible by 11:

$$10a + 9b + 8c + 7d + 6e + 5f + 4g + 3h + 2i + n$$

If $n = 10$, the letter X is used for the check digit.

a. The ninth edition of the book *For All Practical Purposes* was published in the United States by W. H. Freeman, whose publisher number is 4292. Freeman assigned the number 4316 to the hardcover version of the book. What is the check digit? (Use 1 for the country.)

b. Freeman assigned the number 5482 to the softcover version of the book in part a. What is the check digit?

c. Select two adjacent digits of the ISBN for the softcover version of the 9th edition of For All Practical Purposes and transpose them. Is the error detected?

6. The calculations needed detect errors can be tedious. Construct an example using an ISBN to show how matrices can be used to perform the necessary calculations.

7. The ISBN error-detection model uses divisibility by 11 so that it can detect errors caused by transposition of adjacent digits. To see why this works, consider a valid ISBN with digits represented by a through j, j being the check digit.

   a. Write an algebraic expression for the ISBN's check-digit calculation.

   b. Write an algebraic expression for the check-digit calculation if the first and second digits are transposed.

   c. Whenever two numbers are divisible by 11 (or any other positive integer), their sum, difference, and product are also divisible by 11. Write an algebraic expression for the difference between the expressions you wrote in parts a and b.

   d. Use your result to explain why the original and transposed calculations cannot both be divisible by 11.

   e. How does this argument change if the two adjacent digits are not the first and second?

8. Consider a 4-digit number with digits (from left to right) a, b, c, and d. Show that the divisibility test for 11 given in this lesson holds.

9. Show that for a 4-dgit number, if the sum of the digits is divisible by 9, then so is the 4-digit number.

10. Magicians sometimes use divisibility tests and other facts about numbers as the basis for number and card tricks. Here is an example.

    - Have your subjects pick a 3-digit number with different first and last digits.

    - Ask your subjects to write the number backwards and then subtract the smaller 3-digit number from the larger.

- Next, invite your subjects to multiply their result by any positive integer they choose. (There is no limit to the size of the multiplier, but your subjects must do the work!)
- Invite your subjects to scramble the digits of their answer in any order. (Since a zero is never written in a number's first position, they cannot put a zero first.)
- Finally, ask your subjects to circle the leftmost digit of their final answer and read you the remaining digits. You are then able to tell them the digit they have circled.

You are able to do this by finding the digit that makes the sum of the digits divisible by 9. For example, if a subject reads you 6–1–4–2–7, you mentally calculate $6 + 1 + 4 + 2 + 7 = 20$ and $2 + 0 = 2$, which requires a 7 to reach 9. Thus, this person has a 7 circled.

a. Show that the difference between a 3-digit number and the number formed by reversing its digits is divisible by 9.

b. In this trick, your subjects are allowed to scramble the digits of their number. Explain why a number that is divisible by 9 is still divisible by 9 when the digits are reordered.

Projects

11. Research and report on a few error-detecting models that are not discussed in this lesson. Include an explanation of the divisibility test that is used and the types of errors the model detects. Here are a few that you might consider.
    - Postal money orders
    - America Express travelers checks
    - Airline tickets
    - Bank routing numbers
    - Credit card numbers
12. Research and report on codes and check digits used in your area. Possibilities include state driver's license numbers, student identification numbers used by your school, and bar codes used for equipment or textbooks in you school. Discuss the coding model used for each.

Lesson 9.3

Error Correction Models

Sometimes detecting errors isn't enough. Photos from space are subject to severe noise due to radiation and other issues associated with transmitting signals over hundreds of thousands of miles. Errors occur, and often recovering bad images is not an option—errors must be corrected whenever possible. Improvements in technology and the development of error-correction models have resulted in major improvement in the quality of images from space over the years (Figures 9.6 and 9.7)

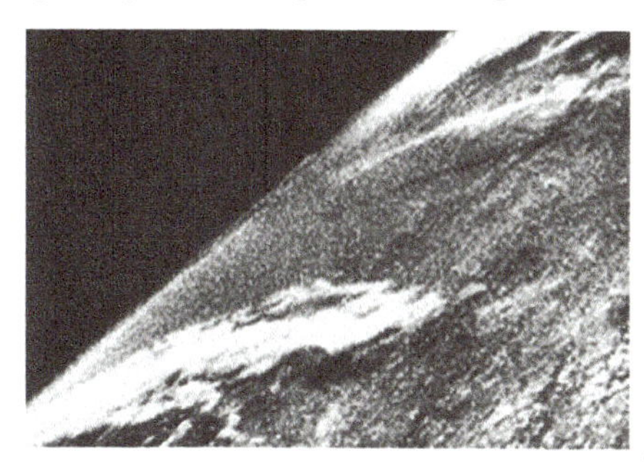

Figure 9.6. The first photo from space. Taken in 1947 from a U. S. rocket, the photo shows a portion of the earth.

Figure 9.7. A 2007 photo from the Hubble space telescope depicts white dwarf star in decay. The star is about 4,000 light years away.

Binary Codes

A **binary code** is one that uses just two characters. For example, the Postnet system is binary. Each pixel of a photo transmitted from space has a binary representation. For convenience, the following example is highly simplified.

Suppose that each pixel of a photo can be only black or white. 1 represents a black pixel, and 0 represents a white pixel. Thus, the binary code that is transmitted for the row of five pixels in Figure 9.8 is 10010.

Figure 9.8 A row of five pixels: black, white, white, black, white

However, if noise causes the signal to be received as 10110, there is no way to know that an error has occurred. There is a simple trick to get around this difficulty: use more than one binary character for a single pixel. That is, build redundancy into the model.

Using two characters isn't enough. Here's why: If 11 represents black and 00 represents white, and one error causes a pixel to be received as 10, it's a 50-50 guess whether the pixel is black or white.

If 111 is black and 000 is white, and a pixel is received as 101, then it seems more likely that the pixel should be black than white. (Of course, this is true only if the probability of a single-digit error is relatively low.)

The amount of redundancy that is needed depends on how probable errors are: the higher the probability of an error, the greater the number of characters needed for a single pixel.

If five characters are sufficient for a single pixel, then the code transmitted for Figure 9.8 is this string of 1s and 0s: 1111100000000001111100000.

The basic idea in this error-correction model is to choose the legal code that is closest to an erroneous one. For example, if a 5-digit code is received as 10110, it misses the legal code for black by 2 characters and the legal code for white by three characters. Thus, in this sense 10110 is closer to 11111 than it is to 00000, so it is decoded as 11111, or black.

Mathematician of Note

Richard Hamming (1915-1998) developed the first error-correction model of the type described here while working at Bell Laboratories.

The idea of measuring the distance between two strings by counting the number of characters in which they are different has become key to information science and is often referred to as Hamming distance. That is, the **Hamming distance** between two binary strings of equal length is the number of positions in which they are different.

Reed-Solomon Codes

Reed-Solomon codes use an error-detection model based on polynomials. The model also uses the trick of building in redundancy. Although the model is binary, it is easier to understand in the familiar base 10, which is used here. Variations on this model are used to correct errors in cell phone transmission and compact discs (Figure 9.9).

The polynomials used in the Reed-Solomon model are of high degree and code large amounts of information. For simplicity, suppose you want to encode just three base-10 numbers: 5.2, 6.8, and 7.9, in that order. A property of polynomial functions that makes them useful for error correction is that a polynomial function of degree $n - 1$ can capture n data points exactly. Thus, to capture 5.2, 6.8, and 7.9 exactly (in that order), form the pairs (1, 5.2), (2, 6.8), and (3, 7.9). Since there are three pairs, there is a polynomial function of degree 2 that fits them exactly. As you know, a polynomial function of degree 2 is called *quadratic*.

Figure 9.9
Some CD players use a Reed-Solomon error-correction model.

There are a variety of ways to find a quadratic function that fits these three pairs. Perhaps the quickest is to perform quadratic regression on a graphing calculator (Figure 9.10), which gives

$$y = -0.25x^2 + 2.35x + 31.$$

L1	L2	L3 1
1	5.2	------
2	6.8	
3	7.9	

L1(4)=

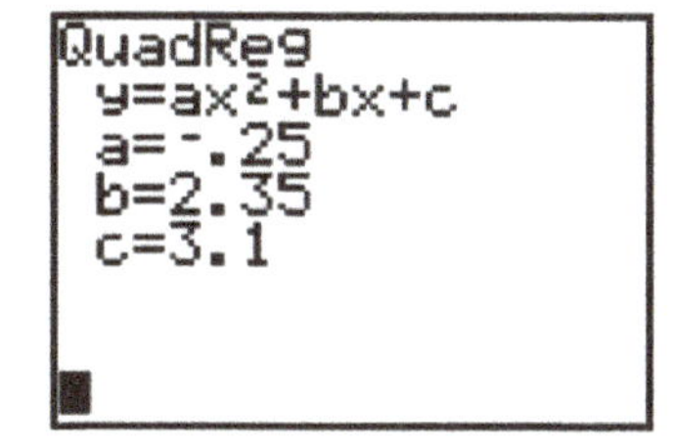

Figure 9.10. Quadratic regression on a calculator.

To build in redundancy, add new values by evaluating the function for, say, 4, 5, 6, 7, 8, 9, and 10. Together with the original three values, this gives

5.2, 6.8, 7.9, 8.5, 8.6, 8.2, 7.3, 5.9, 4.0, 1.6.

These 10 values are transmitted. Now, suppose that because of an error such as a scratch on a CD, two of these numbers—the third and the sixth—are misread:

5.2, 6.8, 9.9, 8.5, 8.6, 6.2, 7.3, 5.9, 4.0, 1.6.

The decoder recognizes that these ten values are not compatible with a quadratic function. Moreover, the decoder can tell that two of them are bad and which two. Although the decoder uses a different method, you can recognize the errors by looking at a graph (Figure 9.11).

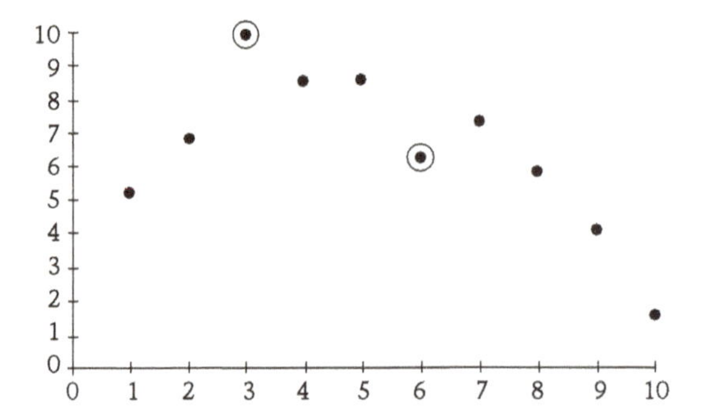

Figure 9.11. Two points do not fit a parabolic shape.

The circled points clearly do not fit the familiar parabolic shape of a quadratic function's graph. The decoder can correct the errors by first performing quadratic regression on any three of the eight good points to get the original quadratic function: $y = -0.25x^2 + 2.35x + 31$.

The errors are then corrected by evaluating this function for $x = 3$ and $x = 6$:

$$-0.25(3)^2 + 2.35(3) + 31 = 7.9;$$
$$-0.25(6)^2 + 2.35(6) + 31 = 8.2.$$

Of course, since $x = 6$ is associated with redundant information, only $x = 3$ needs correcting.

Mathematician of Note

Irving Reed (1923-2012) and Gustave Solomon (1930-1996) developed the error-correction model that bears their names in 1960 while on the staff of MIT.

Exercises

1. Find the Hamming distance between each pair of strings.

   a. 11100111

   b. 11111 11011

   c. 1101011 0110111

   d. 11011 100111

2. Assume that each string was generated with a binary coding model in which all 1s represent a black pixel and all 0s represent a white pixel. Also assume that the probability of an error in a single digit is fairly low. State whether the string is more likely to represent a black pixel or a white pixel.

   a. 1110110

   b. 00100

3. Write a binary code for the row of six pixels shown. Use a string of six 1s for black and a string of 6 0s for white.

4. This binary string was transmitted using the system described in Exercise 3. It contains errors, and the probability of an error in a single digit is low.

   110101111011100000011110000101

   a. Find the errors. Explain your method.

   b. Write the corrected string.

   c. This string represents a row of black and white pixels. Sketch them.

5. Assume that the probability a single digit in a binary code is wrong is 10%. Use your knowledge of probability from Chapter 6 to answer each question.

   a. What is the probability that all of the digits in a string of five digits are wrong?

   b. What is the probability that all of the digits in a string of five are right?

c. What is the probability that exactly four digits in a string of five are right?

d. What is the probability that exactly three digits in a string of five are right?

e. What is the probability that at least three digits in a string of five are right?

f. These probability calculations imply an important modeling assumption about the transmission of digits. What is it?

6. Use your results from Exercise 5 to answer these questions about a model for transmitting black and white pixel information. The model uses five 1s for black and five 0s for white.

a. Do you think this model is a reasonable one if the probability of a single-digit error is 10%? Explain.

b. If the probability of a single-digit error increases to 20%, is the model still reasonably good? Explain.

c. Compare the situation in part b with a similar situation in which six characters are used instead of five.

d. Compare the situation in part b with a similar situation in which seven characters are used instead of 5.

7. A Reed-Solomon coding model is used to transmit the values 1.6, 4.8, 6.4, and 5.2, in that order.

a. What degree polynomial should be used? Explain.

b. The values are coded as ordered pairs: (1, 1.6), (2, 4.8), (3, 6.4), and (4, 5.2). What is the polynomial that codes these pairs?

c. A total of 10 values are to be transmitted: the original four and six redundant values. Make a table showing all of them.

8. A quadratic Reed-Solomon model results in receiving these values: 4.2, 9.8, 7.6, 6.6, 3.8, –5.8, –7.2, and –15.4, in that order.

a. Do you think any of the values are erroneous? Explain.

b. Find the coding quadratic.

c. Use the quadratic you found in part b to correct the errors you identified in part a.

9. Using a graph to identify errors in data encoded with a Reed-Solomon model works well if you are familiar with the shape of the graph of the encoding polynomial. This is seldom the case, but there are other methods. You should be familiar with an alternative method from your work in Chapter 8. Consider these received values that were encoded with a quadratic Reed-Solomon model: 1.6, 5.2, 6.4, 5.8, 1.6, –4.4, –12.8, and –23.6. Write them in a table and discuss how errors might be detected from the table without the use of a graph.

10. Regression is not the only way to fit a polynomial to data. Go back to part b of Exercise 8 and show how the quadratic could be found by solving a system of equations.

A New Use for Bacteria

Time
January 26, 2011

A team of students at Hong Kong's Chinese University may have discovered a way to encrypt large chunks of computer data into a strain of microscopic E. coli bacteria, with one gram capable of holding as much information as 450 2TB hard drives. That's shelves upon shelves of electronic files, distilled down smaller than a petri dish.

The team of 11 students, 2010 gold medalists in the Massachusetts Institute of Technology iGEM biotechnology competition, have successfully found a way to embed text, images, music and video into the bacteria's cells. A giant leap, especially when considering that in 2007, Japanese researchers were only able to embed a cell with a single line of code.

Biostorage presents other technological advantages. "Bacteria can't be hacked," says Allen Yu, a student instructor who participated in the study. "All kinds of computers are vulnerable to electrical failures or data theft. But bacteria are immune from cyber attacks."

Lesson 9.4

Privacy Codes

The codes you met in the first three lessons of this chapter are public. That is, information on how to code and decode is freely available. Codes are also used to keep information private, and thus the coding model is known only to a few. Of course, there are incentives for others to crack these codes and gain access to the secrets that the coding models protect. These kinds of codes are the stuff of movies and novels, but they are also very practical. They protect millions of financial transactions every day.

Cryptography

Cryptography is the science of making and cracking privacy coding models. One of the first uses of cryptography know to history was by Julius Caesar.

Here is the equivalent of Caesar's model using the letters of the modern alphabet. Slide the letters of the alphabet forward a few places. For example, three places:

a	b	c	d	e	f	g	h	i	j	k	l	m	n	o	p	q	r	s	t	u	v	w	x	y	z			
			a	b	c	d	e	f	g	h	i	j	k	l	m	n	o	p	q	r	s	t	u	v	w	x	y	z

Wrap the letters that are left at end back to the beginning:

a	b	c	d	e	f	g	h	i	j	k	l	m	n	o	p	q	r	s	t	u	v	w	x	y	z
x	y	z	a	b	c	d	e	f	g	h	i	j	k	l	m	n	o	p	q	r	s	t	u	v	w

To code a message, replace each of its letters with the associated letter. For example, to code the word *code:*

c → z o → l d → a e → c.

Thus, the encoded form of the word *code* is *zlac.*

To decode *zlac,* the recipient reads the table in reverse.

Because of its history, a model that replaces one letter of the alphabet with another is sometimes called a Caesar cipher. A **cipher** is a coding model that replaces each letter of a message with another letter or other character.

Cracking a Cipher

Caesar ciphers are easy to crack and thus seldom used today. They can be cracked with basic statistics and a trial-and-error approach. In any language, some letters are used more frequently than others. For example, the letter *e* is most common in English passages—even of fairly short length. The histogram in Figure 9.12 shows typical letter frequencies in English.

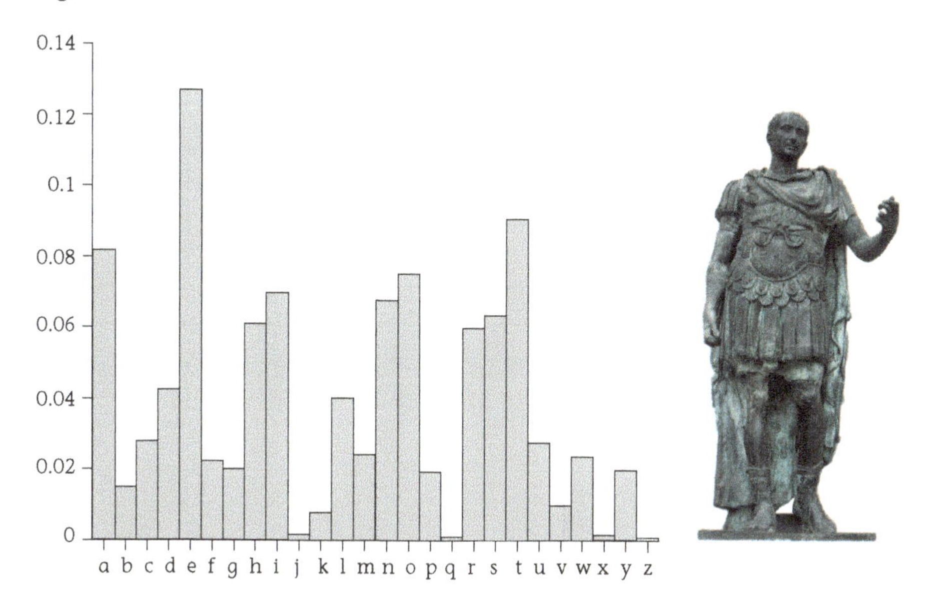

Figure 9.12. English letter frequencies.

A code cracker works from the assumption that the most common character in a coded message represents *e*. Using a trial-and-error process and knowledge about the party who sent the message or the message's likely contents, code crackers can almost always crack simple ciphers.

You may have read descriptions of this type of code cracking in novels. Edgar Allan Poe (in *The Gold Bug*) and Arthur Conan Doyle (in *The Adventure of the Dancing Men*) are two well-known authors who have built plots around statistical code cracking.

A Matrix Coding Model

A good privacy-coding model makes it easy for the right people to code and decode messages, but hard for others to crack. Caesar ciphers do the former, but not the latter. Matrices, which you studied in Chapters 3 and 7, can be used to develop a better model—one in which the character that represents *e* isn't always the same.

This model associates each letter with a number:

A	B	C	D	E	F	G	H	I	J	K	L	M
1	2	3	4	5	6	7	8	9	10	11	12	13
N	O	P	Q	R	S	T	U	V	W	X	Y	Z
14	15	16	17	18	19	20	21	22	23	24	25	26

If you want other characters in your messages, such as a blank space or period, add them to the end of the table and extend the numbering.

If your message is "Meet me at the mall," replace each letter with the associated number:

m	e	e	t	m	e	a	t	t	h	e	m	a	l	l
13	5	5	20	13	5	1	20	20	8	5	13	1	12	12

Next, select a keyword, such as *fish*. Since fish has four letters, write the numeric form of the message in a matrix with four columns: (use 27 for unused matrix elements):

$$\begin{bmatrix} 13 & 5 & 5 & 20 \\ 13 & 5 & 1 & 20 \\ 20 & 8 & 5 & 13 \\ 1 & 2 & 12 & 27 \end{bmatrix}$$

Write a matrix with the same number of rows and columns with the numbers associated with *fish* in each row:

$$\begin{bmatrix} 6 & 9 & 19 & 8 \\ 6 & 9 & 19 & 8 \\ 6 & 9 & 19 & 8 \\ 6 & 9 & 19 & 8 \end{bmatrix}$$

Add the two matrices:

$$\begin{bmatrix} 13 & 5 & 5 & 20 \\ 13 & 5 & 1 & 20 \\ 20 & 8 & 5 & 13 \\ 1 & 2 & 12 & 27 \end{bmatrix} + \begin{bmatrix} 6 & 9 & 19 & 8 \\ 6 & 9 & 19 & 8 \\ 6 & 9 & 19 & 8 \\ 6 & 9 & 19 & 8 \end{bmatrix} = \begin{bmatrix} 19 & 14 & 24 & 28 \\ 19 & 14 & 20 & 28 \\ 26 & 17 & 24 & 21 \\ 7 & 11 & 31 & 35 \end{bmatrix}$$

The coded message is sent in numeric form as
19 14 24 28 19 14 20 28 26 17 24 21 7 11 31 35.

Notice that the first *e* in the message is coded as 14, but the second *e* is coded as 24.

Provided the recipient knows the keyword, the message is easy to decode: follow the same procedure, but use matrix subtraction.

A Binary Coding Model

Internet financial transactions use a variety of models to keep information such as credit card numbers private. One such model uses addition of binary strings.

Binary string addition is based on three basic facts:
$0 + 0 = 0$, $1 + 0 = 1$, and $1 + 1 = 0$. To add two binary strings, write one beneath the other and use the three basic facts:

$$\begin{array}{r} 1010011 \\ \underline{1110101} \\ 0100110 \end{array}$$

This type of binary addition has a feature that makes it useful for transmission of account numbers over the Internet: if you add the same number twice to a given number, you get the given number back again. For example, add the result of the previous addition to the second of the two addends:

$$\begin{array}{r} 1110101 \\ \underline{0100110} \\ 1010011 \end{array}$$

When you make a credit card purchase online, the process works like this:

- Your computer converts your credit card number to a binary string.
- The merchant sends a random binary string to your computer.
- Your computer adds the random string to the binary string that represents your credit card number and sends the result to the merchant.
- The merchant adds the binary string that they sent you to the number they receive and gets your actual credit card number as a result.

Brown Students Decode Roger Williams Shorthand

Boston Globe
December 5, 2012

Lucas Mason-Brown, a senior at Brown University, was an unlikely candidate to help unlock the secrets of a centuries-old New England manuscript.

First, he's a math major.

Second, he's not exactly an expert on 17th-century theologians.

Mason-Brown, 21, is part of a small team of Brown students to crack a previously undeciphered shorthand used by Williams, the religious thinker and founder of Rhode Island—a mystery that had stumped researchers for years.

Mason-Brown used a mix of statistical analysis and historical research to reveal the meaning of some of the theologian's last writings, a series of extensive notes written in the margins of a book.

Williams, who died in 1683, was a Puritan religious thinker who championed the separation of church and state and advocated for the rights of Native Americans.

Exercises

1. a. In developing a Caesar coding model, you must decide what to do with blanks. List a few options.

   b. Use a Caesar model that slides the alphabet forward 5 places to code this message: four score and seven years ago. Give at least two versions based on the options for a blank that you listed in part a.

   c. Does encoding a blank space as a character improve a coding model? Explain.

2. This message was coded with a Caesar model that slides the alphabet forward four places. Decode the message.
   8 13 23 7 22 9 24 9 17 5 24 12 13 23 7 19 19 16

3. In *The Adventure of the Dancing Men*, Sherlock Holmes solves a crime by cracking coded messages in which stick figures are used for letters of the alphabet. Holmes describes his method in the story: "As you are aware, E is the most common letter in the English alphabet, and it predominates to so marked an extent that even in a short sentence one would expect to find it most often. ... It is true that in some cases the figure was bearing a flag, and in some cases not, but it was probable, from the way in which the flags were distributed, that they were used to break the sentence up into words."

   a. Two of the messages are shown here. What character represents E?

   b. Code crackers often know something about the general circumstances surrounding messages they are trying to crack. In *The Adventure of the Dancing Men*, Abe Slaney send the messages to Elsie, the wife of Mr. Hilton Cubitt. Try to crack the two messages in part a.

4. In Edgar Allan Poe's *The Gold Bug*, a key part of the story involves cracking the following coded message. In this message, what character most likely represents the letter e? Explain.

```
53++!305))6*;4826)4+.)4+);806*;48!8`60))85;]8*:+*8!83(88)5*!;46
(;88*96*?;8)*+(;485);5*!2:*+(;4956*2(5*4)8`8*;4069285);)6!8)4++;1
(+9;48081;8:8+1;48!85;4)485!528806*81(+9;48;(88;4(+?34;48)4+;161;
:188;+?;
```

5. a. Use a matrix model with keyword *discrete* to code: when in the course of human events. Leave the blank spaces intact.

   b. Code the message in part a with the same model, but use 27 for a blank space.

6. This message was coded with a matrix model that encodes a blank as a 27th character and uses the keyword *math*. Decode it.

   17 10 39 11 31 6 40 13 40 14 21 28 21 6 33 9 33 10 23 27 40 19 35 11 24 20

7. A binary coding model adds a random string to encode other binary strings. If the random string is 100110111, encode each string:

   a. 110011101

   b. 111011000

8. Here are two strings encoded with a binary model that added the random string 10010111 to each of them. Decode each string.

   a. 00111010

   b. 11101101

9. What happens when a binary string is added to itself?

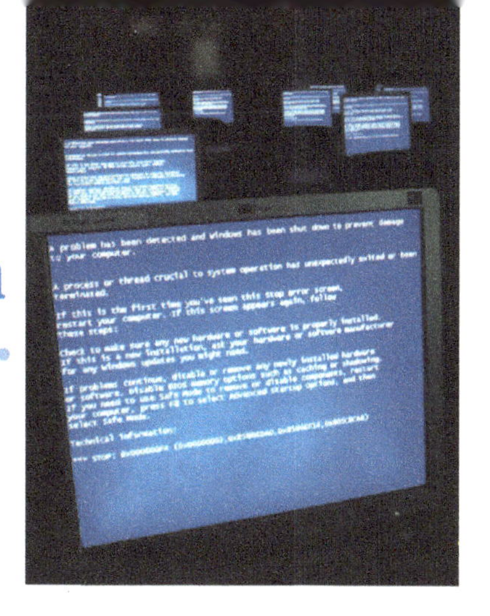

Chapter 9 Extension

A Public Key Model

Privacy coding models have an inherent disadvantage: the sender and receiver must share the coding method, and anyone who knows the method can read their messages. In the 1970s, mathematicians Ronald Rivest, Adi Shamir, and Leonard Adelman devised a way around this issue. They developed a model that allows the method to be made public without comprising the coded messages. With the RSA (for Rivest, Shamir, Adelman) model, knowledge of the key does not help a code cracker.

Mathematicians of Note

Ronald Rivest, Adi Shamir, and Leonard Adelman in 2003

The RSA model uses computers to do calculations with very large numbers, but also takes advantage of the limitations of computer power. It also uses some basic facts about numbers, particularly prime numbers.

Computational Complexity

Algorithms exist that enable computers to do some calculations efficiently, including these that are essential to the RSA model:

- Multiplying large numbers;
- Finding large primer numbers.

Recall that a prime number is one with no divisors other than 1 and the primer number itself: 2, 3, 5, 7, 11, ...

However, there isn't an algorithm that allows computers to find all of the divisors of very large numbers. Although there are a few tricks that can shorten the process, computers must do this by brute force: trying all possible divisors. Moreover, using brute force to find the divisors of a very large number takes prohibitive amounts of computer time.

Modular Arithmetic

Although 9 + 5 =14, it is not so on a clock. If it is 9 o'clock, and you agree to meet someone in 5 hours, you will meet at 2. Thus, on a clock 9 + 5 is 2. Clock arithmetic is a form of modular arithmetic.

A number *a* modulo another number *b* (or just a mod b) is the remainder when a is divided by b. For example, 14 mod 12 is 2 because 2 is the remainder when 14 is divided by 12.

The days of the week are an example of a modulo 7 system. Number the days with 1 for Sunday, 2 for Monday, and so forth. If today is Tuesday (day 3) and you have a deadline in 23 days, the day of the deadline can be found by adding 3 and 23, then dividing by 7. Since 26 divided by 7 has a remainder of 5, the deadline is on a Thursday.

Finding a Public Key

Suppose you would like to use the RSA model to make public a key that others can use to send you private information. Here is a description of the process and an example that uses small numbers instead of the very large numbers that the model actually uses.

Description of process	Small number example
Choose two large prime numbers, p and q	$p = 3$, $q = 5$
Find $n = pq$ and $m = (p - 1)(q - 1)$	$n = 3 \times 5 = 15$, $m = (3 - 1)(5 - 1) = 8$
Find s and t so that $st = 1 \text{ mod } m$	$s = 3$, $t = 3$ ($3 \times 3 = 9 = 1 \text{ mod } 8$)*
Make the values of n and s public.	$n = 15$, $s = 3$ are made public.

*These values are not unique. For example, 3 and 11 also work.

Encoding and Decoding with RSA

With the exception of the size of the numbers involved, the process of encoding and decoding with the public key is fairly simple. This example uses the values in the previous table to encode and decode a single letter.

Description of encoding process	Single letter example
Replace each character with its numeric counterpart.	To encode the letter *m*, replace it with 13 (13th letter of the alphabet)
Raise the numeric value to the *s* power and reduce mod *n*.	$13^3 = 2197$, which is 7 mod 15.
Send this value.	The value that is sent is 7.

Description of decoding process	Single-letter example
Raise the received value to the *t* power and reduce mod *n*.	$7^3 = 343$, which is 13 mod 15.
Replace this value with the corresponding character.	The 13th letter of the alphabet is *m*.

Cracking a message coded with the RSA model requires knowing the value of *t*, which is derived from *m*, which in turn is derived from *p* and *q*. Finding *p* and *q* requires finding the factors of *n*, which, if large enough, cannot be done in reasonable time even on very fast computers.

If an RSA key is made public, there is danger that someone could pose as an imposter, claiming to send messages from another person. RSA provides a way to sign messages to prevent this.

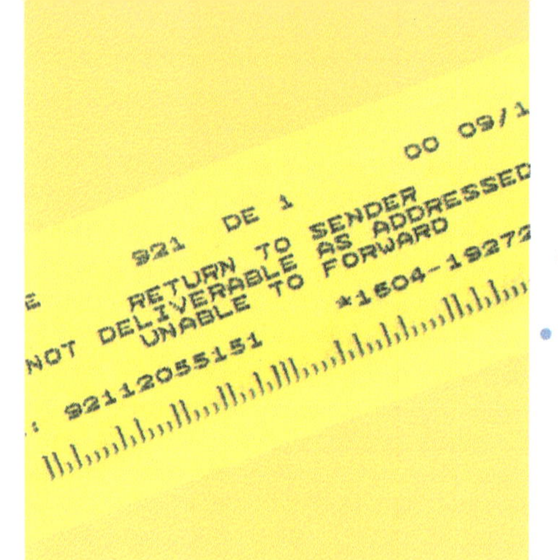

Chapter 9 Review

1. Write a summary of what you think are the important points of this chapter.

2. What is the check digit for the ZIP code 26410–2814?

3. Does this Postnet code represent a legitimate ZIP code? If you think it does, give the numeric ZIP code. If you think it doesn't, what ZIP code do you think it is likely to represent?

4. A manufacturer's UPC identification number is 39124. What is the check digit for this manufacturer's general merchandise product number 18672?

5. Show that for a 4-digit number, if the sum of the digits is divisible by 3, then so is the number.

6. What is the Hamming distance between these two strings: 101101 and 110110?

7. Write a binary code for the five pixels shown. Use a string of four 1s for black and a string of four 0s for white.

8. A binary model is used to transmit black and white pixels. The model uses a string of five 1s for black and a string of five 0s for white. The following string is received for a row of black and white pixels. Assuming the probability of a single error is low, what is the likely black/white configuration of the row of pixels?

001001011100000000101111101000

9. A binary coding model uses a string of seven 1s for a black pixel and seven 0s for a white pixel. If the probability that a single digit is received incorrectly is 15%, what is the probability that at least four digits in a string of seven are received correctly if individual digit errors are independent?

10. A reed-Solomon model is used to transmit the pairs (1, 5.2), (2, 7.1), and (3, 6.8).

    a. What type of mathematical function should be used?

    b. Find the proper function.

    c. A total of six values are to be transmitted: the original three and three redundant values. Find the three redundant values and explain your method.

11. This message was coded with a Caesar cipher that shifts the alphabet 6 characters forward. Blanks were coded as a 27th character. Decode the message.

    14 7 25 26 11 33 19 7 17 11 25 33 29 7 25 26 11

12. This message was coded with a model that replaces each letter of the alphabet and a blank with a number. What number do you think represents the letter e? Explain.

    8 34 13 25 16 12 21 11 34 16 21 34 21 12 12 11 34 16 26 34 8 34 13 25 16 12 21 11 34 16 21 11 12 12 11

13. This message was coded with a matrix model that encodes a blank as a 27th character and uses the keyword moon. Decode it.

    22 42 27 29 35 20 42 18 22 34 18 32 18 35 20 41 26 16 35 22

14. A binary coding model uses the random string 101110110 to encode other binary strings. Use this random string to encode 011101001

Bibliography

Baylis, John. 1998. *Error Correcting Codes: A Mathematical Introduction.* London; New York: Chapman & Hall.

COMAP. 2013. *For All Practical Purposes: Introduction to Contemporary Mathematics.* 9th ed. New York: W. H. Freeman.

Kahn, David. 1996. *The Codebreakers.* New York: Scribner.

Malkevitch, Joseph, Gary Froelich, and Daniel Froelich. 1991. *Codes Galore.* Lexington, MA: COMAP, Inc.

Malkevitch, Joseph, and Gary Froelich. 1993. *Loads of Codes.* Lexington, MA: COMAP, Inc.

Rice, Bart F, and Carroll O. Wilde. 1979. *Error Correcting Codes I.* Lexington, MA: COMAP, Inc.

Sherman, Gary J. 1979. *A Double-Error Correcting Code.* Lexington, MA: COMAP, Inc.

Singh, Simon. 2000. *The Code Book.* New York: Knopf Doubleday.

CHAPTER 10

Additional Topics in Discrete Mathematics

The mathematical topics that are considered to be "discrete" vary from mathematician to mathematician and from textbook to textbook. The broad topics that have been included in Chapters 1 through 9 are what we, the authors, consider to be the most important and interesting for high school students. But in reality, there are a few additional topics that we felt are worthy of attention. So we have included them here in this final chapter.

Lesson 10.1

Logic

Most people who have access to a computer have used a search engine such as Yahoo! or Google to look for something or someone on the Internet. The Internet is an amazing resource of information. It contains so much information that finding what you need can sometimes be a challenge. It is often helpful to take the time to refer to a site's help pages for an in-depth explanation of the kind of searching that can be done.

Learning how to use search engines correctly can help you obtain better search results. Many search engines use what is known as **Boolean logic**. An expression in Boolean logic is simply a statement that is either true or false. For example, the expression "This page contains a picture of Leibniz" is a true statement.

Mathematician of Note

Gottfried Wilhelm Leibniz (1646–1716)

A brilliant German mathematician, Leibniz developed the foundations of modern mathematical logic at the age of 20. Even though Aristotle is considered to be the "founder of logic," Leibniz is considered by many to be the founder of logic as a serious mathematical discipline. During his lifetime, Leibniz made many important contributions to mathematics, but it was many years after his death before his contributions to logic were recognized.

Suppose you want to search the Internet for information about *Mars*. To do so you enter the word *Mars* into your search engine. In turn, the search engine interprets the word as the expression "This Web page contains the word *Mars,*" and it returns a list of pages for which the expression is true.

The terms AND, OR, and NOT can be used to create more complex expressions that can help to narrow or expand your search. For example, if you wish to search for pages that contain the word *Mars*, but not the words *Venus* or *Jupiter*, you can use the expression "Mars AND (NOT Venus) AND (NOT Jupiter)." Most search engines give you advanced search options that help you use this type of logic to customize your search.

Boolean Logic

Is the expression "Mars AND (NOT Venus) AND (NOT Jupiter)" equivalent to the expression "Mars AND NOT (Venus AND Jupiter)"? That is, if you enter these expressions in a search engine, do the searches yield the same results? To answer this question, you need to know about Boolean logic and the words AND, OR, and NOT.

NOT is denoted by the symbol ~. Given a statement p, "$\sim p$" is called the **negation** of p and is read "not p." For example, let p be the statement "Today is Monday." Then $\sim p$ can be interpreted as "Today is not Monday." If a statement p is true, then NOT p is false. If p is false then NOT p is true. This information about the negation relationships is summarized in a **truth table** shown in Table 1.

Point of Interest

Sometimes the symbol $\neg$ is used to indicate the negation of an expression.

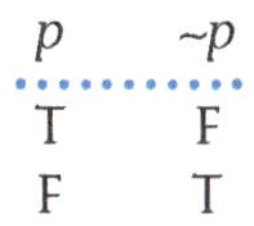

p	$\sim p$
T	F
F	T

Table 1. Truth table for $\sim p$.

AND is denoted by the symbol $\wedge$. Given another statement q, "$p \wedge q$" is called the **conjunction** of p and q. It is read "p and q." The expression $p \wedge q$ is true only when both statements are true. Otherwise, it is false.

OR is denoted by the symbol $\vee$. The expression "$p \vee q$" is called the **disjunction** of p and q. It is read "p or q." To avoid ambiguity about the meaning of the word *or*, mathematicians and logicians have defined "p or q" to mean that the expression is true when at least one of p or q is true. It is false only when both p and q are false.

Tables 2 and 3 summarize the conjunction and disjunction relationships. Notice that the first two columns of each table show all possible truth values of p and q.

p	q	$p \wedge q$
T	T	T
T	F	F
F	T	F
F	F	F

Table 2. Truth table for $p \wedge q$.

p	q	$p \vee q$
T	T	T
T	F	T
F	T	T
F	F	T

Table 3. Truth table for $p \vee q$.

Now that truth values are assigned to the statements $\sim p$, $p \wedge q$, and $p \vee q$, it is possible to examine more complicated expressions. What about expressions such as $p \wedge q \wedge r$? Expressions such as these are considered ambiguous since you do not know which conjunction to perform first: $p \wedge q$ or $q \wedge r$. To avoid confusion, parentheses are used in expressions like these. For example, $p \wedge q \wedge r$ should be written as either $(p \wedge q) \wedge r$ or as $p \wedge (q \wedge r)$.

If you examine the truth tables for $(p \wedge q) \wedge r$ and $p \wedge (q \wedge r)$, you find that they have the same truth value for each possible assignment of Boolean variables. Hence, these two expressions are said to be **logically equivalent**. So, in this case, the way the expression $p \wedge q \wedge r$ is interpreted does not matter.

Point of Interest

Search engines apply Boolean logic in three different ways.

- Full Boolean logic with the words "and," "or," and "not"
- Implied Boolean logic that uses the symbols + to mean "must contain the word" and – to mean "must not contain the word"
- Fill-in templates that use specified words

To determine how a search engine works, consult its Help files.

But this is not always the case. For example, the expression $p \wedge q \vee r$ can be interpreted as either $(p \wedge q) \vee r$ or $p \wedge (q \vee r)$. If you examine the truth tables for these two expressions you find that they are **not** logically equivalent. Therefore, in order to give $p \wedge q \vee r$ meaning, parentheses must be used.

Now it's possible to return to the original computer search question. If you enter the expressions "Mars AND (NOT Venus) AND (NOT Jupiter)" and "Mars AND NOT (Venus AND Jupiter)", do you get the same results? To answer the question, you need to determine whether these two expressions are logically equivalent.

Begin by translating the word expressions into mathematical expressions.

Original statement	Mathematical translation
Mars AND (NOT Venus) AND (NOT Jupiter)	$p \wedge (\sim q) \wedge (\sim r)$
Mars AND NOT (Venus AND Jupiter)	$p \wedge \sim (q \wedge r)$

Next create a truth table for each mathematical expression.

To create a truth table for $p \wedge (\sim q) \wedge (\sim r)$, make six columns with headings p, q, r, $\sim q$, $\sim r$, and $p \wedge (\sim q) \wedge (\sim r)$. (See Table 4.) The bolded columns are helpful in determining the truth value for the last column.

p	q	r	$\sim q$	$\sim r$	$p \wedge (\sim q) \wedge (\sim r)$
T	T	T	F	F	F
T	T	F	F	T	F
T	F	T	T	F	F
T	F	F	T	T	T
F	T	T	F	F	F
F	T	F	F	T	F
F	F	T	T	F	F
F	F	F	T	T	F

Table 4. Truth table for $p \wedge (\sim q) \wedge (\sim r)$.

Look closely at the table and you see that the only time $p \wedge (\sim q) \wedge (\sim r)$ is true is when p is true, q is false, and r is false. In other words, the only time this statement is true in a search is when the word *Mars* appears on a page, the word *Venus* does not appear, and the word *Jupiter* does not appear.

To create a truth table for $p \wedge \sim(q \wedge r)$, make a table with six columns. Label the columns p, q, r, $q \wedge r$, $\sim(q \wedge r)$, and $p \wedge \sim(q \wedge r)$. (See Table 5.)

p	q	r	$q \wedge r$	$\sim(q \wedge r)$	$p \wedge \sim(q \wedge r)$
T	T	T	T	F	F
T	T	F	F	T	T
T	F	T	F	T	T
T	F	F	F	T	T
F	T	T	T	F	F
F	T	F	F	T	F
F	F	T	F	T	F
F	F	F	F	T	F

Table 5. Truth table for $p \wedge \sim(q \wedge r)$.

Notice that the expression $p \wedge (\sim q) \wedge (\sim r)$ is true in three cases: when p is true, q is true, and r is false; when p is true, q is false, and r is true; and when p is true, q is false, and r is false. In other words, it is true when the word *Mars* appears on a page, *Venus* does not appear, and *Jupiter* does not appear. But it is also true when the word *Mars* appears on a page and exactly one of the other two appears. And this is not what you want the search to do.

Since the last columns of the two truth tables differ, you know that $p \wedge (\sim q) \wedge (\sim r)$ is not logically equivalent to $p \wedge \sim(q \wedge r)$. Therefore, you would expect searches performed using these two expressions to return different results.

Exercises

1. Give the negation of each statement.

   a. $5 + 8 = 13$

   b. My favorite singer is Carrie Underwood.

   c. It is 6:00 o'clock in the morning and I am tired.

   d. I do not live in New England.

2. Let p represent the statement "it is warm" and q represent the statement "it is rainy." Represent each statement using logical symbols.

   a. It is warm and it is rainy.

   b. It is cold and it is rainy.

   c. It is warm or it is rainy.

   d. It is cold and it is not rainy.

3. Let p represents the statement "There is life on Mars." Let q represents the statement "There is no life on the moon." Which statement represents $p \wedge q$?

   a. There is life on Mars or no life on the moon.

   b. There is life on both Mars and the moon.

   c. There is life on Mars but not on the moon.

   d. There is life on neither Mars nor the moon.

4. Let p represent the statement "$5 + 2 = 7$" and q represent the statement "$5 + 2 < 7$." Write a statement that represents $\sim p \wedge (\sim q)$.

5. Fill in the columns below to create a truth table for $p \wedge (\sim q)$.

p	q	$\sim q$	$p \wedge (\sim q)$
T	T		
T	F		
F	T		
F	F		

6. Fill in the columns below to create a truth table for $\sim p \wedge q$.

p	q	$\sim p$	$\sim p \wedge q$
T	T		
T	F		
F	T		
F	F		

7. Examine the last columns of the truth tables in Exercises 5 and 6. Are the expressions $p \wedge (\sim q)$ and $\sim p \wedge q$ logically equivalent? Explain.

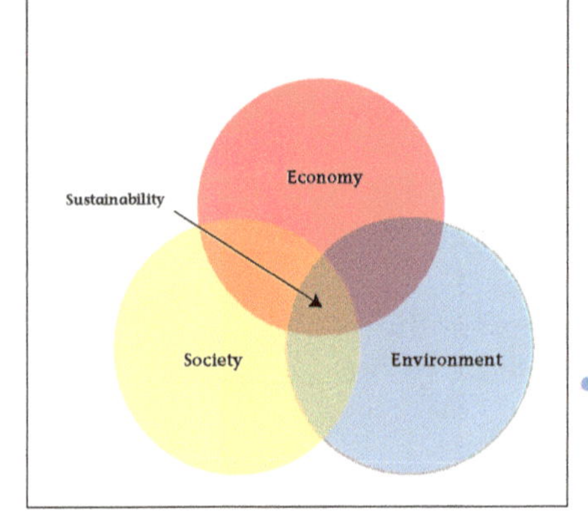

Lesson 10.2

Set Notation

When you think of a **set** in mathematics, you usually think of a collection of objects called "elements" or "members" of the set. For example, as you learned in Lesson 4.3, the edges of the graph in Figure 10.1 can be written as a set: $E = \{AB, BC, BD\}$.

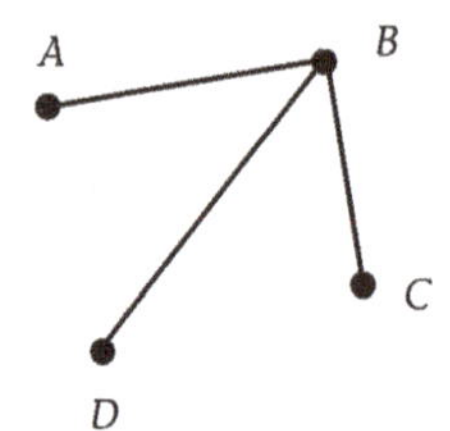

Figure 10.1.

Since edge AB is an element of set E, you can write $AB \in E$ (read "AB is an element of E"). And since edge AD is *not* an element of set E, you can write $AD \notin E$ (read "AB is not an element of E").

Suppose a graph has no edges. That is, its set of edges has no members. This set is called the **empty set** or the **null set**. The symbol $\varnothing$ is used to denote a set with no elements at all.

If every element of set A is also in set B, you can say that A is a **subset** of B. This can be written as $A \subset B$. For example, the set V containing the vertices of the graph in Figure 10.1 can be written: $V = \{A, B, C, D\}$. Suppose $W = \{A, C, D\}$. Since every member of W is also in V, you can say that $W \subset V$.

Operations on Sets

The **union** of sets M and N is the set consisting of all of the elements in M or in N. This union can be written as $M \cup N$. Suppose $M = \{1, 2, 3, 4\}$ and $N = \{4, 5\}$. Then $M \cup N = \{1, 2, 3, 4, 5\}$. Notice that 4 is an element of both sets but is listed only once in the union.

The **intersection** of sets M and N, written as $M \cap N$, consists of only the elements that belong to both M and N. So, $M \cap N = \{4\}$ since 4 is the only element that appears in both sets.

It is possible for two sets to have no elements in common. In that case, the sets are said to be **disjoint** and their intersection is the empty set.

Example 1

$A = \{0, 2, 4, 6, 8\}$; $B = \{3, 4, 5, 6\}$; $C = \{1, 3, 5, 7, 9\}$

Find

a) $A \cap B$ b) $A \cup B$

c) $A \cap C$ d) $B \cup C$

Solution:

a) $A \cap B = \{4, 6\}$ b) $A \cup B = \{0, 2, 3, 4, 5, 6, 8\}$

c) $A \cap C = \varnothing$ (the sets are disjoint) d) $B \cup C = \{1, 3, 4, 5, 6, 7, 9\}$

In Lesson 5.1 the complement of a graph G is defined as a graph with the same vertices as G, but its edges are those not in G. It is also possible to find the **complement** of a set. But before this can be defined, the **universal set** must be defined. The universal set, U, is the set of all of the elements in a particular set theory situation. For example, if the situation involves the number of spots showing when two dice are rolled, then $U = \{2, 3, 4, 5, 6, 7, 8, 9, 10, 11, 12\}$.

The complement of a set A is a set that contains all of the elements of U that are not in A and is denoted $\bar{A}$. For example, if $U = \{2, 3, 4, 5, 7, 8, 9\}$ and $A = \{2, 4, 5, 9\}$, then $\bar{A} = \{3, 7, 8\}$.

Point of Interest

The complement of A is sometime written as A' or $\sim A$.

Venn Diagrams

Relationships among sets can be depicted with Venn diagrams. The shaded regions in Figures 10.2 and 10.3 show the union and the intersection of sets X and Y respectively.

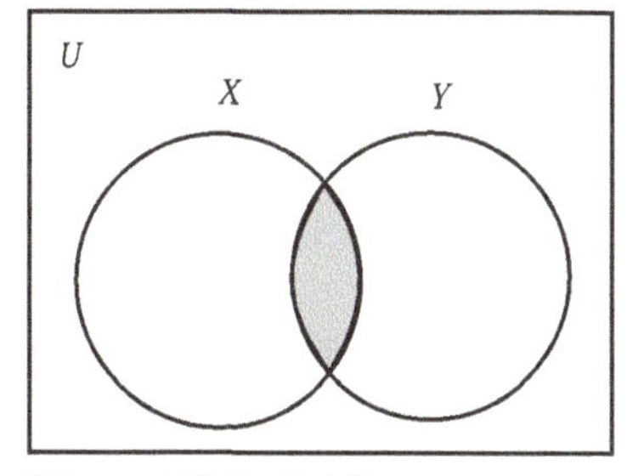

Figure 10.2. $X \cup Y$.

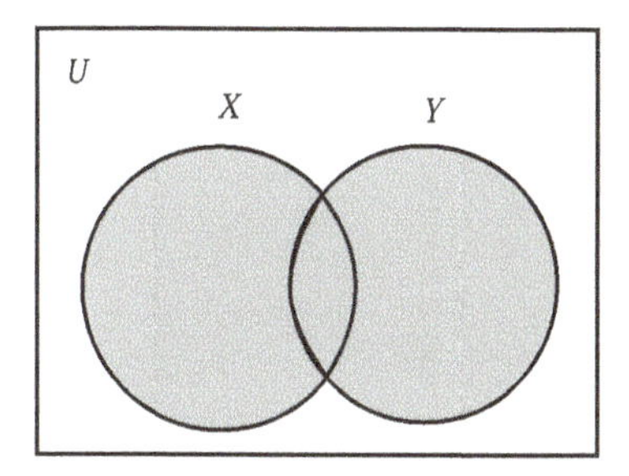

Figure 10.3. $X \cap Y$.

Figure 10.4 shows two intersecting sets, R and S. Figure 5 shows two sets, M and N, that are disjoint.

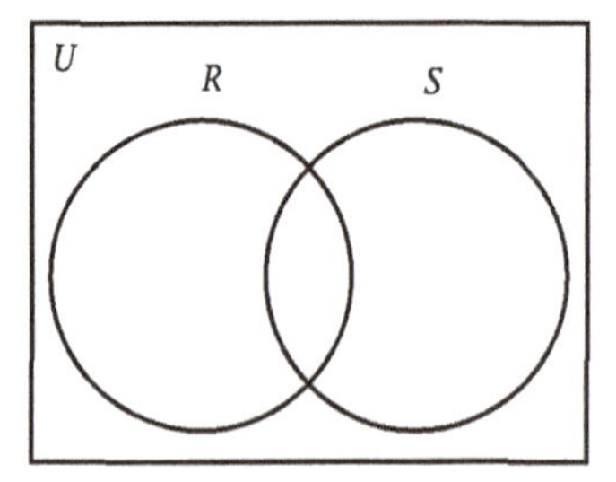

Figure 10.4. Intersecting sets.

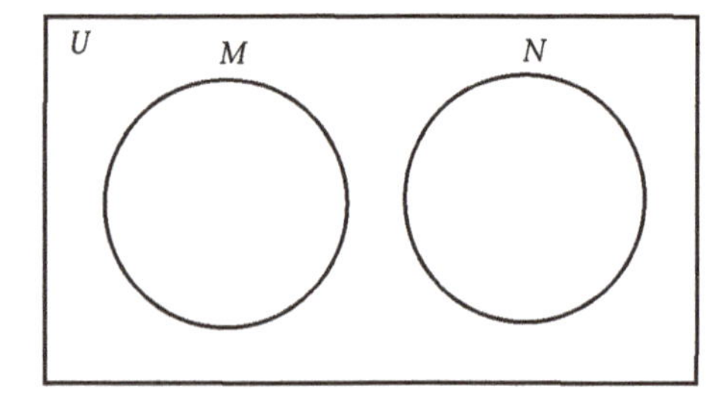

Figure 10.5. Disjoint sets.

The shaded area in Figure 10.6 shows $\bar{A}$.

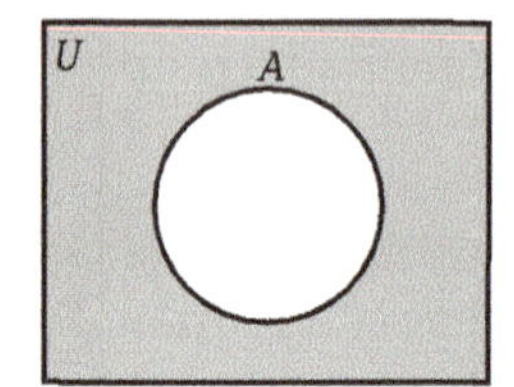

Figure 10.6.

Example 2

Figure 10.7 shows a Venn diagram with two intersecting sets R and S. Which regions of the figure correspond to the following descriptions?

1. R 2. $R \cap S$

3. $R \cup S$ 4. $\overline{R \cup S}$

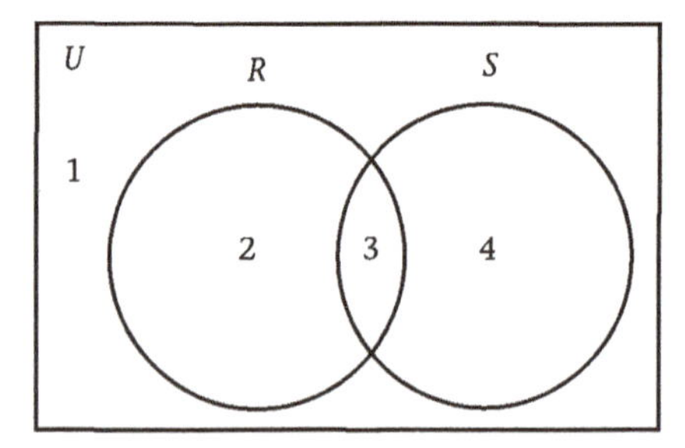

Figure 10.7.

Solution:

1. Regions 2 and 3 2. Region 3

3. Regions 2, 3, and 4 4. Region 1

Exercises

1. Let $A = \{1, 2, 3\}$, $B = \{3, 4\}$, $C = \{1, 2\}$, and $D = \{3\}$. Which of the following statements are true?

   a. $C \subset A$

   b. $B \subset A$

   c. $D \subset A$

   d. $B \subset D$

   e. $D \subset B$

2. For the set $A = \{1, 2, 3, 4, 5\}$, list all of the subsets of A that have exactly two elements.

3. Is $\{a, d, f\}$ a subset of $\{w, a, f, f, l, e\}$? Explain.

4. What do you know if $A \subset B$ and $B \subset A$?

5. Let $R = \{1, 2, 3, 4, 5, 6\}$, $S = \{1, 3, 5\}$, $T = \{2, 4, 6, 8\}$.

   Find

   a. $R \cup T$

   b. $T \cap R$

   c. $T \cup S$

   d. $S \cap T$

6. The universal set, U, is equal to $\{1, 2, 3, 4, 5, 6\}$ and $W = \{1, 3, 5\}$. Find $\overline{W}$.

7. The figure below shows a Venn diagram with two disjoint sets R and S. Which regions of the figure correspond to the given descriptions?

   a. $R \cup S$

   b. $\overline{R \cup S}$

   c. $R \cap S$

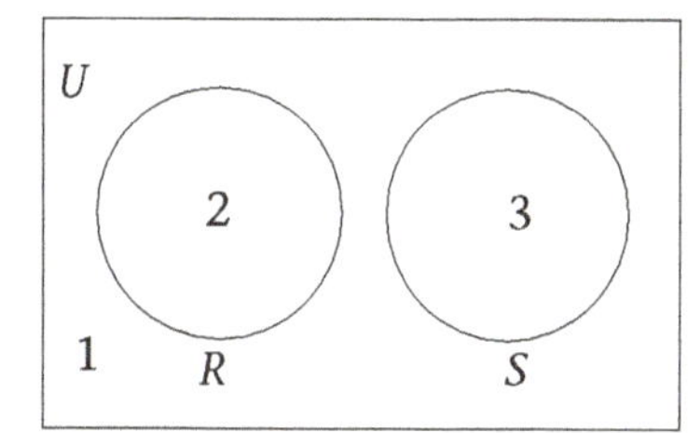

8. A student is using a search engine to look for information about dogs and cats on the Internet. Use the Venn diagrams below to answer the following questions.

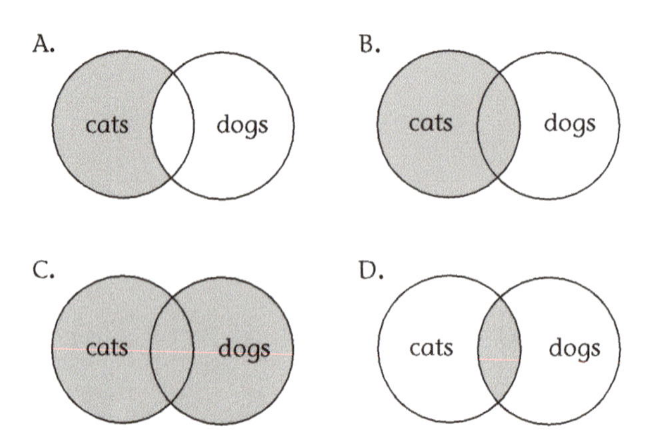

a. Which diagram describes the search if the student wants information on **both** of the terms?

b. Which diagram describes the search if the student wants information on **at least one** of the terms?

c. Which diagram describes the search if the student wants information on cats **not** dogs of the terms?

9. The figure below shows a Venn diagram with three intersecting sets R, S, and T. Which regions of the figure correspond to the given descriptions?

a. $R \cap T$

b. $S \cup T$

c. $R \cup S \cup T$

d. $R \cap S \cap T$

e. $\overline{R \cup S \cup T}$

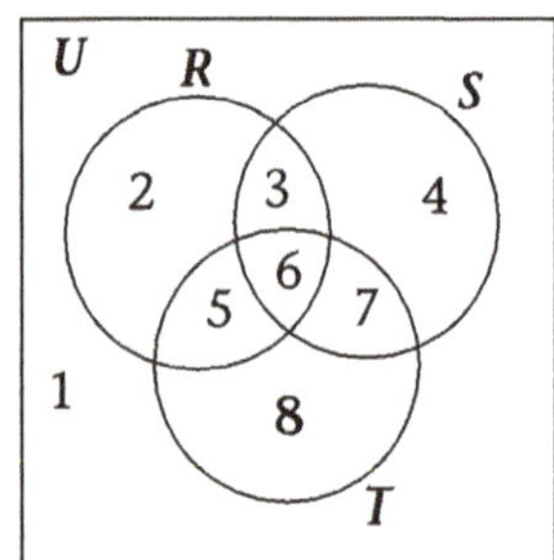

Lesson 10.3

Bin Packing

Many problems in discrete mathematics are optimization problems. That is, they are concerned with maximizing or minimizing something. For example, suppose a carpenter must cut pieces from boards of fixed length for a project such as building a bookcase or siding a house. To control costs, the carpenter wants to minimize the number of boards used. This is an example of a stock-cutting problem.

Stock-cutting problems are also called bin-packing problems. That is because the boards can be considered bins into which pieces of various sizes are packed. Cutting pieces from boards is an example of one-dimensional bin packing. Cutting pieces of plywood from 4′ × 8′ sheets is an example of two-dimensional bin packing.

Solving a stock-cutting problem by hit-and-miss can be wasteful because, for example, keeping track of leftover pieces can be difficult, especially in large-scale problems. Since bin packing has many applications, efficient algorithms are important.

Consider a carpenter who needs pieces of lengths 6, 6, 5, 5, 5, 2, 2, 2, 2, 3, 3, and 3 feet cut from boards 8 feet long.

Next-Fit Algorithms

In a next-fit algorithm, the carpenter cuts pieces in the order they are listed and discards the remainder of a board as soon as a piece can no longer be cut from it. Here are the steps:

1. Cut the piece listed first from a new board.
2. Cut the piece listed next from the previous leftover piece. If the leftover piece is not long enough, discard it and use a new board.
3. Continue in this way. Never go back to a leftover board once it is discarded.

Figure 1 shows how the first five boards are used. The pieces without shading are waste.

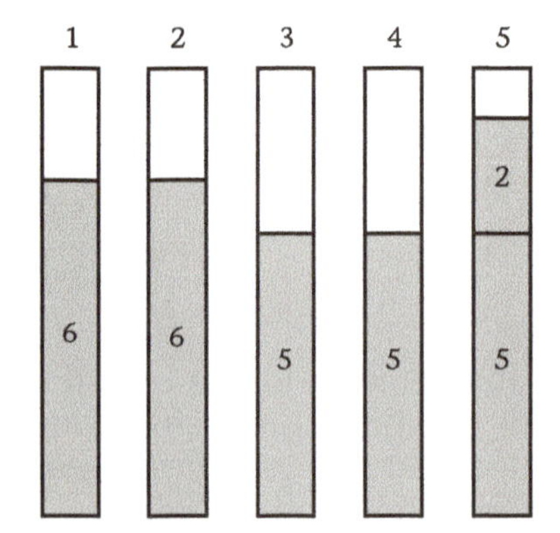

Figure 1.

If the carpenter continues with this algorithm, eight boards are used.

Before applying this algorithm, the carpenter could choose to sort the pieces in increasing order or decreasing order. If the carpenter chooses the latter, the algorithm is known as **next-fit decreasing**. In this example, the list is now 6, 6, 5, 5, 5, 3, 3, 3, 2, 2, 2, 2. If the carpenter applies the next-fit algorithm after sorting, seven boards are used.

First-Fit Algorithms

In a first-fit algorithm, the carpenter lines up waste pieces in the order in which they are cut. To cut a new piece, the carpenter first checks the waste pieces in order until one long enough is found. If none is found, the new piece is cut from a new board.

Figure 2 shows how the pieces are cut in this example. Note that only six boards are required, and the three waste pieces are 1, 1, and 2 feet in length.

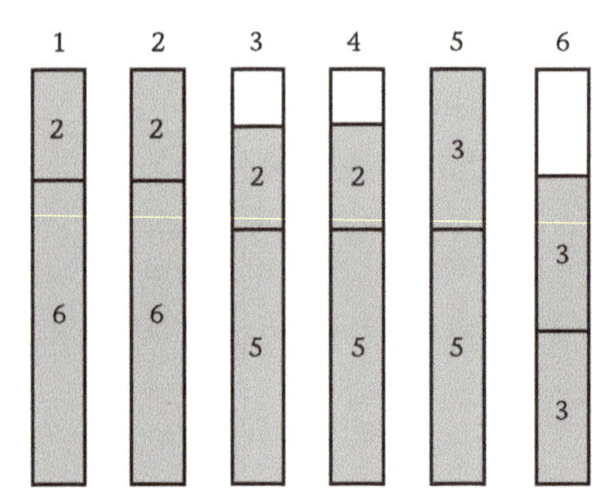

Figure 2.

Again, the first-fit algorithm can be applied in a slightly different way by first sorting the lengths in decreasing order, in which case the algorithm is called **first-fit decreasing**. In this example, first-fit decreasing also uses six boards and wastes a total of 4 feet. But the waste is a single four-foot piece, which might be more useful to the carpenter in another job than the leftover pieces that result from the original first-fit algorithm.

Worst-Fit Algorithms

In a worst-fit algorithm, the carpenter inspects all waste pieces before cutting a new piece and cuts it from the largest remaining waste piece. Of course, if there is no waste piece of sufficient length, the carpenter uses a new board.

Figure 3 shows how the worst-fit algorithm works in this example.

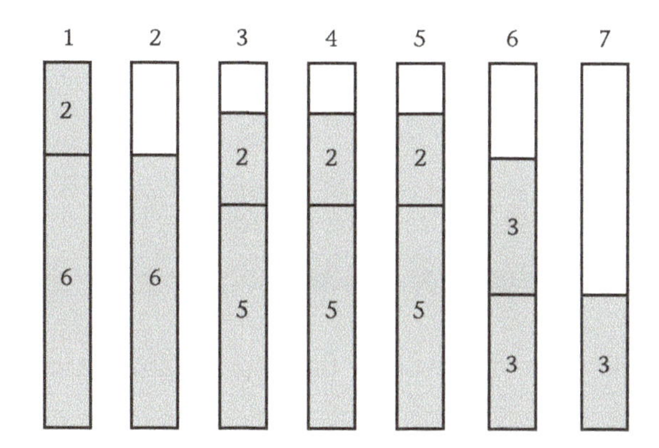

Figure 3.

The carpenter can sort the lengths in decreasing order to make this algorithm a **worst-fit decreasing** algorithm. In this case, worst-fit decreasing yields a solution as good as first-fit decreasing: six boards are used, and there is one waste piece four feet long.

You are probably wondering whether first-fit decreasing and worst-fit decreasing always yield the best possible solution in one-dimensional bin-packing problems. They do not. In fact, mathematicians have not found a single algorithm that produces an optimal solution in all such problems. Moreover, they have discovered that bin-packing algorithms sometimes produce paradoxes.

For example, Ronald Graham has found a bin-packing problem in which first-fit decreasing actually increases the number of bins used when one of the items to be packed is removed!

Mathematician of Note

Ronald Graham (1935–)

Ronald Graham is Chief Scientist at the California Institute for Telecommunication and Information Technology and a member of the Department of Computer Science and Engineering at the University of California-San Diego.

Exercises

1. A plumber needs pieces of pipe of lengths 6, 6, 5, 2, 5, 2, 2, 2, 2, 3, 3, and 3 cut from pipes that are 8 feet long.
   a. Use the first-fit algorithm to determine how many 8-foot pipes are required.
   b. Make a diagram to show how the 8-foot pipes are used.

2. Materials such as paper are often produced in long rows. These long rolls are called *raws*. As customers specify the lengths they need, the raws are then cut into smaller pieces called *finals*. Consider a company that carries 10-foot long rolls of paper and an order of finals of lengths 6, 3, 7, 6, and 4 feet.
   a. Use the next-fit algorithm to determine how many rolls of raws must be cut to fill the order.
   b. Use the next-fit decreasing algorithm to determine how many rolls of raws must be cut to fill the order.

3. Consider a company that carries 10-foot long rolls of paper and an order of finals of lengths 7, 5, 7, 3, 2, and 4 feet.

   a. Use the first-fit algorithm to determine how many rolls of raws must be cut to fill the order.

   b. Use the first-fit decreasing algorithm to determine how many rolls of raws must be cut to fill the order.

4. Consider a company that carries 10-foot long rolls of paper and an order of finals of lengths 7, 5, 7, 3, 2, and 4 feet.

   a. Use the worst-fit algorithm to determine how many rolls of raws must be cut to fill the order.

   b. Use the worst-fit decreasing algorithm to determine how many rolls of raws must be cut to fill the order.

5. Using bins with a capacity 10, apply the next-fit algorithm, the first-fit algorithm, and the worst-fit algorithm to the list 3, 6, 2, 1, 5, 7, 2, 4, 1, and 6. For each algorithm, how many bins are required?

Lesson 10.4

Linear Programming

Linear programming is a mathematical process that uses linear functions to solve certain kinds of optimization problems. The word "programming" is somewhat confusing since neither computer nor calculator programming is involved.

Since linear programming uses continuous linear functions, some people do not consider it a part of discrete mathematics. However, it often appears in courses that include many discrete topics.

Linear programming originated in World War II, when it was used by George Dantzig to solve problems involving allocation of military supplies.

Today linear programming is used in many industries. For example, in oil refining, it is used to solve a variety of problems including creating gasoline blends. Consider a simple example.

Mathematician of Note

George Dantzig (1914–2005) served as Professor of Operations Research and Computer Science at Stanford University.

A manager in charge of gasoline blending must produce at least 60,000 barrels of gasoline with an octane rating of at least 87. To create the blend, the manager combines two formulations. Formulation A has an octane rating of 92 and costs $28 per barrel. Formulation B has an octane rating of 85 and costs $23 per barrel. The manager can purchase no more than 50,000 barrels of each formulation. The manager wants to produce the blend at the lowest possible cost.

Constraints and Objective Function

The conditions under which the manager must minimize cost are called **constraints**. In linear programming, constraints are expressed as linear inequalities. An **objective function** describes the quantity to be optimized; in this case, the cost of the blend.

The constraints are:

- The number of barrels of each formulation is positive.
- Neither the number of barrels of formulation A nor the number of barrels of formulation B can exceed 50,000.
- The total number of barrels of the blend must be at least 60,000.
- The octane rating of the blend must be at least 87.

In words, the objective function is: The cost of the blend is $28 per barrel of formulation A used plus $23 per barrel of formulation B used.

Mathematical Representation

An essential part of a mathematical modeling process is the representation of key aspects of a situation in mathematical terms. In linear programming, representation involves choosing variables, writing symbolic inequalities to represent constraints, and writing an objective function in symbolic form.

First, let A represent the number of barrels of formulation A used in the blend and let B represent the number of barrels of formulation B used in the blend. Let C represent the total cost of the blend.

The constraints can now be expressed as:

- $A \geq 0$; $B \geq 0$.
- $A \leq 50{,}000$; $B \leq 50{,}000$.
- $A + B \geq 60{,}000$ or $B \geq 60{,}000 - A$.
- $\frac{92A + 85B}{A + B} \geq 87$, which simplifies to $B \leq \frac{5}{2}A$.

In symbolic form, the objective function is $C = 28A + 23B$.

Feasible Region

Figure 1 is a graph of the inequalities. Since all constraints must be satisfied, only the portion of the graph that satisfies all inequalities is shaded.

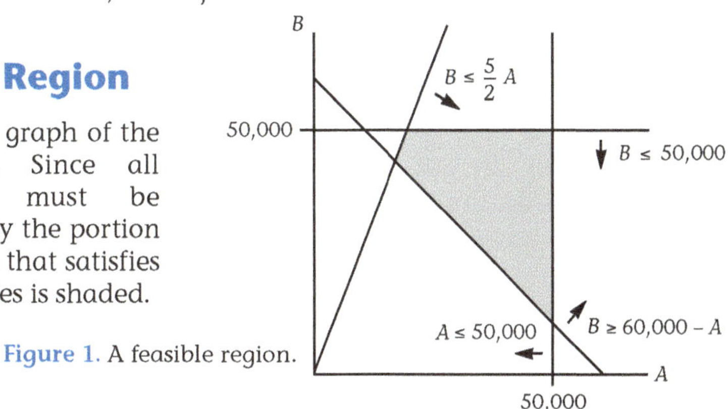

Figure 1. A feasible region.

Since the feasible region contains all pairs (A, B) that satisfy the constraints, a pair that minimizes the objective function must lie in the region.

A fundamental theorem of linear programming says that the pair that minimizes (or maximizes) the objective function under these constraints lies at one of the region's vertices. Therefore, finding the minimum cost requires evaluating the cost function at the region's vertices only. Although the coordinates of some vertices are obvious, finding others requires solving systems of equations. For example, to find the coordinates of this region's leftmost vertex, solve the system

$$B = 60{,}000 - A$$
$$B = \frac{5}{2}A.$$

The following table shows the function's values at each vertex.

A	B	$28A + 23B$
50,000	50,000	2,550,000
50,000	10,000	1,630,000
20,000	50,000	1,710,000
17,143	42,857	1,465,715

A minimum cost of approximately $1,465,700 occurs when approximately 17,143 barrels of formulation A and 42,857 barrels of formulation B are used.

Real-world linear programming often operates with thousands of constraints. Finding and checking all vertices in such situations is time consuming. George Dantzig invented a matrix procedure called the **simplex method** that requires finding and checking only a small fraction of all vertices and that is readily implemented on a computer. In 1984, Narendra Karmarkar invented an algorithm that is considerably more efficient than the simplex algorithm.

Exercises

1. A feasible region has five corners at points (1, 2), (2, 1), (5, 1), (5, 3), and (3, 4). If the objective function is $P = 5x + 2y$, which point maximizes the profit?
2. A feasible region has five corners at points (1, 2), (2, 1), (5, 1), (5, 3), and (3, 4). If the objective function is $P = 5x + 2y$, what is the maximum profit possible?

3. a. Graph the following system of constraints.

   $x \geq 2$

   $y \geq 0$

   $2x + y \leq 12$

   b. Find the coordinates of the three vertices of the region.

   c. Find the minimum and maximum values of the objective function $P = 25x + 30y$ for this region.

4. a. Graph the following system of constraints.

   $x \geq 0$

   $y \geq 0$

   $2x - y \geq -1$

   $x + y \leq 4$

   b. Find the coordinates of the four vertices of the region.

   c. Find the minimum and maximum values of the objective function $C = 5x + 2y$ for this region.

5. Find the minimum and maximum values of the objective function $C = 2x + 6y$.

   Constraints:

   $x \geq 1$

   $y \geq 2$

   $2x + y \leq 12$

6. A crafter makes two sizes of picture frames. A large frame uses 5 feet of framing and a medium frame uses 4 feet. The crafter has 40 feet of framing and wants to make at least 4 large frames. Let x represent the number of large frames and y represent the number of medium frames.

   a. Write three constraints.

   b. Graph the system of constraints.

   c. The profit on a large frame is \$35 and the profit on a medium frame is \$30. Write an objective function for the crafter's total profit P.

   d. How many large frames and how many medium frames should the crafter make in order to maximize profit?

Chapter 10 Review

1. Write a summary of what you think are the important point of the lessons in this chapter.

2. Give the negation of the statement "I am 18 years old and I do not like pretzels."

3. Let p represent the statement "It is hot in Texas." Let q represent the statement "No one lives near Lake Linden."

   a. Write a statement that represents $\sim p$.

   b. Write a statement that represents $p \wedge q$.

   c. Write a statement that represents $p \vee q$.

4. Fill in the columns below to create a truth table for $p \vee (\sim q)$.

p	q	$\sim q$	$p \vee (\sim q)$
T	T		
T	F		
F	T		
F	F		

5. Let $A = \{2, 4, 6, 8, 10\}$, $B = \{1, 3, 5, 7, 9\}$, and $C = \{0, 1, 2, 3, 4, 5\}$

   a. Find $A \cup C$.

   b. Find $A \cap B$.

   c. Find $C \cap B$.

6. The universal set, U, is equal to {2, 4, 6, 8 10, 12} and $A = \{2, 4, 10\}$. Find $\overline{A}$.

7. Which is *not* an example of a “bin-packing” algorithm?

   a. the first-fit algorithm

   b. the next-fit algorithm

   c. the second-fit algorithm

   d. the worst-fit algorithm

8. Consider a company that carries 10-foot long rolls of paper and an order of finals of lengths 6, 3, 7, 6, and 4 feet. Use the first-fit algorithm to determine how many rolls of raws must be cut to fill the order.

9. A feasible region has four corners at points (1, 4), (4, 1), (1, 0), and (0, 2). The objective function is $P = 7x + y$.

   a. Which point maximizes the profit?

   b. What is the maximum profit possible?

10. Find the minimum and maximum values of the objective function $C = 14x - 3y$.

    Constraints:
    $x \geq 1$
    $y \geq 0$
    $y \leq -2x + 6$

Bibliography

COMAP. 2010, *For All Practical Purposes: Mathematical Literacy in Today's World.* 8th ed. New York: W. H. Freeman.

COMAP. 1997 *Principles and Practice of Mathematics.* 1st ed. New York: Springer-Verlag.

Dossey, John, A. Otto, L Spence, and C Vanden Eynden. 2006. *Discrete Mathematics.* 5th ed. Upper Saddle River, NJ: Pearson.

Gass, Saul. 1990. *An Illustrated Guide to Linear Programming.* New York: Dover Publications.

Gass, Saul. 2003. *Linear Programming: Methods and Applications.* New York: Dover Publications.

Graham, R. *Combinatorial Scheduling, in Mathematics Today.* L. Steen (Ed.), Springer-Verlag, New York, 1978, p.183-211.

Henle, James, Garfield, and Thomas Tymoczko. 2011. *Sweet Reason: A Field Guide to Modern Logic.* 2nd ed. Hoboken, NJ: Wiley-Blackwell.

Sriskanddarajah, Jeganathan. 1992. *Optimality Pays: An Introduction to Linear Programming.* Lexington, MA: COMAP, Inc.

Answers to Selected Exercises

CHAPTER 1

Lesson 1.1

11. a. 120 720

    b. The number of schedules possible when there are n choices is n times the number of schedules possible when there are $n - 1$ choices.

12. 8 12 17

Lesson 1.2

6. a.

A	30.8%	69.2%
B	19.2%	0.0%
C	23.1%	0.0%
D	26.9%	30.8%

8. Plurality: B Borda: D Runoff: C Sequential runoff: C

12. $C_n = C_{n-1} - 1$

Lesson 1.3

3. a. A and B

4. a. A

   b. B

8. a. A

   b. C

9. a. From first to last: A, B, C, D

   b. From first to last, the new ranking is B, A, D.

11. a. There are two new comparisons. A total of three comparisons must be made.

    b. There are three new comparisons. A total of six comparisons must be made.

    c.

1	0	0
2	1	1
3	2	3
4	3	6
5	4	10
6	5	15

Lesson 1.4

1. Nondictatorship

2. If the method were repeated, the same ranking might not result. Therefore, condition 5 is violated. Nondictatorship (condition 1) is also violated.

4. Condition 4

7. None

14. a. { } {A} {B} {C} {A, B} {A, C} {B, C} {A, B, C}

    b. { } {A} {B} {C} {D} {A, B} {A, C} {A, D} {B, C} {B, D} {C, D} {A, B, C} {A, B, D} {A, C, D} {B, C, D} {A, B, C, D}

    c. $V_n = 2V_{n-1}$

16. 4 5

17. $V1_n = V1_{n-1} + 1$, or $V1_n = V1_{n-1}$

Lesson 1.5

1. a. The possible coalitions: { ; 0} {A; 3} {B; 2} {C; 1} {A, B; 5} {B, C; 3} {A, C; 4} {A, B, C; 6}

    The winning coalitions: {A, B; 5} {A, C; 4} {A, B, C; 6}

    b. A:3 B:1 C:1

    c. A:2 B:2 C:0

6. $C_n = 2C_{n-1}$

7. { } {A} {B} {C} {D} {A, B} {A, C} {A, D} {B, C} {B, D} {C, D} {A, B, C} {A, B, D} {A, C, D} {B, C, D} {A, B, C, D}

8. a. The winning coalitions are: {A, B; 51%} {A, C; 51%} {A, B, C; 76%} {A, B, D; 75%} {A, C, D; 75%} {B, C, D; 74%} {A, B, C, D; 100%}. Of these, A is essential to 5, B to 3, C to 3, and D to 1.

   b. The winning coalitions are: {A, B; 88%} {A, C; 54%} {A, D; 52%} {A, B, C; 95%} {A, B, D; 93%} {A, C, D; 59%} {B, C, D; 53%} {A, B, C, D; 100%}. Of these, A is essential to 6, B to 2, C to 2, and D to 2.

   c. In part a, D has 24% of the stock and one-twelfth of the power. In part b, D has 5% of the stock, but two-twelfths of the power.

Chapter 1 Review

2. a. D

   b. B

   c. A

   d. E

   e. C

   f. C

3. 19, 42, 89

4. a. It can occur in either the runoff or sequential runoff model.

5. a. Wilson; no

   b. They ranked him last.

   c. The voters in the last group could have switched to Roosevelt, their second choice, and thereby prevented Wilson from winning.

   d. Borda, runoff, and Condorcet give the election to Roosevelt.

6. Conditions 2, 3, and 5

7. Arrow's fourth condition

8. Arrow proved that no group ranking model that ranks three or more choices always adheres to his five fairness conditions.

9. Yes.

10. a. Clinton: .43 + .2 × .38 + .35 × .19 = .5725 or about 57%
Bush: .38 + .15 × .43 + .3 × .19 = .5015, or about 50%
Perot: .19 + .3 × .43 + .2 × .38 = .395, or about 40%

12. This is a 5-3-1 Borda point system.

13. a. The plurality winner is A in each case.

b. No.

14. a. {A, B, C, D; 12}, {A, B, C; 10}, {B, C, D; 8}, {A, C, D; 9}, {A, B, D; 9}, {A, B; 7}, {A, C; 7}

b. A: 5, B: 3, C: 3, D: 1

c. No, A's power is disproportionately high, while D's is low.

d. All voters now have equal power.

15. The indices of members A, B, and E approximately reflect their share of the total vote. Member C's index is higher than it should be, at the expense of member D.

CHAPTER 2

Lesson 2.2

3. a. $40,000 $35,000

b. $80,000 – $40,000 = $40,000

c. $40,000

d. $35,000 $42,500 $37,500

f. Marmaduke receives $40,000.

4. a.

	Amy	Bart	Carol
Final settlement	4,788.89	4,988.89	4,722.23

7. a. $\begin{bmatrix} 13{,}000 & 121{,}000 & 0 \\ 11{,}000 & 127{,}000 & 0 \\ 12{,}000 & 123{,}000 & 0 \end{bmatrix}$

b. The value to Alan of the items that Betty receives

Lesson 2.3

1. b. 42.857

   c. Sophomore quota: 10.83

   d. Sophomore seats: 11; junior seats: 6; senior seats: 4

2. a. Sophomore adjusted ratio: $464 \div 11 = 42.18$; junior adjusted ratio: $240 \div 6 = 40$

   b. Sophomore seats: 11; junior seats: 6; senior seats: 4

4. a.

4.545	4.762
10.454	10.952

   When the ideal ratio is 22, the decimal part of the 100-member class is larger than the decimal part of the 230-member class. The situation is reversed when the ideal ratio drops to 21.

   b. For a small class

Lesson 2.4

1. a.

10.83	10	10	11	11
5.6	5	5	6	6
4.57	4	4	5	5

   b. Sophomores: 11; juniors: 6; seniors: 4

   c.

43.63	43.82
43.56	43.83

   d. 43.56 (Sr.) 43.63 (Jr.) 44.19 (So.)
   Sophomore seats: 11; junior seats: 6; senior seats: 4

   e. 43.82 (Jr.) 43.83 (Sr.) 44.24 (So.)
   Sophomore seats: 11; junior seats: 5; senior seats: 5

   f. The model favored by a given class can be seen in the following table of final apportionment results:

	Hamilton	Jefferson	Webster	Hill
Sophomore	11	11	11	11
Junior	6	6	6	5
Senior	4	4	4	5

4. a. 50

   b. 3.7 4
      2.6 3
      1.6 2

   c. 52.8571
      52
      53.3333

   d. Freshman: 21; sophomore: 4; junior: 3; senior: 2

Lesson 2.5

1. a. Ava feels that she has exactly one-third.

   b. Bert and Carlos each could feel that he receives more than one-third.

4. One-sixth or 0.16

5. b. One-sixth

   c. Four-sixths or 0.67

   d. One-third or 0.33

8. a. $2 \times 3 = 6$

   b. $k(k + 1)$ or $k^2 + k$

9. a. Yes. No.

   b. Yes. Yes.

   c. Probably not. Yes.

Lesson 2.6

1. a. $k + 1$ $k - 1$

   b. $k + 2$ k $2k + 1$ $2k - 1$

2. a. New handshakes: 3 Total handshakes: 6

   b. New handshakes: 4 Total handshakes: 10

3. a. 7

   b. k

   c. $H_n = H_{n-1} + (n - 1)$

4. a. 45

b. $\frac{k(k-1)}{2}$ $\frac{2k(2k-1)}{2}$ $\frac{(k+1)k}{2}$

c. $\frac{k(k-1)}{2}$

d. k

e. $\frac{(k+1)k}{2}$

f. $\frac{k(k-1)}{2} + k = \frac{k^2-k}{2} + \frac{2k}{2} = \frac{k^2+k}{2} = \frac{(k+1)k}{2}$

6. a. $V_{k+1} = 2V_k$

b. $V_n = 2^n$

c. $V_{k+1} = 2^{k+1}$

Chapter 2 Review

2. Answers are rounded to the nearest dollar.

	Joan	Henry	Sam
Fair share	$9,213	$8,933	$8,767
Items received	Lot	Computer, stereo	Boat
Cash	$1,213	$6,333	$2,067
Final settlement	$9,676	$9,396	$9,229

3. Answers are rounded to the nearest dollar.

	Anne	Beth	Jay
Fair share	$3,200	$2,967	$3,017
Items received	Car, computer		Stereo
Cash	–$5,600	$2,967	$2,017
Final Settlement	$3,406	$3,172	$3,222

4. Answers are rounded to the nearest dollar.

	Lynn	Pauline	Tim
Fair share	$4,225	$2,150	$2,163
Items received	Car, kayak	Guitar, watch	
Cash	–$2,475	–$750	$2,163
Final settlement	$4,756	$2,416	$2,428

5. a. 10
   b. 64.7, 24.7, 10.6
   c. 65, 25, 10
   d. 64, 24, 10
   e. 9.9538, 9.8800, 9.6364
   f. 65, 25, 10
   g. 65, 25, 11
   h. 10.0310, 10.0816, 10.0952
   i. 64, 25, 11
   j. 65, 25, 11
   k. 10.0313, 10.0837, 10.1067
   l. 64, 25, 11
   m. 64, 25, 11
   n. State A gains population and state C loses population, but A loses a seat to C.
6. B would strongly favor the Jefferson model, while A would strongly oppose it.
7. Balinski and Young proved that any apportionment model sometimes produces one of three undesirable results: violation of quota, the loss of a seat when the size of the legislative body increases even if population doesn't decrease, and the loss of a seat by one state whose population increases to another whose population decreases.
8. Arnold and Betty
9. Have each of the original four divide his or her piece into five pieces that he or she considers equal. Have the new person select a piece from each of the others.
10. $k^2 - k$ or $(k - 1)$
11. $\frac{k^2 - k}{2}$ or $\frac{k(k-1)}{2}$

CHAPTER 3

Lesson 3.1

3.

	Jacket	Shirt	Pants
Shop 1	25	75	75
Shop 2	30	50	50
Shop 3	20	40	35

4. a. $A_{21} = \$1.09$, $A_{12} = \$10.86$, $A_{32} = \$3.89$

   b. A_{21} represents the cost of drinks at Vin's.
   A_{12} represents the cost of pizza at Toni's.
   A_{32} represents the cost of salad at Toni's.

5. S_3 represents the cost of pizza at Sal's.

9. $$A+B = \begin{bmatrix} 10.10 + 1.15 & 10.86 + 1.10 & 10.65 + 1.25 \\ 3.69 + 0.00 & 3.89 + 0.45 & 3.85 + 0.50 \end{bmatrix}$$

$$C = \begin{array}{r} \\ \text{Pizza} \\ \text{Salad} \end{array} \begin{array}{c} \begin{array}{ccc} \text{Vin's} & \text{Toni's} & \text{Sal's} \end{array} \\ \begin{bmatrix} \$11.25 & \$11.96 & \$11.90 \\ \$3.69 & \$4.34 & \$4.35 \end{bmatrix} \end{array}$$

12. Decrease in statistics is shown with a negative sign.

14. b. Commutative property:

$$A+B=B+A=\begin{bmatrix} 5 & 1 \\ 1 & 6 \end{bmatrix}$$

Associative property:

$$A+(B+C)=\begin{bmatrix} 4 & -2 \\ 3 & 1 \end{bmatrix} + \begin{bmatrix} 3 & 7 \\ -1 & 4 \end{bmatrix} = \begin{bmatrix} 7 & 5 \\ 2 & 5 \end{bmatrix}$$

$$(A+B)+C=\begin{bmatrix} 5 & 1 \\ 1 & 6 \end{bmatrix} + \begin{bmatrix} 2 & 4 \\ 1 & -1 \end{bmatrix} = \begin{bmatrix} 7 & 5 \\ 2 & 5 \end{bmatrix}$$

16. a. $$A+O=\begin{bmatrix} 4 & -2 \\ 3 & 1 \end{bmatrix} + \begin{bmatrix} 0 & 0 \\ 0 & 0 \end{bmatrix} = \begin{bmatrix} 4 & -2 \\ 3 & 1 \end{bmatrix} = A$$

$$O+A=\begin{bmatrix} 0 & 0 \\ 0 & 0 \end{bmatrix} + \begin{bmatrix} 4 & -2 \\ 3 & 1 \end{bmatrix} = \begin{bmatrix} 4 & -2 \\ 3 & 1 \end{bmatrix} = A$$

$$A-A=\begin{bmatrix} 4 & -2 \\ 3 & 1 \end{bmatrix} - \begin{bmatrix} 4 & -2 \\ 3 & 1 \end{bmatrix} = \begin{bmatrix} 0 & 0 \\ 0 & 0 \end{bmatrix} = O$$

Lesson 3.2

1. a. T represents the cost of four pizzas with additional toppings and four salads with a choice of two dressings from each of the three pizza houses.

   b. $47.60

   c. T_{12} represents the cost of four pizzas at Toni's. T_{21} represents the cost of four salads at Vin's.

2. a., b.

$$J = \begin{array}{c} \\ \text{Pearl} \\ \text{Jade} \end{array} \begin{array}{c} \begin{array}{cccc} e & p & n & b \end{array} \\ \begin{bmatrix} 16 & 8 & 12 & 10 \\ 40 & 20 & 24 & 18 \end{bmatrix} \end{array}$$

   c. 24 jade necklaces

   d. J_{21} represents the number of jade earrings that the jeweler expects to sell in June. J_{12} represents the number of pearl pins that the jeweler expects to sell in June.

9. $$\text{Rate} \overset{\begin{array}{ccc} \text{CD} & \text{CU} & \text{Bond} \end{array}}{\begin{bmatrix} 0.073 & 0.065 & 0.075 \end{bmatrix}} \begin{array}{c} \text{CD} \\ \text{CU} \\ \text{Bonds} \end{array} \overset{\text{Amt Invest}}{\begin{bmatrix} \$10{,}000 \\ 17{,}000 \\ 12{,}000 \end{bmatrix}} = \$2{,}735$$

10. a. The transpose of a row matrix is a column matrix.

    b. The transpose of a column matrix is a row matrix.

    c. $$M^T = \begin{array}{c} e \\ p \\ n \\ b \end{array} \overset{\begin{array}{cc} p & j \end{array}}{\begin{bmatrix} 8 & 20 \\ 4 & 10 \\ 6 & 12 \\ 5 & 9 \end{bmatrix}}$$

11. a. Hours

	e	p	n	b
Hours	2	1	2.5	1.5

b. [42.5   93.5]

c. Hours

	p	j
Hours	42.5	93.5

d. It takes the jeweler 42.5 hours to make the pearl jewelry and 93.5 hours to make the jade jewelry.

Lesson 3.3

1. a. $AB = \begin{bmatrix} 0 & 5 & 5 \\ -14 & 8 & 15 \end{bmatrix}$

2. a.

	Mike	Liz	Kate
Mike	$261,000	$0	$250
Liz	$235,000	$0	$215
Kate	$255,000	$0	$325

b. The entries in row 1 represent the value to Mike of the items that he, Liz, and Kate received. Row 2 represents the values to Liz and row 3 the values to Kate.

3. a. Matrix Q:

	Burger	Special	Potato	Fries	Shake
Rosa	0	1	0	1	1
Max	1	0	1	0	1

b. Matrix C:

	Cal	Fat	Chol
Burger	450	40	50
Special	570	48	90
Potato	500	45	25
Fries	300	30	0
Shake	400	22	50

c. The dimensions of Q are 2 by 5, and those of C are 5 by 3.

d. The dimensions of the product QC are 2 by 3.

e. The dimensions of C can be described as Foods by Contents, and the dimensions of Q times C can be described as Persons by Contents.

f.
$$R = \begin{array}{c} \\ \text{Rosa} \\ \text{Max} \end{array} \begin{array}{c} \begin{array}{ccc} \text{Cal} & \text{Fat} & \text{Chol} \end{array} \\ \begin{bmatrix} 1{,}270 & 100 & 140 \\ 1{,}350 & 107 & 125 \end{bmatrix} \end{array}$$

9. The diagram below shows the polygons plotted for parts a through h.

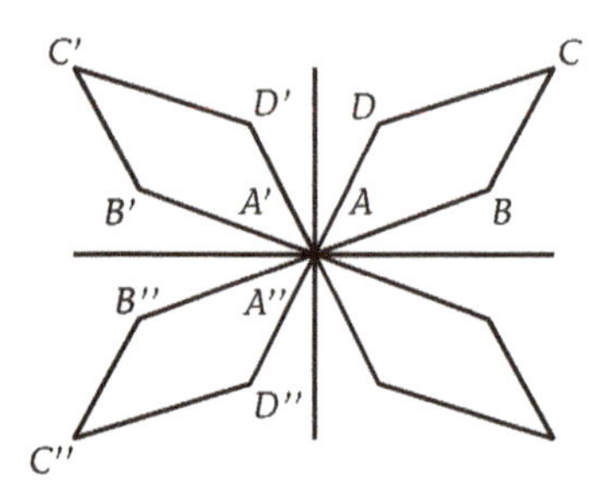

a. $T_1P = \begin{bmatrix} 0 & -6 & -8 & -2 \\ 0 & 2 & 6 & 4 \end{bmatrix}$

c. Polygon $A'B'C'D'$ is the reflection of polygon $ABCD$ in the y-axis.

d. $T_2P = \begin{bmatrix} 0 & -6 & -8 & -2 \\ 0 & -2 & -6 & -4 \end{bmatrix}$

e. Polygon $A''B''C''D''$ is the reflection of polygon $A'B'C'D'$ in the x-axis.

f. $T_2T_1 = \begin{bmatrix} -1 & 0 \\ 0 & -1 \end{bmatrix} = R$

$RP = \begin{bmatrix} 0 & -6 & -8 & -2 \\ 0 & -2 & -6 & -4 \end{bmatrix}$

The effect of R on P is to rotate $ABCD$ 180 degrees about the origin.

Lesson 3.4

1. a. 18.97 newborn rats

   b. 9.96, 8.1, 7.29, 9.36, 2.4

   c. Distribution after 6 months: 18.97, 9.96, 8.1, 7.29, 9.36, 2.4; Total 56 rats

   d. Distribution after 9 months: 18.32, 11.38, 8.96, 7.29, 5.83, 5.62; Total 57 rats

   Distribution after 12 months: 18.02, 10.99, 10.24, 8.06, 5.83, 3.5; Total 57 rats

2. a. 118.4 newborn deer

   b. 30, 24, 21.6, 21.6, 8.4, 0

   c. The product is the number of newborn deer after 1 cycle.

3. a. $\begin{bmatrix} 50 & 30 & 24 & 24 & 12 & 8 \end{bmatrix} \begin{bmatrix} 0.6 \\ 0 \\ 0 \\ 0 \\ 0 \\ 0 \end{bmatrix}$

   b. $\begin{bmatrix} 50 & 30 & 24 & 24 & 12 & 8 \end{bmatrix} \begin{bmatrix} 0 \\ 0.8 \\ 0 \\ 0 \\ 0 \\ 0 \end{bmatrix}$

4. a. Distribution after 3 months: 0, 21, 0, 0, 0, 0; Total 21 rats

   b. Distribution after 3 months: 11, 3, 4.5, 4.5, 4, 3; Total 30 rats

Lesson 3.5

1. a. $P_5 = P_0L^5 = [19.47 \quad 10.81 \quad 9.89 \quad 9.22 \quad 6.45 \quad 3.50]$

   b. Total Population: 59.35 rats

   c. $P_7 = [20.47 \quad 12.11 \quad 10.51 \quad 8.76 \quad 7.12 \quad 4.43]$

   Total Population: 63.41 rats

2. To reach 250 members:

	Cycles	Population	Years
a.	61	253.2	15.25
b.	69	250.9	17.25

3. a. 4 cycles: 56.65 −1.31%
   5 cycles: 59.35 4.77%
   6 cycles: 61.76 4.06%

   b. The population appears to decline then increase again.

   c. $P_{25} = 108.488$, $P_{26} = 111.789$, $P_{27} = 115.191$. The growth rate in each case is 3.04%.

4. a. $P_{25} = 86.054$, $P_{26} = 88.674$, $P_{27} = 91.372$
   $P_{25} = 67.098$, $P_{26} = 69.142$, $P_{27} = 71.247$
   $P_{25} = 54.823$, $P_{26} = 56.491$, $P_{27} = 58.210$
   $P_{25} = 154.512$, $P_{26} = 159.215$, $P_{27} = 164.060$

Chapter 3 Review

2. a. 6 elements

   b. $C_{12} = -2$; $C_{21} = -1$

3. a. $A + C = \begin{bmatrix} 6 & -2 \\ 3 & 7 \\ 2 & -1 \end{bmatrix}$

   b. It is not possible to subtract these matrices because their orders are not the same.

c. $(A + C) - D = \begin{bmatrix} 5 & 1 \\ 1 & 9 \\ -1 & 0 \end{bmatrix}$

d. $2A + D = \begin{bmatrix} 5 & -3 \\ 10 & 12 \\ 1 & 5 \end{bmatrix}$

4. a. $L = \begin{array}{c} \\ \end{array} \begin{matrix} \text{Mex} & \text{Chips} & \text{Salsa} & \text{Drinks} \end{matrix}$

$L = \begin{bmatrix} 35 & 6 & 6 & 12 \end{bmatrix}$

b. L_2 = number of bags of chips ordered

L_4 = number of six-packs of drinks

c. Cost = Number $\begin{matrix} \text{Mex} & \text{Chips} & \text{Salsa} & \text{Drinks} \\ [\ 35 & 6 & 6 & 12\ ] \end{matrix} \times \begin{bmatrix} \text{Mex} \\ \text{Chips} \\ \text{Salsa} \\ \text{Drinks} \end{bmatrix} \begin{bmatrix} \$4.50 \\ \$1.97 \\ \$2.10 \\ \$2.89 \end{bmatrix}$

= [$216.60]

5. a. C =

	Lodging	Food	Rec
Crystal	$13.00	$20.00	$5.00
Springs	$12.50	$19.50	$7.50
Bear	$20.00	$18.00	$0.00
Beaver	$40.00	$0.00	$0.00

b. C_{22} = $19.50

C_{43} = $0.00

c. C_{13} = cost for recreation at Crystal Lodge

C_{31} = cost for lodging at Bear Lodge

6. a.

	System	Cartridge	Case
Z-Mart	$39.50	$24.50	$8.50
Base	$49.90	$29.95	$12.50

b.

	System	Cartridge	Case
Z-Mart	$35.55	$22.05	$7.65
Base	$39.92	$23.96	$10.00

c.

$$\begin{array}{l} \\ \text{Z-Mart} \\ \text{Base} \end{array} \begin{array}{c} \begin{array}{ccc} \text{System} & \text{Cartridge} & \text{Case} \end{array} \\ \begin{bmatrix} \$3.95 & \$2.45 & \$0.85 \\ \$9.98 & \$5.99 & \$2.50 \end{bmatrix} \end{array}$$

d.

$$4\begin{bmatrix} \$35.55 & \$22.05 & \$7.65 \\ \$39.92 & \$23.96 & \$10.00 \end{bmatrix} =$$

$$\begin{array}{l} \\ \text{Z-Mart} \\ \text{Base} \end{array} \begin{array}{c} \begin{array}{ccc} \text{System} & \text{Cartridge} & \text{Case} \end{array} \\ \begin{bmatrix} \$142.20 & \$88.20 & \$30.60 \\ \$159.68 & \$95.84 & \$40.00 \end{bmatrix} \end{array}$$

7. a. 3×2

b. 4×3

c. Not possible. The number of columns of Q does not equal the number of rows of S.

d. 4×3

8. a.

$$\text{No.} \begin{array}{c} \begin{array}{ccc} \text{Plate} & \text{Large} & \text{Small} \end{array} \\ \begin{bmatrix} 5 & 3 & 7 \end{bmatrix} \end{array}$$

b.

$$\begin{array}{l} \\ \text{Plate} \\ \text{Large} \\ \text{Small} \end{array} \begin{array}{c} \begin{array}{cccc} \text{Ebony} & \text{Walnut} & \text{Rose} & \text{Maple} \end{array} \\ \begin{bmatrix} 100 & 800 & 600 & 400 \\ 200 & 1{,}200 & 1{,}000 & 800 \\ 50 & 500 & 450 & 400 \end{bmatrix} \end{array}$$

c.

$$\begin{array}{cccc} \text{Ebony} & \text{Walnut} & \text{Rose} & \text{Maple} \\ [1{,}450 & 11{,}000 & 9{,}150 & 7{,}200] \end{array}$$

d.

$$\text{No.} \begin{array}{c} \text{Weeks} \\ \begin{bmatrix} 15 & 12 & 14 \end{bmatrix} \end{array}$$

Total time = 15 + 12 + 14 = 41 weeks

9. \$16.225

10.

$$\$ \begin{array}{c} \begin{array}{ccc} \text{Jazz} & \text{Symp} & \text{Orch} \end{array} \\ \begin{bmatrix} 300.00 & 335.00 & 373.50 \end{bmatrix} \end{array}$$

11. a. $AB = \begin{bmatrix} 8 & -2 & 4 \\ 23 & 38 & -13 \\ -1 & 19 & -11 \end{bmatrix}$

b. $BA = \begin{bmatrix} 2 & -1 \\ 29 & 33 \end{bmatrix}$

c. CA is not defined. The number of columns of C does not equal the number of rows of A.

d. $DA + E = [12 \quad 10]$

12. $A^T = \begin{bmatrix} 4 & 5 \\ 2 & 1 \\ 6 & 3 \end{bmatrix}$

13. a. $M = \begin{bmatrix} 1 & 1 \\ 1 & 1 \end{bmatrix}$, $M^2 = \begin{bmatrix} 2 & 2 \\ 2 & 2 \end{bmatrix}$, $M^3 = \begin{bmatrix} 4 & 4 \\ 4 & 4 \end{bmatrix}$,

$M^4 = \begin{bmatrix} 8 & 8 \\ 8 & 8 \end{bmatrix}$

b. $M^5 = \begin{bmatrix} 16 & 16 \\ 16 & 16 \end{bmatrix}$

c. $M^n = \begin{bmatrix} 2^{n-1} & 2^{n-1} \\ 2^{n-1} & 2^{n-1} \end{bmatrix}$ where n is a natural number

d. The conjecture is true for $n = 1$ since

$$M^1 = \begin{bmatrix} 2^{1-1} & 2^{1-1} \\ 2^{1-1} & 2^{1-1} \end{bmatrix} = \begin{bmatrix} 2^0 & 2^0 \\ 2^0 & 2^0 \end{bmatrix} = \begin{bmatrix} 1 & 1 \\ 1 & 1 \end{bmatrix};$$

Assume true for n = any natural number k, that is,

$$M^k = \begin{bmatrix} 2^{k-1} & 2^{k-1} \\ 2^{k-1} & 2^{k-1} \end{bmatrix}.$$

To complete the proof, we need to show that the conjecture is true for $n = k + 1$. That is, we must show that

$$M^{k+1} = \begin{bmatrix} 2^k & 2^k \\ 2^k & 2^k \end{bmatrix}$$

$$M^{k+1} = MM^k = \begin{bmatrix} 1 & 1 \\ 1 & 1 \end{bmatrix}\begin{bmatrix} 2^{k-1} & 2^{k-1} \\ 2^{k-1} & 2^{k-1} \end{bmatrix} = \begin{bmatrix} 2^{k-1}+2^{k-1} & 2^{k-1}+2^{k-1} \\ 2^{k-1}+2^{k-1} & 2^{k-1}+2^{k-1} \end{bmatrix}$$

$$= \begin{bmatrix} 2\cdot 2^{k-1} & 2\cdot 2^{k-1} \\ 2\cdot 2^{k-1} & 2\cdot 2^{k-1} \end{bmatrix} = \begin{bmatrix} 2^k & 2^k \\ 2^k & 2^k \end{bmatrix}.$$

Therefore the conjecture is true for all natural numbers n.

e. a. $M = \begin{bmatrix} 1 & 0 \\ 2 & 3 \end{bmatrix}$, $M^2 = \begin{bmatrix} 1 & 0 \\ 8 & 9 \end{bmatrix}$, $M^3 = \begin{bmatrix} 1 & 0 \\ 26 & 27 \end{bmatrix}$,

$$M^4 = \begin{bmatrix} 1 & 0 \\ 80 & 81 \end{bmatrix}$$

b. $M^5 = \begin{bmatrix} 1 & 0 \\ 242 & 243 \end{bmatrix}$

c. $M^n = \begin{bmatrix} 1 & 0 \\ 3^n - 1 & 3^n \end{bmatrix}$ where n is a natural number

d. The conjecture is true for $n = 1$ since

$$M^1 = \begin{bmatrix} 1 & 0 \\ 3^1 - 1 & 3^1 \end{bmatrix} = \begin{bmatrix} 1 & 0 \\ 2 & 3 \end{bmatrix}.$$

Assume true for n = any natural number k, that is,

$$M^k = \begin{bmatrix} 1 & 0 \\ 3^k - 1 & 3^k \end{bmatrix}.$$

To complete the proof, we need to show the conjecture is true for $n = k + 1$. That is, we must show that

$$M^{k+1} = \begin{bmatrix} 1 & 0 \\ 3^{k+1} - 1 & 3^{k+1} \end{bmatrix}$$

$$M^{k+1} = MM^k = \begin{bmatrix} 1 & 0 \\ 2 & 3 \end{bmatrix} \begin{bmatrix} 1 & 0 \\ 3^k - 1 & 3^k \end{bmatrix} = \begin{bmatrix} 1 & 0 \\ 2 + 3 \cdot 3^k - 3 & 3 \cdot 3^k \end{bmatrix} =$$

$$\begin{bmatrix} 1 & 0 \\ 3^{k+1} - 1 & 3^{k+1} \end{bmatrix}.$$

Therefore the conjecture is true for all natural numbers n.

14. identity; square matrix with ones along the diagonal and zeros elsewhere

15. a. Yes. $AB = BA = I$

    b. Yes. $AB = BA = I$

    c. No. These are not square matrices. $AB \neq BA \neq I$.

16. a.

	Male	Female
A	189	196
B	176	180
C	251	254

    b.

A	385
B	356
C	505

    Note: There may be a slight difference in the totals matrix due to rounding.

17. a. 24 months

    b. $L = \begin{bmatrix} 0.0 & 0.6 & 0.0 & 0.0 & 0.0 & 0.0 \\ 0.5 & 0.0 & 0.8 & 0.0 & 0.0 & 0.0 \\ 1.1 & 0.0 & 0.0 & 0.9 & 0.0 & 0.0 \\ 0.9 & 0.0 & 0.0 & 0.0 & 0.8 & 0.0 \\ 0.4 & 0.0 & 0.0 & 0.0 & 0.0 & 0.6 \\ 0 & 0.0 & 0.0 & 0.0 & 0.0 & 0.0 \end{bmatrix}$

    c. 10%

    d. After the 17th cycle, or about 35 years

CHAPTER 4

Lesson 4.1

1.

Start 0 A 5 B 6 D 4 G 10 Finish
B 6 E; A 5 C 4 E; E 8 G; C 4 F 4 G

2.

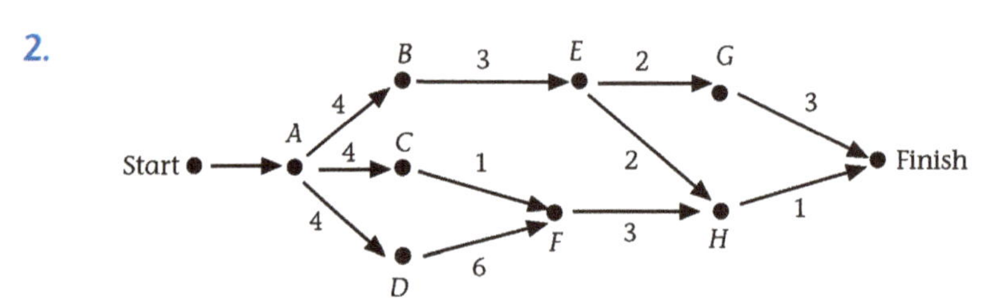

7. a. *A*: 2, none; *B*: 4, none; *C*: 3, *A*; *D*: 3, *C* and *B*; *E*: 2, *B*; *F*: 1, *E*;
G: 4, *D* and *F*

Lesson 4.2

1. EST for *C* through *G*: 7, 10, 11, 16, 23
Minimum Project Time: 26
Critical Path: Start–*ACEFG*–Finish

3. EST for *D* through *L*: 4, 5, 6, 6, 6, 6, 9, 9, 15
Minimum Project Time: 20
Critical Path: Start–*ADHKL*–Finish

5. a.

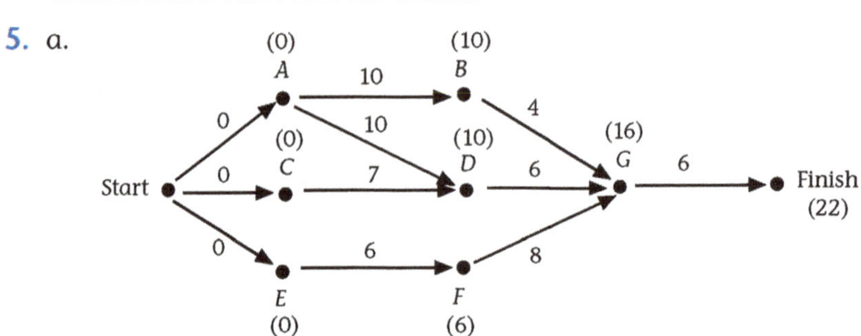

b. Minimum Project Time: 22

c. Critical Path: Start–*ADG*–Finish

d. The minimum time is reduced to 21 days, to 20 days.

e. No, below 8 days *A* is no longer on the critical path.

9. a. Day 16, Day 17, Day 18. Both task *G* and the project will be delayed.

b. Day 11

c. Day 5, Day 6, Day 5

Lesson 4.3

1.

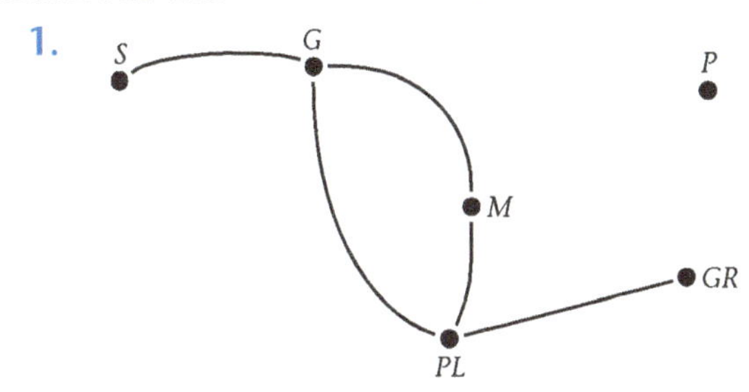

4. a. Not adjacent: *A*&*F*, *B*&*F*, *B*&*C*, *A*&*C*, *A*&*D*, *A*&*E*, *B*&*E*, *B*&*D*, *F*&*C*, *F*&*D*

b. *F*, *E*, *D*, *C*

c. No, there is no path from *A* or *B* to the vertices *C*, *D*, *E*, or *F*.

d. No, not every pair of vertices is adjacent. For example, *B* and *C* are not adjacent.

7. a.

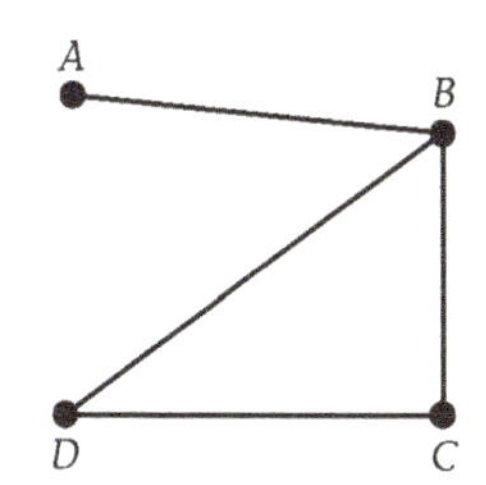

8. a. $\begin{bmatrix} 0 & 1 & 0 & 1 & 1 \\ 1 & 0 & 1 & 1 & 0 \\ 0 & 1 & 0 & 1 & 0 \\ 1 & 1 & 1 & 0 & 1 \\ 1 & 0 & 0 & 1 & 0 \end{bmatrix}$

11. $\deg(V) = 3$; $\deg(X) = 2$; $\deg(Y) = 2$; $\deg(Z) = 1$

13. a. $\deg(B) = 2$; $\deg(C) = 6$, $\deg(D) = 3$, $\deg(E) = 2$

b. $\begin{bmatrix} 1 & 0 & 1 & 0 & 0 \\ 0 & 0 & 2 & 0 & 0 \\ 1 & 2 & 0 & 1 & 2 \\ 0 & 0 & 1 & 1 & 0 \\ 0 & 0 & 2 & 0 & 0 \end{bmatrix}$

14.

K_4	4	12	$T_4 = T_3 + 6$
K_5	5	20	$T_5 = T_4 + 8$
K_6	6	30	$T_6 = T_5 + 10$

Recurrence relation: $T_n = T_{n-1} + 2(n - 1)$

Lesson 4.4

1. a. Both. The degrees of all vertices are even.

3. New Circuit: *S, b, e, a, g, f, S*
Final Circuit: *S, e, f, a, b, c, S, b, e, a, g, f, S*

4. Answers may vary. Sample circuit: *e, d, f, h, d, c, h, b, c, g, a, h, g, f, e*

9. a. Sample digraph:

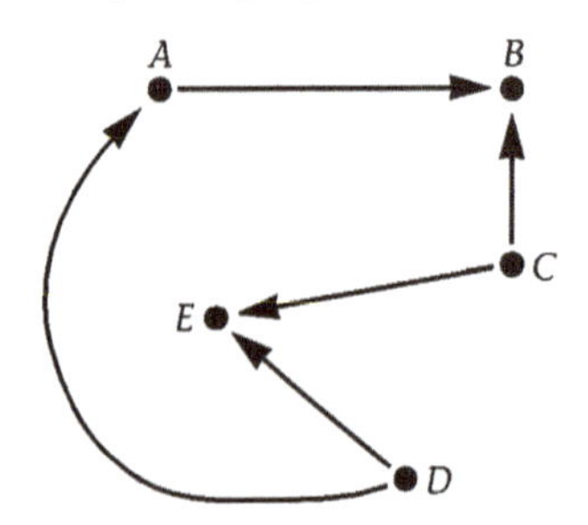

11. a. Yes.

b. No.

13. a.

$$\begin{array}{c|ccccc} & a & b & c & d & e \\ \hline a & 0 & 0 & 1 & 0 & 1 \\ b & 1 & 0 & 0 & 1 & 0 \\ c & 0 & 1 & 0 & 0 & 0 \\ d & 0 & 0 & 1 & 0 & 0 \\ e & 0 & 1 & 0 & 1 & 0 \end{array}$$

Lesson 4.5

1. Graphs a and d have Hamiltonian circuits. For graphs b and c, the theorem does not apply.

9.

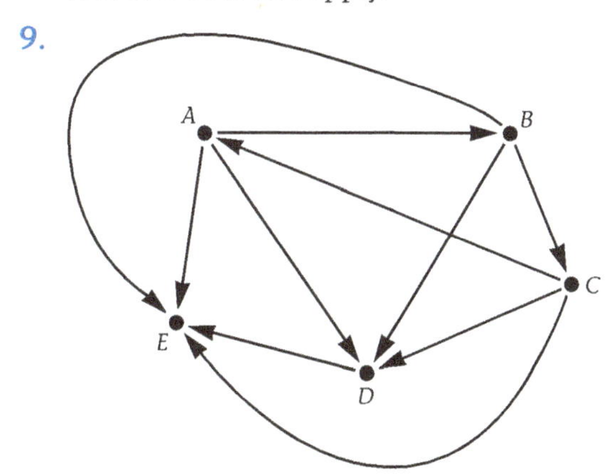

12.

4	6
5	10
6	15

$S_n = S_{n-1} + (n - 1)$

16. a. $M = \begin{bmatrix} 0 & 0 & 1 & 1 & 1 \\ 1 & 0 & 1 & 1 & 0 \\ 0 & 0 & 0 & 1 & 1 \\ 0 & 0 & 0 & 0 & 1 \\ 0 & 1 & 0 & 0 & 0 \end{bmatrix}$

b. $M^2 = \begin{bmatrix} 0 & 1 & 0 & 1 & 2 \\ 0 & 0 & 1 & 2 & 3 \\ 0 & 1 & 0 & 0 & 1 \\ 0 & 1 & 0 & 0 & 0 \\ 1 & 0 & 1 & 1 & 0 \end{bmatrix}$

c. The winner would be B.

Lesson 4.6

1. a. 4

   b. 2

3. a. List the vertices in order from the ones with the greatest degree to the ones with the least.

   b. Those not adjacent to it or adjacent to one with that color

6. a. 2, 3, 4, 5

10. 4

13. a. Sample coloring:

Chapter 4 Review

2.

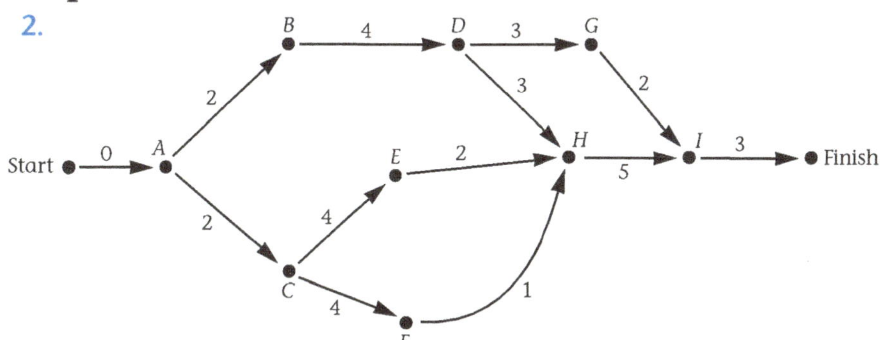

3. a. $A,(0)$; $B,(4)$; $C,(4)$; $D,(7)$; $E,(6)$; $F,(11)$; $G,(10)$; $H,(15)$; $I,(16)$; $J,(18)$

   b. Minimum Project Time: 23

4. a.

Task	Time	Prerequisites
Start	0	—
A	2	None
B	3	None
C	4	*A*
D	4	*A*, *B*
E	2	*B*
F	3	*C*
G	5	*D*, *E*
H	7	*F*, *G*
Finish		

5. a. $A,(0)$; $B,(2)$; $C,(2)$; $D,(6)$; $E,(6)$; $F,(6)$; $G,(9)$; $H,(9)$; $I,(14)$

   b. Critical Path: Start–*ABDHI*–Finish
   Minimum Project Time: 17

6. a.

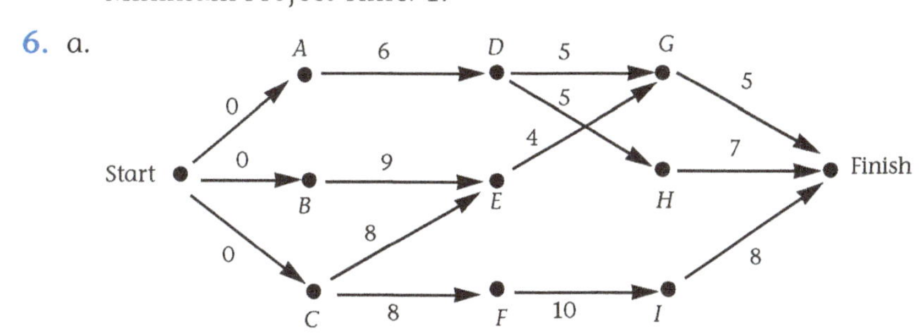

b. A,(0); B,(0); C,(0); D,(6); E,(9); F,(8); G,(13); H,(11); I,(18)

c. Minimum Project Time: 26

d. Critical Path: Start–*CFI*–Finish

7. a. Yes, a path exists from each vertex to every other vertex.

b. No, not every pair of vertices is adjacent.

c. A, D, or C

d. *BCDE* or *BCAE*

e. Deg(C) = 4

f.
$$\begin{array}{c} \\ A \\ B \\ C \\ D \\ E \end{array} \begin{array}{c} \begin{array}{ccccc} A & B & C & D & E \end{array} \\ \begin{bmatrix} 0 & 0 & 1 & 0 & 1 \\ 0 & 0 & 1 & 0 & 0 \\ 1 & 1 & 0 & 1 & 1 \\ 0 & 0 & 1 & 0 & 1 \\ 1 & 0 & 1 & 1 & 0 \end{bmatrix} \end{array}$$

8. a. Sample graph:

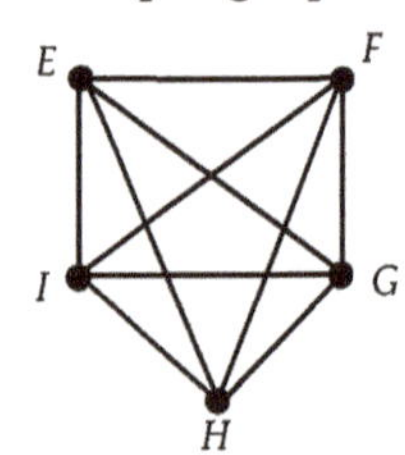

b. 4

c. Yes, the degrees of all vertices are even.

9. Sample graph:

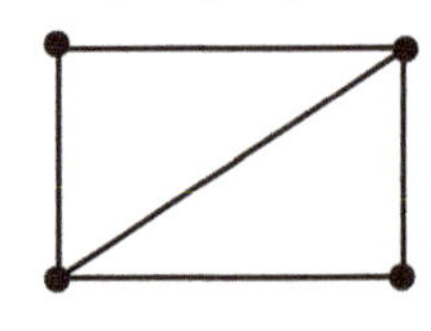

10. a. Euler path. Two vertices have odd degrees, and the remaining vertices have even degrees.

    b. Euler circuit. All vertices have even degrees.

11. a.

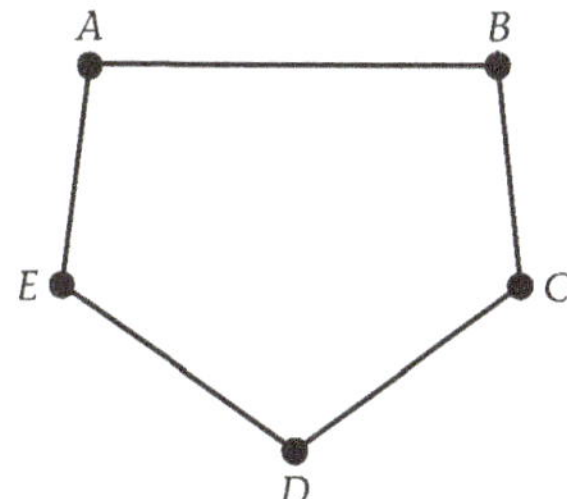

    b.

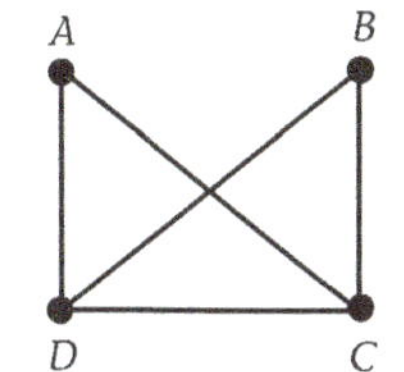

12. a. No.

    b. Yes.

    c.

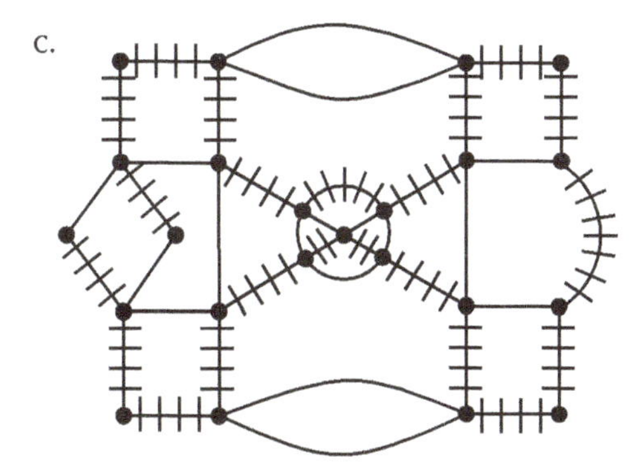

13. a. Yes.

    b. No, the graph has exactly two odd vertices. You would have to begin at one of the vertices with an odd degree and end at the other.

14.

	A	B	C	D	E	F
A	0	1	0	0	0	0
B	0	0	1	0	0	1
C	0	0	0	1	0	0
D	0	1	0	0	1	0
E	0	0	0	0	0	1
F	1	0	0	0	0	0

15. a.

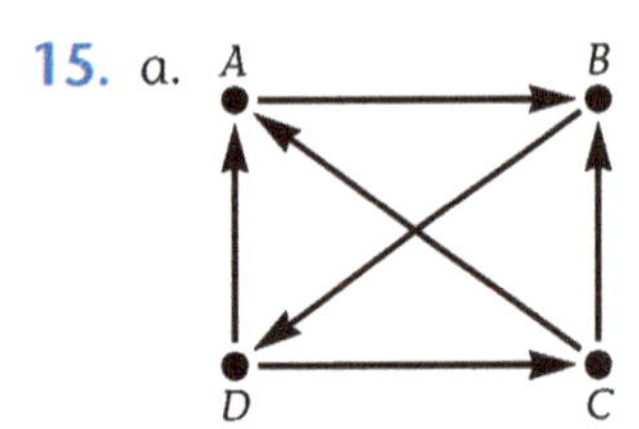

b. There is no Condorcet winner because none of the candidates can beat all of the other candidates in a one-on-one race.

c. *D, C, A, B; C, A, B, D; C, B, D, A; B, D, C, A; A, B, D, C*

d. If the Hamiltonian path, *B, D, C, A* is chosen for a pairwise voting scheme, *B* wins. The path shows that for *B* to "survive," you need first to pair *A* and *C*. *C* wins. Next pair the winner, *C*, against *D*, and *D* wins. Finally, pair the winner, *D*, against *B*, and *B* is the final winner of the election.

16. Sample graph:

17. a.

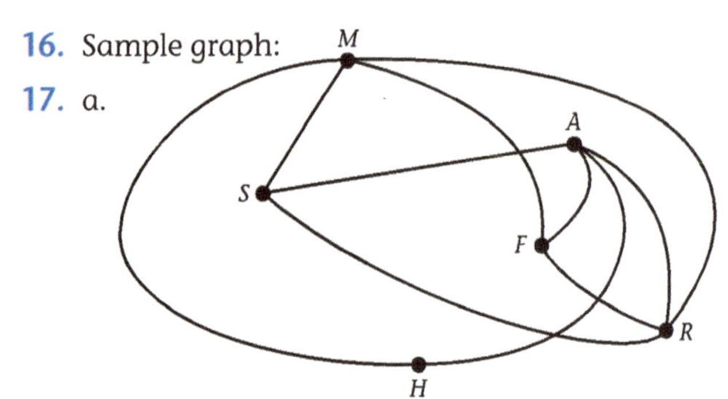

b. Three time slots. Sample schedule:

Time 1—Math & Art, Time 2—Reading & History,
Time 3—Science & French

18. a.

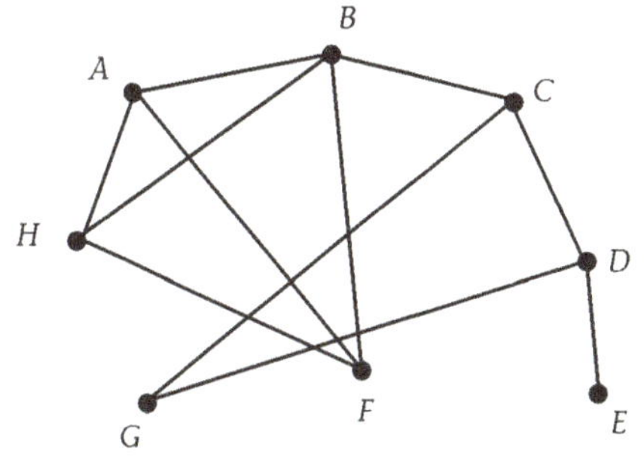

b. 4 frequencies

19. a. No, the outdegrees are not equal to the indegrees at each vertex.

b. Yes, the outdegree equals the indegree at all vertices but two. At one of those two vertices, the indegree is one greater than the outdegree and at the other vertex, the outdegree is one greater than the indegree.

Answers will vary, but the paths must begin at B and end at D.

20. a., b. Sample graph and coloring:

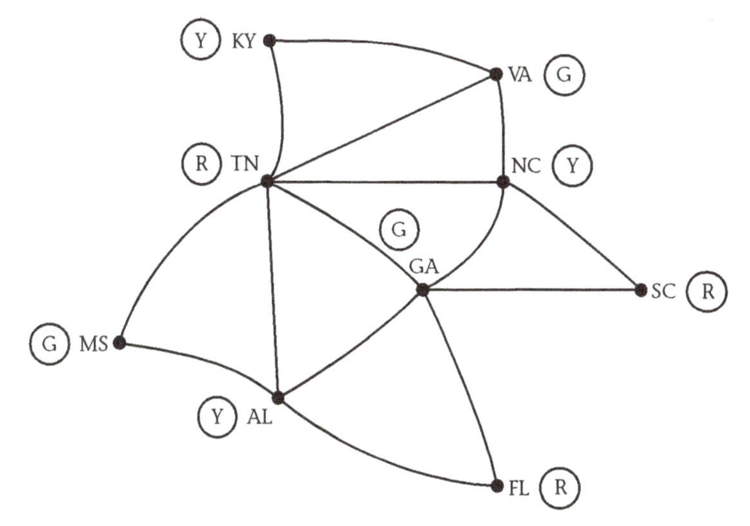

c. Three colors

CHAPTER 5

Lesson 5.1

1. Planar

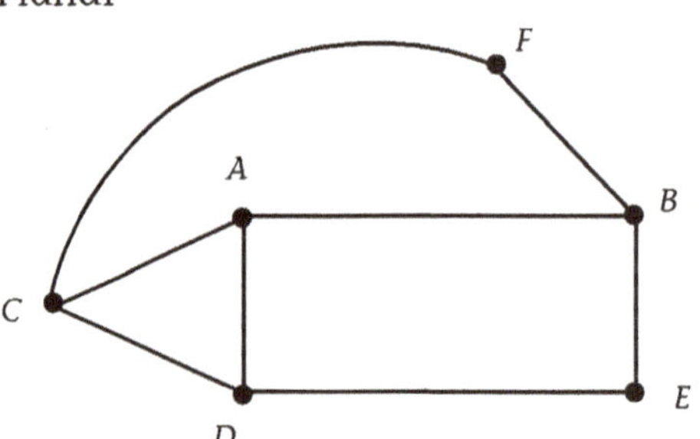

5. Hint: move *A* and *B* around.

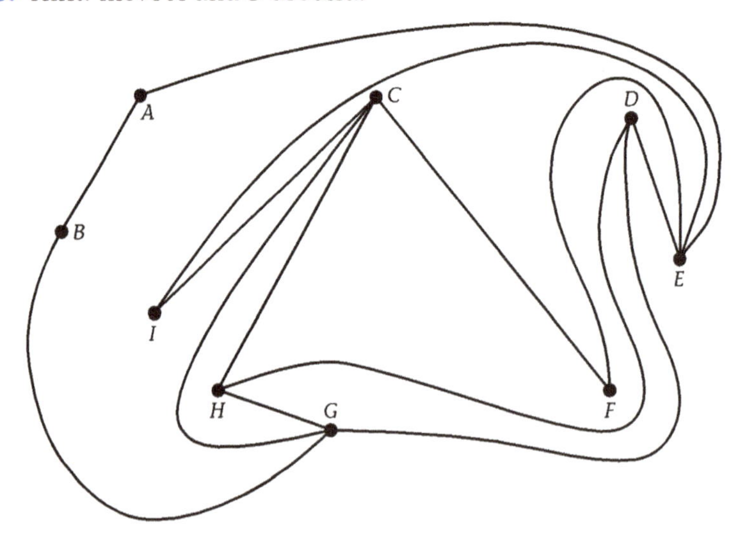

8.

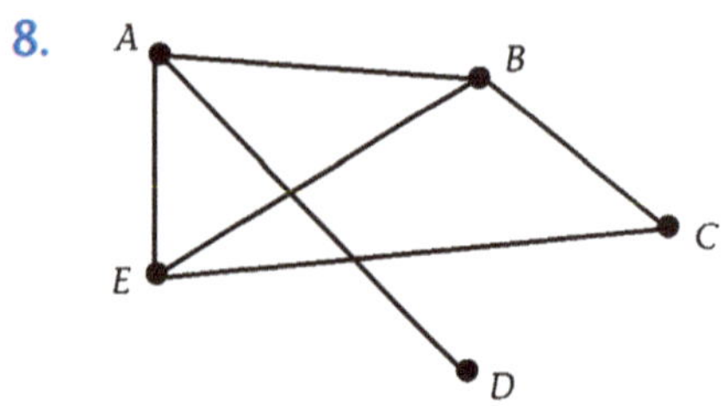

11. a.

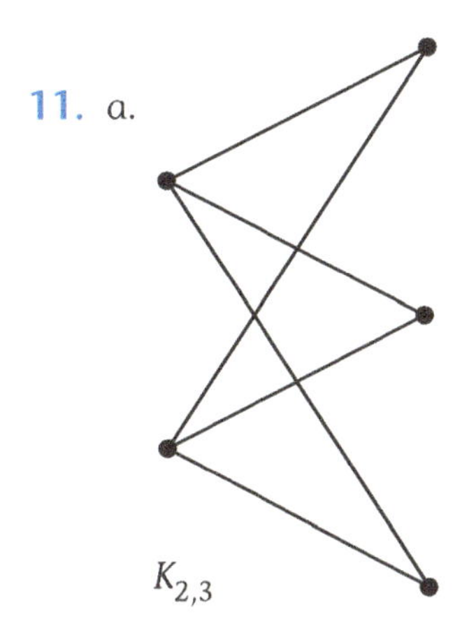

12. a. {*A*, *B*, *C*, *D*, *E*, *F*} and {*G*}

15. 6, 12, *mn*

18. 30 handshakes, bipartite

21. a, d; because all of the edges and vertices of the original graph are in a and d

23. No. No.

Lesson 5.2

1. a., b.

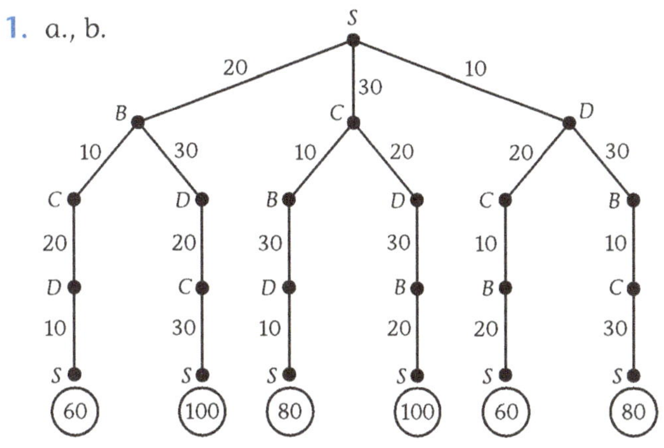

c. *SDCBS*

d. *SDCBS*

e. Yes.

5. a. 0.36 seconds, about 24 hours

7. Shortest possible circuit: *SACBS*. Total distance: 18.75 mm

Lesson 5.3

1. 6, 12, 11, *C*, *BC*. The shortest path from *A* to *E* is *AHGE*.

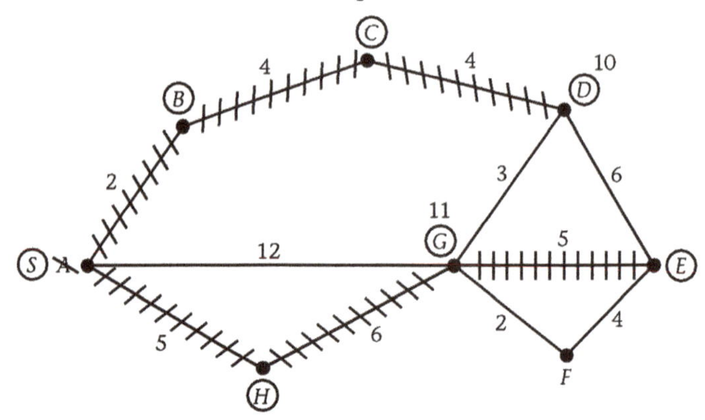

2. *ABECDF* (11)

5. a. Albany, *C*, *E*, *H*, Ladue

   b. Albany, *B*, *D*, Fenton, *G*, *K*, Ladue. This problem yields a different solution than part a, because you have to find two solutions and then add them: First you must find the shortest path from Albany to Fenton, then a path with Fenton as Start.

9.

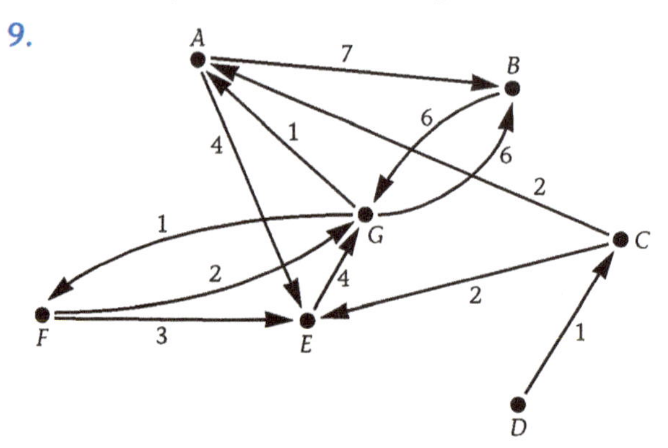

Shortest path: *DCAB*. Least charge: $1 + 2 + 7 = 10$

Lesson 5.4

1. *BCEFB, CDEC, BCDEFB, BCFB, CEFC, CDEFC*

3. a. 5 vertices

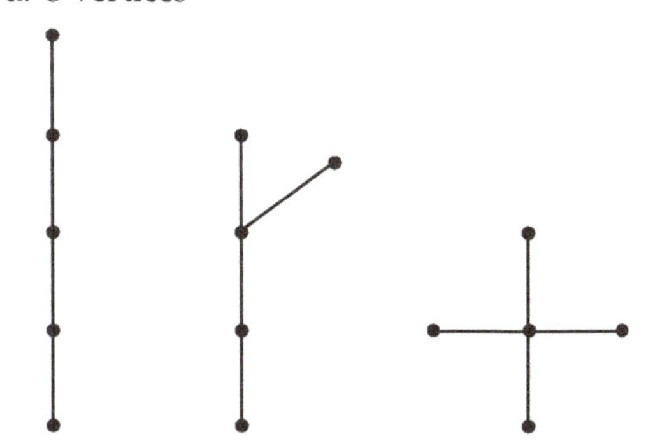

4.

Number of Vertices	Number of Edges
1	0
2	1
3	2
4	3
n	$n - 1$

   a. 18 edges

   b. 16 vertices

   c. The number of edges = the number of vertices – 1.

7. a.

Number of Vertices	Sum of the Degrees of the Vertices	Recurrence Relation
1	0	$S_1 = 0$
2	2	$S_2 = S_1 + 2$
3	4	$S_3 = S_2 + 2$
4	6	$S_4 = S_3 + 2$
5	8	$S_5 = S_4 + 2$
6	10	$S_6 = S_5 + 2$

   b. $S_n = S_{n-1} + 2$

9. Answers will vary. Possible forest:

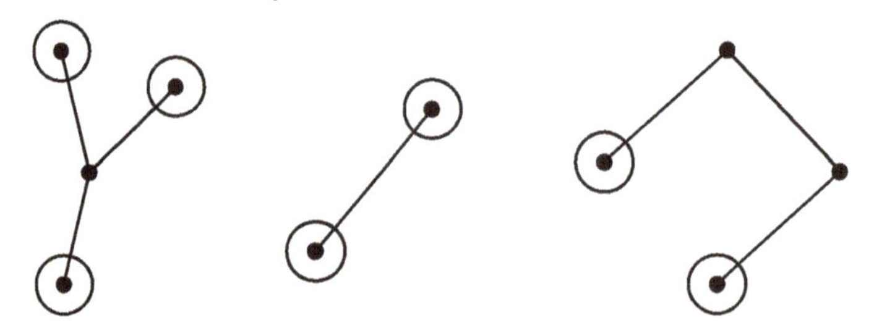

12. Rhombus

Lesson 5.5

1. One possible spanning tree:

4. One possible spanning tree:

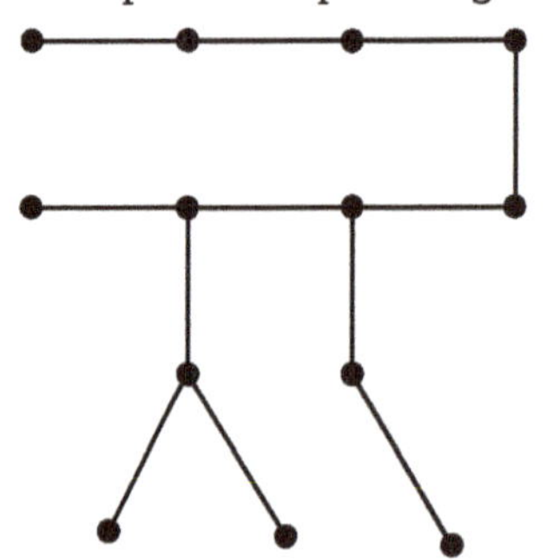

7. a. Yes.

   b. *E* and *J*

   c. *DE* and *DJ*

d. *F, K, I*

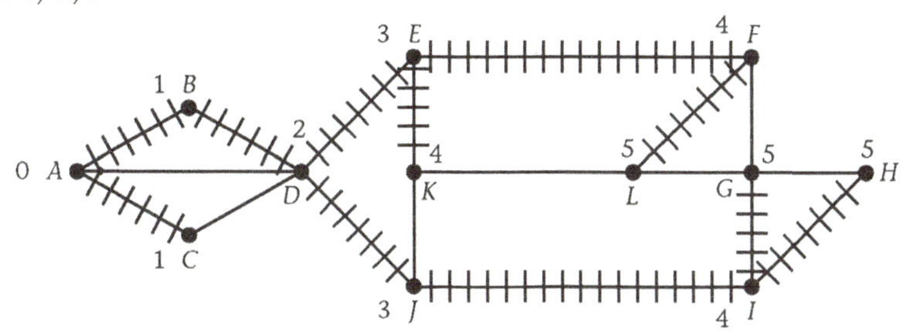

8. This is one of many possibilities.

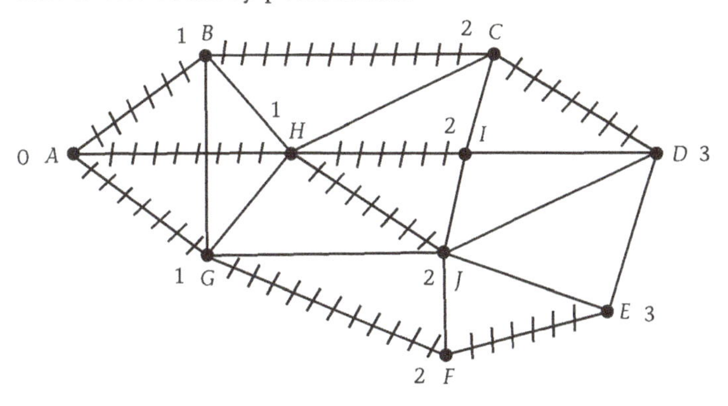

11. The minimum weight is 10.

15. $2,100

20. a. Weight 28

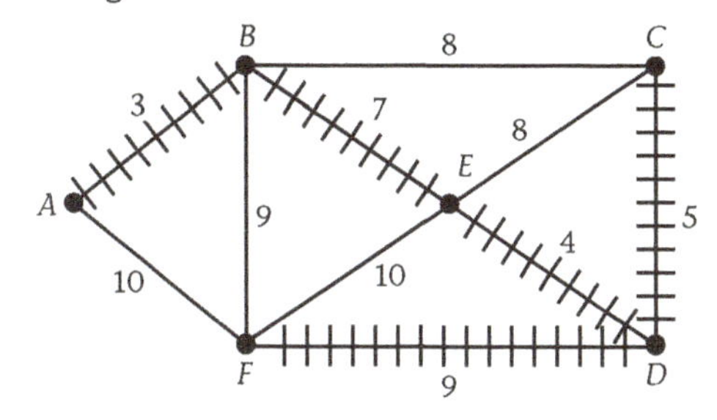

b. A to F is 10; A to B is 3; A to C is 11; A to E is 10; A to D is 14.

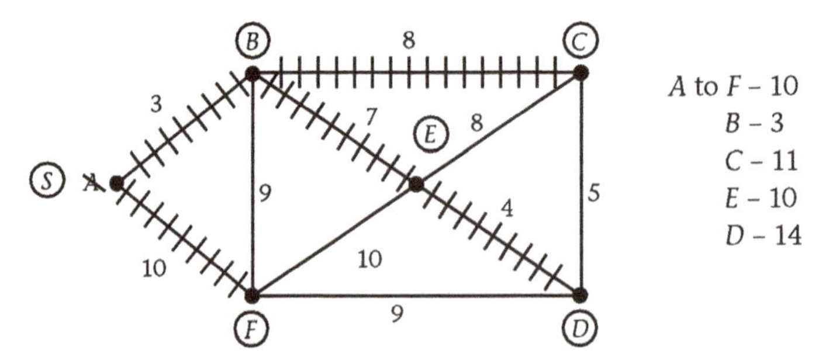

c. No. It is a spanning tree, but in this example it is not minimal. Its total weight is 32, which is greater than 28.

Lesson 5.6

1.

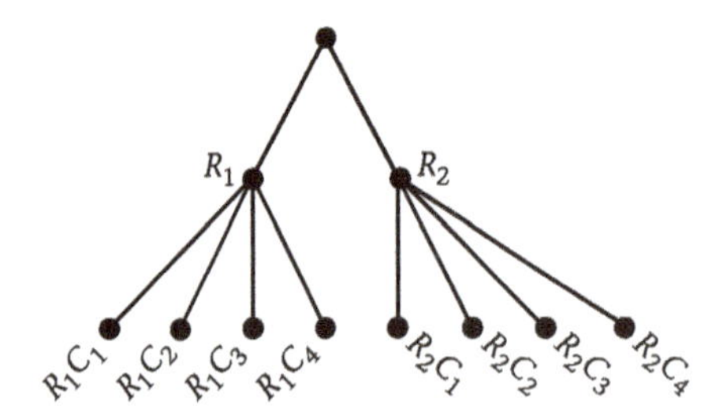

3. Binary Tree

   a. V is level 2.

   b. C is the parent.

   c. G and H are children.

7. There are 17 questions in the book.

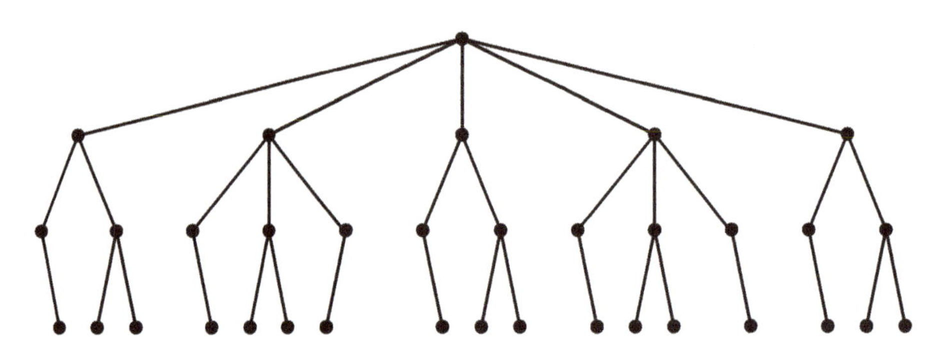

11.

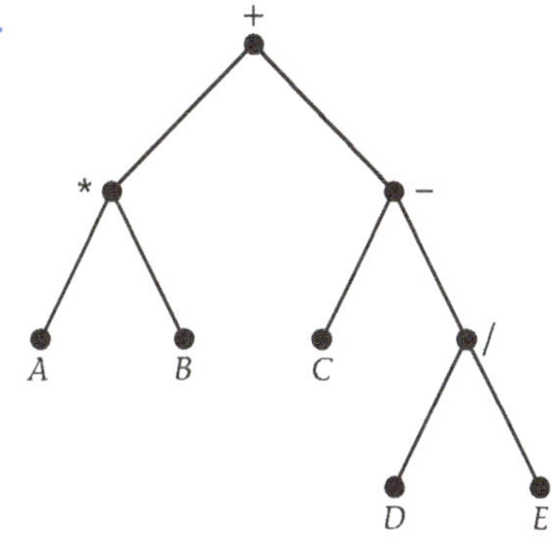

12. 3 2 * 8 2 3 * − +

16. a. 33

b. 7

c. 9

d. 14

17. a. 2 3 6 * + 4 1 + −

20. *ABDEGHCFI*

22. 17

23. a. 8

Chapter 5 Review

2. Planar;

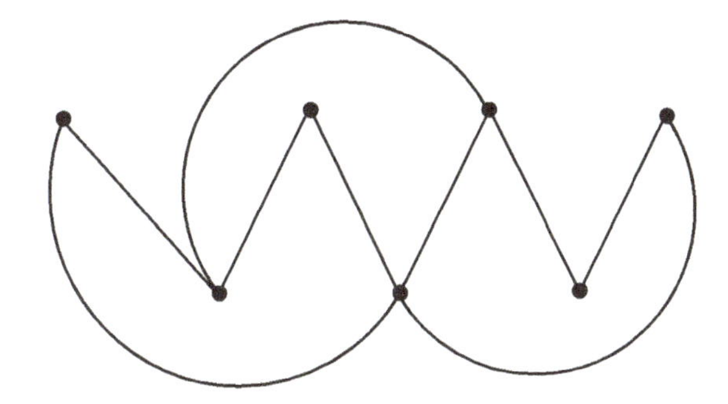

3.

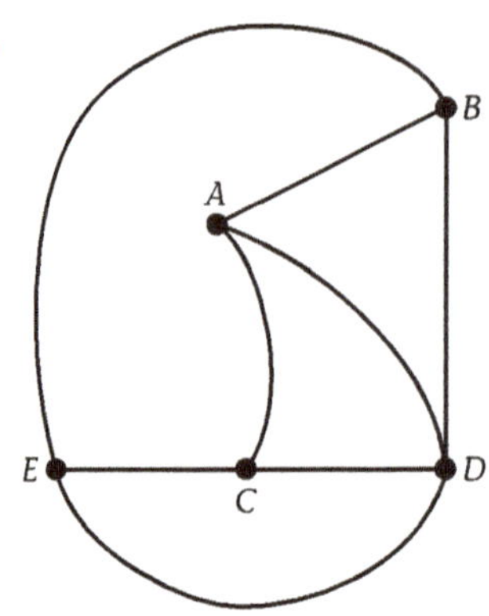

4. a. 2

   b. 2

   c. 2

   d. 2

5. a. The vertices of the graph can be divided into two sets so that each edge of the graph has one endpoint in each set.

   b. Yes, all possible edges from one set of vertices to the other are drawn.

   c. Yes.

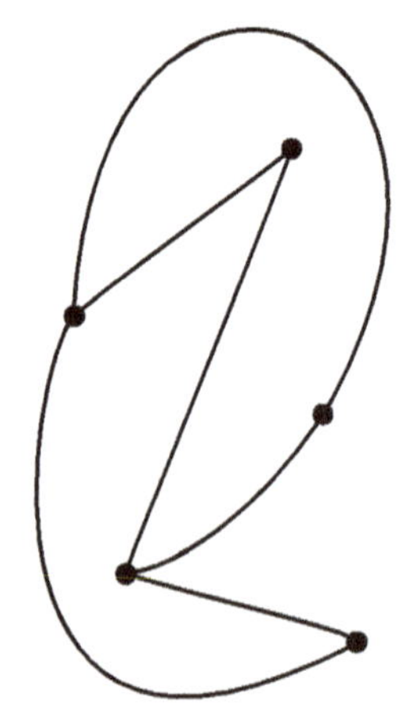

   d. 2

6. One possible solution:

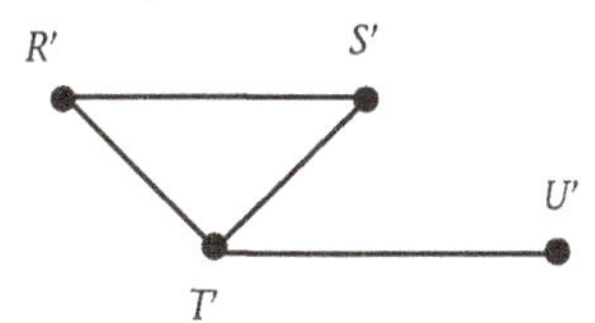

7. a. O-SCM-O

   b. 314 ft

   c. Hamiltonian circuit

8. Total weight: 9 + 23 +31 + 35 + 28 = 126

9.

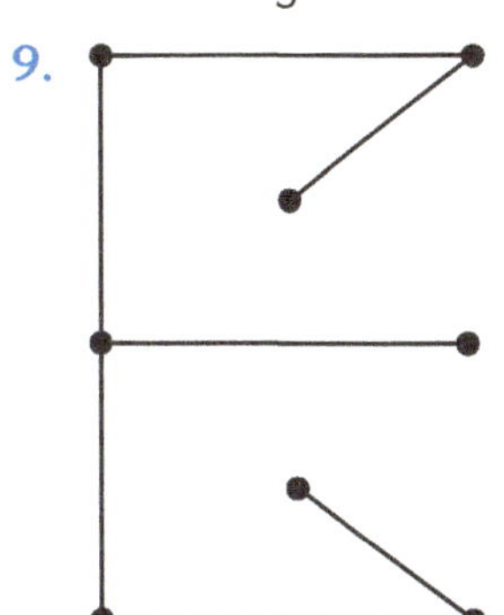

10. 3, 4, 5, n

11. The circuit does not include all the vertices of the graph.

12. a. Home-T-P-G-H-BB

    b. 14 miles

    c.

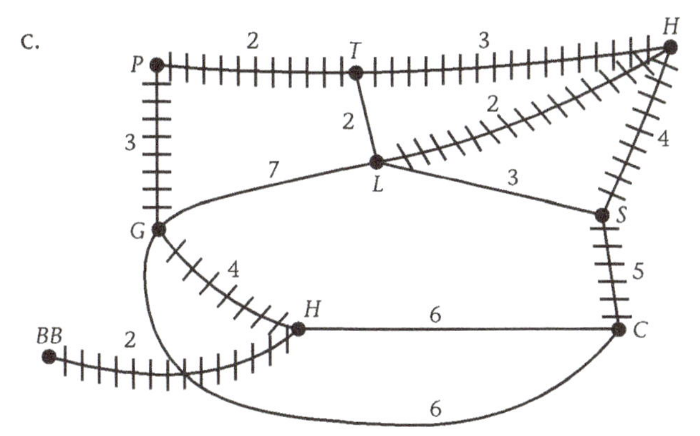

13. The total weight of a minimum spanning tree for the graph is 23 miles.

14. a. Yes, it is a connected graph with no cycles.

b. Yes, it is a connected graph with no cycles.

c. No, the graph contains a cycle.

d. No, the graph is not connected.

15. The graph will no longer be a tree because it will contain a cycle, multiple edges, or a loop.

16.

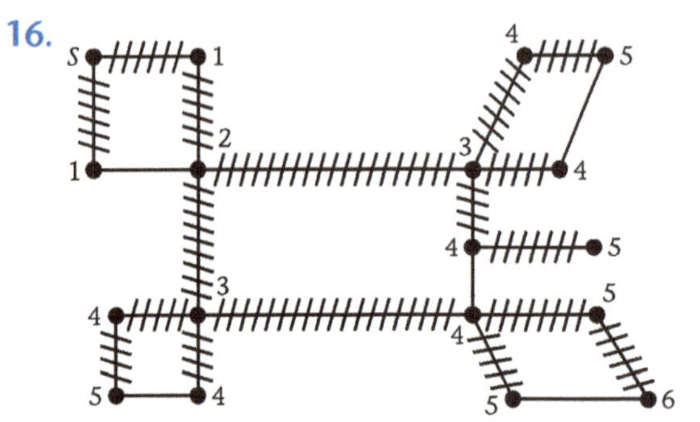

17. a.

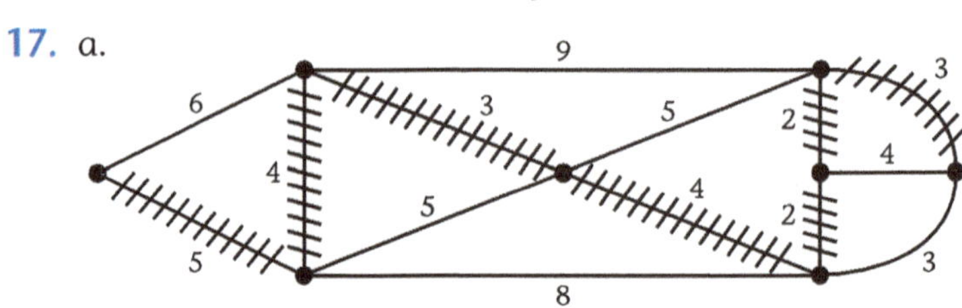

b. Total cost = $23,000

18. One possible solution

19. Problems similar to those in Lesson 5.5

20.

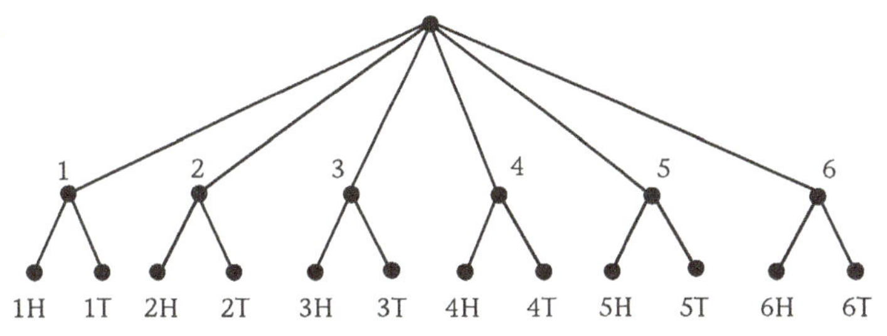

21.

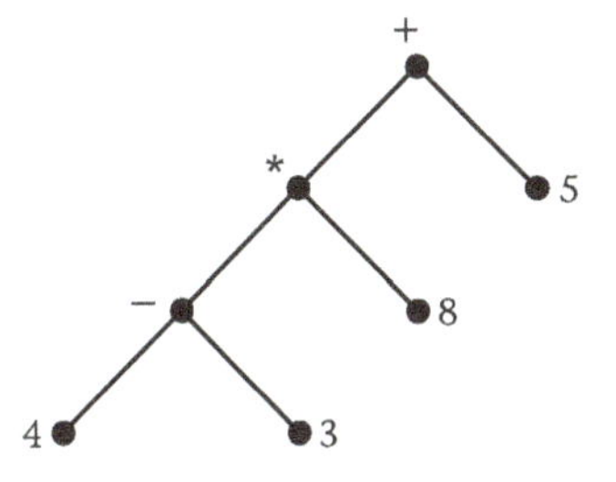

22. 4 6 3 – + 5 2 * +

23. 18

24. Any expression is possible. The following is just one example:

The expression: 3 * (2 + 6) – 5

The expression tree:

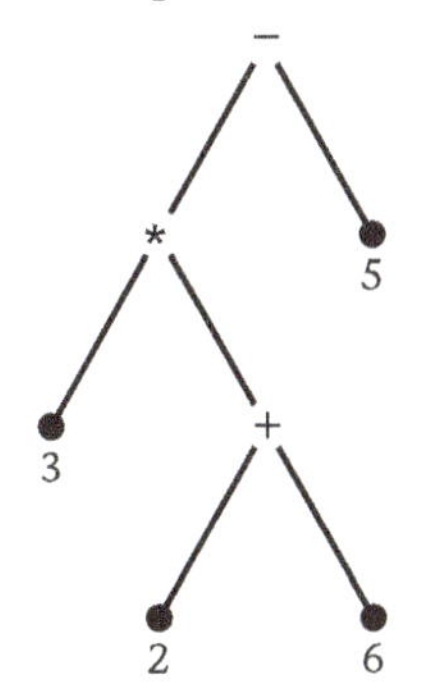

The postorder listing: 3 2 6 + * 5 –

CHAPTER 6

Lesson 6.1

1. *li, lo, ln, ls, il, io, in, is, ol, oi, on, os, nl, ni, no, ns, sl, si, so, sn*
2. 5, 4, $5 \times 4 = 20$, $6 \times 5 = 30$
3. (1, 2) (1, 3) (1, 4) (1, 5) (1, 6) (1, 7) (1, 8) (1, 9) (2, 3) (2, 4) (2, 5) (2, 6) (2, 7) (2, 8) (2, 9) (3, 4) (3, 5) (3, 6) (3, 7) (3, 8) (3, 9) (4, 5) (4, 6) (4, 7) (4, 8) (4, 9) (5, 6) (5, 7) (5, 8) (5, 9) (6, 7) (6, 8) (6, 9) (7, 8) (7, 9) (8, 9)
4. 9, 8, $9 \times \frac{8}{2} = 36$

13. 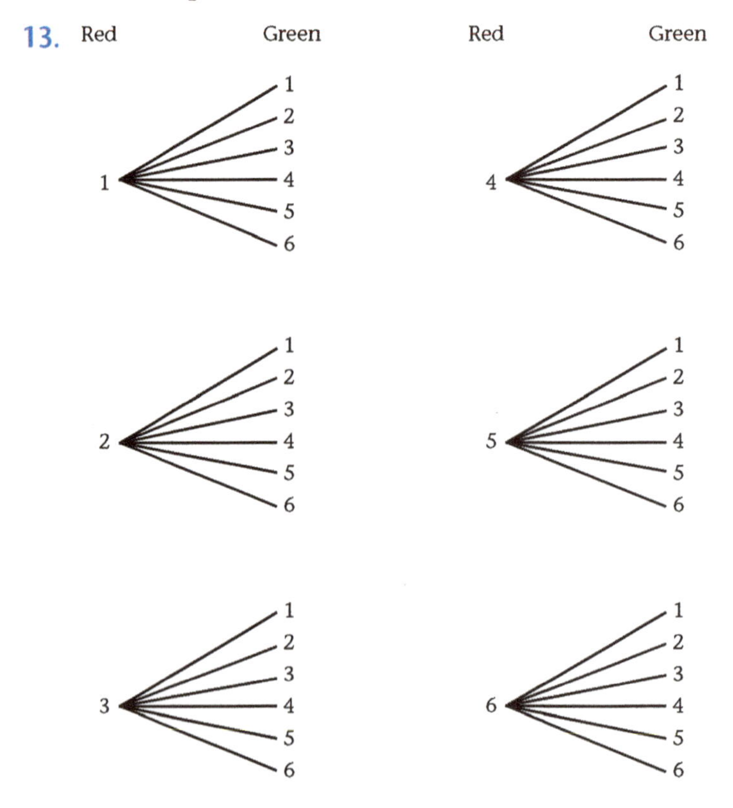

15. a. Win \$1: 10 ways, win \$2: 1 way, lose \$1: 25 ways

Lesson 6.2

1. $\frac{10!}{6!}$ 5040
3. a. 15

   b. A particular front sprocket and a particular rear sprocket
5. a. 12,167,000

   b. $\frac{1}{12,617}$
7. a. 72

   b. $\frac{1}{2}$

   c. 0

   d. 1
10. a. 2,704

    b. 2,652
13. a. 30! or about 2.6525×10^{32}

    b. About 1.0827×10^{28}; the number of seating arrangements is about 24,500 times as large.
14. a. 6

    b. A road from Claremont to Upland and a road from Upland to Pasadena

    c. 9
18. a. 100,000 manufacturers

    b. 100,000 products
19. a. 1,048,576

    b. About 10 years; about 262 feet

Lesson 6.3

3. a. 672

   b. 504

   c. 2,380. They are the same.

4. a. 210

b. 210

c. 1,024

5. a. 1,326

b. 325

c. $\frac{325}{1,326}$ or about 0.245

8. a. 10

b. 10

9. a. 7,059,052

b. About 245 weeks, or a little less than 5 years

c. About 353 feet

d. 13,983,816

e. $\frac{80,000}{13,983,816}$ or about .00572

f. The probability of winning today is a little more than half (about 0.6) what it was before the change.

10. a. $C(6, 5) \times C(47, 1) = 282$

b. 16,215

c. 324,300

14. 511

20. a. 7

b. 21

c. 28

e. $\frac{7}{28}$ or $\frac{1}{4}$

f. 91

Lesson 6.4

1. a. $\frac{520}{1,000} = .52$

   b. $\frac{196}{360}$, or about .544

   c. No, but they are fairly close.

   d. $\frac{360}{1,000} = .36$

   e. $\frac{196}{1,000} = .196$. The product is .1872.

   f. $.544 \times .36 = .196$. They are the same.

   g. No.

3. a. $\frac{684}{1,000} = .684$

   b. $.36 + .52 = .88$, which is larger than .684

   c. No.

5. a.

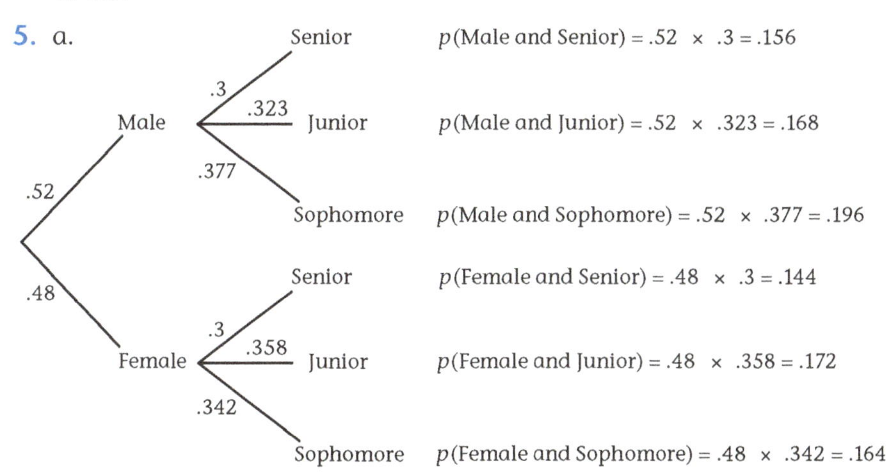

   b. 1

9. a. .63

   b. That the outcomes of the two games are independent

10. a.

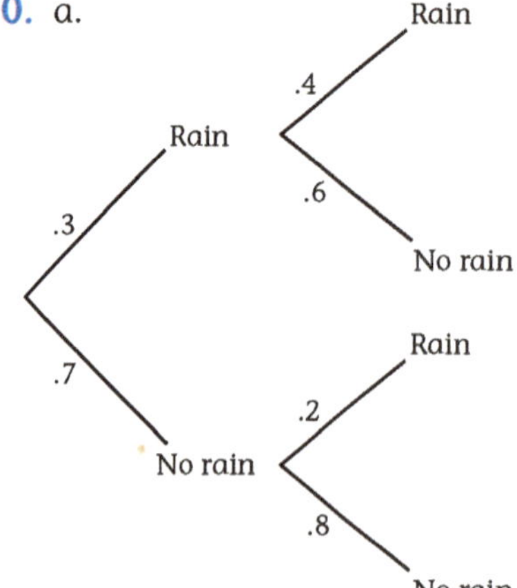

p(Rain and Rain) = .3 × .4 = .12

p(Rain and No rain) = .3 × .6 = .18

p(No rain and Rain) = .7 × .2 = .14

p(No rain and No rain) = .7 × .8 = .56

b. .12

c. .32

d. .56

e. No.

13. $\frac{1}{216}$

16. a. $.977^6$, or about .870

b. The first is .870, and the second is $1 - \frac{1}{6} = \frac{5}{6}$, or about .83.

c. .000529

d. .999471, .99683

19. a. $\frac{1}{80}$

b. That the two events are independent

c. The probability of selecting a man with red hair is .1; the probability of selecting a man who owns a blue car is .125; and the probability of selecting a man who has red hair and owns a blue car is about .0124. The product of the first two probabilities is .0125, so the events are quite close to being independent.

20. a. 36

b. $\frac{1}{36}$

c. $\frac{1}{1{,}296}$

d. $\frac{35}{36}$

e. $\frac{1{,}225}{1{,}296}$

f. $\left(\frac{35}{36}\right)^{21}$, or about .5534

Lesson 6.5

1. a. The probability of each outcome is $\frac{1}{2}$, and successive applications are independent.

   b. The probability of choosing the third answer is $\frac{1}{2}$, while the probability of choosing each of the others is $\frac{1}{4}$; successive applications are independent.

   c. Each possibility probably has a $\frac{1}{4}$ chance of occurring, but successive applications may not be independent if, for example, one finger is injured.

   d. Each possibility has the same chance of occurring, and successive applications are independent.

2. a. 10

   b. .3125

   c. .03125, .15625, .3125, .3125, .15625, .03125

   d. .00243, .02835, .1323, .3087, .36015, .16807

3. a. .2051

   b. .1172

   c. .0439

   d. .0098

   e. .00098

   f. .3770

5. a. .2765

   b.

Number of Women	0	1	2	3	4	5	6
Probability	.0467	.1866	.3110	.2765	.1382	.0369	.0041

8. a. $\frac{1}{7{,}059{,}052}$; $\frac{7{,}059{,}051}{7{,}059{,}052}$

   b. \$2.82, but this assumes that the jackpot is not shared with another party.

   c.

Amount won	27,000,000	-5,000,000
Probability	$\frac{5{,}000{,}000}{7{,}059{,}052}$	$\frac{2{,}059{,}052}{7{,}059{,}052}$

   The expectation is \$17,665,933, but this assumes the jackpot is not shared.

10. a.

Amount Won	-1	1	20
Probability	$\frac{21}{36}$	$\frac{14}{36}$	$\frac{1}{36}$

   b. \$0.36

   c. No, the council loses about 36 cents per play. One way of correcting this is to give no prize for matching a single number. In fact, the jackpot could then be increased but should be kept under \$35.

13. a. $\frac{1}{4}$

   b.

Amount Won	-\$0.50	\$1.00
Probability	$\frac{3}{4}$	$\frac{1}{4}$

   c. -\$0.13

   d. Yes, about 87 cents per play

Chapter 6 Review

2. a. $\frac{7{,}900}{46{,}900}$, or about .168

   b. $\frac{2{,}300}{13{,}700}$ or about .168

   c. $\frac{13{,}700}{46{,}900}$ or about .292

   d. $\frac{2{,}300}{46{,}900}$ or about .049

   e. $\frac{19{,}300}{46{,}900}$ or about .412

   f. Yes.

   g. No.

3. a. 45,697,600

4. $C(5, 1) = 5$, $C(5, 2) = 10$, $C(5, 3) = 10$, $C(5, 4) = 5$, $C(5, 5) = 1$

5. a. $\frac{1}{8}$

   b. $\frac{1}{8}$

   c. $\frac{1}{4}$

6. a. 720

   b. 48

   c. $\frac{1}{15}$

   d. The math books can be in positions 1 and 2, or in positions 2 and 3, or in positions 3 and 4, or in positions 4 and 5, or in positions 5 and 6. $5 \times 48 = 240$.

   e. $\frac{1}{3}$

7. a.

Amount Won	\$2	\$1	–\$1
Probability	$\frac{1}{4}$	$\frac{1}{4}$	$\frac{1}{2}$

   b. $\frac{1}{4}$

   c. Win about \$25

8. a. 255

9. a. 720

   b. About 28

10. There are $C(39,5) = 575{,}757$ different winning tickets possible in the first and $C(36,6) = 1{,}947{,}792$ ways of winning in the second, so the probability of winning in the first is between three and four times as great as in the second. About five in the first and one or two in the second.

11. a. $\frac{1}{4}$

   b. $\frac{4}{17}$

   c. $\frac{1}{17}$

d.

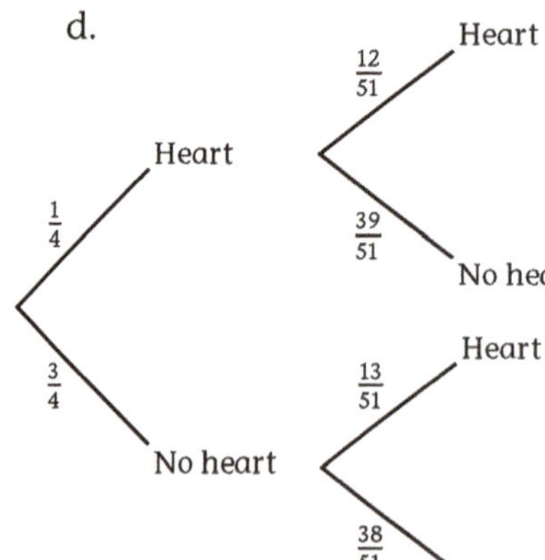

p(Heart and Heart) $= \frac{1}{4} \times \frac{12}{51} = \frac{12}{204}$

p(Heart and No heart) $= \frac{1}{4} \times \frac{39}{51} = \frac{39}{204}$

p(No heart and Heart) $= \frac{3}{4} \times \frac{13}{51} = \frac{39}{204}$

p(No heart and No heart) $= \frac{3}{4} \times \frac{38}{51} = \frac{114}{204}$

e. No.

12. a. 43

13. a. 165

b. 75

c. $\frac{75}{165}$

d. 135

e. $\frac{135}{165}$

14. a.

Good driver (.9)
- .05 Ticket — p(Good driver and Ticket) = .045
- .95 No ticket — p(Good driver and No ticket) = .855

Bad driver (.1)
- .7 Ticket — p(Bad driver and Ticket) = .07
- .3 No ticket — p(Bad driver and No ticket) = .03

b. 5,750

c. 3,500

d. $\frac{3{,}500}{5{,}750}$

15. a. $\frac{1}{1{,}000}$

    b. $\frac{729}{1{,}000}$

    c. $\frac{504}{1{,}000}$

16. a. 100

    b. 90

    c. 45

17. a. Approximately .161

    b. Approximately .965

18. a. 62%

    b. Approximately .871

    c. Approximately 16%

19. 15

20. a. .2

    b. .1

    c. .6

    d. .2

    e. No.

    f. Yes.

22. a. Approximately 5%

    b. Approximately 93%

    c. About 1

23. Mutually exclusive: rolling a number divisible by 5 and rolling a number divisible by 3, rolling a number divisible by 5 and rolling a number divisible by 2. Independent: rolling a number divisible by 2 is $\frac{1}{2}$ and rolling a number divisible by 3.

24. a. No.

    b. No.

25. a. $\frac{1}{175{,}223{,}510}$

    b. This probability is about 168 times the probability of winning the jackpot.

    c. No sooner than about a third of a minute before the drawing

CHAPTER 7

Lesson 7.1

1.

Total Production Units	Units Used Internally	Units for External Sales
500	0.05(500) = 25	500 – 0.05(500) = 475
900	45	855
2,000	100	1,900
5,000	250	4,750
2,500	125	2,375
7,500	375	7,125
P	$0.05P$	$P - 0.05P$

2. a. 2%

   b. C .02

   c. $D = P - 0.02P$

   d. $P \approx \$20{,}408$

3. a. .01 .02 .03 Chips Computers .20

   b.

	Chips	Computers
Chips	0.02	0.20
Computers	0.01	0.03

   c. \$20 chips; \$10 computers

   d. \$150 computers, \$1,000 chips

4.

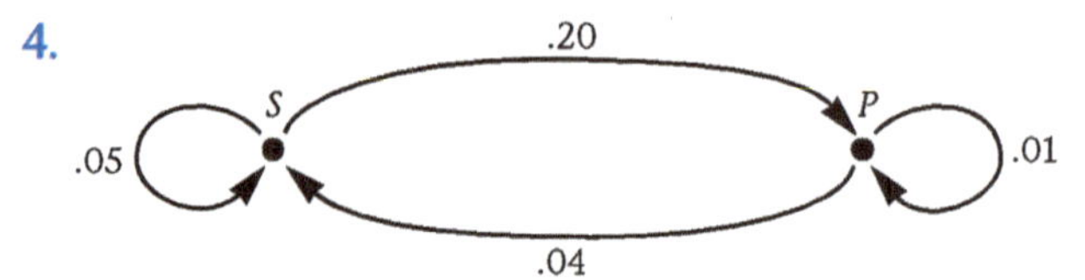

5. a. 5; 4

   b. 1; 20

6. a. \$1 million used within service

   \$0.8 million needed from production

   c. Available for service demands: 20 – 1 – 8 = \$11 million

   Available for production demands: \$38.8 million

Lesson 7.2

2. a. $P = \begin{matrix} \text{Chips} \\ \text{Computers} \end{matrix} \begin{bmatrix} \$40{,}000 \\ \$50{,}000 \end{bmatrix}$

   b. $CP = \begin{matrix} \text{Chips} \\ \text{Computers} \end{matrix} \begin{bmatrix} \$10{,}800 \\ \$1{,}900 \end{bmatrix}$

   c. $D = P - CP = \begin{matrix} \text{Chips} \\ \text{Computers} \end{matrix} \begin{bmatrix} \$29{,}200 \\ \$48{,}100 \end{bmatrix}$

5. a.

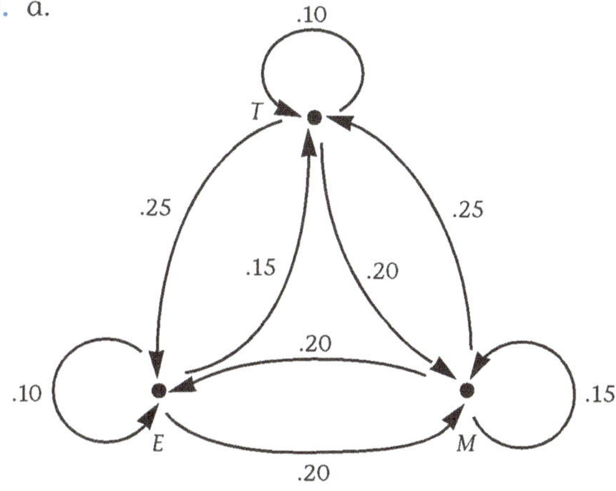

   b. $C = \begin{matrix} & \begin{matrix} \text{T} & \text{E} & \text{M} \end{matrix} \\ \begin{matrix} \text{T} \\ \text{E} \\ \text{M} \end{matrix} & \begin{bmatrix} 0.10 & 0.25 & 0.20 \\ 0.15 & 0.10 & 0.20 \\ 0.25 & 0.20 & 0.15 \end{bmatrix} \end{matrix}$

   The entries in the matrices for parts c through f represent millions of dollars.

c. $P = \begin{matrix} T \\ E \\ M \end{matrix} \begin{bmatrix} 150 \\ 200 \\ 160 \end{bmatrix}$

d. $CP = \begin{matrix} T \\ E \\ M \end{matrix} \begin{bmatrix} 97.0 \\ 74.5 \\ 101.5 \end{bmatrix}$

e. $D = \begin{matrix} T \\ E \\ M \end{matrix} \begin{bmatrix} 53.0 \\ 125.5 \\ 58.5 \end{bmatrix}$

f. $P = \begin{matrix} T \\ E \\ M \end{matrix} \begin{bmatrix} 218.60 \\ 195.24 \\ 239.65 \end{bmatrix}$

6. e. $P = (I - C)^{-1}D = \begin{matrix} S \\ M \\ A \end{matrix} \begin{bmatrix} 11.03 \\ 11.61 \\ 9.21 \end{bmatrix}$

Lesson 7.3

1. a. $D_0 = [0.75 \quad 0.25]$, $D_1 = [0.625 \quad 0.375]$, $D_2 = [0.5875 \quad 0.4125]$
   $D_3 = [0.57625 \quad 0.42375]$, $D_4 = [0.572875 \quad 0.427125]$
   b. $D_{10} = [0.571430 \quad 0.428570]$, $D_{15} = [0.571429 \quad 0.428571]$
3. a. $[0.571429 \quad 0.428571]$
4. $D_0 = [1 \quad 0]$, $D_1 = [0.7 \quad 0.3]$, $D_2 = [0.61 \quad 0.39]$,
   $D_3 = [0.583 \quad 0.417]$, $D_4 = [0.5749 \quad 0.4251]$
   $D_{10} = [0.571431 \quad 0.428569]$; $D_{15} = [0.571429 \quad 0.428571]$
   After several weeks, about 57% of the students will be eating in the cafeteria.
5. a. No. The sum of the entries in row 2 is greater than 1.
   c. No. Entries must be probabilities (between 0 and 1 inclusive).
   e. Yes.

6. a. 0.42

 b. 0.08

 c. $D_0 = [0.6 \quad 0.4]$

 d. $T = \begin{bmatrix} .7 & .3 \\ .2 & .8 \end{bmatrix}$

 e. $D_7 = D_0(T^7) = [0.40 \qquad 0.60]$

 f. 40% chance of rain

13. a. $D_0 = [1 \quad 0 \quad 0]$

 b. $D_4 = D_0T^4 = [0.4748 \quad 0.2912 \quad 0.2340]$

 After four days there is a 47% probability that the rat will be well, 29% probability that it will be ill, and 23% probability that it will dead.

Lesson 7.4

1.

	Best Strategies Player 1	Best Strategies Player 2	Strictly Determined	Saddle Point
a.	row 1	column 2	yes	8
c.	row 1or 2	column 1	no	
e.	row 3	column 3	yes	4

2. a. Best strategies: Row 2, column 2. Saddle point is 3.

 b. Best strategies stay the same. This adds 4 to the saddle point. (3 + 4) = 7.

4. a. Every other row dominates row B. Eliminate row B.

 Column E dominates column G. Eliminate column G.

 Best strategies: Row C and column F. The saddle point is 3.

7. a.

		Gretchen 1	2	3	
Jon	1	10	−10	−10	[−10]
	2	−20	20	−20	−20
	3	−30	−30	30	−30
		[10]	20	30	

No saddle point

 b. This is not a strictly determined game. There is no saddle point.

Lesson 7.5

1. Payoff matrix is:

$$\begin{array}{cc} & \text{Tina} \\ & \begin{array}{cc} \text{Heads} & \text{Tails} \end{array} \\ \text{Sol} \begin{array}{c} \text{Heads} \\ \text{Tails} \end{array} & \begin{bmatrix} 4 & -2 \\ -3 & 1 \end{bmatrix} \end{array}$$

b. The probability that both Sol and Tina show heads is .15.

c. The probability that Tina shows tails when Sol shows heads is .35.

d. The probability that Tina shows heads when Sol shows tails is .15.

e. The probability that both Sol and Tina show tails is .35.

f.

Outcome	HH	HT	TH	TT
Probability	.15	.35	.15	.35
Amount won	4	−2	−3	1

g. Sol's payoff expectation for this game is −0.20. Sol will lose 2 pennies every 10 plays.

h. It is the same.

2. a. $A = [.75 \quad .25]$, $B = \begin{bmatrix} 4 & -2 \\ -3 & 1 \end{bmatrix}$, $C = \begin{bmatrix} .3 \\ .7 \end{bmatrix}$,

then $ABC = [.75 \quad .25]\begin{bmatrix} 4 & -2 \\ -3 & 1 \end{bmatrix}\begin{bmatrix} .3 \\ .7 \end{bmatrix} = [-0.2]$

b. Sample answer:

$A = [.25 \quad .75]$. $ABC = [.25 \quad .75]\begin{bmatrix} 4 & -2 \\ -3 & 1 \end{bmatrix}\begin{bmatrix} .3 \\ .7 \end{bmatrix} = [-0.2]$

c. Sample answer:

Let $C = \begin{bmatrix} .25 \\ .75 \end{bmatrix}$. $ABC = [.4 \quad .6]\begin{bmatrix} 4 & -2 \\ -3 & 1 \end{bmatrix}\begin{bmatrix} .25 \\ .75 \end{bmatrix} = -0.2$

6. a. Payoff matrix:

$$\begin{array}{cc} & \text{Player 2} \\ & \begin{array}{cc} 1 & 2 \end{array} \\ \text{Player 1} \begin{array}{c} 1 \\ 2 \end{array} & \begin{bmatrix} 2 & -3 \\ -3 & 4 \end{bmatrix} \end{array}$$

b. Best strategies: Both row and column players play 1 finger seven-twelfths of the time and 2 fingers five-twelfths of the time.

c. The row player can expect to lose 1 cent in 12 plays or about 8 cents in 100 plays.

d. This is not a fair game since the row player will lose. In a fair game the expected payoff for both players is the same.

8. a. The group against should send out mailings three-fourths of the time and go door-to-door one-fourth of the time.

The group in favor should send out mailings one-half of the time and go door-to-door one-half of the time.

b. Let $A = [.75 \quad .25]$, $B = \begin{bmatrix} 300 & 200 \\ 100 & 400 \end{bmatrix}$, and $C = \begin{bmatrix} .5 \\ .5 \end{bmatrix}$.
Then $ABC = [250]$.

So, the opposing group only gets 250 signatures, which is not enough to get the issue on the ballot.

Chapter 7 Review

2. a.

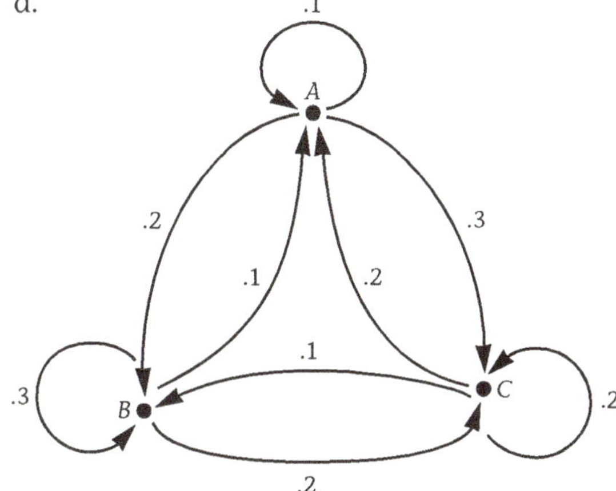

b. $CP = \begin{matrix} A \\ B \\ C \end{matrix} \begin{bmatrix} 7.7 \\ 7.4 \\ 5.8 \end{bmatrix}$ $\quad D = P - CP = \begin{matrix} A \\ B \\ C \end{matrix} \begin{bmatrix} 0.3 \\ 4.6 \\ 9.2 \end{bmatrix}$

c. $P = (I - C)^{-1}D = \begin{bmatrix} 18.6 \\ 20.4 \\ 22.2 \end{bmatrix}$

3. This game is strictly determined. The saddle point for this matrix is –2. Mike's best strategy is not to bluff.

4. a.

$$\text{Mike}\quad\begin{matrix} & \begin{matrix}\text{Brit} \\ \text{Black} \quad \text{Red}\end{matrix} \\ \begin{matrix}\text{Black} \\ \text{Red}\end{matrix} & \begin{bmatrix} 7 & -6 \\ -4 & 3 \end{bmatrix}\end{matrix}$$

b. Mike's best strategy is to play his black card seven-twentieths of the time and his red card thirteen-twentieths of the time.

Brit's best strategy is to play her black card nine-twentieths of the time and her red card eleven-twentieths of the time.

d.

Outcome	BB	BR	RB	RR
Probability	$\frac{63}{400}$	$\frac{77}{400}$	$\frac{117}{400}$	$\frac{143}{400}$
Amount won	7	–6	–4	3

e. Mike can expect to lose an average of 3 cents every twenty hands played.

5. a. The probability of another quiz on Friday is about 29%.

b. Students should expect that the teacher will start class with a quiz one-third of the time.

c. She will start class with a quickie review 40% of the time.

6. a.

$$C = \begin{matrix} & \begin{matrix} O & I & B \end{matrix} \\ \begin{matrix} O \\ I \\ B \end{matrix} & \begin{bmatrix} .20 & .25 & .55 \\ .45 & .35 & .20 \\ .20 & .25 & .55 \end{bmatrix} \end{matrix}$$

b. Probability of another Italian sandwich on Wednesday is 28.5%.

c. In the long run:

$$\begin{bmatrix} 0 & 1 & 0 \end{bmatrix}\left(\begin{bmatrix} .20 & .25 & .55 \\ .45 & .35 & .20 \\ .20 & .25 & .55 \end{bmatrix}\right)^{50} = [.2694\ldots \quad .2777\ldots \quad .4527\ldots];$$

O: 27%, I: 28%, B: 45%

7. a.

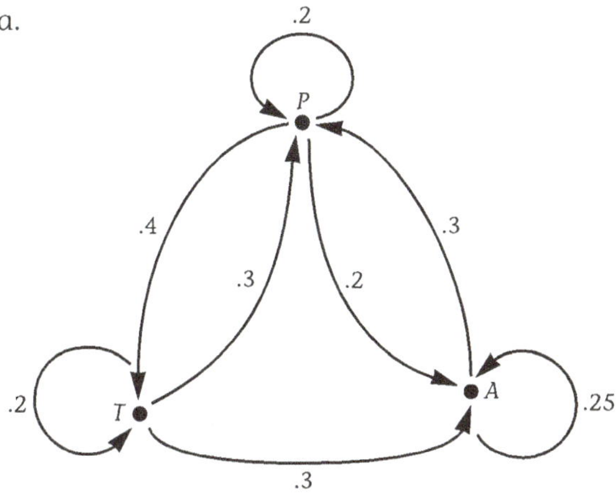

b. $C = \text{From} \begin{array}{c} \\ \text{T} \\ \text{P} \\ \text{A} \end{array} \begin{array}{c} \text{To} \\ \begin{array}{ccc} \text{T} & \text{P} & \text{A} \end{array} \\ \left[\begin{array}{ccc} 0.2 & 0.3 & 0.3 \\ 0.4 & 0.2 & 0.2 \\ 0.0 & 0.3 & 0.25 \end{array}\right] \end{array}$

c. Transportation is most dependent on petroleum (0.4) and least dependent on agriculture (0.0).

d. \$1.08 million from petroleum. \$1.35 million from agriculture.

e. $CP = \begin{array}{c} \text{T} \\ \text{P} \\ \text{A} \end{array} \begin{bmatrix} 16.00 \\ 16.00 \\ 11.25 \end{bmatrix} \qquad D = \begin{bmatrix} 4.00 \\ 9.00 \\ 3.75 \end{bmatrix}$ (in millions of dollars)

f. $P = \begin{bmatrix} 16.4 \\ 17.5 \\ 11.0 \end{bmatrix}$ (in millions of dollars)

8. The best strategy for both companies is to focus on school district A.

9. a. The Democrats' best strategy is to go with strategy A one-fourth of the time and strategy B three-fourths of the time. The Republicans' best strategy is to go with strategy C one-half of the time and strategy D one-half of the time.

b. Expected payoff for the Democrats: 45% of undecided voters will join them.

10. a.

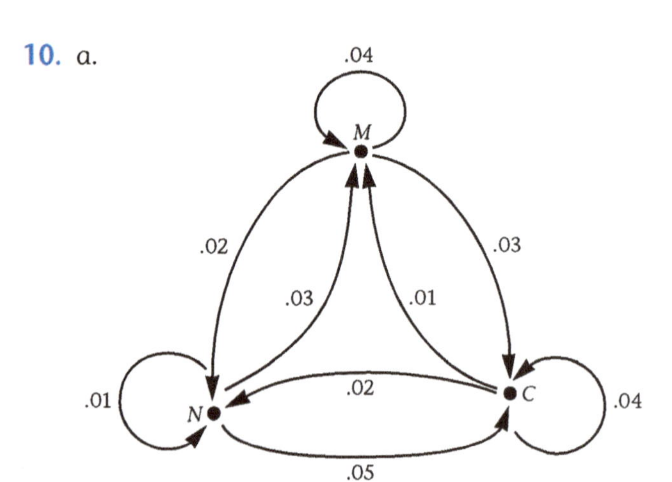

b. $P = \begin{bmatrix} \$54{,}136 \\ \$34{,}112 \\ \$42{,}941 \end{bmatrix}$

c. $CP = \begin{bmatrix} \$4{,}136 \\ \$4{,}112 \\ \$2{,}941 \end{bmatrix}$

11. Store A: Lower prices. Store B: No change.

CHAPTER 8

Lesson 8.1

1. b. 1, 4, 9

c.

Number of Couples	Number of Handshakes	Recurrence Relation
1	1	—
2	4	$H_2 = H_1 + 3$
3	9	$H_3 = H_2 + 5$
4	16	$H_4 = H_3 + 7$
5	25	$H_5 = H_4 + 9$

2. a.

Couples	Handshakes
1	0
2	2
3	6
4	12

b. $2n - 2$

c. $H_n = H_{n-1} + 2n - 2$

3. a. i. $H_n = H_{n-1} + 3$

ii. $H_n = (H_{n-1}) \times 2$

b. 16, 64, 36, 5,040

5. a. 4

c. 0

7. a.

Term Number	Number of Handshakes	First Differences	Second Differences
1	0	—	—
2	1	1	—
3	3	2	1
4	6	3	1
5	10	4	1
6	15	5	1
7	21	6	1
8	28	7	1

10. a.

Term Number	Number of Bees
0	5,000
1	5,600
2	6,272
3	7,024.64
4	7,867.60

b. $B_n = 1.12B_{n-1}$

c. After 27 years, in 2014

Lesson 8.2

2. a. $H_n = 2n^2 - 5n$

   b. $H_n = 0.29 + 0.23(n - 1)$

   d. $H_n = 3^{n-1}$

3. a. $T_n = T_{n-1} + n - 1$

   b. 0

4. a. 1, –2, –11, –38, –119, –362

   b. 2.5

6. a.

Row Number	Number of Seats	Total Seats
1	24	24
2	26	50
3	28	78
4	30	108
5	32	140
6	34	174

   b. $S_n = S_{n-1} + 2$

   c. $S_n = 24 + 2n - 2$

   d. 37

   f. $T_n = T_{n-1} + 24 + 2n - 2$

   g. $T_n = n^2 + 23n$

Lesson 8.3

1. b. i. $H_n = H_{n-1} + 3$

      iii. $H_n = (H_{n-1}) \times 1.2$

      iv. $H_n = H_{n-1} + H_{n-2}$

   c. i. $H_n = 2 + 3(n - 1)$

      iii. $H_n = 10(1.2^{n-1})$

5. c. $5,755.11

   g. $5,772.76

6.

Year	4.8% Monthly	5% Yearly
0	$5,000.00	$5,000.00
1	$5,245.35	$5,250.00
2	$5,502.74	$5,512.50
3	$5,772.76	$5,788.13

7. a. 74; 585

   b. 63.25; 817.5

10. $117,463.15

11. Approximately 6.5%

15. a. About $71,300

    b. A little over $22,200

16. a. 1.14471

    b. 2,318

Lesson 8.4

1. a. $93,070.22

   b. $48,000

   c. $45,070.22

   d. $202,107.52, $93,070.22

5. a.

T (in months)	T_n
0	12,000
1	11,804
2	11,606.63
3	11,407.87

   b. $A_n = (A_{n-1}) \times 1.007 - 286$

   c. It takes 52 months.

7. a. $M_n = (M_{n-1}) \times 1.1 - 2000$

   b. At the end of the third year, $9,352

   c. 10%

   d. The fixed point is $20,000.

8. $14,837.25

Lesson 8.5

1. a. –3; $T_n = 4(2^{n-1}) - 3$; 2.5353×10^{30}

   b. 3.5; $T_n = 1.5(3^{n-1}) + 3.5$; 2.5769×10^{47}

2. a. $B_n = (B_{n-1})\left(1+\frac{0.08}{12}\right) + 150$

   b.

Month	Balance
0	150
1	301
2	453.01
3	606.03
4	760.07

   c. –22,500

   d. $22{,}801\left(1+\frac{0.08}{12}\right)^{n-1} - 22{,}500$, or $22{,}650\left(1+\frac{0.08}{12}\right)^{n} - 22{,}500$

   e. \$225,194.28

   f. 472 months

   g. Approximately \$333

3. a. $B_n = (B_{n-1}) \times 1.01 - 260$

   c. 26,000

   e. \$8,223.71

   f. 62 months

Lesson 8.6

1. a.

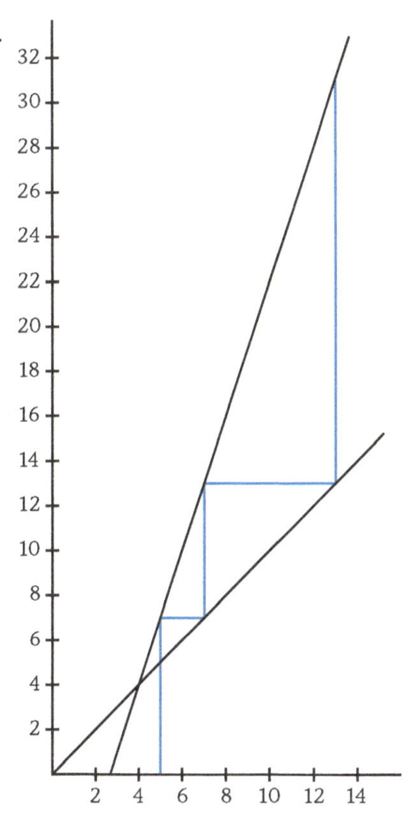

b.

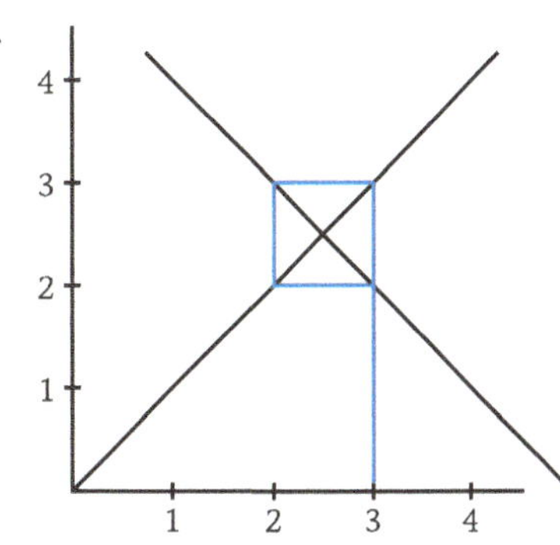

c.

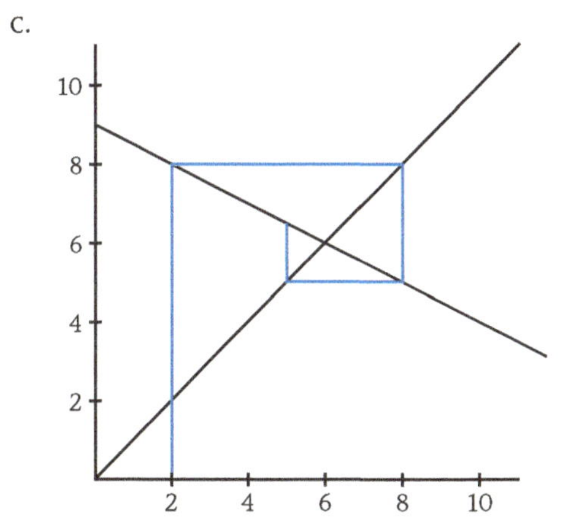

d.

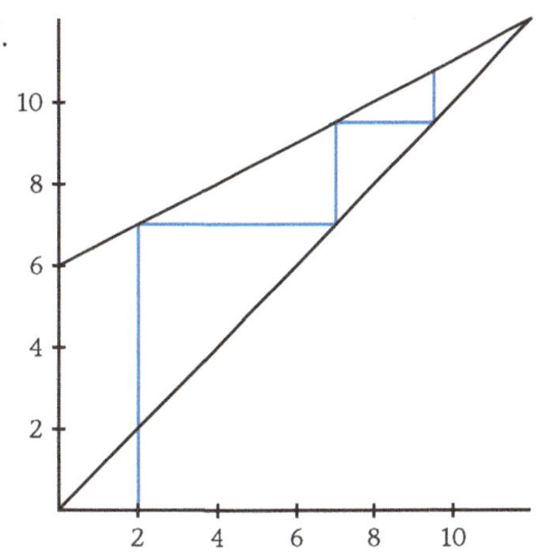

4. a. 1

b. 3

c.

n	t_n
1	1
2	5
3	13

d. $t_n = 2t_{n-1} + 3$

6. b.

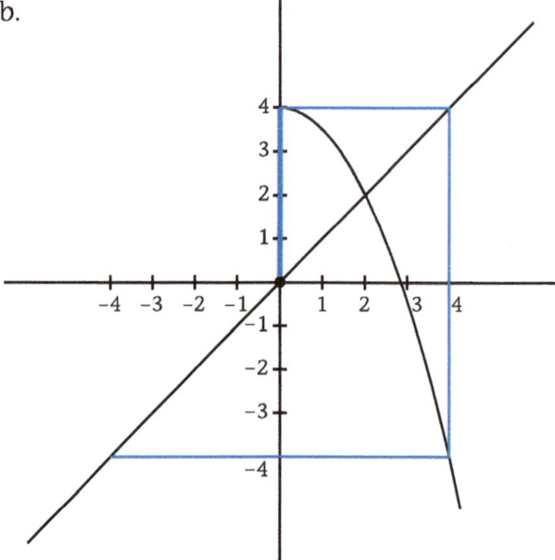

c. The behavior is unpredictable.

f. 2, −4

Chapter 8 Review

2. a. $H_n = H_{n-1} + 4$; $H_n = 2 + 4(n - 1)$; 398

b. $H_n = 3H_{n-1} - 1$; $H_n = \frac{5}{2}3^{n-1} + \frac{1}{2}$; 4.29×10^{47}

c. $H_n = (H_{n-1}) \times 2$; $H_n = 3(2^{n-1})$; 1.9014759×10^{30}

d. $H_n = 2H_{n-1} + 3$; $H_n = 2^n - 3$; 1.2676506×10^{30}

3. a. Arithmetic
   b. Neither
   c. Geometric
   d. Neither
4. a. 0; 0, 0, 0, 0
   b. –0.75; –0.75, –0.75, –0.75, –0.75
   c. No fixed point
   d. 1; 1, 1, 1, 1
5. a. $H_n = 2(5^{n-1})$; 3.1554436×10^{69}
   b. $H_n = 2.75(5^{n-1}) - 0.75$; 4.338735×10^{69}
   c. $H_n = 2 + (-3)(n - 1)$; –295
   d. $H_n = (-2)^{n-1} + 1$; -6.338253×10^{29}
6. a. Geometric
   b. Neither
   c. Arithmetic
   d. Neither
7. a. $O_n = 1.03\ O_{n-1}$
   b. $C_n = 0.982\ C_{n-1}$
8. a.

N	S_n	First Differences	Second Differences	Third Differences
1	1	—	—	—
2	5	4	—	—
3	14	9	5	—
4	30	16	7	2
5	55	25	9	2

   c. $\frac{n^3}{3} + \frac{n^2}{2} + \frac{n}{6}$; 204
9. Second degree; $H_n = n^2 - 6$

10. a.

Day	Gifts That Day	Total Gifts
1	1	1
2	3	4
3	6	10
4	10	20
5	15	35
6	21	56

b. $G_n = G_{n-1} + n$

$T_n = T_{n-1} + \frac{n^2}{2} + \frac{n}{2}$

c. $G_n = \frac{n^2}{2} + \frac{n}{2}$

$T_n = \frac{n^3}{6} + \frac{n^2}{2} + \frac{n}{3}$

11. a. $P_n = P_{n-1} + 0.21$

b. $P_n = 0.49 + 0.21(n - 1)$

12. a.

Month	Balance
0	\$1,000
1	\$1,004
2	\$1,008.02
3	\$1,012.05

b. $B_n = 1.004(B_{n-1})$

c. $B_n = 1{,}000(1.004)^n$

d. 14 years, 6 months

13. a.

Month	Balance
0	\$5,000
1	\$5,126.67
2	\$5,254.01
3	\$5,382.03

b. $B_n = (B_{n-1})\left(1+\frac{0.064}{12}\right) + 100$

c. $B_n = 23{,}750\left(1+\frac{0.064}{12}\right)^n - 18{,}750$

d. \$13,928.99

e. 200 months

14. a. $C_n = C_{n-1} + 3.50$

b. $C_n = 50 + 3.5n$

c. \$4.00

15. a. $B_n = 1.008(B_{n-1}) - 230$

b. $B_n = -17,750(1.008)^n + 28,750$

c. 61 months

d. \$3,030

e. \$352.88

16. a. $R_n = 0.9R_{n-1}$

b. $R_n = 1,000(0.9^n)$

c. About 6.58 minutes or 6 minutes, 35 seconds

17. a. $V_n = 1.5V_{n-1} + 4,000$

b. $V_n = 16,000(1.5)^{n-1} - 8,000$

18.

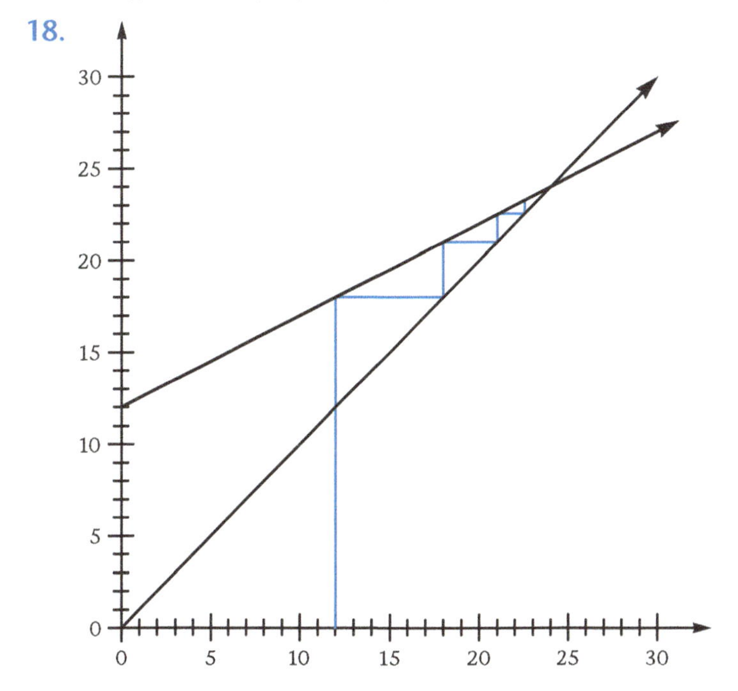

19. a. $A_n = 0.4A_{n-1} + 500$

b. The amount of medication in the body stabilizes at 833 mg.

c. The cobweb would be attracted to the point (833.33, 833.33), which is the intersection of $y = x$ and $y = 0.4x + 500$, as shown in this figure.

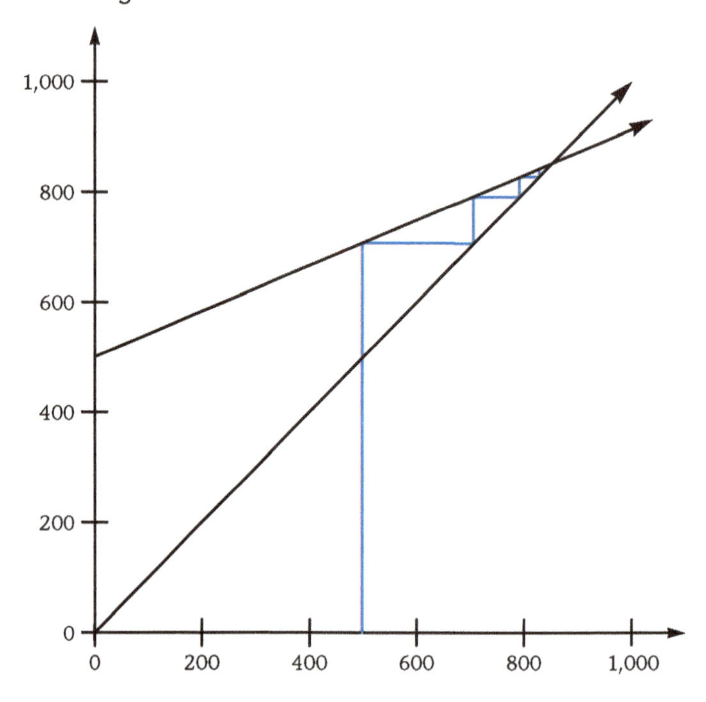

d. The amount in the body reaches the stable value (833 mg, in this case) more quickly. The stable value is probably near the optimal dosage of the drug.

20. a. $C_n = 0.8C_{n-1} + 1$

b. The concentration quickly exceeds the recommended maximum of 3 ppm and gradually approaches 5 ppm.

CHAPTER 9

Lesson 9.1

6. The extra digit for 01824 is 5.
8. a. The ZIP code is 01730–1459.

Lesson 9.2

1. The check digit is 0.
4. b. The check digit is 2.
5. a. The check digit is 3.

Lesson 9.3

1. a. 2
4. b. 11111111111000000111111000000
5. a. .00001
7. b. $y = -0.2x^3 + 0.4x^2 + 3.4x - 2$

Lesson 9.4

1. b. If blanks are left as blanks, the first six characters are: 11 20 26 23 B 24 (B represents a blank). If 27 represents a blank, the first six characters are 11 20 26 23 32 24.
5. a. The first twelve characters are: 27 17 24 17 B 14 34 B 24 17 24 B (B represents a blank).
7. a. 010101010

Chapter 9 Review

2. The check digit is 2.
4. The check digit is 7.
6. 4
7. 11110000111100001111

8. The likely configuration is white–black–white–white–black–white.
9. About 99%
10. a. A second-degree polynomial

    b. $y = -1.1x^2 + 5.2x + 1.1$

    c. 4.3, –0.4, and –7.3
11. Haste makes waste
12. Probably 12
13. The message is: I love discrete math.
14. 110011111

CHAPTER 10

Lesson 10.1

1. d. I live in England.
2. c. $p \vee q$

Lesson 10.2

2. {1, 2}, {1, 3}, {1, 4}, {1, 5}, {2, 3}, {2, 4}, {2, 5}, {3, 4}, {3, 5}, {4, 5}
5. b. {2, 4, 6}
7. a. Regions 2 and 3
9. b. Regions 3, 4, 5, 6, 7, and 8

Lesson 10.3

1. a. 6 pipes are needed.
3. a. 4 raws
5. next-fit: 5 bins

Lesson 10.4

2. $31
3. b. (2, 0), (6, 0), and (2, 8)
6. a. $x \geq 4$, $y \geq 0$, and $5x + 4y \leq 40$

Chapter 10 Review

2. I am not 18 years old and I like pretzels.
3. a. It is not hot in Texas.

   b. It is hot in Texas and no one lives near Lake Linden.

   c. It is hot in Texas or no one lives near Lake Linden.
4. Answer:

p	q	$\sim q$	$p \vee \sim q$
T	T	F	T
T	F	T	T
F	T	F	F
F	F	T	T

5. a. {0, 1, 2, 3, 4, 5, 6, 8, 10}

   b. ∅

   c. {1, 3, 5}
6. $\bar{A} = \{6, 8, 12\}$
7. c
8. 3 raws
9. a. (4, 1)

   b. $29
10. The minimum value C is 2. The maximum value C is 42.

Illustration/Photo Credits

Cartoons on paages 74, 93, 121, 156, 183, 241, 306, 334, 401, 410, 445, and 470 are by Tom Durfee

Page 2: Tom Arthur Creative Commons Attribution-Share Alike 2.0 Generic license.

Page 7: By Pontificia Universidad Católica de Chile [CC-BY-SA-2.0 (http://creativecommons.org/licenses/by-sa/2.0)], via Wikimedia Commons

Page 8: Jean-Charles de Borda. Image is in the public domain because its copyright has expired. Wikimedia Commons

Page 18: By Linda Bartlett (Photographer) [Public domain], via Wikimedia Commons

Page 19: Marquis de Condorcet. Image is in the public domain because its copyright has expired. Wikimedia Commons

Page 27: By Rwickham (Own work) [CC0], [Public domain], via Wikimedia Commons

Page 28: By Linda A. Cicero / Stanford News Service (Stanford News Service) [CC-BY-3.0 (http://creative commons.org/licenses/by/3.0)], via Wikimedia Commons.

Page 35: By Almonroth (Own work) [CC-BY-SA-3.0 (http://creativecommons.org/licenses/by-sa/3.0)], via Wikimedia Commons.

Page 36: By Joyce N. Boghosian, White House photographer (The White House Blog) [Public domain], via Wikimedia Commons

Page 37: John Banzhaf III. By Self-portrait (Wikipedia:Contact us/Photo submission) [CC-BY-SA-3.0 (http://creativecommons.org/licenses/by-sa/3.0)], via Wikimedia Commons

Page 43: Cartoon. Copyright © by Thaves. Distributed from www.thecomics.com

Page 45: Daniel Schwen [GFDL (http://www.gnu.org/copyleft/fdl.html), CC-BY-SA-3.0 (http://creative commons.org/licenses/by-sa/3.0/) or CC-BY-SA-2.5 (http://creativecommons.org/licenses/by-sa/2.5)], via Wikimedia Commons

Page 47:By Andrew (Tawker) (Own work) [GFDL (http://www.gnu.org/copyleft/fdl.html), CC-BY-SA-3.0 (http://creativecommons.org/licenses/by-sa/3.0/) or CC-BY-2.5 (http://creativecommons.org/licenses/by/2.5)], via Wikimedia Commons

Page 48: Bayernnachrichten.de at de.wikipedia [Attribution], from Wikimedia Commons

Page: 51 Donald Saari http://www.math.uci.edu/~dsaari/

Page 56: Public Domain. Work of the Federal Government. Wikimedia Commons

Page 57: Visitor7 Own Work, Creative Commons Attribution-Share Alike 3.0 Unported.

Page 58: By Evan-Amos (Own work) [CC0], via Wikimedia Commons

Page 61: By Littleinfo (Own work) [Public domain], via Wikimedia Commons

Page 64: FOR BETTER OR WORSE © 1992 Lynn Johnston Productions. Dist. By Universal Press Syndicate.

Reprinted with permission. All rights reserved.

Page 65: James Brown, publicdomainpictures.net

Page 67: House. By Scott Worsley (Own work) [CC-BY-SA-3.0 (http://creativecommons.org/licenses/by-sa/3.0)], via Wikimedia Commons

Page 67: Boat. By Mick from Northamptonshire, England (Boats At Jones Boat Yard) [CC-BY-2.0 (http://creativecommons.org/licenses/by/2.0)], via Wikimedia Commons

Page 67: Mazda Automobile. By NRMA Motoring and Services from Sydney, Australia (2012 Mazda 3 - NRMA New Cars Uploaded by FAEP) [CC-BY-2.0 (http://creativecommons.org/licenses/by/2.0)], via Wikimedia Commons

Page 69: National Park Service. Public Domain. Work of the Federal Government.

Page 72-73: Hamilton / Jefferson. Public Domain, Wikimedia Commons

Page 72: Congress DP 2-10 By Scrumshus (Own work) [Public domain], via Wikimedia Commons

Page 82: Jake DeGroot [CC-BY-SA-3.0 (http://creativecommons.org/licenses/ by-sa/3.0)], via Wikimedia Commons

Page 88: Source:http://rangevoting.org /BalinskiL.html

Page 88: Source: http://econ.jhu.edu/directory/ h-peyton-young

Page 91: Cake By Ardfern (Own work) [CC-BY-SA-3.0 (http://creativecommons.org/licenses/by-sa/3.0) or GFDL (http://www.gnu.org/copyleft/fdl.html)], via Wikimedia Commons

Page 96: Peanuts © reprinted by permission of United Feature Syndicate, Inc.

Page 97: dominos/pennys. G. Ward, Comp, Inc.

Page 100: Dominos: Honza Groh (Jagro) Creative Commons Attribution-Share Alike 3.0 Unported

Page109: Source: http://www.uschillel.org/uschillel/board/board-dr-solomon-golomb/

Page 111: By David [Public domain or Public domain], via Wikimedia Commons

Page 111: Hugo Steinhaus http://www.matematycy.interklasa.pl/cytaty/steinhaus.php

Page 112: Steven Brams http://en.wikipedia.org/wiki/Steven_ Brams.

Alan Taylor Source:http://www.math.union.edu/people/faculty/taylora.html

Page 114: NASA and in the Public Domain. Wikimedia Commons.

Page 119: Stan Musial public domain picture. Wikimedia Commons

Page 120: By Jakob Dettner (de:User:Jdettner), Rainer Zenz (de:User:Rainer Zenz), SoothingR (en:User:SoothingR) (Own work) [CC-BY-SA-2.0-de (http://creativecommons.org/licenses/by-sa/2.0/de/deed.en)], via Wikimedia Commons

Page 129: Salad. Work of the U.S. Government. Public domain, wikimedia Commons

Page 138: Calendar. By photo originally taken by Tomasz "Polimerek" Ganicz, design of the calendars by Johanna Pung; all pictures visible on the calendars are copyrighted by their respective authors and have been released under the Creative Commons Attribution Share-Alike 3.0 licence. (Own work) [CC-BY-SA-3.0 (http://creativecommons.org/licenses/by-sa/3.0)], via Wikimedia Commons

Page 139: Salad. This work is in the public domain in the United States because it is a work prepared by an officer or employee of the United States Government as part of that person's official duties under the terms of Title 17, Chapter 1, Section 105 of the US Code.

Page 149: Rat. By Oskila (Own work) [CC-BY-SA-3.0 (http://creativecommons.
org/licenses/by-sa/3.0)], via Wikimedia Commons.

Page 151: Copyright (c) 1995 by Thaves. Distributed from www.thecomics.com.

Page 154: Rat Photograph. By Dezidor (Own work) [CC-BY-3.0 (http://creativecommons.org/licenses/by/3.0)], via Wikimedia Commons

Page 159: Word of the U.S. government Dept of Agricultureal Research and in the public domain. Wikimedia Commons

Page 162: By Jefferson liffey (Lifey College) [GFDL(http://www.gnu.org/copyleft/fdl.html) or CC-BY-SA-3.0 (http://creativecommons.org/licenses/by-sa/3.0)], via Wikimedia Commons

Page 170: By KlickingKarl (Own work) [CC-BY-SA-3.0 (http://creativecommons.
org/licenses/by-sa/3.0) or GFDL (http://www.gnu.org/copyleft/fdl.html
)], via Wikimedia Commons

Page 171: By U.S. Navy photo by Photographer's Mate 2nd Class Michael Winter. [Public domain], via Wikimedia Commons

Page 175: Polaris.Work of the U.S. federal government, all EPA images are in the public domain

Page 177 tower By Gellerj (Own work) [CC-BY-SA-3.0 (http://creative commons.org/licenses/by-sa/3.0) or GFDL (http://www.gnu.org/copyleft/fdl.html)], via Wikimedia Commons

Page 182: Test tube. By Linda Bartlett (Photographer) [Public domain or Public domain], via Wikimedia Commons

Page 190: Pencil By Juliancolton (Own work) [Public domain], via Wikimedia Commons

Page 191: Euler. Jakob Emanuel Handmann [Public domain], via Wikimedia Commons

Page 201: Fire hydrant By brdavids (originally posted to Flickr as Hydrant) [CC-BY-2.0 (http://creativecommons.org/licenses/by/2.0)], via Wikimedia Commons

Page 202: Sir William Hamilton. This image is in the public domain because its copyright has expired. Wikimedia Commons.

Page 210: Checkmate photo by Alan Light, licensed under the Creative Commons Attribution-Share Alike 3.0 Unported license.

Page 218: Postman By Chong Fat (Own work (Self took photo)) [Public domain], via Wikimedia Commons

Page 221: Mapping. By Maersk Line (Adjusting the route Uploaded by russavia) [CC-BY-SA-2.0 (http://creativecommons.org/licenses/by-sa/2.0)], via Wikimedia Commons.

Page 232: DM5 02Slovenija_2007_4col.svg Public domain, via Wikimedia Commons.

Page 241: Salesman Illustration. Comap, Inc.

Page 247: Rt. 495 Map By Sswonk [Public domain], via Wikimedia Commons

Page: 247: By Hamilton Richards [GFDL (http://www.gnu.org /copyleft/fdl.html) or CC-BY-SA-3.0 (http://creativecommons.org/licenses/by-sa/3.0/)], via Wikimedia Commons.

Page 253: By Photo by Jocelyn Augustino/FEMA (This image is from the FEMA Photo Library.) [Public domain], via Wikimedia Commons

Page 254: By Jorge Barrios (Own work) [GFDL (http://www.gnu.org/ copyleft /fdl.html) or CC-BY-3.0 (http://creativecommons.org/licenses/by/3.0)], via Wikimedia Commons.

Page 255: Tree Illustration. Comap, inc.

Page 262: Dave Barber, Comap, inc.

Page 263: Earthquake Map. By USGS/Joan Gomberg and Eugene Schweig [Public domain], via Wikimedia Commons.

Page 275: Coin flip. G. Ward, Comap, Inc.

Page 281: HP 50g, Wikimedia Commons

Page 286: By Bill Branson (Photographer) [Public domain], via Wikimedia Commons

Page 289: By Matt Britt [CC-BY-2.5 (http://creativecommons.org/ licenses/by/2.5)], via Wikimedia Commons

Page 298: By U.S. Navy photo by Mass Communication Specialist 2nd Class Ron Kuzlik [Public domain], via Wikimedia Commons.

Page 298: Cartoon. The Wizard of Id by Brant Parker and Johnny Hart. Reprinted by permission of John L. Hart FLP, and Creators Syndicate, Inc.

Page 305: Students. By Carrie E. David Ford (SMDC/ARSTRAT) (United States Army) [Public domain], via Wikimedia Commons

Page 306: Comap, Inc.

Page 308: Bicycle gears By Thegreenj

(Own work) [GFDL (http://www.gnu.org/copyleft/fdl.html) or CC-BY-SA-3.0 (http://creativecommons.org/licenses/by-sa/3.0/)], via Wikimedia Commons

Page 309: Woman's basketball. By «Marylandstater» «reply». [Public domain], via Wikimedia Commons

Page 317: Cards. By Imager Visioner.Imager Visioner at en.wikipedia [Public domain], from Wikimedia Commons

Page 320: Cards. Bicycle Cards.com

Pafe 322: Cartoon. Frank and Ernest © reprinted by permission of Newspaper Enterprise Association, Inc.

Page 322: Band. By English: Staff Sgt. Clinton Firstbrook [Public domain], via Wikimedia Commons

Page 326: Bike lock. Gary Froelich. Comap, Inc.

Page 327: By Jarrett Campbell from Cary, NC, USA (IMG_1622) [CC-BY-2.0 (http://creativecommons. org/licenses/by/2.0)], via Wikimedia Commons

Page 328: Cards. Liko81 [CC-BY-3.0 (http://creativecommons.org/licenses/by/3.0)], via Wikimedia Commons

P329: Die By Stephen Silver (Open Clip Art Library) Public Domain, via Wikimedia Commons

Page 334: Comap, Inc.

Page 336: Cassini Space craft. By Cassini_spacecraft_de.jpg: NASA derivative work: Nova13 (talk) derivative work: Nova13 (Cassini_spacecraft_de.jpg) [Public domain or Public domain], via Wikimedia Commons

Page 342: American roulette wheel By Film8ker at en.wikibooks [Public domain], from Wikimedia Commons

Page 343: Challenger By Kennedy Space Center [Public domain], via Wikimedia Commons

Page 344: Orion module By NASA (http://twitpic.com/5gbbh (modified to remove text)) [Public domain], via Wikimedia Commons

Page 347: Thomas Bayes. This work is in the public domain in the United States, and those countries with a copyright term of life of the author plus 100 years or less.

Page 348: Air hockey L6.5 opener By U.S. Air Force photo by Airman 1st Class Alexxis Pons Abascal [Public domain], via Wikimedia Commons

Page 356: Gavel Photo by Chris Potter. This file is licensed under the Creative Commons Attribution 2.0 Generic license. Wikimedia commons.

Page 357: Cell Photograph. By OpenStax College [CC-BY-3.0 (http://creativecommons.org/ licenses/by/3.0)], via Wikimedia Commons

P364 Dime. Work of U.S Govenment and in the Public Domain. Wikimedia Commons.

Page 364: Stanislaw Ulam. By Originally uploaded by Deer*lake (Transferred by Deer*lake) (Originally uploaded on en.wikipedia) [Public domain], via Wikimedia Commons

Page 367: Mercedes. By OSX (Own work) [Public domain], via Wikimedia Commons.

Page 368: Cartoon. Peanuts © reprinted by permission of United Feature Syndicate, Inc.

Page 378: This work is in the public domain in the United States because it is a work prepared by an officer or employee of the United States Government as part of that person's official duties.

Page 378: Wassily Leontief. Wikimedia Commons.

Page 386: Nissan Leaf battery. By Tennen-Gas (Own work) [CC-BY-SA-3.0 (http://creativecommons.org/ licenses/by-sa/3.0)], via Wikimedia Commons

Page 387: Battery. By Frettie (Own work) [GFDL (http://www.gnu.org/copyleft/fdl.html) or CC-BY-3.0 (http://creativecommons.org/licenses/by/3.0)], via Wikimedia Commons

Page 398: By U.S. Department of Agriculture (20111019-FNS-RBN-1772) [CC-BY-2.0 (http://creativecommons.org/licenses/by/2.0)], via Wikimedia Commons

Page 398: Andrei A. Markov. [Public domain], via Wikimedia Commons

Page 401: Comap, Inc.

Page 405: G. Ward, Comap, Inc.

Page 409: By Symode09 (Own work) [Public domain], via Wikimedia Commons.

Page 409: John Neumann. By wikispaces [CC-BY-SA-3.0 (http://creativecommons.org/licenses/by-sa/3.0) or CC-BY-SA-3.0 (http://creativecommons.org/licenses/by-sa/3.0)], via Wikimedia Commons

Page 410: Comap, Inc.

Page 410: Nash. By Elke Wetzig (Elya) (Own work) [GFDL (http://www.gnu.org/copyleft/fdl.html) or CC-BY-SA-3.0 (http://creativecommons.org/licenses/by-sa/3.0/)], via Wikimedia Commons

Page 418: Penny. By Daniel Schwen (Own work) [Public domain or CC-BY-SA-2.5 (http://creativecommons.org/licenses/by-sa/2.5)], via Wikimedia Commons

Page 431: By KUSHI (自分で撮影) [Public domain], via Wikimedia Commons

Page 435: © 1997 Greg Howard. Reprinted with special permission of King Features Syndicate.

Page 436: By NASA image courtesy Jeff Schmaltz, MODIS Rapid Response Team,Goddard Space Flight Center [Public domain], via Wikimedia Commons.

Page 444: By Tobias Wolter (Own work) [GFDL http://www.gnu.org/copyleft/fdl.html, CC-BY-SA-3.0 (http://creativecommons.org/licenses/by-sa/3.0/) or CC-BY-SA-2.5-2.0-1.0 (http://creativecommons.org/licenses/by-sa/2.5-2.0-1.0)], via Wikimedia Commons.

Page 445: Comap, Inc.

Page: 450 http://jeff560.tripod.com/images/fibonacci.jpg

Page 453: By Rama (Own work) [CC-BY-SA-2.0-fr (http://creative commons.org/licenses/by-sa/2.0/fr/deed.en)], via Wikimedia Commons

Page 455:By User:Greg L (English Wikipedia) [GFDL (www.gnu.org/copyleft/fdl.html) or CC-BY-SA-3.0 (http://creativecommons.org/licenses/by-sa/3.0/)], via Wikimedia Commons.

Page 461: By Nightscream (Own work) [CC-BY-3.0 (http://creativecommons.org/licenses/by/3.0)], via Wikimedia Commons.

Page 462:By USDA photo by Scott Bauer [Public domain], via Wikimedia Commons.

Page 470: Comap, Inc.

Page 471: By Mohler Addison, U.S. Fish and Wildlife Service [Public domain], via Wikimedia Commons

Page 472: By permission of John L. Hart FLP, and Creators Syndicate, Inc.

Page 473: Comap, Inc.

Page 476: By Merzperson at en.wikipedia [Public domain], from Wikimedia Commons.

Page 477: From the box cover of the original Tower of Hanoi Puzzle. http://www.cs.wm.edu/~pkstoc/toh.html

Page 487: By Gilibaumer (Own work) [CC-BY-SA-3.0 (http://creativecommons.org/licenses/by-sa/3.0)], via Wikimedia Commons

Page 495: Copyright © 1998 by Thaves. Distributed by www.thecomics.com

Page 496: I, Luc Viatour [GFDL (http://www.gnu.org/copyleft/fdl.html), CC-BY-SA-3.0 (http://creativecommons.org/licenses/by-sa/3.0/) or CC-BY-SA-2.5-2.0-1.0 (http://creativecommons.org/licenses/by-sa/2.5-2.0-1.0)], via Wikimedia Commons

Page 504: Julia set spiral.png. Wikimedia Commons.

Page 508: By Guillaume Jacquenot Gjacquenot (Own work) [GFDL http://www.gnu.org/copyleft/fdl.html or CC-BY-SA-3.0-2.5-2.0-1.0 (http://creativecommons.org/licenses/by-sa/3.0)], via Wikimedia Commons

Page 509: By sacratomato_hr (Moo) [CC-BY-SA-2.0 (http://creativecommons.org/licenses/by-sa/2.0)], via Wikimedia Commons.

Page 517: Picture taken in the military aviation museum, Dubendorf by Audrius Meskauskas, Audriusa 14:44, 10 February 2006 (UTC). GNU Free Documentation License. Wikimedia Commons.

Page 518: By Robert Kaufmann (This image is from the FEMA Photo Library.) [Public domain], via Wikimedia Commons

Page 518: Zip Code Map. By Denelson83 (Own work, based on Image:ZIP_code_zones.png) [Public domain], via Wikimedia Commons

Page 519: Mr. zip USPS and in the public domain

Page 522: Hand Scanner by G. Ward, Comap, Inc.

Page 529: Hubble Space Telescope. By NASA [Public domain], via Wikimedia Commons

Page 529: 1 Moon Photo Source:http://commons.wikimedia.org/wiki/File:First_photo_from_space.jpg

Page 529: Hubble Photo. Source: http://hubblesite.org/gallery/album/nebula/pr2007009a/

Page 530: Richard Hamming. http://www.computerhistory.org/collections/accession/102632901

Page 532: Reed Solomon. http://reed-solomon.tripod.com

Page 536: ATM. By Count Iblis (Own work) [Public domain], via Wikimedia Commons.

Page 537: Figure 9.12. English letter frequencies. Source: Wikimedia Commons, http://en.wikipedia.org/wiki/File:English_letter_frequency_%28alphabetic%29.svg

Page 537: Julius Caessar. By Georges Jansoone (JoJan) (Own work (own photo)) [GFDL (http://www.gnu.org/copyleft/fdl.html) or CC-BY-SA-3.0-2.5-2.0-1.0 (http://creativecommons.org/licenses/by-sa/3.0)], via Wikimedia Commons

Page 539: With permission. © W.Carson/Distributed by Universal Press Syndicate.

Page 543: By Grj23 (Own work) [CC0], via Wikimedia Commons.

Page 543: Mathematicians of Note, Ronald Rivest, Adi Shamir, and Leonard Adelman in 2003. Source: http://www.usc.edu/dept/molecular-science/RSA-2003.htm

Page 546: By Scanned and processed by Michael Romanov (Personal collection) [CC-BY-3.0 (http://creativecommons.org/licenses/by/3.0)], via Wikimedia Commons

Page 549: By DARPA (Defense Advanced Research Projects Agency (DARPA)) [Public domain], via Wikimedia Commons

Page 550: Female Student on computer. By Jeff Billings [CC-BY-SA-3.0 (http://creativecommons.org/licenses/by-sa/3.0)], via Wikimedia Commons

Page 550: Gottfried Wilhelm Leibniz.By Johann Friedrich Wentzel d. Ä. [Public domain], via Wikimedia Commons

Page 554: Mars. By NASA and The Hubble Heritage Team (STScI/AURA) [Public domain], via Wikimedia Commons.

Page 556: Venn Diagram. By Sustainability Hub [CC-BY-SA-3.0 (http://creativecommons.org/licenses/by-sa/3.0)], via Wikimedia Commons. Redrawn by Comap, Inc.

Page 561: Table saw. By Patrick Thor (Own work) [CC-BY-SA-3.0 (http://creativecommons.org/licenses/by-sa/3.0)], via Wikimedia Commons

Page 564: Ronald Graham. By Cheryl Graham (http://math.ucsd.edu/~fan/ron/jug.html) [CC-BY-3.0 (http://creativecommons.org/licenses/by/3.0)], via Wikimedia Commons.

Page 566: Gasoline truck. By Graham Richardson from Plymouth, England (Asda DK07OLX Uploaded by oxyman) [CC-BY-2.0 (http://creativecommons.org/licenses/by/2.0)], via Wikimedia Commons

Page 566: Geo Dantzig. Source: http://news.stanford.edu/news/2006/june7/memldant-060706.html

Page 570: Sunrise. By Wing-Chi Poon [CC-BY-SA-3.0 (http://creativecommons.org/licenses/by-sa/3.0)], via Wikimedia Commons

Index

A

D

E

H

I

J

K

L

M

N

O

P

T

U

V

W

Y

Z